Sensor Data Fusion Analysis for Broad Applications

Sensor Data Fusion Analysis for Broad Applications

Editor

Natividad Duro Carralero

MDPI

Basel • Beijing • Wuhan • Barcelona • Belgrade • Novi Sad • Cluj • Manchester

Editor
Natividad Duro Carralero
Department of Computer
Sciences and Automatic
Control
UNED
Madrid
Spain

Editorial Office
MDPI AG
Grosspeteranlage 5
4052 Basel, Switzerland

This is a reprint of articles from the Special Issue published online in the open access journal *Sensors* (ISSN 1424-8220) (available at: https://www.mdpi.com/journal/sensors/special_issues/sensor_data_fusion).

For citation purposes, cite each article independently as indicated on the article page online and as indicated below:

Lastname, A.A.; Lastname, B.B. Article Title. *Journal Name* **Year**, *Volume Number*, Page Range.

ISBN 978-3-7258-1589-0 (Hbk)
ISBN 978-3-7258-1590-6 (PDF)
doi.org/10.3390/books978-3-7258-1590-6

Contents

About the Editor

Natividad Duro Carralero

Natividad Duro Carralero has a degree in Physical Sciences from the Complutense University of Madrid (1995) and a PhD in Sciences from the National University of Distance Education (2002). She is currently a member of the Computer and Automation Department of the UNED, and she has worked as a full-time associated professor since 2008. Her research activity covers different aspects in the field of systems engineering and automation, such as process control (the subject in which she carried out her PhD), modeling and simulation of photobioreactors, machine learning techniques for thermonuclear fusion databases, big data, and virtual and remote laboratories applied for teaching. From all of these lines of research, publications of which she is the author or co-author have emerged. These publications include more than 40 publications with an impact index (JCR), three books on university teaching subjects, one chapter in a research-oriented book, and multiple publications in scientific conference proceedings. In addition, she has also participated in more than 10 research projects regarding competitive public calls in relation to these issues.

sensors

MDPI

Editorial

Sensor Data Fusion Analysis for Broad Applications

Natividad Duro

Department of Computer Sciences and Automatic Control, Universidad Nacional de Educación a Distancia (UNED), C/Juan del Rosal, 16, 28040 Madrid, Spain; nduro@dia.uned.es

1. Introduction

Sensor data fusion analysis plays a pivotal role in a variety of fields by integrating data from multiple sensors to produce more accurate, reliable, and comprehensive information than that achieved by individual sensors alone. Advances in technology have allowed us to obtain relevant information from this analysis and many applications take advantage of this fact from different points of view. Not only can greater accuracy and reliability be achieved by combining data from multiple sources, but equally, better decision-making is executed, supported by informed decisions and predictive analytics from data fusion, which can be crucial for the operation of many industrial applications.

Therefore, efficiently integrating and analyzing data from multiple sources directly impacts the attainment of higher productivity and more efficient operations in industry. In this sense, resource optimization can be achieved by mitigating the need for additional sensors and reducing redundancy by leveraging data from existing sensors more effectively.

Different emerging areas gain relevant benefits from sensor data fusion analysis. These include industrial applications, medical or biomedical applications, robotics, monitoring systems, transportation systems, information systems, or control processes.

Applications in all of these areas require advanced algorithms and techniques to analyze and interpret the data from various sensors. To this end, it is still necessary to investigate several challenges such as reducing the complexity and computational load, especially in real-time applications, ensuring privacy and security, and maximizing data interoperability, among others.

2. Overview of Published Papers

All submissions were judged on technical merit and relevance, and sixteen high-quality papers were ultimately accepted to appear in this Special Issue.

Below, a list of accepted contributions is provided, each accompanied by a brief description. We believe that such papers offer updates and key insights into the research areas that this Special Issue focuses on and may therefore inspire future work in this area.

In the first contribution, the authors use RGB thermal fusion data as a robust neural network-based neonatal face detection method. They show that it is possible to achieve accurate sensor fusion for short distances using a ToF camera as an additional sensor. Its solution offers good precision, increasing data efficiency and economizing the process.

In the second paper, the authors present a novel methodology that provides an early fusion module capable of introducing the required reliability in a next-generation lightweight object detector in the event of sensor failure as well as extreme weather conditions. The article demonstrates that together, the early melt detector and the multimodal marker work robustly and transparently. Additionally, integrating a GPU enables the system to perform exceptionally well in real time.

The third study presents a new odometry system implemented in an autonomous wheelchair. It uses LSTM neural networks to estimate the speed of the robot using an encoder sensor. Real-time retraining allows the system to self-calibrate and adapt to changes in the defined model and also reduces the influence of some unsystematic errors. The paper

Citation: Duro, N. Sensor Data Fusion Analysis for Broad Applications. *Sensors* **2024**, *24*, 3725. https://doi.org/10.3390/s24123725

Received: 31 May 2024
Revised: 3 June 2024
Accepted: 6 June 2024
Published: 7 June 2024

shows that the proposed wheelchair localization is a robust solution for state estimation in challenging environments.

The fourth article addresses the great interest that the academic and scientific community has in the development of technologies related to geospatial and terrestrial phenomena. It reflects a strong trend of using the SVM method to evaluate large volumes of textual and geospatial data, as well as the use of Weka software. The results present the implementation of an innovative, practical, and systematic approach for information extraction and the recommendation of relevant knowledge. The method uses calibration parameters for each object and achieves results with high precision.

The fifth study implements a real-time PPC for MLS subject to dynamic uncertainty and external disturbances. A modified GFTSMM function has been developed based on the errors of the proposed PPC, so the tracking error variables quickly converge to the equilibrium point. Thanks to the designed observer, the approximate value of all uncertainty is known, which reduces vibration and improves control performance. The combination of GFTSMC, PPC, and MTOSMO is a novel strategy that guarantees a stable position of the controlled ball and the possibility of orbital tracking with good performance in real time.

The sixth article presents a solution in the field of health, safety, and fire to obtain temporally synchronous data from high-resolution sensors. The authors develop an energy-conserving multi-sensor fusion framework that powers low-power model-driven microcontrollers using machine learning. Likewise, they apply optimization techniques using anomaly detection modes to provide real-time information on demand that saves lives. The paper presents the application and results in a real-life healthcare scenario.

The seventh paper mainly focuses on designing a highly accurate trainable EKF-based localization framework using inertial measurement units for an autonomous ground vehicle with dead reckoning. The goal is to fuse it with a laser image for simultaneous detection and localization and for mapping estimation based on range sensors that improve performance. Convolution neural networks, backpropagation algorithms, and gradient descent methods are implemented in the system to optimize the parameters. Additionally, a unique cost function is used to train the models and improve the accuracy. The research is generic and applicable to various robot localization models assisted by inertial measurement units.

Paper eight proposes an Echo State Network architecture that exploits sensor data fusion to detect failures in a scaled replica of an industrial plant. Thanks to their sparse weight structure, Echo State Networks function as recurrent neural network models, exhibiting low complexity and memory consumption, making them suitable for deployment on an Edge device. By analyzing vibration and current signals, the proposed model can correctly detect most of the faults that occur in the plant. The experimental results demonstrate the feasibility of the proposed approach and present a comparison with other approaches, where it is shown that the presented methodology is the best trade-off in terms of precision, recall, F1-score, and inference time.

The ninth study presents new data fusion approaches for the characterization of musts and wines based on biogenic amine and elemental composition. The paper applies inductively coupled plasma techniques to determine a wide range of elements. The authors obtain good descriptive models to describe the different compositions of the wines and musts using data fusion.

The tenth paper shows a multiblock approach to fuse processes and near-infrared sensors for online prediction of polymer properties. The main goal of the study is to explore the feasibility of multiblock regression methods to build real-time monitoring models that predict two quality properties of acrylonitrile butadiene styrene by fusing data from near-infrared and process sensors. Several prediction models taking advantage of sensor measurements have been created, which have provided good prediction results and allowed for the identification of the most relevant block data for the prediction of quality parameters.

Paper eleven proposes an optimization framework for the large-scale field placement of optical sensors to improve border protection. Compared to the frequently used maximum area coverage approach, this method minimizes undetected passages in the monitored area. The paper takes into consideration both natural and built environmental coatings. The optimization is performed using a bacterial evolutionary algorithm. Therefore, the developed simulation framework based on ray tracing provides an excellent opportunity to optimize large areas.

In the twelfth study, different types of artificial neural networks are investigated to estimate the arrival time in acoustic emission signals. Convolutional neural network models and a novel capsule neural network are proposed. The models have been tested with experimental data acquired in the framework of a localization problem to identify targets with known coordinates on a square aluminum plate. The models have been shown to have excessive precision at significant noise levels.

Paper thirteen evaluates the performances of five clustering algorithms: k-means, fuzzy C-means (FCM), hierarchical, mean shift, and density-based spatial clustering of applications with density-based noise. The paper analyzes the impacts of input data format and feature selection on the quality of management zone delineation. It uses key soil fertility attributes collected with an online visible and near-infrared spectrometer, demonstrating that k-means is the optimal clustering method for management zone delineation.

Paper fourteen proposes the use of the Delicar system, a self-driving product delivery vehicle that can drive the vehicle on the road and report the current geographical location to the authority in real time through a map. A camera captures the road image and transfers it to the computer using socket server programming. The system's infrastructure is also low-cost and easy to install.

The fifteenth paper proposes a recent iterative learning algorithm for sensor data fusion to detect pitch actuator failures in wind turbines. The development of this proposed approach is based on iterative learning control and Lyapunov's theories. Numerical experiments have been carried out to support the study. These experiments use a well-known model of a wind turbine hydraulic pitch actuator with some common faults, such as high oil content in the air, hydraulic leaks, and pump wear.

The last paper presents a detailed review of state-of-the-art data fusion solutions for data storage and indexing from various types of sensors, feature engineering, and multimodal data integration. The review aims to serve as a guide for the early stages of an analytic pipeline of manufacturing prognosis. The reviewed literature showed that in fusion and preprocessing, the methods chosen to be applied in this sector are beyond the state of the art. Existing weaknesses and gaps that will lead to future research were also identified.

3. Conclusions

In summary, the sixteen papers collected in this Special Issue represent a good example of the uses of data sensor fusion in industrial applications. The papers show a wide interest in the research area.

Future research on sensor data fusion analysis will be focused on the miniaturization of sensors and components, as well as an increased use of multi-sensor systems and wireless and autonomous radio sensors. Developing smaller and smaller sensors will allow us to understand the world as we know it more easily and accurately. On the other hand, one of the main challenges in sensor fusion using mobile devices will be the quality of the data collected. Mobile sensors are easily affected by noise, interference, calibration errors, and outliers.

Conflicts of Interest: The authors declare no conflicts of interest.

List of Contributions

1. Gleichauf, J.; Hennemann, L.; Fahlbusch, F.B.; Hofmann, O.; Niebler C.; Koelpin, A. Sensor Fusion for the Robust Detection of Facial Regions of Neonates Using Neural Networks. *Sensors* **2023**, *23*, 4910.
2. Neves, F.S.; Claro, R.M.; Pinto, A.M. End-to-End Detection of a Landing Platform for Offshore UAVs Based on a Multimodal Early Fusion Approach. *Sensors* **2023**, *23*, 2434.
3. Fariña, B.; Acosta, D.; Toledo, J.; Acosta, L. Improving Odometric Model Performance Based on LSTM Networks. *Sensors* **2023**, *23*, 961.
4. Silva, L.A.; Mendes, A.S.; Blas, H.S.S.; Bastos, L.C.; Gonçalves, A.L.; de Moraes, A.F. Active Actions in the Extraction of Urban Objects for Information Quality and Knowledge Recommendation with Machine Learning. *Sensors* **2023**, *23*, 138.
5. Truong, T.N.; Vo, A.T.; Kang, H.-J. Real-Time Implementation of the Prescribed Performance Tracking Control for Magnetic Levitation Systems. *Sensors* **2022**, *22*, 9132.
6. D'souza, O.; Mukhopadhyay, S.C.; Sheng, M. Health, Security and Fire Safety Process Optimisation Using Intelligence at the Edge. *Sensors* **2022**, *22*, 8143.
7. Milam, G.; Xie, B.; Liu, R.; Zhu, X.; Park, J.; Kim, G.; Park, C.H. Trainable Quaternion Extended Kalman Filter with Multi-Head Attention for Dead Reckoning in Autonomous Ground Vehicles. *Sensors* **2022**, *22*, 7701.
8. Bruneo, D. De Vita, F. Detecting Faults at the Edge via Sensor Data Fusion Echo State Networks. *Sensors* **2022**, *22*, 2858.
9. Mir-Cerdà, A.; Granell, B.; Izquierdo-Llopart, A.; Sahuquillo, À.; López-Sánchez, J.F.; Saurina, J.; Sentellas, S. Data: Fusion Approaches for the Characterization of Musts and Wines Based on Biogenic Amine and Elemental Composition. *Sensors* **2022**, *22*, 2132.
10. Strani, L.; Vitale, R.; Tanzilli, D.; Bonacini, F.; Perolo, A.; Mantovani, E.; Ferrando, A.; Cocchi, M. A Multiblock Approach to Fuse Process and Near-Infrared Sensors for On-Line Prediction of Polymer Properties. *Sensors* **2022**, *22*, 1436.
11. Kovács, S.; Bolemányi, B.; Botzheim, J. Placement of Optical Sensors in 3D Terrain. Using a Bacterial Evolutionary Algorithm. *Sensors* **2022**, *22*, 1161.
12. Zonzini, F.; Bogomolov, D.; Dhamija, T.; Testoni, N.; De Marchi, L.; Marzani, A. Deep Learning Approaches for Robust Time of Arrival Estimation in Acoustic Emission Monitoring. *Sensors* **2022**, *22*, 1091.
13. Javadi, S.H.; Guerrero, A.; Mouazen, A.M. Clustering and Smoothing Pipeline for Management Zone Delineation Using Proximal and Remote Sensing. *Sensors* **2022**, *22*, 645.
14. Chy, M.K.A.; Masum, A.K.M.; Sayeed, K.A.M.; Uddin, M.Z. Delicar: A Smart Deep Learning Based Self Driving Product Delivery Car in Perspective of Bangladesh. *Sensors* **2022**, *22*, 126.
15. Acho, F.; Pujol-Vázquez, G. Data Fusion Based on an Iterative Learning Algorithm for Fault Detection in Wind Turbine Pitch Control Systems. *Sensors* **2021**, *21*, 8437.
16. Tsanousa, A.; Bektsis, E.; Kyriakopoulos, C.; González, A.G.; Leturiondo, U.; Gialampoukidis, I.; Karakostas, A.; Vrochidis S.; Kompatsiaris, I. A Review of Multisensor Data Fusion Solutions in Smart Manufacturing: Systems and Trends. *Sensors* **2022**, *22*, 1734.

sensors

MDPI

Article

Sensor Fusion for the Robust Detection of Facial Regions of Neonates Using Neural Networks

Johanna Gleichauf [1,*], Lukas Hennemann [1], Fabian B. Fahlbusch [2,3], Oliver Hofmann [1], Christine Niebler [1] and Alexander Koelpin [4]

1 Nuremberg Institute of Technology, 90489 Nuremberg, Germany; christine.niebler@th-nuernberg.de (C.N.)
2 Division of Neonatology and Pediatric Intensive Care, Department of Pediatrics and Adolescent Medicine, Friedrich-Alexander-Universität Erlangen-Nürnberg, 91054 Erlangen, Germany; fabian.fahlbusch@uk-erlangen.de
3 University Children's Hospital Augsburg, Neonatal and Pediatric Intensive Care Unit, 86156 Augsburg, Germany
4 Hamburg University of Technology, 21073 Hamburg, Germany; alexander.koelpin@tuhh.de
* Correspondence: johanna.gleichauf@th-nuernberg.de; Tel.: +49-911-5880-1093

Abstract: The monitoring of vital signs and increasing patient comfort are cornerstones of modern neonatal intensive care. Commonly used monitoring methods are based on skin contact which can cause irritations and discomfort in preterm neonates. Therefore, non-contact approaches are the subject of current research aiming to resolve this dichotomy. Robust neonatal face detection is essential for the reliable detection of heart rate, respiratory rate and body temperature. While solutions for adult face detection are established, the unique neonatal proportions require a tailored approach. Additionally, sufficient open-source data of neonates on the NICU is lacking. We set out to train neural networks with the thermal-RGB-fusion data of neonates. We propose a novel indirect fusion approach including the sensor fusion of a thermal and RGB camera based on a 3D time-of-flight (ToF) camera. Unlike other approaches, this method is tailored for close distances encountered in neonatal incubators. Two neural networks were used with the fusion data and compared to RGB and thermal networks. For the class "head" we reached average precision values of 0.9958 (RetinaNet) and 0.9455 (YOLOv3) for the fusion data. Compared with the literature, similar precision was achieved, but we are the first to train a neural network with fusion data of neonates. The advantage of this approach is in calculating the detection area directly from the fusion image for the RGB and thermal modality. This increases data efficiency by 66%. Our results will facilitate the future development of non-contact monitoring to further improve the standard of care for preterm neonates.

Keywords: non-contact monitoring; neonates; sensor fusion; neural network; face detection

Citation: Gleichauf, J.; Hennemann, L.; Fahlbusch, F.B.; Hofmann, O.; Niebler, C.; Koelpin, A. Sensor Fusion for the Robust Detection of Facial Regions of Neonates Using Neural Networks. *Sensors* **2023**, *23*, 4910. https://doi.org/10.3390/s23104910

Academic Editor: Natividad Duro Carralero

Received: 17 April 2023
Revised: 14 May 2023
Accepted: 16 May 2023
Published: 19 May 2023

1. Introduction

Neonates on the Neonatal Intensive Care Unit (NICU), especially those born preterm, require continuous monitoring (e.g., via electrocardiogram (ECG), pulse oximeter or temperature probes) of their vital signs (heart rate, respiratory rate and body temperature). Most of these techniques, however, rely on direct skin contact which might pose a clinical problem particularly for immature neonates with commonly increased skin sensitivity. Clinically, this can lead to pressure marks, eczema, skin irritations and even epidermal abrasion [1]. Hence, current research aims to develop non-contact monitoring approaches to improve neonatal comfort. Technically, monitoring of vital signs requires a reliable detection of the head and facial regions such as the nose. Unfortunately, the most common face detection methods are only utilized on adults and cannot be readily applied to the body proportions of neonates [2]. Moreover, sufficient open-source data of neonates on the NICU is lacking. In the clinical setting, non-contact monitoring is challenged by incubator-related issues such as the varying quality of lighting and the positioning of detectors in close proximity to the area of interest. Therefore, our aim is to collect thermal and Red Green Blue (RGB) camera data of incubator-nursed (preterm) neonates on the NICU. Thermal cameras measure temperature gradients and display it within a false color image and are light independent. RGB cameras collect visible light and display it how a human would perceive it. RGB camera images hold more features than thermal camera images. The idea is to fuse the different image modalities to allow a robust face detection under all conditions. A neural network shall be trained with the fused data.

1.1. State of the Art

In this chapter, we describe the State of the Art for thermal-RGB-fusion approaches as well as neural networks for face detection.

1.1.1. Thermal-RGB-Fusion

Direct and indirect approaches for the fusion of an RGB with a thermal camera will be presented in the following section.

Direct Extrinsic Calibration

St. Laurent et al. use the angles and the ratio of the field of view (FoV) of the thermal and the RGB camera [3]. The calibration is facilitated using parallel or convergent optical axes of the cameras. Shivakumar et al. use a projection of the RGB coordinates into the 3D space using a depth image which is determined by stereo depth calculations [4]. Their RGB camera is a Stereolabs Zed Mini (Stereolabs, San Francisco, CA, USA). The 3D coordinate is then projected into the thermal camera image. For the intrinsic calibration, a checkered pattern is used. Yang et al. first detect the centers of the circles of the calibration target within the RGB and the thermal camera image [5]. The root mean square error (RMSE) of the coordinates of the detected circle centers is calculated. Finally, normalized cross correlation (NCC) is used as an image registration procedure.

Indirect Extrinsic Calibration

Since it is difficult to achieve a high precision using a direct extrinsic calibration, approaches for an improved extrinsic calibration using a third sensor will be discussed. Krishnan and Saripalli apply a cross calibration of an RGB camera with a thermal camera with a light detection and ranging (LIDAR) sensor [6]. First, the RGB camera is calibrated with the LIDAR. The transformations between the edges of the checkered calibration target within the camera image and the 3D point cloud are determined. The extrinsic calibration between the thermal camera and the LIDAR is determined in the same manner using a different calibration target which is visible within the thermal camera image. Gleichauf et al. employed a thermal-RGB-laser fusion for a project with the German railway [7]. In this case, no direct fusion between the RGB and thermal camera takes place, but the single systems are fused with the 3D laser scanner by themselves. The points of the laser scanner are projected into the two image modalities.

Thermal-ToF-Fusion

Tisha et al. fuse a thermal camera with a Kinect2. First, the intrinsic calibration takes place and in the next step the homography of both cameras is calculated (intrinsic times extrinsic). The multiplication of the homographies with a pixel point of the camera delivers the pixel point within the other camera [8]. Van Baar et al.'s extrinsic calibration approach is based on the publication by Yang et al. [9]. They fuse an RGB image with a depth image. For the thermal camera a checkered pattern with heated resistors is used [10]. Cao et al. fuse a thermal with a Red, Green, Blue -Depth (RGB-D) camera (Kinect2) using the thermal-guided iterative closest point (T-ICP) algorithm [11]. A thermal-3D-pointcloud registration is applied. The resulting error is defined by a nonlinear least-squares objective and minimized by the Gauss–Newton method. Pfitzner fuses a thermal camera with a Kinect via the infrared sensor frames of the Kinect to the thermal camera and the Kinect2.

For the fusion the RGB images of the Kinect camera systems are used [12]. The method by Rocco Spremolla is very similar, where first the intrinsic calibrations of the thermal and the Kinect2 cameras are determined and then the relative poses of thermal/RGB (Kinect) to the infrared (IR)/Depth of the Kinect. The transformations can then be calculated [13].

RGB-ToF-Fusion

Similarly fusion approaches between ToF and RGB cameras exist. Salinas et al. recommend using a depth dependent homography lookup table instead of calculating the extrinsic parameters of the ToF and RGB camera. In both modalities, point correspondences need to be found [14]. Kim et al. use five RGB and three ToF cameras which are set up in a semicircle. The intrinsic and extrinsic calibration can be calculated using the intensity image of the ToF camera but inaccurate registrations of the depth images can occur [15]. The aim is to achieve a precise 3D reconstruction of surfaces.

Fusion Using Neural Networks

There are several approaches using neural networks for generating fusion images. In each case, the images are taken from a long distance [16]. The network used by Alexander et al. is used for civil infrastructure applications and states that the robustness increases for fusion images [17]. Jung et al. use a neural network for the creation of fusion images out of near infrared (NIR) and noisy RGB images [18]. Wang et al. apply a multi-channel convolutional neural network (CNN) for infrared and visible image fusion using images of the same scene and position [19]. Another approach uses salience detection and a CNN for fusion image generation [20]. Using neural networks for the sensor fusion requires a lot of data to train the networks. As we demonstrated in this article, all of these fusion approaches are only applicable for long distance environments.

1.1.2. Face Detection Using Neural Networks and Image Processing

For adults, there are a lot of scientific contributions which address the detection of body regions. The field of application can differ and is not restricted to the monitoring of vital signs. The detection of the head and face became more popular after the publication of the WiderFace dataset [21]. Face detection developed rapidly and mainly focused on extreme and real variation problems including scaling, pose, occlusion, lighting conditions, blur, etc. Neural networks are mainly used for the common object detection and trained for the special case of face detection with the WiderFace dataset and adjusted. Known methods are YOLO5Face [22], RetinaFace [23], MTCNN [24], Mask R-CNN [25] and an implementation of the Faster R-CNN [26]. These are all adaptions of the original architecture (trained on RGB data of the WiderFace dataset) especially designed to overcome the problems of the face detection. Some were extended such that the pose or the orientation of the face can be detected. The dataset also holds images of babies but they are neither newborns nor recorded within the clinical field. Most of the dataset consists of images of adults. Therefore, nets which were trained on the WiderFace dataset are not suitable for detecting the face region of neonates within the incubator. There are many other models which were trained with the WiderFace dataset. An overview can be found on the WiderFace website [21].

There is little research on the head detection within thermal images. Cheong et al. [27] use the Otsu threshold method to convert thermal into binary images. Using the horizontal projection of the images the global minimum can be determined which helps to identify the height and width of the head region.

Kopaczka et al. [28] use a so-called Histogram of oriented gradients detector for the evaluation of thermal images. They also compare other machine learning methods and classic approaches within image processing. They conclude that machine learning methods deliver better results than traditional approaches.

Silva et al. [29] use deep learning methods such as the neural network YOLOv3 by Redmon et al. [30]. They apply transfer learning in order to implement a network trained on RGB data to detect heads within thermal images.

A similar approach can be found with Vuković et al. [31]. They use a R-CNN which delivers very good results but has no real-time capability.

Mucha et al. [32] implement the SCRFD DL architecture by Guo et al. [33] which returned satisfactory precision values for the WiderFace Challenge. SCRFD DL was adapted such that faces can be detected within thermal images.

1.1.3. Face Detection for Neonates

As the proportions of adult faces differ from the faces of neonates the previously presented methods cannot be used for neonates without further adaptions [2]. Chaichulee et al. use a self designed multi-task CNN which segments the visible skin of the neonate (clinical field) within the RGB image [34]. In most cases the torso and the head are made visible. These regions are used to determine the respiratory and heart rate.

Green et al. [35] extended the network such that the head, the torso and the nappy area can be detected. It is based on the Faster R-CNN [36]. The whole network is made up of a core network based on VGG16 [37], three branches for the patient detection and skin segmentation with movement control and a branch for the detection of the body regions. The average precision achieved for the "head" class is 98.2. The data set holds 2269 images of 15 subjects of different ethnic backgrounds and sex. The recordings were taken within an incubator during the stay on the NICU.

Kyrollos et al. use the RetinaNet [38], to detect the thorax and face region of neonates in an RGB video stream [39]. A model was trained with a transfer learning approach by applying weights which were pre-trained on the ImageNet dataset [40]. Three different core nets were tested: ResNet50, ResNet101 and ResNet152. There were only small differences between the three models so it was decided to use the fastest model, ResNet50. A mean average precision of 56.9 was achieved. The results of the detection are used for the calculation of the respiratory rate. The RGB images (200 images per subject with 5 subjects) were not taken within an incubator.

A CNN with pyramidal hierarchic features is presented by Lu et al. [41]. First, an adaption of VGG16 [37] is used to extract the implicit features of the normalized image. Multi-scale feature maps are selected to predict and detect different sizes of facial features of neonates. The third part of the system contains two parallel branches: a classification branch of facial and non-facial regions and a regression branch for the position of the face. The dataset holds 3600 images with different perspectives, gestures, facial expressions and backgrounds. If the images were taken on a NICU is unknown. An average precision of 92.7 for the facial region was achieved.

An often-used model for the detection of neonates is YOLOv3 [30] and YOLOv5[42]. In Nagy et al.'s work, YOLOv3 is the basis for the detection of head, torso and hands of a nurse or parents [43]. In combination with a LSTM (Long Short-Term Memory) the static objects within the RGB video can be detected. The pulse and respiratory rate are determined from the results of the classification block (YOLOv3 plus LSTM). The author's dataset holds 96 h of labeled RGB video data recorded on a NICU. The achieved precision of the object detection is not stated in the publication. The results of the classification block made up of YOLOv3 and LSTM show a sensitivity of 97.9, specificity of 97.5 and an F1 score of 98.1. Therefore, the object detection should also deliver high levels of precision.

Khanam et al. [44] use Redmon's YOLOv3 for the face detection of preterm neonates in the incubator. The head needs to be detected during normal lighting conditions as well as during phototherapy with ultraviolet (UV) light. The dataset holds 473 images from the internet. A transfer learning approach was chosen. The starting point were the weights of a YOLOv3 net trained on the MS COCO dataset. The authors do not state the precision of the face detection.

By training YOLOv5 on a proprietary RGB dataset with labeled faces of newborns in the clinical field Hausmann et al. almost reach real-time face detection [2]. The results show a precision of 68.7. For the dataset, the University of South Florida Multimodal Neonatal

Pain Assessment Dataset (USF-MNPAD-I) [45] was used. In comparison, a net was also trained with the WiderFace dataset, which led to a precision of 7.37.

Dosso et al. [46] also use YOLOv5. They compare different face detection models such as RetinaFace [23] and YOLO5Face [22] which were trained with the WiderFace dataset. By using transfer learning approaches with their own dataset they reach a precision of 81.45. The best results were achieved by the fine-tuned YOLO5Face which they called NICUFace. The dataset contained 33 subjects which were filmed over 153 h. They also use a thermal camera but its data were not used for their work.

Antink et al. [47] follow another approach by extending an encoder decoder method through a modified version of the ResNet50 encoder. Their aim is to segment different regions of the body of preterm neonates. They use a freely available dataset holding segmentation data of adults. A second dataset containing NIR and RGB images of preterms on the NICU was used for transfer learning. The head is well segmented within the RGB images (Intersection over Union (IoU) of 82.0). For the torso, an IoU of 41.0 is reached. Within the NIR images the detection delivered less accurate results. By applying methods of data augmentation and generating artificial NIR images an IoU of 62.0 for the head detection and of 14.0 for the torso was achieved. Voss et al. extended this research by using a U-Net architecture for training with RGB, NIR and fusion images [48]. Their fusion does not take place on the images itself, but on the feature level. They reach a mean IoU of 85.0 with their fusion model. The results for the RGB model is similar with 85.0. Within the NIR images the accuracy drops significantly to a mean IoU of 75.0. In all modalities, the segmentation of the torso performed worst.

Beppu et al. research the detection of different body regions within thermal images [49]. They trained YOLOv5 [42] to detect six body regions such as the head, torso, arms and legs. For the dataset, 4820 thermal images of 26 different subjects extracted out of 1872 h of video material were used. The images were taken within an incubator on a NICU. The head detection reached a precision of 94.8, whereas the torso an average precision (AP) of 95.8. They use the detected regions to determine the body temperature of preterms.

Besides the deep learning methods there are other approaches for the face detection. Awais et al. [50] solves the problem of the face detection using the CIE L*a*b intensity based detection by Neophytou et al. [51] and Fairchild [52] within their framework NFAR (Neonatal Face Attributes Recognition) for RGB images. NFAR reaches a precision of up to 94.9.

No literature regarding the face detection of neonates with fused images currently exists.

1.1.4. Face Detection Using Fused Images

Two sensor fusion approaches for the face detection based on thermal and visual image data are introduced by Bebis et al. [53]. The first approach is pixel based which is applied within the wavelet plane, the other one is feature based in the Eigenspace. For the wavelet method, differences within the resolution are detected, for the Eigenspace global image features are fused. The Eigenspace method is based on a Principal Component Analysis (PCA). The aim is to calculate a fused image out of the infrared and the RGB data and to use the most significant features in both spectra. The heat energy which is emitted by the face can be measured by the thermal camera. The reflection of light from the facial surface is detected by the visual camera. A simple pixel to pixel fusion has the disadvantage that spacial information is lost. If the fusion takes place for different resolution planes, features with different spacial dimensions can be used, especially the most significant ones. Both approaches were tested by Bebis et al. with the Equinox dataset [54] with long wave infrared (LWIR) and RGB camera images. The dataset only holds head-on images of adults faces. In our case, it has to be trained with facial data of neonates. In comparison, the fusion within the wavelet space delivers better results, but the computational expense is higher. In the future, the authors want to use a hair-wavelet-transformation [53]. Both of these methods have only been tested on adults and do not necessarily use the fused images for the face detection or the vital sign detection.

Chen et al. propose a neural network for creating fusion images for face recognition [55]. They state that the recognition improves when fusing data. The network is only trained on adults.

1.1.5. Neural Networks Using Fused Image Data

There are a few approaches using neural networks with fused image data as input. The applications lie within the perception for automated driving [56] and pedestrian detection (Faster R-CNN) [57]. In both cases, the fusion images are also created by a neural network. Shopovska et al. propose that the fusion generates more informative images [57].

There are so-called RGB-T trackers which track objects in the foreground using neural networks within RGB thermal images. Thus, in contrast to image fusion, the fusion precision is less important [58,59].

There are no neural networks using fused image data for the face detection for neonates. We will now present a method addressing this gap.

2. Materials and Methods

We now describe all parts which are required for our sensor fusion as well as the training of the neural networks for neonatal face detection. First, the theoretical approach is presented, and then the hardware and software are described in more detail. The measurement series we recorded will be presented in the last step.

2.1. Concept and Theoretical Approach

In this section, the concept of our fusion approach, the architecture of our neural networks and the theory behind the sensors used is presented.

2.1.1. Sensors

First, the theory behind the thermal, RGB and the 3D-ToF camera are described. A detailed description of the hardware used will be presented in the following section.

Thermal cameras are 2D infrared thermometers which detect the emitted heat of objects. Within the thermal camera image temperature gradients are displayed as false color images. RGB cameras collect visible light and display the image how a human being would perceive it [60].

Within time-of-flight cameras two principles can apply: the phase difference method and the impulse time-of-flight method. In our ToF camera the phase difference method is used [61]. This is based on the phase shift caused by the reflected modulated signal. Using the phase shift ϕ and the wavelength λ of the modulated signal the distance to the object can be calculated:

$$d = \frac{\phi}{2 \cdot \pi} \cdot \frac{\lambda}{2} \tag{1}$$

2.1.2. Intrinsic Calibration

Fusion between the different sensors requires prior intrinsic calibration of the thermal and RGB camera. During the rectification the tangential and radial distortion coefficients are calculated and removed. It relates to the projection of the chip plane onto the image plane of the camera sensor such that the transformation between the sensor's pixel coordinates and the world coordinate frame is given. Both cameras can be modelled as a pinhole camera model represented by this formula:

$$\begin{pmatrix} u \\ v \\ 1 \end{pmatrix} = \begin{pmatrix} f_x & 0 & u_0 \\ 0 & f_y & v_0 \\ 0 & 0 & 1 \end{pmatrix} \begin{pmatrix} X/Z \\ Y/Z \\ 1 \end{pmatrix} \tag{2}$$

with f_x and f_y as focal length, u_0 and v_0 as center of the camera sensors. u and v are the x- and y-coordinate within the image. Z is a scaling factor.

Images of the calibration target have to be processed from different positions with different distances and skew so that the distortion coefficients can be calculated as precisely as possible. The rectification can be calculated using the following formula:

$$\begin{pmatrix} u \cdot (1 + k_1 r^2 + k_2 r^4 + k_3 r^6) + 2p_1 v + p_2 \cdot (r^2 + 2u^2) \\ v \cdot (1 + k_1 r^2 + k_2 r^4 + k_3 r^6) + 2p_1 \cdot (r^2 + 2v^2) + p_2 u \end{pmatrix} \tag{3}$$

with k_1, k_2 and k_3 as radial distortion coefficients and p_1 and p_2 as tangential distortion coefficients. r is the distance between u and v.

2.1.3. Sensor Fusion

In the following, the concept of our sensor fusion approach will be described. The precision of direct fusion between the thermal and RGB camera is limited to its specific calibration distance [3]. This is not ideal due to the varying sizes of the neonates the distance to the cameras can differ. Therefore, it is our aim to develop an indirect fusion approach using the Time of Flight camera of our sensor setup from [62] as the third sensor. The advantage of the 3D ToF camera is that it can take the depth into account. The following steps are necessary for the indirect thermal-RGB-fusion:

- Detect circles of the calibration target within the ToF mono image and calculate the corresponding depth points.
- Detect circles within RGB and thermal image.
- Calculate transformation between RGB and ToF camera using the circle centers.
- Calculate transformation between thermal and ToF camera using the circle centers.
- Project RGB points into the thermal image (at the position of the ToF points).

2.1.4. Neural Networks RetinaNet and YOLOv3

Now the architecture of the neural networks RetinaNet and YOLOv3 will be presented.

RetinaNet

RetinaNet is a CNN developed by Facebook AI Research (FAIR) and published by Lin et al. in [38]. It is known as one of the best one-stage object detection models that is proven to work well with dense and small scale objects.

RetinaNet is composed of several networks, as displayed in Figure 1. On top of a Residual Network (ResNet) (a) [63] backbone a Feature Pyramid Network (FPN) (b) [64] is used. ResNet utilizes skip connections to extract deep features [63]. An FPN is used on top of ResNet for constructing a rich multiscale feature pyramid from one single resolution input image [64]. Lateral connections allow RetinaNet to merge feature maps with the same resolution. The feature pyramid is included to improve the accuracy of detection for objects of a different scale. In the original paper [38], five different scales are used. On top of the FPN, two subnets are added: first, a classification subnet (c) which predicts the probability of the presence of an object at each position for each anchor and object class. The second subnet is the bounding box regression subnet (d) with the purpose of regressing the offset of an anchor box to a nearby ground truth object.

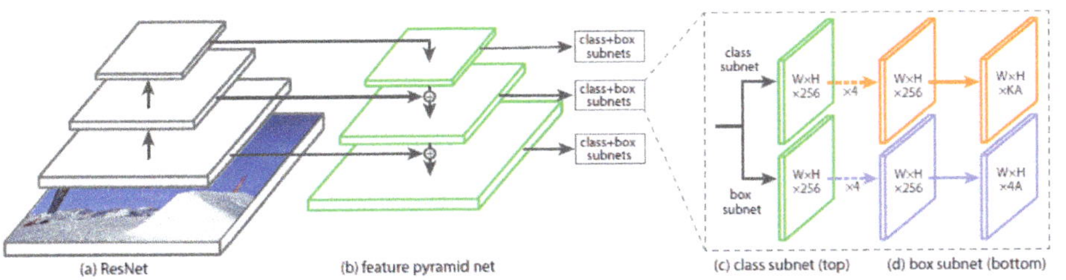

Figure 1. Structure of RetinaNet from the original paper [38]. Reprinted with permission from [38]. 2018, Lin et al.

It has been shown that there is an extreme imbalance between the foreground and background classes in a single-stage detector [38]. To overcome this problem, RetinaNet uses a special loss function, called Focal Loss. It is used on the output of the classification subnet. By using Focal Loss, less loss is produced by "easy" negative samples (such as the background), so that the loss is focused on "hard" samples (real objects we want to detect), which improves prediction accuracy. Focal Loss reduces the cost of simple detections and increases the cost of correcting misclassified objects.

RetinaNet achieves state-of-the-art performance and outperforms most one-stage and two-stage detectors, such as Fast Region-Based Convolutional Neural Networks (R-CNN) [65] or Faster R-CNN [36].

YOLOv3

"You only look once", or YOLO, is one of the faster object detection algorithms found in the literature. YOLOv3 is the third revised version of YOLO published by Redmon et al. in 2018 [30].

The structure of YOLOv3 relies on similar elements as RetinaNet. The backbone is Darknet-53, a fully convolutional network with 53 layers. Like ResNet, it uses so-called skip connections to bypass individual layers during training in order to enable a deeper architecture. Darknet is used for feature extraction. Compared with that of ResNet, the network structure used by Darknet-53 better utilizes the graphical processing unit (GPU), making it more efficient to evaluate and thus faster. For the task of detection, 53 additional layers are added on top of Darknet, so that YOLOv3 is based on a fully convolutional architecture with 106 layers. The overall structure is shown in Figure 2.

A major innovation of YOLOv3 in contrast to its predecessors [66] is that predictions are performed at three different scales. The 82nd layer performs the first detection. During the passage of the previous 81 layers, the input image is down-sampled by a factor of 32.

After a few convolutional layers, the 79th layer is deeply concatenated with the 61st layer sampled upward by two. The resulting layer is then subjected to further convolutional layers until another detection is performed in layer 94. The prediction is thus performed on a feature map that is smaller than the initial image by a factor of 16.

The last detection is then performed in layer 106, which corresponds to a down-sampling factor of 8. Similarly to the previous prediction head, a deep convolution is applied. The 91st layer is up-sampled by a factor of two and then used to convolute layer 36.

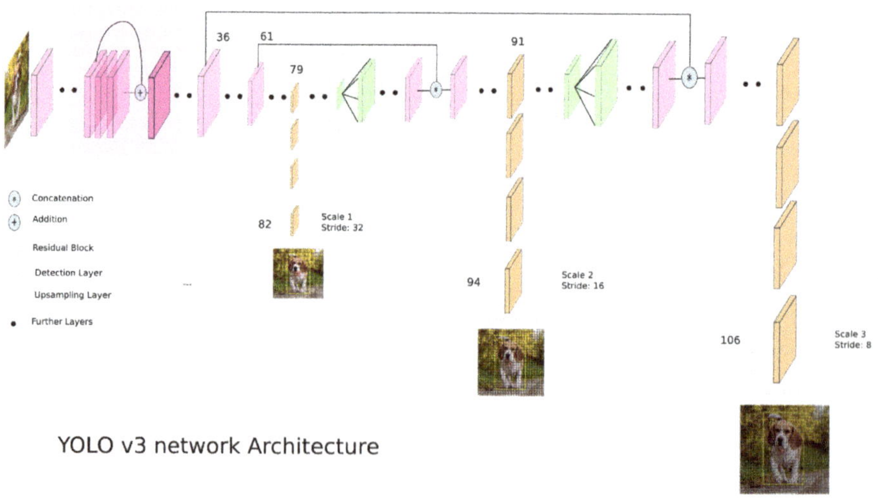

Figure 2. YOLOv3 structure from Ayoosh Kathurias blog post in Towards Data Science [67]. Reprinted with permission from [67]. 2018, Ayoosh Kathuria.

In their paper, Redmon et al. tried to use Focal Loss as a loss function. However, this did not bring any further improvements. YOLOv3, on the other hand, relies on logistic regression to determine the object confidence and class predictions [30].

YOLOv3 is currently considered to be one of the fastest object detection algorithms and is a good choice if real-time detection is needed, with negligible loss of accuracy.

2.1.5. Data Augmentation

In general, data augmentation methods can be divided into two categories: traditional and advanced methods. While advanced methods such as Generative Adversial Networks (GANs) are computationally intensive, traditional methods have a low complexity time [68]. In this paper, we focus on the use of different traditional data augmentation techniques due to their ease of use. All data sets are subjected to data augmentation. The parameters were empirically adjusted to the specific properties of the image types.

Mirroring

The mirroring is performed by swapping the individual pixel values on the basis of the symmetry axes. Either a single mirroring on x- or y-axis or a double mirroring on both is accomplished. In addition to the single images, the bounding box is also transformed to prevent relabeling. Obtained images are shown in Figures 3b and 4b.

Rotation

The rotation operation uses degree increments of 90° between 0° and 360°. Other angles were neglected here, as it would have entailed relabeling. To maintain the proportions of the image content, the rotated images are then resized back to their original size. For the thermal and fused images, this corresponds to a size of 640×480, for the RGB images the total size is 1146×716. This can be seen in Figures 3c and 4c.

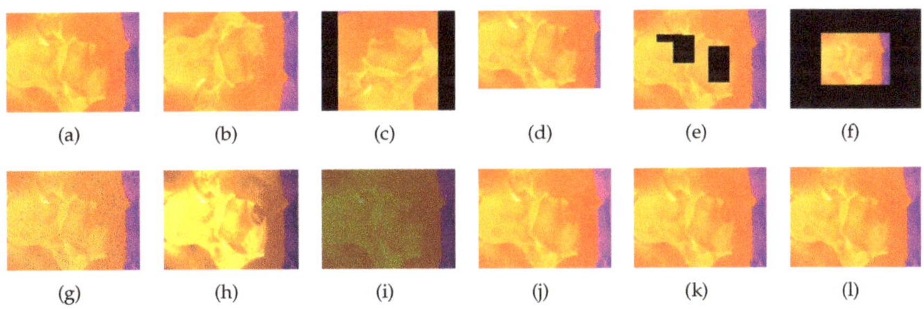

Figure 3. Augmented thermal images: (**a**) original; (**b**) vertical mirroring; (**c**) rotation; (**d**) random crop; (**e**) erasing; (**f**) zooming; (**g**) salt/pepper; (**h**) histogram equalization; (**i**) contrast; (**j**) saturation; (**k**) blurring; (**l**) sharpening [69].

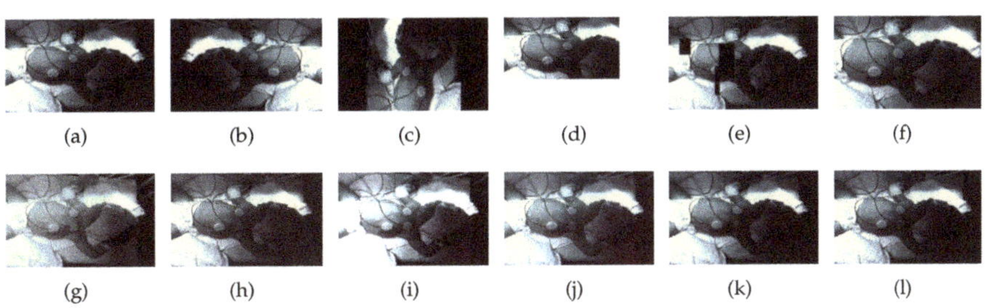

Figure 4. Augmented RGB images: (**a**) original; (**b**) horizontal mirroring; (**c**) rotation; (**d**) random crop; (**e**) erasing; (**f**) zooming; (**g**) histogram equalization; (**h**) Gaussian; (**i**) contrast; (**j**) saturation; (**k**) blurring; (**l**) sharpening [69].

Zooming

This transformation is dedicated to zooming in or out to the center of an image. It returns an enlarged or shrunk view of the image without changing the image dimensions. The zoom factor is set at random between the values 0.5 and 1.5. This corresponds to a reduction or enlargement of 50%. Zooming is also applied on the bounding boxes to avoid relabeling. Images transformed with zooming can be seen in Figures 3f and 4f.

Random Crop

Random crop is similar to zooming. The difference is that instead of zooming in or out of the center of the image, a randomly generated section of the image is selected. This is then used as the new image. The bounding box is either recalculated or discarded if less than 75% of the area is in the image section. The image dimensions are retained. The image section is moved to the upper left corner of the new image. The rest of the image is filled in with black (see Figures 3d and 4d).

Erasing

This transformation randomly erases rectangular areas of different sizes within an image. If bounding boxes with an overlap of more than 50% are deleted, they are discarded. Up to three of these rectangles are applied per image. Sample images can be viewed in Figures 3e and 4e.

Noise

Several genres of noise can be utilized for data augmentation. In this case, either salt-and-pepper (Figure 3g) or Gaussian noise (Figure 4h) is added to the original image. Salt-and-pepper noise is known as impulse noise. It presents itself as sparsely occurring white and black pixels. The transformation is irrelevant for the bounding boxes, so they do not need to be changed.

Contrast and Saturation Changing

The contrast of individual images is changed by multiplying all pixel values by a random constant between 0.5 and 0.75 (see Figures 3i and 4i). By adding a random constant between 0.1 and 0.5, the saturation and brightness of the images are modified. Before this, the images are converted from RGB to HSV color space. The resulting images are shown in Figures 3j and 4j.

Blurring and Sharpening

For blurring, a Gaussian filter is applied to the image with a standard deviation between 2 and 4. Sharpening is performed by using a 3×3 filter kernel with randomly generated weights.

Histogram Equalization

When using the histogram equalization (see Figures 3j and 4j) the intensity distribution of the image is transformed to an equal distribution. It is applied to improve the contrast of an image. The aim is to stretch the intensity values for the display of the image. This improves the visibility of details within the image which otherwise would be hard to be seen. In this paper, the histogram equalization implementation of openCV is used. Before applying it the image has to be converted to the YCbCr color space.

2.2. Hardware Setup

In this section, the hardware setup for recording the data, performing the sensor fusion and training the neural networks is presented.

2.2.1. Thermal Camera

We used the thermal camera Optris PI 640 (by the company Optris, Berlin, Germany) with a field of view of 90° × 64°. The camera has an optical resolution of 640 × 480 pixels. It works within the wavelength spectrum between 7.5 µm and 13 µm. We use the measurement range of −20 °C to 100 °C. The thermal sensitivity is 75 mK. The ROS package *optris_drivers* is used as driver [70].

2.2.2. RGB Camera

The RGB camera used is the JAI GO-2400C-USB Camera with a Fish-Eye lens FUJINON FE185C086HA-1 by Sony (Tokyo, Japan). It has 7.5 µm square pixels and a field of view of 185° × 185°. The driver used was *ros_cvb_camera_driver* [71]. The resulting image has a size of 1146 × 716 pixels.

2.2.3. 3D-Time-of-Flight Camera

We use a 3D time-of-flight camera called CamBoard pico flexx by the company pmd (Siegen, Germany). The camera has a field of view of 62° × 45° and a resolution of 224 × 171 (38k) pixels. The measurement range is 0.1–4 m. A frame rate of up to 45 frames per second (fps) (3D frames) is delivered. We set a frame rate of 35 fps in our case. The camera can be connected via USB to an Ubuntu 18.04 computer with Robot Operating System (ROS) Melodic installed. The Royale SDK provided by the manufacturer as well as the ROS package *pico_flexx_driver* are the drivers used [72].

2.2.4. Computer with GPU

An OMEN HP computer with Ubuntu 18.04.6 bionic operating system and a GeForce RTX 1080 Ti GPU from NVIDIA (NVIDIA, Santa Clara, CA, USA) (11 GB Memory) is used. We used the NVIDIA drivers 465.19.01, CUDA 11.3, cuDNN. We had the Melodic ROS version installed for recording the camera data.

2.3. *Software Algorithms*

All software algorithms for the sensor fusion as well as the face detection of our approach will be described in the following.

2.3.1. Calibration

For the intrinsic calibration the ROS-package *camera_calibration* was used [73]. There is a special calibration program for the fish eye optics of the RGB camera [74].

Thermal Camera

The calibration target we used was made up of a 2 mm thick aluminum plate and a 1 mm thick high density fibreboard from which 6 × 20 circles with a diameter of 1 cm were cut out with a laser cutter. The distance between the circle centers is 13.96 mm. The calibration target was placed in the image with a horizontal orientation such that the long edge lies horizontally and the short edge vertically within the image. In total, 144 images were read. For the calibration the ROS package *camera_calibration* was used.

Concomitantly, *rqt_reconfigure* needs to be started so that the temperature threshold can be set. In this case, the temperatures are coded in black and white. Depending on the temperature of the calibration target the suitable threshold varies in order to make all circles of the calibration target visible.

This resulted in a very good rectification of the image (see Figure 5). The cushion effects at the edges of the image are completely removed and the calibration target is visualized as rectangle within the thermal camera image. This was tested at different areas within the image. Using a baby doll, the occurrence of proximity-related distortions was excluded.

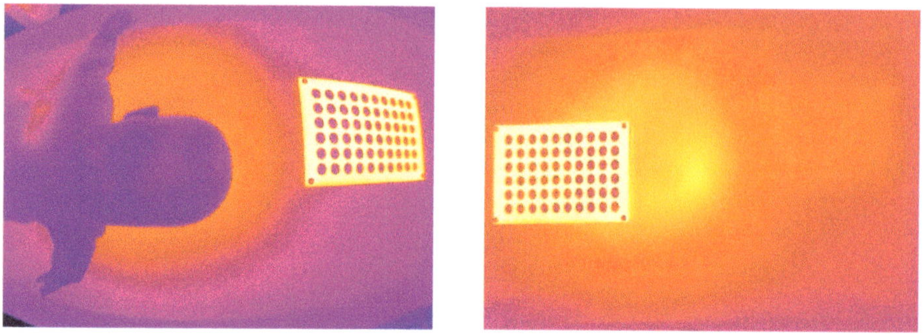

Figure 5. Intrinsic calibration of the thermal camera with 144 images. Rectified image (**right**) and non-rectified image (**left**). The cushion effect is sufficiently removed.

RGB Camera

First the RGB image was cropped to a size of 1146 × 716 pixels. The calibration was then performed with a 5 × 9 chess board with squares of size 14 mm. The result can be seen in Figure 6.

In the last step, the values of the calibration were manually adapted such that the cushion effect decreases. It was empirically determined how the image changes when changing the projection as well as the distortion matrix. The step by step adaption can be seen in Figure 7.

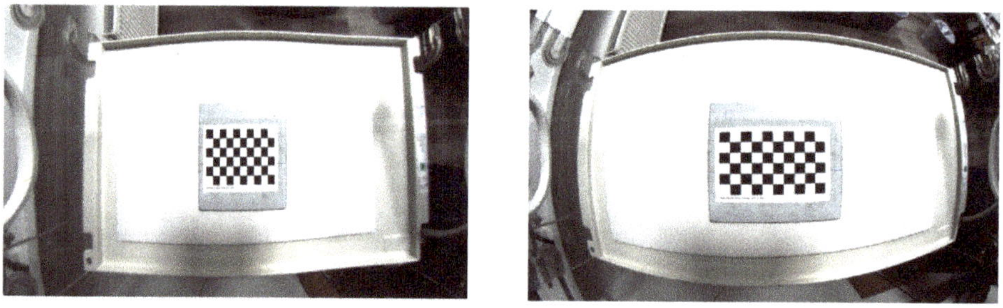

Figure 6. Intrinsic calibration with the cropped RGB camera image using 196 images (**left**). The calibration target is a 5 × 9 chess board with squares of size 14 mm. Non-rectified image (**right**).

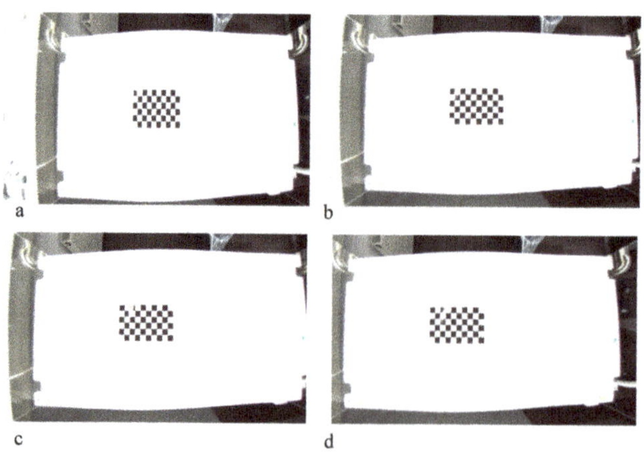

Figure 7. *Cont.*

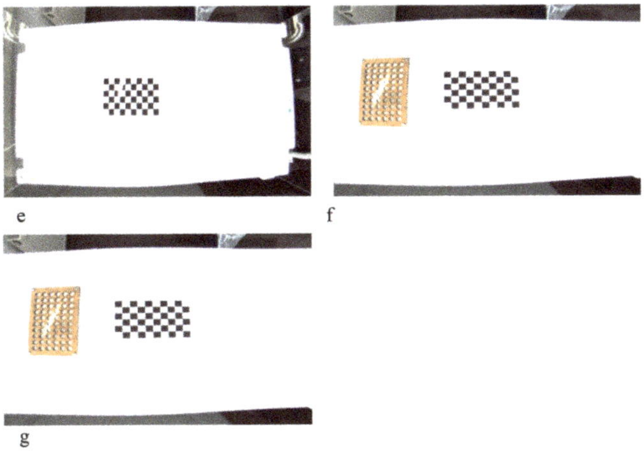

Figure 7. Empirical adaption of the intrinsic calibration of the RGB camera. (**a**) original image; (**b**) change the first value of the projection matrix from 337.7501220703125 to 400; (**c**) change the first value of the projection matrix from 337.7501220703125 to 410; (**d**) change the second value of the projection matrix from 567.37183155695675 to 500; (**e**) change the third value of the projection matrix from 456.1886901855469 to 470; (**f**) projection matrix with first value 560, second value 500 and third value 470; (**g**) change distortion matrix to [−0.2282522216659725, 0.03337792115699235, 0.0010099526756546235, −0.0010636239771667497, 0] and previous projection matrix.

2.3.2. Thermal-RGB-Fusion

Circle Detection

For the detection of the circle centers, blob detection was used. Blob detection delivers so called keypoints which hold the center coordinates of the blob [75,76]. The results can be seen in Figure 8.

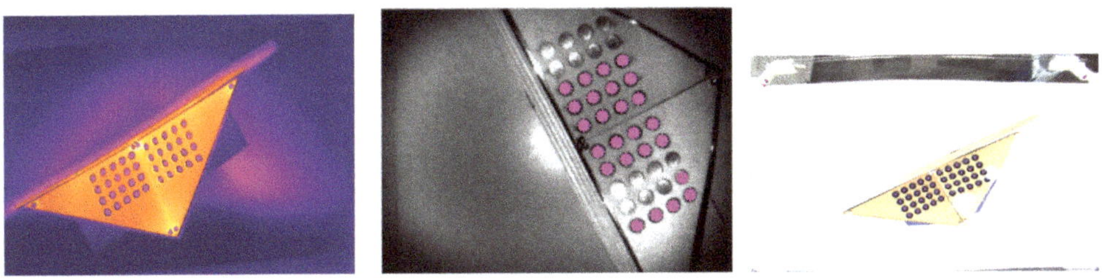

Figure 8. Blob detection within the thermal (**left**), ToF (**middle**) and RGB camera image (**right**). For the RGB image the contrast had to be increased by placing a blue paper sheet underneath.

RGB-ToF and Thermal-ToF Fusion

The pixel indices of the ToF camera image and the point cloud correspond to each other due to the underlying internal fusion. By ordering the point cloud the corresponding 3D point of each pixel can be determined. In order to calculate the transformation, the 3D ToF points and the 2D thermal and RGB points are inserted into the function *SolvePnP*. The function calculates the orientation and position of the object from the given object points (3D ToF points) and image projections (2D thermal and RGB points), as well as the camera matrix and the distortion coefficients [77]. Internally, the Direct Linear Transform (DLT) algorithm is applied which calculates the matrix projecting the 3D ToF points into the 2D image plane.

In order to receive the rotation matrix from *rvec* the openCV function *Rodrigues* can be used [78]. It calculates the corresponding matrix from the vector. For applying the calculated transformation on the ToF data and to project the points into each image the function *projectPoints* is applied [79].

In the following, the point correspondences which were used to calculate the RGB-ToF fusion (26 correspondences, see Figure 9) and the thermal-ToF fusion (15 correspondences, see Figure 10) are marked and the projected points into the images are shown. The most suitable combination of point correspondences was found empirically.

In order to achieve a very high fusion precision significant points from the video of the neonate can be included as point correspondences. In our case, individual adaptions had to be made for each subject.

Once the extrinsic transformations are known, the thermal-RGB-fusion can be calculated. For each ToF point within the point cloud the corresponding RGB and thermal pixels are determined (see Figure 11). Then, the color information of the RGB pixel is inserted at the corresponding point within the thermal image, delivering the final fusion image. The size of the fusion image is the same as the thermal camera image.

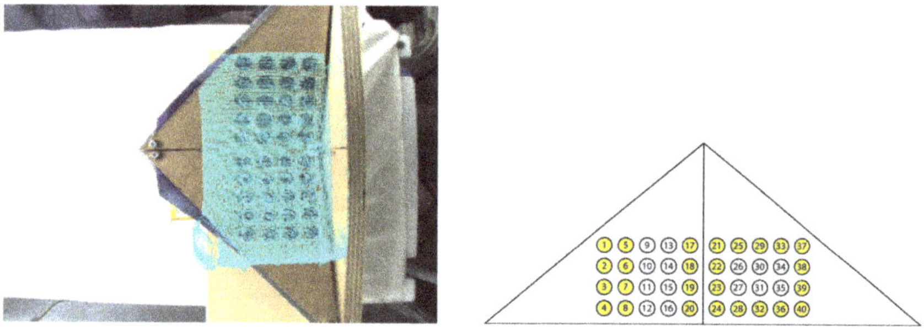

Figure 9. Extrinsic calibration of the RGB with the ToF camera; 26 point correspondences were used (**right**). The turquoise points are the projected ToF points (**left**).

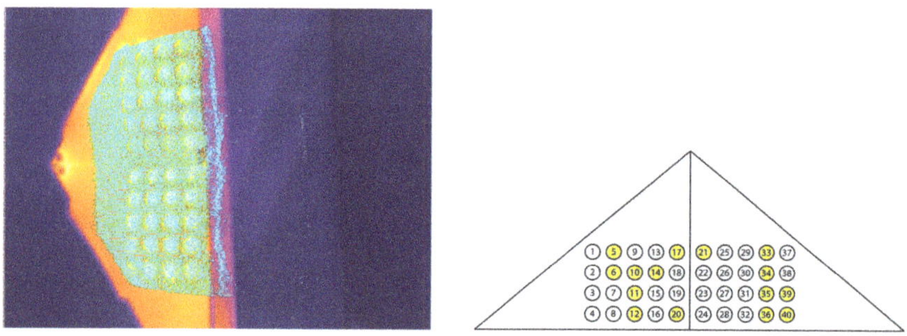

Figure 10. Extrinsic calibration of the thermal with the ToF camera; 15 point correspondences were used (**right**). The turquoise points are the projected ToF points (**left**).

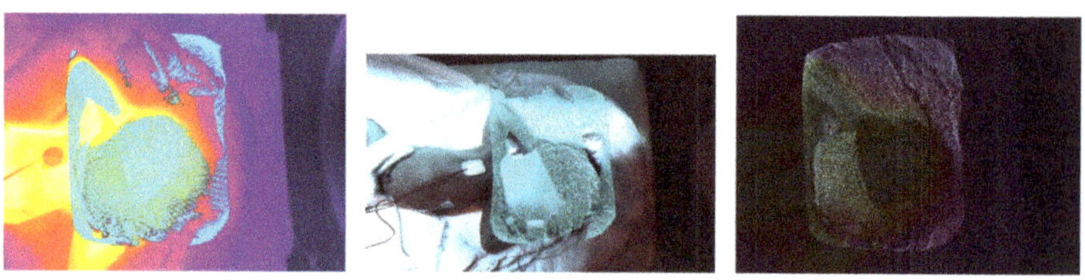

Figure 11. Thermal-ToF-fusion (**left**), RGB-ToF fusion (**middle**) and thermal-RGB-fusion camera image (**right**).

2.3.3. Neural Networks RetinaNet and YOLOv3

RetinaNet

In this paper, the RetinaNet implementation of the Git repository of fizyr is used [80], supporting *keras* and *tensorflow* (v2.4. were employed). For functional GPU support, some software packages needed to be installed. This includes NVIDIA drivers (driver 465.19.01), *CUDA* 11.3, *cuDNN* and *OpenCV* 4.5.254.

YOLOv3

In this paper, the Git repository of AlexeyAB is used [81]. For a smooth use of YOLOv3, several software requirements need to be met (details see above).

2.4. Measurement Series

The datasets holding thermal, RGB and ToF data were recorded at the Department of Pediatrics and Adolescent Medicine, University of Erlangen-Nürnberg, Germany. The study was conducted in accordance with the Declaration of Helsinki, and approved by the Ethics Committee of the Friedrich-Alexander Universität Erlangen-Nürnberg (protocol code: 197_20B and date of approval: 28 April 2021, amendment protocol code: 20-197_1-B, approved on 22 December 2022).

2.4.1. Subjects

The subjects included in this study (see Tabel 1) were neonates of a minimum gestational age of 34 + 0 (late pre-term) that were clinically stable from a cardio-respiratory perspective. All subjects required phototherapy due to jaundice. The recordings took place in parallel to phototherapy.

Table 1. List of subjects. Term neonates were > 37 + 0 weeks of gestation.

Subject	Gestational Age	Age during Study	Sex	Weight
01	34 + 0	2 days	male	1745 g
02	term	5 days	female	3650 g
03	term	4 days	female	2330 g
04	term	13 days	male	3300 g
05	term	2 days	female	2750 g

2.4.2. Training, Validation and Test Datasets

Our training datasets contain 1400 labeled images plus 1400 augmented images per each modality from subject 01 to 04. Our validation dataset contains 300 images from each modality of subject 01 to 04 which were not included in the training dataset. Table 2 shows the number of instances per class per modality.

Table 2. Number of instances per class per modality within the validation dataset.

Modality	Head	Nose	Torso	Intervention
RGB	1193	575	1129	160
Thermal	1199	305	1183	123
Fusion	1190	8	1055	80

The validation dataset was used to calculate the levels of average precision during the training process. With the test dataset we evaluate the performance of our trained networks. As test dataset 65 random images of each image modality of subject 05 were used which have not been seen by the networks yet. It is important that the distribution of the test dataset is similar to the training dataset.

3. Results

In this section, the training results for each modality and network are presented. Each network was trained for 64 epochs and the epoch with the best precision was determined. First, we show the quantitative results, where we compare the average precision per class. In the second step we present the test results with our test dataset.

3.1. RGB Dataset

In this section, we describe the results for the RGB dataset.

3.1.1. RetinaNet

Figure 12 shows the average precision vs. epochs for the different classes within the RGB images. The best results are achieved in epoch 25. In this epoch, we obtain an average precision of 1.0 for the head, 0.9937 for the nose, 0.99 for the torso and 0.94 for the intervention.

We used the weights of epoch 25 to evaluate our test dataset. We picked two representative images to visualize the detections (see Figure 13). The corresponding confidence scores are shown within the image. We achieved APs of 0.9255, 0.0, 0.7841 and 0.5884 for the classes head, nose, torso and intervention, respectively, over the whole test dataset.

3.1.2. YOLOv3

Figure 14 shows the average precision vs. epochs for the different classes within the RGB images for YOLOv3. The best precision is reached in epoch 64. In this epoch we receive APs of 1.0, 0.9885, 0.9991 and 0.9821 for head, nose, torso and intervention, respectively.

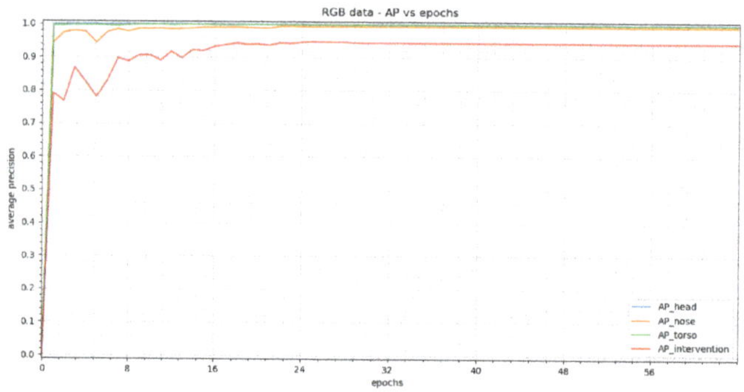

Figure 12. The training results for the RGB images achieved by RetinaNet can be seen in this figure. The average precision (AP) vs. epochs for the different classes are displayed. The trend shows a steep rise at the beginning which results in a stable plateau in epoch 25.

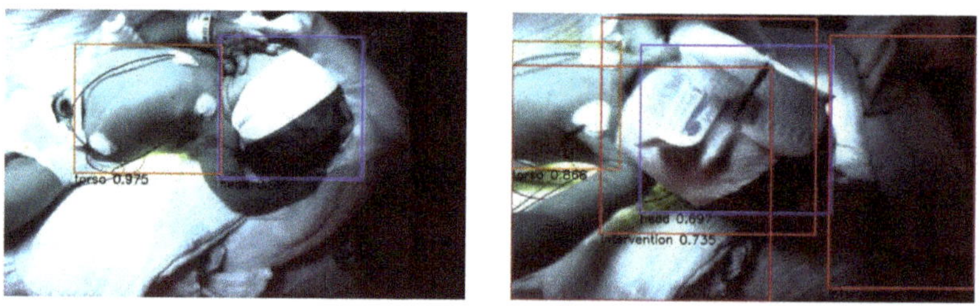

Figure 13. Visual evaluation of two example RGB images with RetinaNet. Good case (**left**) with confidence scores 0.992 for head and 0.975 for torso. Bad case (**right**) where the head is also detected as intervention with confidence scores 0.697 for head, 0.868 for torso and 0.735 for intervention. The false detection probably occurs due to the shorter distance between the face and the camera. This case is not sufficiently represented within the training dataset.

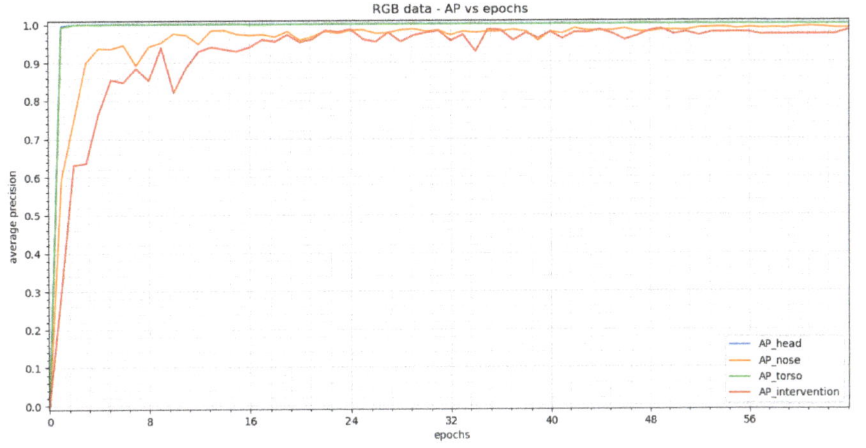

Figure 14. The training results for the RGB images reached by YOLOv3 can be seen in this figure. The average precision (AP) vs. epochs for the different classes are displayed. The trend shows a steep rise at the beginning and an oscillating behavior for class "nose" and "intervention". The best precision values for all classes were achieved in epoch 64.

The weights of epoch 64 for evaluating our test dataset were used. Figure 15 shows two representative images to visualize the detections. The corresponding confidence scores can be found within the image description. We reach APs of 0.8563, 0.0, 0.8301 and 0.536 for the classes head, nose, torso and intervention, respectively, over the complete test dataset.

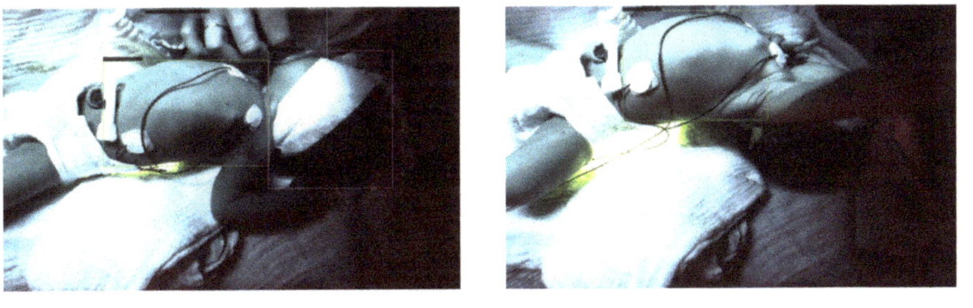

Figure 15. Visual evaluation of two example RGB images with YOLOv3. Good case (**left**) with confindence scores of 0.95 for the head, 0.99 for torso and 0.58 for the intervention. Bad case (**right**) where the head is not detected caused by the occlusion by the arm. Another factor is the twisted pose of the neonate's body. This leads to a confidence score of 0.30 for the torso.

3.2. Thermal Dataset

In this section, the results for the thermal dataset are presented.

3.2.1. RetinaNet

In Figure 16 the average precision vs. epochs for the different classes within the thermal images can be seen. Epoch 28 delivers the best results. We achieve an average precision of 0.9969, 0.9864, 0.9862 and 0.8695 for head, nose, torso and intervention, respectively.

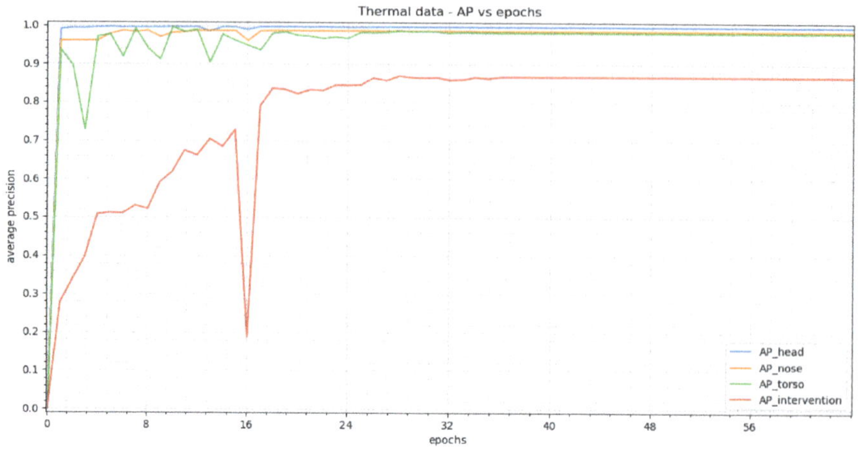

Figure 16. The resulting average precision values for the thermal images trained with RetinaNet are shown within the figure. First, a steep incline can be seen with the exception of the class "intervention" which shows a slower rise. In epoch 16, a collapse can be noted. By epoch 28 the precision values result in stable plateaus.

To evaluate our test dataset the weights of epoch 28 were used. Figure 17 shows two representative images to visualize the detections. The corresponding confidence scores can be found within the image. For the evaluation of our whole dataset we receive APs of 0.9227, 0.0, 0.7816 and 0.4856 for the classes head, nose, torso and intervention, respectively.

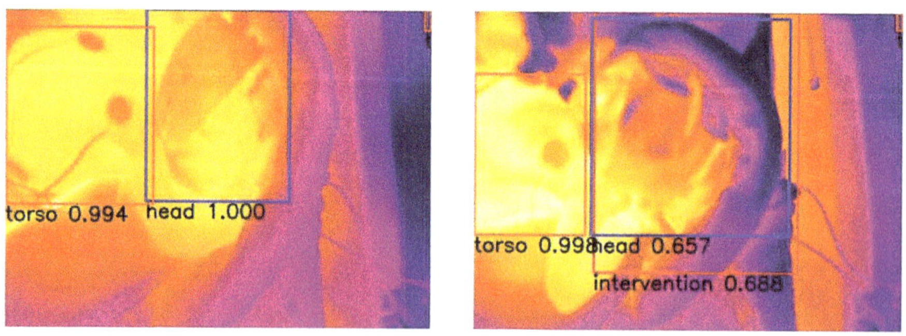

Figure 17. Visual evaluation of thermal images with RetinaNet. Good case (**left**) with confidence scores 1.0 for head and 0.994 for the torso. Bad case (**right**) where the head is also detected as intervention. The false detection occurs as the cloth surrounding the neonate's face has a similar temperature as an adult's, and is therefore detected as intervention. This leads to confidence scores of 0.657 for the head, 0.998 for the torso and 0.688 for the intervention.

3.2.2. YOLOv3

Figure 18 displays the average precision vs. epochs for the different classes within the thermal images for YOLOv3. The best results occur in epoch 61. There APs of 0.9983 for the head, 0.9993 for the nose, 0.9963 for the torso and 0.9225 for the intervention are achieved.

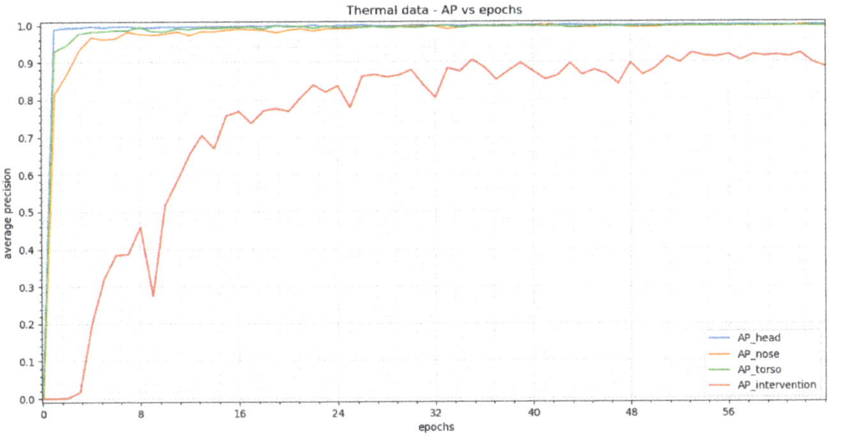

Figure 18. This figure shows the results for the training of YOLOv3 with thermal images. A steep increase can be seen at the beginnig for all classes other than "intervention". "Intervention" shows an oscillating behavior.

We used the weights of epoch 61 to evaluate our test dataset. Figure 19 displays two representative images to visualize the detections. The corresponding confidence scores are shown within the image description. We receive APs of 0.8924, 0.0, 0.8611 and 0.5706 for the classes head, nose, torso and intervention, respectively, over the complete test dataset.

3.3. *Fusion Dataset*

In this section, we describe the results for the fusion dataset.

3.3.1. RetinaNet

Figure 20 shows the average precision vs. epochs for the different classes within the fusion images. In this case, the best epoch is number 38. We receive average precision values of 0.9949 for the head, 0.0 for the nose, 0.9934 for the torso and 0.7683 for the intervention.

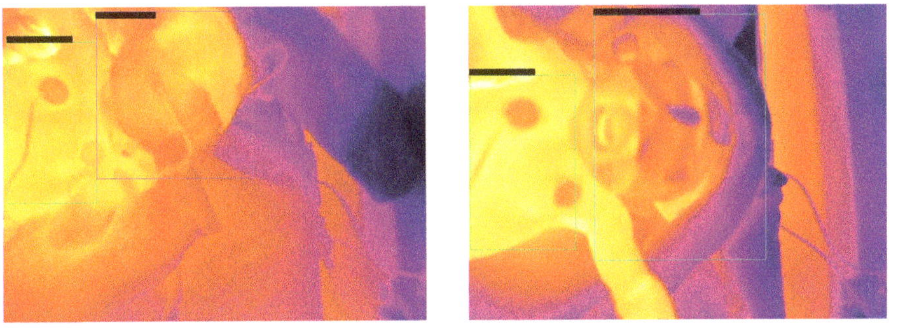

Figure 19. Visual evaluation of thermal images with YOLOv3. Good case (**left**) with confidence scores of 1.0 for the head and 0.95 for the torso. Bad case (**right**) where the head is detected as intervention with confidence scores of 0.99 for the torso and 0.25 for the intervention. The reasons are the same as mentioned for RetinaNet.

Figure 21 displays two representative images to visualize the detections. For the evaluation of our test dataset the weights of epoch 38 were used. The corresponding confidence scores can be seen within the image. We reached APs of 0.9958, 0.0, 0.6863 and 0.6574 for the classes head, nose, torso and intervention, respectively, over the whole test dataset.

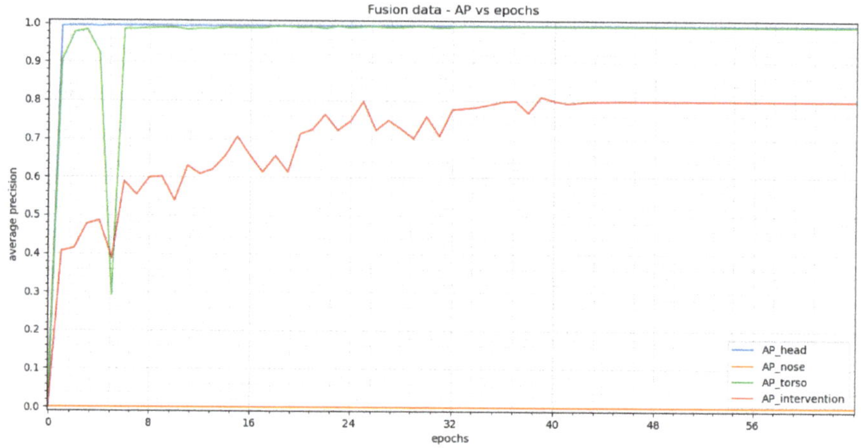

Figure 20. This figure displays the training results of RetinaNet with fusion images. The trend of "head" and "torso" has a steep incline at the beginning, resulting in a stable plateau in epoch 7. Nonetheless, there is a strong decline in epoch 5 for "torso". "Intervention" shows a slower rise, resulting in a plateau in epoch 38. Due to the insufficient amount of instances of "nose", the results cannot be taken into account. A further discussion of this issue can be found in Section 4.

3.3.2. YOLOv3

In Figure 22, the average precision vs. epochs for the different classes within the fusion images for YOLOv3 can be seen. Epoch 56 delivers the most precise findings. In this epoch, we obtain average precision values of 0.9949 for the head, 0.3274 for the nose, 0.9948 for the torso and 0.8390 for the intervention.

Figure 23 shows two representative images to visualize the detections. Our test dataset was evaluated using the weights of epoch 56. The corresponding confidence scores can be found within the image description. We reach APs of 0.9455, 0.0, 0.7864 and 0.5199 for the classes head, nose, torso and intervention, respectively, over the complete test dataset.

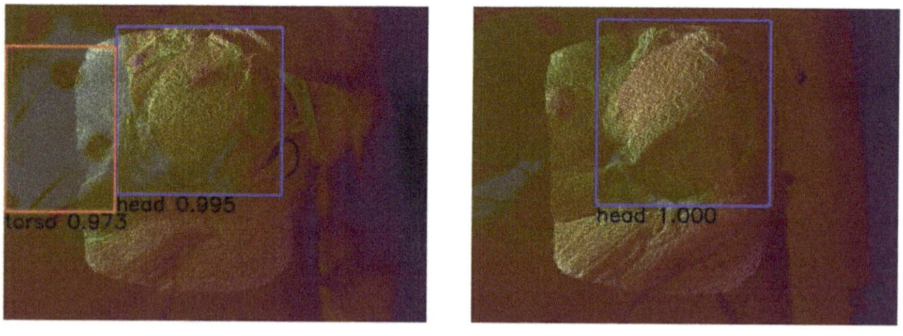

Figure 21. Visual evaluation of two fusion images with RetinaNet. Good case (**left**) with confidence scores of 0.995 for the head and 0.973 for the torso. Bad case (**right**) where the torso is not detected with a confidence score of 1.0 for the head. This is most likely caused by the small overlap of the torso and the ToF pointcloud due to a twisted pose of the neonate. Those images were not represented in our training dataset.

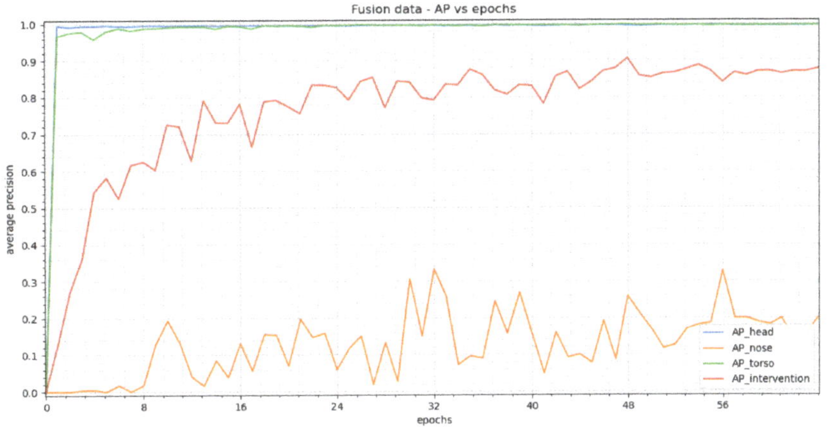

Figure 22. The results of YOLOv3 trained with fusion images can be seen in this figure. At the beginning a steep rise can be seen for "head" and "torso", resulting in a stable plateau in epoch 24. "Nose" and "intervention" both show an oscillating behavior. "Intervention" has overall higher precision values than "nose" and its oscillation has a smaller amplitude. The precision values for "nose" cannot be taken into account due to an insufficient amount of class instances.

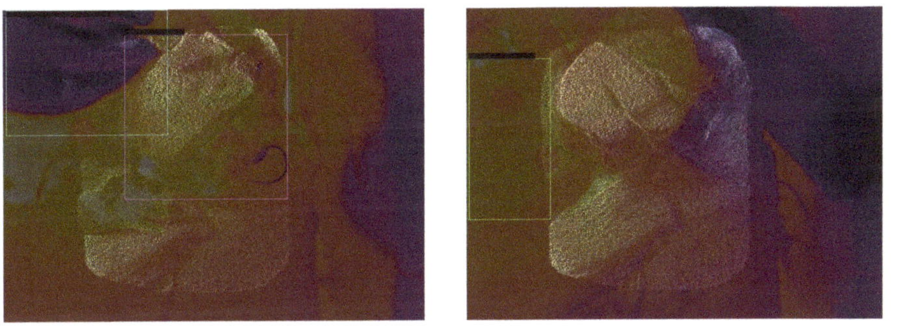

Figure 23. Visual evaluation of two sample fusion images with YOLOv3. Good case (**left**) with confidence scores of 0.95 for the head and 0.94 for the intervention. Bad case (**right**) where the head is not detected. The confidence score for the torso is 0.40.

4. Discussion

In this section, we discuss the results of the sensor fusion as well as the training results. Finally, we compare our results to the state of the art.

4.1. Comparison of Theoretical Approach for the Sensor Fusion

When we performed our sensor fusion approach we expected that the procedure only has to be undertaken once and that the extrinsic transformation can be applied to all subjects. This was not the case, most probably as small changes of the neonatal head and body position have a great impact due to the proximity of the cameras to the subject.

4.2. Discussion of Training Results

When discussing our training results we refer to the average precision values in Table 3.

Table 3. Average precision values for different classes using different modalities and networks (validation dataset).

	Modality	Epoch	Average Precision			
			Head	Nose	Torso	Intervention
RetinaNet	RGB	25	1.0	0.9937	0.99	0.94
	Thermal	28	0.9969	0.9864	0.9862	0.8695
	Fusion	38	0.9949	0.0 *	0.9934	0.7683
YOLOv3	RGB	64	1.0	0.9885	0.9991	0.9821
	Thermal	61	0.9983	0.9993	0.9963	0.9225
	Fusion	56	0.9949	0.3274 *	0.9948	0.8390

* The precision values for "nose" cannot be taken into account due to an insufficient number of class instances within the fusion dataset.

4.2.1. RetinaNet

When comparing the average precision with its trend of class "head" for all modalities almost identical results can be seen (smaller than 1%, see Table 4). The trend of the class "torso" within the RGB image is very similiar to the class "head" curve as can be seen in Figure 12. The same holds true for the fusion with the exception of one epoch where the precision drops significantly (see Figure 20). As this is only the case for one epoch this might be due to a wrong adaption of the weights at the beginning of the training. The thermal data shows an oscillating behavior until it reaches a stable value. This is due to the adjustment of the learning rate during the training process. The optimizer used adapts the step length such that it decreases the closer the optimum. A similar collapse can be seen for "intervention" within the thermal images (see Figure 16).

Table 4. Difference between fusion and RGB/thermal in average precision values per class for RetinaNet (validation dataset).

	Modality	Epoch	Average Precision			
			Head	Nose	Torso	Intervention
RetinaNet	Fusion—RGB	13	−0.0051	−0.9937 *	0.0034	−0.1717
	Fusion—Thermal	10	−0.002	−0.9864 *	0.0072	−0.1012

* The precision values for "nose" cannot be taken into account due to an insufficient number of class instances within the fusion dataset.

When looking at the trend of the "intervention" over all modalities we can see major differences in the precision compared to the "head" and "torso" (see Table 4). Furthermore we observed a relationship between the number of instances per class within the validation dataset and the reached precision i.e., 160 instances within RGB images leads to an AP of 0.94, 123 instances within thermal images gives an AP of 0.8695 and 80 instances within fusion images results in an AP of 0.7683. Therefore, we expect that an increase of instances of "intervention" within the thermal and fusion images will lead to higher precision. The same behavior can be seen for the "nose" within the fusion images where only right instances of the nose were included. Therefore, the precision values for the "nose" should not be taken into account. For the thermal and RGB images 300 "nose" instances within the dataset are enough to reach a similar precision as for the "head" and the "torso".

As the fusion images hold more information than the RGB and thermal images by themselves the training takes more time (around 10 more epochs) to achieve the same precision.

The results generated by the test dataset show an overall decrease in the average precision. The reason for this lies in the complexity of some of the test images which were not present in the training dataset. This can especially be seen for the torso where often only the back or the side was visible due to the neonate lying in a twisted pose. This case was not covered by the training dataset. The training dataset also did not contain new forms of intervention such as the appearance of a nursing bottle. Additionally, the number of instances of the class "intervention" in the training dataset is low, leading to an overall worse performance. As before, the number of instances of the class "nose" is too low, resulting in an AP of 0.0 for all modalities. Clinically the most significant class is the head as the temperature and heart rate detection will depend on it. There were only small differences in the average precision values (improvement of 0.9% for the fusion, decrease of 4.42% for the thermal images and 7.45% for the RGB images). This shows that the fusion performs best. Due to the small number of test subjects included in the study a slight overfitting tendency was observed. We expect the average precision values to reach values comparable with the validation dataset once more subjects are included in the training process.

4.2.2. YOLOv3

For YOLOv3 the classes "head" and "torso" deliver identical precision values for all modalities (difference less than 1%, see Table 5). As with the RetinaNet analysis, a direct relationship was observed in the validation images between the number of instances for the "intervention" class and the achieved average precision values. This means 160 instances within RGB images leads to an AP of 0.9821, 123 instances within thermal images gives an AP of 0.9225 and 80 instances within fusion images results in an AP of 0.8390. As before there are not enough instances of class "nose" within the fusion dataset to achieve precise detection. Therefore, these precision values should not be taken into account. The trends for "intervention" and "nose" show a fluctuating behavior. This can be explained by the usage of the Stochastic Gradient Descent (SGD) optimizer and its learning rate. (The optimizer used adapts the step length such that it decreases the closer the optimum).

Table 5. Difference between fusion and RGB/thermal in average precision levels per class for YOLOv3 (validation dataset).

	Modality	Epoch	Average Precision			
			Head	Nose	Torso	Intervention
YOLOv3	**Fusion—RGB**	−8	−0.0051	−0.6611 *	−0.0043	−0.1431
	Fusion—Thermal	−5	−0.0034	−0.6719 *	−0.0015	−0.0835

* The precision values for "nose" cannot be taken into account due to an insufficient number of class instances.

Similar to RetinaNet an overall decrease in the average precision with the test dataset can be observed for YOLOv3. As the same test dataset was used the same issues occur as described before. These arise due to the complexity of the test images and new forms of intervention which were not taken into account before (e.g., nursing bottle). Another problem is the low number of instances of the classes "nose" and "intervention". For the class "head" a decrease of 14.4% for the RGB, a decrease of 10.6% for the thermal images and 4.94% for the fusion images in the average precision values were observed. The data exhibit an overfitting trend due to the small number of test subjects. As mentioned before, we expect our average precision values to reach values similar to those of the validation dataset when using more test subjects for training.

4.2.3. Comparison between RetinaNet and YOLOv3

When comparing results from RetinaNet and YOLOv3 for the validation fusion data, almost identical average precision values are reached for the classes "head" (difference of 0%) and "torso" (difference of 0.14%). For class "intervention" YOLOv3 has an increase in precision of around 7% compared to RetinaNet (see Table 6). Even though YOLOv3 reaches an average precision of 0.3274 for the "nose", this result should be discounted, as it is caused by the small number of instances within the class. RetinaNet needs fewer epochs to achieve the highest precision values than YOLOv3. When taking into account the oscillating behavior of YOLOv3, as well as distinguishing between the different classes, we see that a similar number of epochs lead to stable precision values for RetinaNet when compared with YOLOv3

Table 6. Comparison of average precision values between classes and modalities RetinaNet—YOLOv3 (validation dataset).

	Modality	Epoch	Average Precision			
			Head	Nose	Torso	Intervention
RetinaNet—YOLOv3	RGB	−39	0.0	0.0052	−0.0091	−0.0421
	Thermal	−33	−0.0014	−0.0066	−0.0101	−0.053
	Fusion	−18	0.0	−0.3274 *	−0.0014	−0.0707

* The precision values for "nose" cannot be taken into account due to an insufficient number of class instances.

As previously stated, both networks exhibit less favourable performances when used on the test dataset compared with the validation dataset due to the aforementioned reasons. Depending on the class one of the two networks performs slightly better (see Section 3).

Redmond et al. [30] and Hennemann [69] mention that the computing time per image is around three times greater for RetinaNet than YOLOv3. A similar difference was observed in our study. We reached approx. 58 ms and approx. 14 ms for RetinaNet and YOLOv3 on our hardware setup.

4.2.4. Summary

In summary, the fusion network is able to deliver the same precision values as the RGB and thermal networks. This is the case if there is an equal number of images and instances per class for each modality and class. If the extrinsic transformation between the thermal and RGB camera is given, the position of the detected region can automatically be calculated for the RGB and the thermal image. A necessary condition is that the region of interest (ROI) lies within the ToF pointcloud. This results in the great advantage that only one network needs to be trained and, therefore, no labeling of the RGB and thermal data is needed. This simplifies the generation of datasets with respect to time and cost (we save 66%).

As shown by Hennemann, better results over all classes can be achieved by training with a greater number of test subjects [69].

When comparing the average computing time of the two networks, YOLOv3 performs slightly better in our setup. When taking the average precision values into account, we observed only marginal differences in average precision values between classes. We therefore believe that both networks are adequate for neonatal face detection.

4.3. Comparison with State of the Art

First we compare our sensor fusion approach to the methods mentioned in Section1.1. We describe how our results using a neural networks training approach compare with other neonatal face detection networks.

4.3.1. Sensor Fusion

To our knowledge, our sensor fusion approach is the first which is explicitly designed for the short distances commonly encountered in the setting of incubator nursing of premature neonates on the NICU. At the time of completion of this work, no literature regarding the use of ToF camera as third sensor for indirect fusion was found.

4.3.2. Neural Network for Face Detection

In this section, we compare our results with the best performing findings in the literature for each class mentioned in Section 1.1. Green et al. reach an average precision of 0.982 for the detection of the head [35]. Our fusion approach achieves a precision of 0.9949 (RetinaNet) and 0.9949 (YOLOv3). For the test dataset we reach 0.9958 and 0.9455 for

RetinaNet and YOLOv3, respectively. This shows that we achieve similarly high precision values for instances of the "head" class within data the neural network has not seen before. For "nose" and "intervention" there are no precision values available within the literature. Beppu et al. receive an AP of 0.985 for the torso [49]. We observe precision values of 0.9934 (RetinaNet) and 0.9948 (YOLOv3) with the fusion network. When considering our test dataset we achieve APs of 0.6863 (RetinaNet) and 0.7864 (YOLOv3). As mentioned before, we expect these values to increase to similar precision values when training with more test subjects. As we can directly calculate the detections for the RGB and thermal images, we will also receive such high precision. So far, we are the first to report the use of thermal-RGB-fused images of neonates to train a neural network.

We hope our results will facilitate the development of non-contact neonatal monitoring techniques to improve the standard of care for this vulnerable group of patients.

5. Conclusions

In this contribution, we investigated the use of thermal-RGB-fusion data for a robust neonatal face detection method based on neural networks. We showed that it is possible to achieve a precise sensor fusion for short distances using a ToF camera as third sensor. We trained two different neural networks (RetinaNet and YOLOv3) with thermal, RGB and fusion images and evaluated the precision values achieved. Our fusion networks deliver the same level of precision as the RGB and thermal networks as long as an equal number of instances per class and modality are available (at least 300) within the dataset. Compared to the state of the art we achieve comparable average precision values for the head. Based on the known extrinsic calibration between our cameras we can easily calculate the detection within the thermal and RGB images from the fusion network. Therefore, this increases data efficiency (training of a single network, reduction of data labelling by 66%) and economizes the process. Both RetinaNet and YOLOv3 could be used for detection of the facial region in fused images of neonates in the clinical setting. For the improvement of the generalization of the neural network we intend to train with more subjects in the future.

Author Contributions: Conceptualization, J.G. and L.H.; methodology, J.G. and L.H.; software, J.G. and L.H.; validation, J.G. and L.H.; investigation, J.G. and L.H.; resources, F.B.F.; clinical supervision, F.B.F.; writing—original draft preparation, J.G. and L.H.; writing—review and editing, J.G., L.H., F.B.F., O.H., C.N. and A.K.; visualization, J.G. and L.H.; supervision, J.G., O.H., C.N. and A.K. All authors have read and agreed to the published version of the manuscript.

Funding: This research was funded by the Federal Ministry of Education and Research (BMBF), Germany. Funding reference: 13FH546IX6.

Institutional Review Board Statement: The study was conducted in accordance with the Declaration of Helsinki, and approved by the Ethics Committee of the Friedrich-Alexander Universität Erlangen-Nürnberg (protocol code: 197_20B and date of approval: 28 April 2021, amendment protocol code: 20-197_1-B, approved on 22 December 2022).

Informed Consent Statement: Informed consent was obtained from the parents of all subjects involved in the study.

Acknowledgments: We would like to thank our participants and their parents who agreed to take part in the study. We would like to thank the nursing staff at the neonatal unit and Antje Neubert for her clinical supervision. We would also like to thank Sarra Babba for her support with the labeling of our data.

Conflicts of Interest: The authors declare no conflict of interest.

Abbreviations

The following abbreviations are used in this manuscript:

AP	average precision
BPM	breaths per minute
CNN	Convolutional Neural Network
DLT	Direct Linear Transform
ECG	Electrocardiogram
FFT	Fast Fourier Transform
FoV	field of view
FPN	Feature Pyramid Network
fps	frames per second
GAN	Generative Adversial Network
GPU	graphical processing unit
IoU	Intersection over Union
IR	infrared
LIDAR	light detection and ranging
LSTM	Long Short-Term Memory
LWIR	long wave infrared
NCC	normalized cross correlation
NICU	Neonatal Intensive Care Unit
NIR	near infrared
PCA	Principal Component Analysis
PCL	Point Cloud Library
R-CNN	Region-Based Convolutional Neural Networks
ResNet	Residual Network
RGB	Red Green Blue
RGB-D	Red, Green, Blue -Depth
RMSE	root mean square error
ROI	region of interest
ROS	Robot Operating System
SGD	Stochastic Gradient Descent
SNR	Signal-to-noise ratio
T-ICP	thermal-guided iterative closest point
ToF	time-of-flight
UV	ultraviolet
YOLO	you only look once

References

1. Larsen, R. Pädiatrische Intensivmedizin. In *Anästhesie und Intensivmedizin für die Fachpflege*; Springer: Berlin/Heidelberg, Germany, 2016; pp. 920–949. [CrossRef]
2. Hausmann, J.; Salekin, M.S.; Zamzmi, G.; Goldgof, D.; Sun, Y. Robust Neonatal Face Detection in Real-world Clinical Settings. *arXiv* **2022**, arXiv:2204.00655. [CrossRef]
3. St-Laurent, L.; Prévost, D.; Maldague, X. Fast and accurate calibration-based thermal/colour sensors registration. In Proceedings of the 2010 International Conference on Quantitative InfraRed Thermography, Quebec, QC, Canada, 24–29 June 2010. [CrossRef]
4. Shivakumar, S.S.; Rodrigues, N.; Zhou, A.; Miller, I.D.; Kumar, V.; Taylor, C.J. PST900: RGB-Thermal Calibration, Dataset and Segmentation Network. In Proceedings of the IEEE International Conference on Robotics and Automation, Paris, France, 31 May–31 August 2020; pp. 9441–9447. [CrossRef]
5. Yang, M.D.; Su, T.C.; Lin, H.Y. Fusion of infrared thermal image and visible image for 3D thermal model reconstruction using smartphone sensors. *Sensors* **2018**, *18*, 2003. [CrossRef] [PubMed]
6. Krishnan, A.K.; Saripalli, S. Cross-Calibration of RGB and Thermal Cameras with a LIDAR for RGB-Depth-Thermal Mapping. *Unmanned Syst.* **2017**, *5*, 59–78. [CrossRef]
7. Gleichauf, J.; Vollet, J.; Pfitzner, C.; Koch, P.; May, S. Sensor Fusion Approach for an Autonomous Shunting Locomotive. In Proceedings of the Informatics in Control, Automation and Robotics, Paris, France, 7–9 July 2020; Gusikhin, O., Madani, K., Eds.; Springer: Cham, Switzerland, 2020; pp. 603–624.
8. Tisha, S.M. *LSU Digital Commons Thermal-Kinect Fusion Scanning System for Bodyshape Inpainting and Estimation under Clothing*; Louisiana State University and Agricultural & Mechanical College: Baton Rouge, LA, USA, 2019.
9. Yang, Q.; Yang, R.; Davis, J.; Nistér, D. Spatial-depth super resolution for range images. In Proceedings of the IEEE Computer Society Conference on Computer Vision and Pattern Recognition, Minneapolis, MN, USA, 17–22 June 2007. [CrossRef]

10. Van Baar, J.; Beardsley, P.; Pollefeys, M.; Gross, M. Sensor fusion for depth estimation, including TOF and thermal sensors. In Proceedings of the 2nd Joint 3DIM/3DPVT Conference: 3D Imaging, Modeling, Processing, Visualization and Transmission, 3DIMPVT 2012, Zurich, Switzerland, 13–15 October 2012; pp. 472–478. [CrossRef]
11. Cao, Y.; Xu, B.; Ye, Z.; Yang, J.; Cao, Y.; Tisse, C.L.; Li, X. Depth and thermal sensor fusion to enhance 3D thermographic reconstruction. *Opt. Express* **2018**, *26*, 8179. [CrossRef] [PubMed]
12. Pfitzner, C. Visual Human Body Weight Estimation with Focus on Medical Applications. Ph.D. Thesis, Universität Würzburg, Würzburg, Germany, 2018.
13. Rocco Spremolla, I.; Antunes, M.; Aouada, D.; Ottersten, B. RGB-D and Thermal Sensor Fusion - Application in Person Tracking . *VISIGRAPP* **2016**, *3*, 610–617. [CrossRef]
14. Salinas, C.; Fernández, R.; Montes, H.; Armada, M. A new approach for combining time-of-flight and RGB cameras based on depth-dependent planar projective transformations. *Sensors* **2015**, *15*, 24615–24643. [CrossRef]
15. Kim, Y.M.; Theobalt, C.; Diebel, J.; Kosecka, J.; Miscusik, B.; Thrun, S. Multi-view image and ToF sensor fusion for dense 3D reconstruction. In Proceedings of the 2009 IEEE 12th International Conference on Computer Vision Workshops, ICCV Workshops 2009, Kyoto, Japan, 27 September–4 October 2009; pp. 1542–1546. [CrossRef]
16. Tang, L.; Yuan, J.; Zhang, H.; Jiang, X.; Ma, J. PIAFusion: A progressive infrared and visible image fusion network based on illumination aware. *Inf. Fusion* **2022**, *83–84*, 79–92. [CrossRef]
17. Alexander, Q.G.; Hoskere, V.; Narazaki, Y.; Maxwell, A.; Spencer, B.F. Fusion of thermal and RGB images for automated deep learning based crack detection in civil infrastructure. *AI Civ. Eng.* **2022**, *1*, 3. [CrossRef]
18. Jung, C.; Zhou, K.; Feng, J. Fusionnet: Multispectral fusion of RGB and NIR images using two stage convolutional neural networks. *IEEE Access* **2020**, *8*, 23912–23919. [CrossRef]
19. Wang, H.; An, W.; Li, L.; Li, C.; Zhou, D. Infrared and visible image fusion based on multi-channel convolutional neural network. *IET Image Process.* **2022**, *16*, 1575–1584. [CrossRef]
20. Wang, Z.; Wang, F.; Wu, D.; Gao, G. Infrared and Visible Image Fusion Method Using Salience Detection and Convolutional Neural Network. *Sensors* **2022**, *22*, 5430. [CrossRef] [PubMed]
21. Yang, S.; Luo, P.; Loy, C.C.; Tang, X. WIDER FACE: A Face Detection Benchmark. *arXiv* **2015**, arXiv:1511.06523. [CrossRef]
22. Qi, D.; Tan, W.; Yao, Q.; Liu, J. YOLO5Face: Why Reinventing a Face Detector. *arXiv* **2021**, arXiv:2105.12931. [CrossRef]
23. Deng, J.; Guo, J.; Zhou, Y.; Yu, J.; Kotsia, I.; Zafeiriou, S. RetinaFace: Single-stage Dense Face Localisation in the Wild. *arXiv* **2019**, arXiv:1905.00641. [CrossRef]
24. Kaipeng, Z.; Zhanpeng, Z.; Zhifeng, L.; Yu, Q. Joint Face Detection and Alignment Using Multitask Cascaded Convolutional Networks. *IEEE Signal Process. Lett.* **2016**, *23*, 1499–1503. [CrossRef]
25. Yudin, D.; Ivanov, A.; Shchendrygin, M. Detection of a human head on a low-quality image and its software implementation. *ISPRS—Int. Arch. Photogramm. Remote Sens. Spat. Inf. Sci.* **2019**, *XLII-2/W12*, 237–241. [CrossRef]
26. Jiang, H.; Learned-Miller, E. Face Detection with the Faster R-CNN. *arXiv* **2016**, arXiv:1606.03473. [CrossRef]
27. Cheong, Y.K.; Yap, V.V.; Nisar, H. A novel face detection algorithm using thermal imaging. In Proceedings of the 2014 IEEE Symposium on Computer Applications and Industrial Electronics (ISCAIE), Penang, Malaysia, 7–8 April 2014; pp. 208–213. [CrossRef]
28. Kopaczka, M.; Nestler, J.; Merhof, D. Face Detection in Thermal Infrared Images: A Comparison of Algorithm- and Machine-Learning-Based Approaches. In Proceedings of the Advanced Concepts for Intelligent Vision Systems, Antwerp, Belgium, 18–21 September 2017; Blanc-Talon, J., Penne, R., Philips, W., Popescu, D., Scheunders, P., Eds.; Springer: Cham, Switzerland, 2017; pp. 518–529.
29. Silva, G.; Monteiro, R.; Ferreira, A.; Carvalho, P.; Corte-Real, L. Face Detection in Thermal Images with YOLOv3. In Proceedings of the Advances in Visual Computing, Lake Tahoe, NV, USA, 7–9 October 2019; Bebis, G., Boyle, R., Parvin, B., Koracin, D., Ushizima, D., Chai, S., Sueda, S., Lin, X., Lu, A., Thalmann, D., et al., Eds.; Springer: Cham, Switzerland, 2019; pp. 89–99.
30. Redmon, J.; Farhadi, A. YOLOv3: An Incremental Improvement. *arXiv* **2018**, arXiv:1804.02767.
31. Vuković, T.; Petrović, R.; Pavlović, M.; Stanković, S. Thermal Image Degradation Influence on R-CNN Face Detection Performance. In Proceedings of the 2019 27th Telecommunications Forum (TELFOR), Belgrade, Serbia, 26–27 November 2019; pp. 1–4. [CrossRef]
32. Mucha, W.; Kampel, M. Depth and thermal images in face detection—A detailed comparison between image modalities. In Proceedings of the 2022 the 5th International Conference on Machine Vision and Applications (ICMVA), New York, NY, USA, 18–20 February 2022.
33. Jia, G.; Jiankang, D.; Alexandros, L.; Stefanos, Z. Sample and Computation Redistribution for Efficient Face Detection. *arXiv* **2021**, arXiv:2105.04714 .
34. Chaichulee, S.; Villarroel, M.; Jorge, J.; Arteta, C.; Green, G.; McCormick, K.; Zisserman, A.; Tarassenko, L. Multi-Task Convolutional Neural Network for Patient Detection and Skin Segmentation in Continuous Non-Contact Vital Sign Monitoring. In Proceedings of the 2017 12th IEEE International Conference on Automatic Face Gesture Recognition (FG 2017), Washington, DC, USA, 30 May–3 June 2017; pp. 266–272. [CrossRef]
35. Green, G.; Chaichulee, S.; Villarroel, M.; Jorge, J.; Arteta, C.; Zisserman, A.; Tarassenko, L.; McCormick, K. Localised photoplethysmography imaging for heart rate estimation of pre-term infants in the clinic. In Proceedings of the Optical Diagnostics and Sensing XVIII: Toward Point-of-Care Diagnostics, San Francisco, CA, USA, 30 January–2 February 2018; Coté, G.L., Ed.; SPIE: Bellingham, WA USA, 2018; p. 26. [CrossRef]

36. Ren, S.; He, K.; Girshick, R.; Sun, J. Faster R-CNN: Towards Real-Time Object Detection with Region Proposal Networks. *arXiv* **2015**, arXiv:1506.01497.
37. Simonyan, K.; Zisserman, A. Very Deep Convolutional Networks for Large-Scale Image Recognition. *arXiv* **2014**, arXiv:1409.1556. [CrossRef]
38. Lin, T.Y.; Goyal, P.; Girshick, R.; He, K.; Dollár, P. Focal Loss for Dense Object Detection. *arXiv* **2017**, arXiv:1708.02002.
39. Kyrollos, D.G.; Tanner, J.B.; Greenwood, K.; Harrold, J.; Green, J.R. Noncontact Neonatal Respiration Rate Estimation Using Machine Vision. In Proceedings of the 2021 IEEE Sensors Applications Symposium (SAS), Sundsvall, Sweden, 23–25 August 2021; pp. 1–6. [CrossRef]
40. Russakovsky, O.; Deng, J.; Su, H.; Krause, J.; Satheesh, S.; Ma, S.; Huang, Z.; Karpathy, A.; Khosla, A.; Bernstein, M.; et al. ImageNet Large Scale Visual Recognition Challenge. *arXiv* **2014**, arXiv:1409.0575. [CrossRef]
41. Lu, G.; Wang, S.; Kong, K.; Yan, J.; Li, H.; Li, X. Learning Pyramidal Hierarchical Features for Neonatal Face Detection. In Proceedings of the 2018 14th International Conference on Natural Computation, Fuzzy Systems and Knowledge Discovery (ICNC-FSKD), Huangshan, China, 28–30 July 2018; pp. 8–13. [CrossRef]
42. Jocher, G.; Stoken, A.; Borovec, J.; NanoCode012; ChristopherSTAN; Changyu, L.; Laughing ; tkianai; Hogan, A.; lorenzomammana; et al. *ultralytics/yolov5: V3.1—Bug Fixes and Performance Improvements*; Zenodo: Genève, Switzerland, 2020. [CrossRef]
43. Nagy, Á.; Földesy, P.; Jánoki, I.; Terbe, D.; Siket, M.; Szabó, M.; Varga, J.; Zarándy, Á. Continuous camera-based premature-infant monitoring algorithms for NICU. *Appl. Sci.* **2021**, *11*, 7215. [CrossRef]
44. Khanam, F.T.Z.; Perera, A.G.; Al-Naji, A.; Gibson, K.; Chahl, J. Non-contact automatic vital signs monitoring of infants in a Neonatal Intensive Care Unit based on neural networks. *J. Imaging* **2021**, *7*, 122. [CrossRef]
45. Salekin, M.S.; Zamzmi, G.; Hausmann, J.; Goldgof, D.; Kasturi, R.; Kneusel, M.; Ashmeade, T.; Ho, T.; Sun, Y. Multimodal neonatal procedural and postoperative pain assessment dataset. *Data Brief* **2021**, *35*, 106796. [CrossRef]
46. Dosso, Y.S.; Kyrollos, D.; Greenwood, K.J.; Harrold, J.; Green, J.R. NICUface: Robust neonatal face detection in complex NICU scenes. *IEEE Access* **2022**, *10*, 62893–62909. [CrossRef]
47. Antink, C.H.; Ferreira, J.C.M.; Paul, M.; Lyra, S.; Heimann, K.; Karthik, S.; Joseph, J.; Jayaraman, K.; Orlikowsky, T.; Sivaprakasam, M.; et al. Fast body part segmentation and tracking of neonatal video data using deep learning. *Med. Biol. Eng. Comput.* **2020**, *58*, 3049–3061. [CrossRef]
48. Voss, F.; Brechmann, N.; Lyra, S.; Rixen, J.; Leonhardt, S.; Hoog Antink, C. Multi-modal body part segmentation of infants using deep learning. *Biomed. Eng. Online* **2023**, *22*, 28. [CrossRef]
49. Beppu, F.; Yoshikawa, H.; Uchiyama, A.; Higashino, T.; Hamada, K.; Hirakawa, E. Body part detection from neonatal thermal images using deep learning. In *Lecture Notes of the Institute for Computer Sciences, Social Informatics and Telecommunications Engineering*; Lecture notes of the Institute for Computer Sciences, Social Informatics and Telecommunications Engineering; Springer International Publishing: Cham, Switzerland, 2022; pp. 438–450.
50. Awais, M.; Chen, C.; Long, X.; Yin, B.; Nawaz, A.; Abbasi Saadullah, F.; Akbarzadeh, S.; Tao, L.; Lu, C.; Wang, L.; et al. Novel Framework: Face Feature Selection Algorithm for Neonatal Facial and Related Attributes Recognition. *IEEE Access* **2020**, *8*, 59100–59113. [CrossRef]
51. Neophytou, N.; Mueller, K. Color-Space CAD: Direct Gamut Editing in 3D. *IEEE Comput. Graph. Appl.* **2008**, *28*, 88–98. [CrossRef]
52. Fairchild, M.D. *Color Appearance Models*, 3rd ed.; The Wiley-IS&T Series in Imaging Science and Technology; John Wiley & Sons: Nashville, TN, USA, 2013.
53. Bebis, G.; Gyaourova, A.; Singh, S.; Pavlidis, I. Face recognition by fusing thermal infrared and visible imagery. *Image Vis. Comput.* **2006**, *24*, 727–742. [CrossRef]
54. Selinger, A.; Socolinsky, D.A. *Appearance-Based Facial Recognition Using Visible and Thermal Imagery: A Comparative Study*; EQUINOX Corp.: New York, NY, USA, 2006.
55. Chen, X.; Wang, H.; Liang, Y.; Meng, Y.; Wang, S. A novel infrared and visible image fusion approach based on adversarial neural network. *Sensors* **2022**, *22*, 304. [CrossRef] [PubMed]
56. Vadidar, M.; Kariminezhad, A.; Mayr, C.; Kloeker, L.; Eckstein, L. Robust Environment Perception for Automated Driving: A Unified Learning Pipeline for Visual-Infrared Object Detection. In Proceedings of the 2022 IEEE Intelligent Vehicles Symposium, Aachen, Germany, 4–9 June 2022; pp. 367–374. [CrossRef]
57. Shopovska, I.; Jovanov, L.; Philips, W. Deep visible and thermal image fusion for enhanced pedestrian visibility. *Sensors* **2019**, *19*, 3727. [CrossRef] [PubMed]
58. Zhang, H.; Zhang, L.; Zhuo, L.; Zhang, J. Object tracking in RGB-T videos using modal-aware attention network and competitive learning. *Sensors* **2020**, *20*, 393. [CrossRef]
59. Zhang, X.; Ye, P.; Peng, S.; Liu, J.; Gong, K.; Xiao, G. SiamFT: An RGB-Infrared Fusion Tracking Method via Fully Convolutional Siamese Networks. *IEEE Access* **2019**, *7*, 122122–122133. [CrossRef]
60. What Is a Visible Imaging Sensor (RGB Color Camera)? Available online: https://www.infinitioptics.com/glossary/visible-imaging-sensor-400700nm-colour-cameras (accessed on 24 January 2023).
61. pmd FAQ. Available online: https://pmdtec.com/picofamily/faq/ (accessed on 19 November 2020).
62. Gleichauf, J.; Herrmann, S.; Hennemann, L.; Krauss, H.; Nitschke, J.; Renner, P.; Niebler, C.; Koelpin, A. Automated Non-Contact Respiratory Rate Monitoring of Neonates Based on Synchronous Evaluation of a 3D Time-of-Flight Camera and a Microwave Interferometric Radar Sensor. *Sensors* **2021**, *21*, 2959. [CrossRef]

63. He, K.; Zhang, X.; Ren, S.; Sun, J. Deep Residual Learning for Image Recognition. *arXiv* **2015**, arXiv:1512.03385.
64. Lin, T.Y.; Dollár, P.; Girshick, R.; He, K.; Hariharan, B.; Belongie, S. Feature Pyramid Networks for Object Detection. *arXiv* **2016**, arXiv:1612.03144.
65. Girshick, R. Fast R-CNN. *arXiv* **2015**, arXiv:1504.08083.
66. Redmon, J.; Farhadi, A. YOLO9000: Better, Faster, Stronger. *arXiv* **2016**, arXiv:1612.08242.
67. Kathuria, A. Available online: https://towardsdatascience.com/yolo-v3-object-detection-53fb7d3bfe6b (accessed on 2 March 2023).
68. Mikołajczyk, A.; Grochowski, M. Data augmentation for improving deep learning in image classification problem. In Proceedings of the 2018 International Interdisciplinary PhD Workshop (IIPhDW), Swinoujscie, Poland, 9–12 May 2018; pp. 117–122. [CrossRef]
69. Hennemann, L. Realisierung und Optimierung der Detektion von Körperregionen Neugeborener zur Kontaktlosen und Robusten Überwachung der Vitalparameter mittels eines Neuronalen Netzes. Master's Thesis, Nuremberg Institute of Technology, Nuremberg, Germany, 2023.
70. May, S. optris_drivers. Available online: https://wiki.ros.org/optris_drivers (accessed on 24 January 2023).
71. Hartmann, C.; Gleichauf, J. ros_cvb_camera_driver. Available online: http://wiki.ros.org/ros_cvb_camera_driver (accessed on 7 June 2019).
72. For Artificial Intelligence University of Bremen, I. pico_flexx_driver. Available online: https://github.com/code-iai/pico_flexx_driver (accessed on 29 April 2020).
73. camera_calibration. Available online: https://wiki.ros.org/camera_calibration (accessed on 14 January 2019).
74. Ocana, D.T. image_pipeline. Available online: https://github.com/DavidTorresOcana/image_pipeline (accessed on 7 June 2019).
75. openCV. How to Detect Ellipse and Get Centers of Ellipse. Available online: https://answers.opencv.org/question/38885/how-to-detect-ellipse-and-get-centers-of-ellipse/ (accessed on 12 May 2022).
76. opencv 3, Blobdetection, The Function/Feature Is Not Implemented () in detectAndCompute. Available online: https://stackoverflow.com/questions/30622304/opencv-3-blobdetection-the-function-feature-is-not-implemented-in-detectand (accessed on 12 May 2022).
77. openCV. solvePnP. Available online: https://docs.opencv.org/3.4/d9/d0c/group__calib3d.htmlga549c2075fac14829ff4a58bc931c033d (accessed on 12 May 2022).
78. openCV. Rodrigues. Available online: https://docs.opencv.org/3.4/d9/d0c/group__calib3d.htmlga61585db663d9da06b68e70cfbf6a1eac (accessed on 12 May 2022).
79. openCV. projectPoints. Available online: https://docs.opencv.org/3.4/d9/d0c/group__calib3d.htmlga1019495a2c8d1743ed5cc23fa0daff8c (accessed on 12 May 2022).
80. Fizyr. Keras-Retinanet. Available online: https://github.com/fizyr/keras-retinanet (accessed on 10 May 2023).
81. AlexeyAB. Darknet. Available online: https://github.com/AlexeyAB/darknet (accessed on 10 May 2023).

sensors

MDPI

Article

End-to-End Detection of a Landing Platform for Offshore UAVs Based on a Multimodal Early Fusion Approach

Francisco Soares Neves [1,2], Rafael Marques Claro [1,2] and Andry Maykol Pinto [1,2,*]

[1] Faculty of Engineering, University of Porto (FEUP), 4200-465 Porto, Portugal
[2] Centre for Robotics and Autonomous Systems—INESC TEC, 4200-465 Porto, Portugal
* Correspondence: andry.m.pinto@inesctec.pt

Abstract: A perception module is a vital component of a modern robotic system. Vision, radar, thermal, and LiDAR are the most common choices of sensors for environmental awareness. Relying on singular sources of information is prone to be affected by specific environmental conditions (e.g., visual cameras are affected by glary or dark environments). Thus, relying on different sensors is an essential step to introduce robustness against various environmental conditions. Hence, a perception system with sensor fusion capabilities produces the desired redundant and reliable awareness critical for real-world systems. This paper proposes a novel early fusion module that is reliable against individual cases of sensor failure when detecting an offshore maritime platform for UAV landing. The model explores the early fusion of a still unexplored combination of visual, infrared, and LiDAR modalities. The contribution is described by suggesting a simple methodology that intends to facilitate the training and inference of a lightweight state-of-the-art object detector. The early fusion based detector achieves solid detection recalls up to 99% for all cases of sensor failure and extreme weather conditions such as glary, dark, and foggy scenarios in fair real-time inference duration below 6 ms.

Keywords: object detection; sensor fusion; early-fusion; computer vision; RGB camera; thermal camera; 3D LiDAR

Citation: Neves, F.S.; Claro, R.M.; Pinto, A.M. End-to-End Detection of a Landing Platform for Offshore UAVs Based on a Multimodal Early Fusion Approach. *Sensors* **2023**, *23*, 2434. https://doi.org/10.3390/s23052434

Academic Editor: Natividad Duro Carralero

Received: 28 January 2023
Revised: 19 February 2023
Accepted: 20 February 2023
Published: 22 February 2023

1. Introduction

Object detection is a perception task that provides the autonomous systems the necessary awareness of the surrounding environment. By extracting features from raw sensory information, it produces meaningful high-level knowledge about surrounding objects (e.g., other vehicles, pedestrians, goal marks, road lanes). In the end, object detection provides semantic representations about the surrounding environment [1]. For detection, the most common on-board sensors used by autonomous systems are two-dimensional/three-dimensional LiDAR, visual cameras, radar, and thermographic cameras. Each sensor has its own advantages and disadvantages that are directly affected by the environmental conditions. Visual cameras alone are the traditional sources of information, however, in recent years, there is a concern for diversifying the set of sensors to increase the robustness of the system for the case of sensor failure against different environmental conditions [2–4]. For example, thermal cameras discriminate really well different objects emitting different temperature intensities. However, on extreme heated environments (e.g., sunny days, hot rooms), the background and foreground becomes indistinguishable and homogeneous. Visual cameras provide colour and texture information, but for foggy, glary, or dark environments, the signal becomes corrupted and noisy. LiDAR sensors extract depth information by analyzing light energy reflected from objects in the surroundings. Nevertheless, in rainy and foggy environments, the reflected information becomes worthless. Hence, for an autonomous system to be reliable, it is a requirement to endow it with a diverse and distinct set of sensor equipment.

Adopting sensor fusion is one of the building blocks to build reliable perception systems. A system of this kind is still able to meaningfully perceive the environment, not only when there are individual failure, but also when environmental conditions corrupt any sensory information making it not reliable. At greater length, sensor fusion is described by combining the information produced by various sensors representing different modalities (e.g., visual, thermographic, RADAR, LiDAR) into a joint representation to produce a less uncertain input. Sensor fusion can happen in three different stages of the detection pipeline [5]: (i) early fusion refers to the combination of multiple input sources into a unique feature vector before feature extraction; (ii) intermediate fusion refers to the process of extracting correlation between features from the joint combination of sensory information inside the perception module and before classification or regression; and (iii) late fusion happens after the classification/regression procedure.

Extracting and combining relevant features from multiple modalities is a challenging process that requires a lot of domain knowledge and it strongly depends on the requirements of the task to solve. Effectively, deep learning technology is already capable of tackling this process [6–8] autonomously. Apart from this deep approaches, when significant correlation across modalities is smoothly inferred, early fusion should be considered [4]. If it is applied, early fusion produces early joint representations directly from raw data and anticipates correlation and redundancy. Several studies have been conducted to find several pair combinations between point cloud, visual, and thermographic into single concatenated representations [9–12]. However, there is a lack of research on combining the three.

In this paper, we contribute by presenting a novel multimodal early-fusion-based perception system that combines visual, thermal, and three-dimensional LiDAR data information to produce reliable detection capabilities against demanding operating conditions such as extreme weather or modality failure. According to the authors' knowledge, there is no equivalent early-fusion method combining point clouds, thermographic, and visual information. This combination of sensors is implemented by a multimodal fiducial ArUco marker called ArTuga (Patent pending (Portuguese Patent Request (PPP) nr. 118328, and European Patent Request (EP) nr. 22212945.4)) proposed by Claro et al. [13]. It enables multimodal detection against several weather conditions for robotic solutions endowing heterogeneous perception systems comprised by visual, thermographic, and LiDAR-based devices. Therefore, it is not possible to replace the marker with an ordinary object given the constructive characteristics of the ArTuga that provide a spatial alignment of certain elements that facilitate the precise and robust detection of the marker. The contributions of this article include:

- Real-time multimodal marker detection that is deployable onboard a UAV;
- Resilient and high-accuracy detection based on early-fusion approaches against sensor failure integrated in the YOLOv7 framework;
- Robustness against demanding weather and operating conditions for extensive experiments using real UAVs landing on a floating platform;
- A new multimodal dataset collected in real offshore and onshore environments during a UAV landing operation, comprised of diverse joint representations of visual, infrared and point-cloud images.

The paper is organized as follows: In Section 2, there is a review of the literature about object detection and sensor fusion. In Section 3, a novel early-fusion methodology is proposed. Section 4 describes the experimental setup. Section 5 exposes the achieved results. Lastly, Section 6 provides the conclusions.

2. Related Work

Object detection (OD) is a computer vision task described by recognizing, identifying, and locating objects of interest within a picture with a certain degree of confidence. OD has gained popularity since it started to fuse with the strong capabilities of neural network technology. OD is mainly subdivided between single-stage and two-stage detection. These two categories compete by finding the optimal trade-off between detection speed and accuracy [14]. Real-time systems require fast and accurate predictions, therefore building real-time object detectors demands solving the former mentioned trade-off.

Two-stage detectors, such as the RCNN [15], fast-RCNN [16], and faster-RCNN [17], follow region proposal principles that focus on the localization of regions of interest in the image (e.g., where objects of interest are located) before performing detection. Firstly, these regions are estimated. Lastly, from these regions, detection is performed. Depending on the detected objects, correspondent confidence probabilities are extracted. These methods focus on achieving high detection accuracy, however, at the cost of being slow. Hence, two-stage detection is not viable for real-time demands. On the other hand, single-stage detectors, such as the you only look once (YOLO) [18] and the single-shot multibox detector (SSD) [19] perform detection and classification in a singular common step. Single-stage detection prioritizes inference speed over accuracy. Naturally, it is suitable to address real-time constraints. Despite inference speed prioritization, these methods achieve at least similar accuracy to two-staged detectors [14]. Within the available single-stage detectors, YOLO is a suitable choice: (i) it is more popular, accessible, and has a stronger documentation than any other method in the literature; (ii) it achieves highly fair detection speeds (up to 45 FPS), and thus it has at least similar accuracy when compared with other two-stage detectors.

Further, multimodal fusion [20] aims to find the combination of apparently disparate multi-domain data (e.g., visible light, infra-red light, sound, laser) sources to produce a more robust and more rich fused signal. One of the hot topics of the current sensor fusion literature is choosing the level in which the fusion takes place in the detection pipeline [5,21,22]. Sensor fusion happens in three levels, such as: (i) *early fusion* as the pixel-level fusion; (ii) *intermediate fusion* as the feature-level fusion; (iii) *late fusion* as the decision-level fusion. Several studies have been conducted by fusing sensory data in different levels of the perception pipeline. Farahnakian and Heikkonen (2020) [10] achieved state-of-the-art performance to detect marine vessels by applying intermediate fusion techniques using thermal and RGB cameras. However, reasonable real-time inference durations (mostly in the order of seconds) were not achieved. Additionally, the authors even suggested the possibility of adding a LiDAR source to the fused input to explore possible improvements. Liu et al. (2022) [12] proposed an intermediate fusion approach using LiDAR and visual camera data for car detection. Choi and Kim (2021) [11] followed an early fusion approach by combining an infrared camera with a three-dimensional LiDAR sensor, achieving reliable performance in impractical environmental conditions for vision-based sensors. Azam et al. (2019) [23] suggested another early fusion approach that fuses three-dimensional LiDAR with thermal images, ensuring reliable performance for both day and night light conditions. Bhanushali et al. (2020) [24] achieved reliable real-time object detection by training an end-to-end SSD detector on the KITTI dataset [25], merging early and late fusion principles. The former study also suggested the addition of other sensors, such as radar, to increase the robustness of the proposed model.

3. Multimodal Early Fusion Approach for Fiducial Marker Detection

Offshore robotic applications operate in challenging weather conditions where corrupt sensor information or even sensor failure situations are expected. Therefore, robotic solutions need to be resilient and endow redundant and heterogeneous perception systems [13]. Thus, based on the available multimodal sources, this research proposes a simple methodology that produces an expressive and redundant joint representation over a multimodal feature space. The proposed system comprises an (i) early-fusion module that applies a novel early-fusion technique and a (ii) lightweight YOLO-based detector that is fed with an early fused input. A system high-level perspective is depicted by the Figure 1.

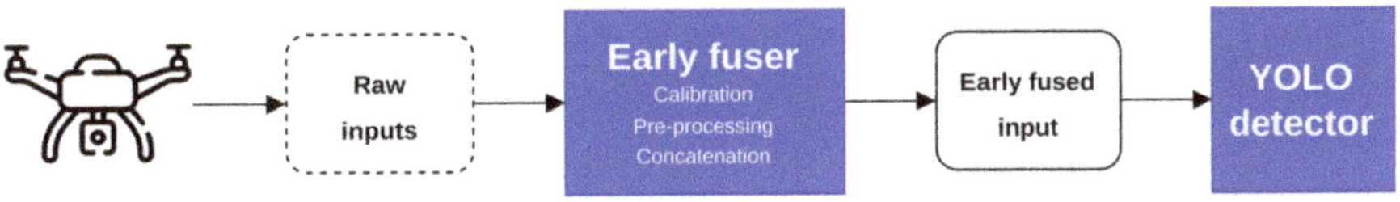

Figure 1. The early-fusion detection system high-level view.

3.1. Early Fuser

Based on spatiotemporal alignment of data streams, the early fuser produces a concatenated 3-channel RGB input containing a joint representation of visual, infrared, and LiDAR modalities. The aim is to anticipate redundancy across input streams and facilitate the detector's feature extraction process. The early fuser procedure is described by: (i) a calibration step where modalities are aligned into a common coordinate system, (ii) a pre-processing step where relevant features for each modality are extracted, and (iii) a final concatenation step where processed modalities are aggregated into a redundant RGB representation.

3.1.1. Calibration

To aggregate sensory information into a common three-channel image, spatial relationships between sensors must be obtained. The visual camera is chosen to be the main frame of reference. Hence, the remaining infrared and LiDAR information are projected into the visual camera image coordinate system. Before projecting the three-dimensional LiDAR point-cloud into the visual image coordinate system, the point-cloud is transformed into the visual camera coordinate system. The three-dimensional spatial relationship between each sensor is described by an extrinsic matrix. Using Zhang's method [26], an extrinsic calibration operation is performed to obtain the extrinsic transformation matrix $E_{l,v} = \begin{bmatrix} R & T \end{bmatrix}$. $E_{l,v}$ is a rigid body transformation described by a rotation

$$R = \begin{bmatrix} r_{11} & r_{12} & r_{13} \\ r_{21} & r_{22} & r_{23} \\ r_{31} & r_{32} & r_{33} \end{bmatrix}$$

and a translation $T = \begin{bmatrix} t_x & t_y & t_z \end{bmatrix}^T$ between the visual camera coordinate system v and the LiDAR coordinate system l, where LiDAR points $P_l = \begin{bmatrix} x_l & y_l & z_l \end{bmatrix}^T$ are converted into visual camera points $P_v = \begin{bmatrix} x_v & y_v & z_v \end{bmatrix}^T$ as follows:

$$\begin{gathered} P_v = E_{l,v} P_v \\ \begin{bmatrix} x_v \\ y_v \\ z_v \end{bmatrix} = \begin{bmatrix} r_{11} & r_{12} & r_{13} & t_x \\ r_{21} & r_{22} & r_{23} & t_y \\ r_{31} & r_{32} & r_{33} & t_z \end{bmatrix} \begin{bmatrix} x_l \\ y_l \\ z_l \end{bmatrix}. \end{gathered} \tag{1}$$

Following, to transform P_v into pixel points $U_v = \begin{bmatrix} u_v & v_v & 1 \end{bmatrix}^T \in \Re^2$, an intrinsic calibration for the visual camera is performed using the Zhang's method to obtain the intrinsic matrix defined by:

$$K_v = \begin{bmatrix} f_x & \gamma & c_x \\ 0 & f_y & c_y \\ 0 & 0 & 1 \end{bmatrix},$$

where (f_x, f_y) are the focal lengths, (c_x, c_y) is the image center and γ is the skew between x and y directions. Therefore, a perspective projection is applied to obtain U_v as follows:

$$\alpha \begin{bmatrix} u_v \\ v_v \\ 1 \end{bmatrix} = K_v \begin{bmatrix} x_v \\ y_v \\ z_v \end{bmatrix},$$

Finally, similar to a stereoscopic camera, an homografy matrix was extracted to map corresponding points between the thermal image and the visual image as carried out by [13].

3.1.2. Pre-Processing

Specific pre-processing techniques are applied for each data stream. The visual stream is undistorted using OpenCV library, to remove natural lens distortion described by K_v. There is no need for rectification because it is a monocular camera. To be represented in a single-channel shape, the visual source, acquired in a RGB representation, is finally converted into a grayscale (single-channel) form. The infrared stream is already acquired in a grayscale representation. Similarly to the visual images, the infrared images undergo undistortion operations by resorting to the OpenCV library using lens distortion described by its intrinsic matrix.

The infrared sensor is properly pre-configured to capture distinct temperature contrast between the ArTuga and the background information. The temperature contrast results in a colour contrast. Availing this natural contrast, a binary threshold filter is applied as an enhancement operation and a filter operation to filter surrounding background noise. The binary operation is a pixel-wise non-linear threshold operation described by:

$$p_{ij}^{binarized} = \begin{cases} 255 & p_{ij}^{orig} > thresh \\ 0 & \text{otherwise} \end{cases}, \quad (2)$$

where 255 is the maximum color value, $thresh = 128$ is the threshold applied, $p_{i,j}^{orig}$ is any original pixel and $p_{i,j}^{binarized}$ is any transformed pixel. The binary threshold operation is depicted in the Figure 2.

Figure 2. The binary threshold operation applied to the infrared images.

As it can be observed, the binarization results in a more recognizable object by enhancing colour contrast and cleaning the background while preserving edge and corner properties.

The LiDAR stream is acquired in a three-dimensional point-cloud representation. Before intrinsic/extrinsic transformations, this stream is pre-processed to filter and enhance specific inherent information. Excessive and needless point-cloud information is filtered using a voxel downsampling operation. Voxel downsampling is a spatial operation that iteratively buckets points into three-dimensional voxels. Each voxel is compressed and generates a unique three-dimensional point by averaging every inner point. This operation reduces the size of the cloud and throws out excessive information while retaining the overall geometric structure. The used voxels have a leaf size of 5 cm. In addition, more points are removed according to their intensity. The ArTuga has a retro-reflective tape in its borders and white-bit coding area in its interior resulting in disparate values of intensity in comparison to the rest of the point cloud. Points with intensity below a maximum intensity of I are suppressed. Hence, the resulting points mostly belong to the aruco object. A two-dimensional grayscale image is generated by assigning a colour to each resultant pixel, depending on the laser depth, according to a 8-bit scaling operation. The colour c for each pixel is computed as follows:

$$c = 255\left(1 - \frac{d}{m_d}\right),$$

where d is the laser distance measurement and m_d is the maximum depth reached during the UAV flight. A final dilation operation is applied to compensate the sparsity of the image, by applying a 30×30 squared kernel as depicted by the Figure 3.

Figure 3. Dilation over the sparsed 2D points.

3.1.3. Concatenation

As a final step, the pre-processed single-channel streams are concatenated into a three-channel RGB image, where the red, green, and blue channels correspond to the LiDAR, infrared, and visual streams, respectively. The channels are presented in the Figure 4.

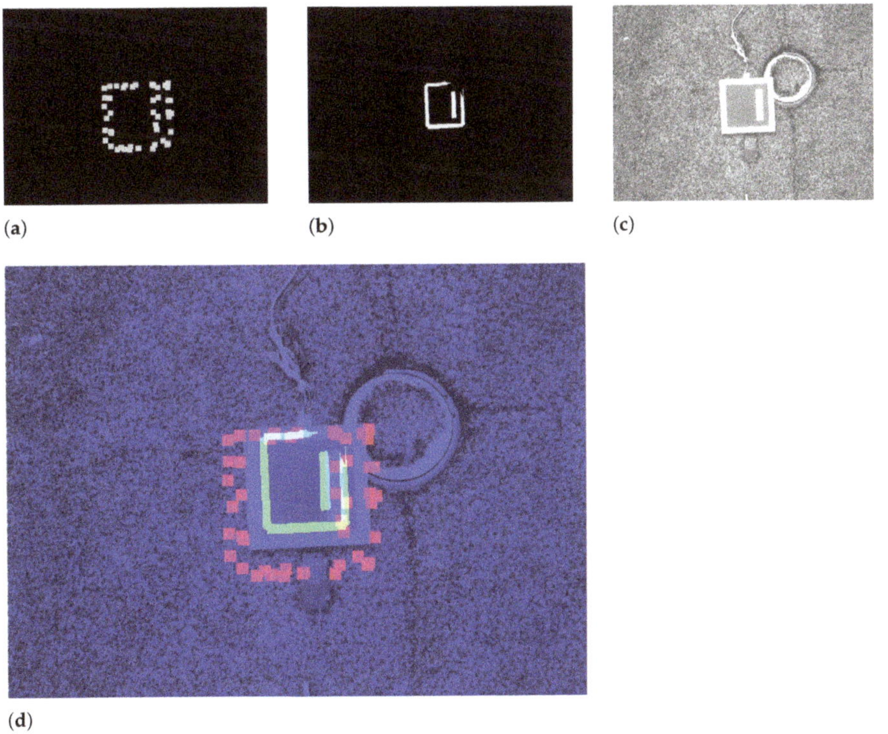

Figure 4. The individual channels and the aggregated RGB image. (**a**) The LiDAR channel. (**b**) The thermal channel. (**c**) The visual channel. (**d**) The aggregated image.

The concatenation finalizes the early fusion pipeline resulting in an input that is more understandable and correlated across modalities to feed the YOLO detector. The pipeline is depicted in the Figure 5.

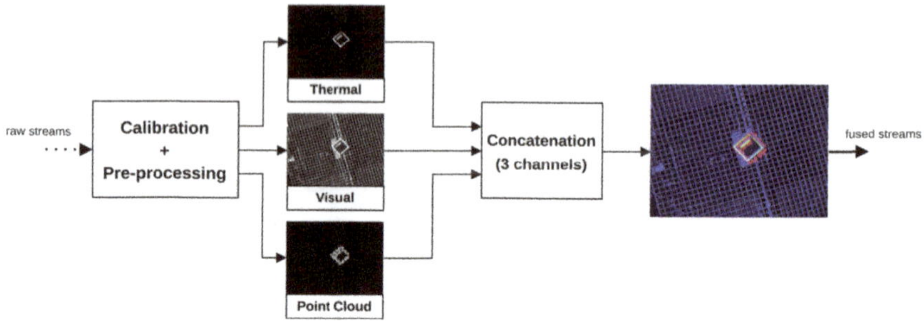

Figure 5. The early fuser pipeline.

3.2. Detector

Nowadays YOLO-based detectors are a solid and popular choice for fast but still accurate detection. Impressively, it is used to tackle real-world tasks (e.g., autonomous cars and UAVs) [27–29], addressing challenging weather conditions [30,31] and real-time speed constraints [32,33]. Nowadays, the most recent versions of YOLO, the YOLOv5 [34], YOLOv6 [35], and the YOLOv7 [36] are the state-of-the-art detectors for fast and accurate detection. This work uses a detector based on the YOLOv7 architecture. Considering real-time demands, the feasibility of the smallest version of the YOLOv7 is explored: the *tiny* version. The aim is to scrutinize the feasibility of these lightweight, faster, and simpler detectors considering the weaker accuracies when compared to the bigger versions. Hence, the proposed early fusion method aims to counteract lightweight limitations by facilitating the input information.

More particularly, YOLO addresses the detection task as a regression problem because it outputs bounding box coordinates (location of objects in the image) and the probability of detection (the confidence in the prediction) for each detected object. YOLO takes a full image as an input, diving it into a into a $n \times n$ grid. For each cell in the grid, it estimates N possible bounding boxes correspondent to N objects. Each detected object is classified by a label l (e.g., $l \in \{pedestrian, truck, traffic\ light\}$). Each bounding box b is defined as a vector as follows:

$$b = \begin{bmatrix} x_{min} & y_{min} & x_{max} & y_{max} & c & l \end{bmatrix},$$

where (x_{min}, y_{min}) and (x_{max}, y_{max}) are the minimum and maximum pixel coordinates, respectively, with respect to the top-left corner of the image, c is the confidence of the prediction containing an object and l is the object class label. Ultimately, YOLO outputs a $N \times 6$ dimensional tensor, where only the most trusted predicted bounding boxes remain, by applying a technique called Non-Maximum-Suppression.

4. Experimental Setup

A UAV named CROW (copter robot for offshore wind-farms), based on a quadcopter frame (with a wingspan of 0.7 m, and a maximum payload of 2 Kg), was remotely operated to execute a few minutes duration flight around a maritime platform that contains the ArTuga marker in its center. The CROW endows a perception system comprised of a (i) three-dimensional LiDAR, a (ii) visual camera, and an (iii) infrared camera. The detection performance is highly dependent on the sensor characteristics. Thus, the sensor choice must provide high resolution capabilities. The set of sensors have the following specifications:

- Visual Camera—*The Imaging Source DFM 37UX273-ML*—Frame Rate: 15 Hz, Resolution: 1440 × 1080 pixels, Field of View: 45° horizontal;

- Thermal Camera—*FLIR Boson 640 Radiometry*—Frame Rate: 15 Hz, Resolution: 640 × 512 pixels, Field of View: 50° horizontal, Temperature Measurement Accuracy: ±5 °C;
- 3D LiDAR—*Ouster OS1-64*—Frame Rate: 10 Hz, Resolution: 64 × 1024 channels, Range: 120 m, Accuracy: ±0.05 m, Field of View: 360° horizontal and 45° vertical.

During the flight, the sensor data is recorded to be further processed to generate the datasets for model training. There was a concern to produce a heterogeneous dataset, therefore the flight operation was conducted considering different spatial perspectives of the platform. Despite the aim of this application being offshore, onshore samples were also acquired to promote heterogeneity. The high variability of an heterogeneous dataset ensures that a model is robust against unexpected and different image perspectives. Naturally, robustness is a major priority to avoid overfitting while training. The CROW UAV executing a landing operation is depicted by the Figure 6.

Figure 6. The conducted real experiment of a landing procedure using the CROW UAV.

4.1. Datasets

The performance of the early fusion detector is going to be compared with three unimodal detectors correspondent to each sensor stream. The unimodal detectors function as fine-tuned baseline detectors for each modality. Hence, the aim is to use them as a performance reference for the multimodal early fusion detector. Accordingly four datasets are generated: a visual, a thermal, a LiDAR and a fusion datasets. The datasets are available in a Google Drive public repository [37].

4.1.1. Visual Dataset

The visual dataset comprises 1449 images representing onshore and offshore visual samples from different spatial points of view of the ArTuga with different backgrounds (e.g., landing platform, water, ground). Besides the inherent variability of the data, some data augmentation techniques are applied for this dataset to prevent overfitness. Brightness variation is a clear augmentation technique to train the model against brighter and dark scenarios. The motivation for this technique is to prepare the detector for sunny and night settings. In more depth, for every image, by increasing up to 25% or decreasing up to 90% the brightness, a darker image is a pixel-wise operation, such as $p_{i,j}^{dark} = p_{i,j}^{orig}(1-\delta), \forall i \in [0, h-1] \subset N, j \in [0, w-1] \subset N$ and a brighter image such as $p_{i,j}^{bright} = p_{i,j}^{orig}(1+\beta), \forall i \in [0, h-1] \subset N, j \in [0, w-1] \subset N$, where $\delta \in]0, 0.9], \beta \in]0, 0.25]$, w, h are the image width and height, respectively, $p_{i,j} \in [0, 255] \subset N$ is a pixel in the ith row and jth column of the image, and 0 and 255 are the minimum and maximum colour values. In addition, flipping and rotation augmentation techniques are applied. The images are resized from the acquisition size 1440 × 1080 to a final size of 640 × 480 pixels. The Figure 7 depicts some samples from the visual dataset.

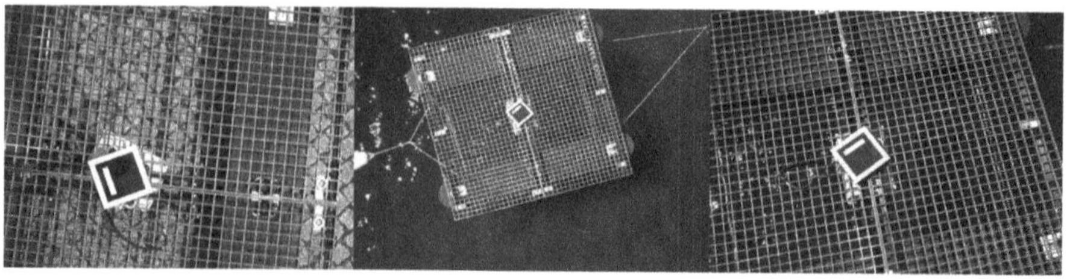

Figure 7. Samples from the visual dataset.

4.1.2. Thermal Dataset

The thermal dataset comprises 441 pre-processed binarized infrared images as already described in the Section 3.1.2. The data augmentation techniques applied for this dataset are flipping and rotation. The images are resized from the acquisition size 1440 × 1080 to a final size of 640 × 480 pixels. The Figure 8 depicts some samples from the thermal dataset.

Figure 8. Samples from the thermal dataset.

4.1.3. LiDAR Dataset

The LiDAR dataset comprises 316 pre-processed two-dimensional point cloud projections already described in the Section 3.1.2. The data augmentation techniques applied for this dataset are flipping and rotation. The images are resized from 1440 × 1080 to a final size of 640 × 480 pixels. Figure 9 depicts some samples from the LiDAR dataset.

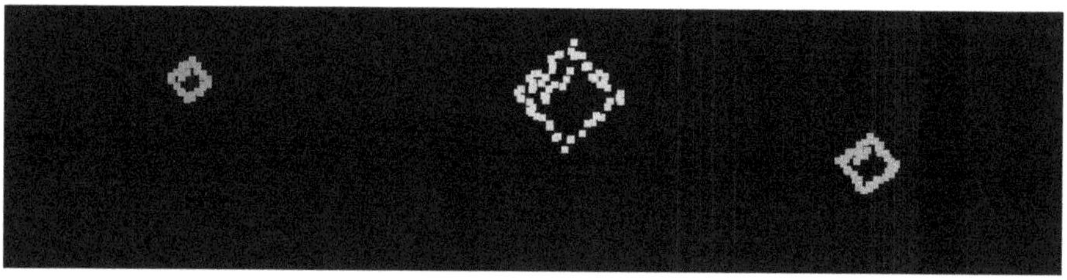

Figure 9. Samples from the LiDAR dataset.

4.1.4. Fusion Dataset

The fusion dataset is constituted by 2158 pre-processed and concatenated RGB images produced by the early fuser described in the Section 3.1. This dataset implements a brightness augmentation technique for the visual channel and flipping and rotation techniques for all channels. The images are resized from 1440 × 1080 to a final size of 640 × 480 pixels. The Figure 10 depicts some samples from the the fusion dataset. As it can be observed, the dataset contains both multimodal samples representing cases when all modalities are available; and unimodal samples representing sensory failure cases when specific modalities are deactivated. Training with unimodal samples intends to promote resilience against sensory failure.

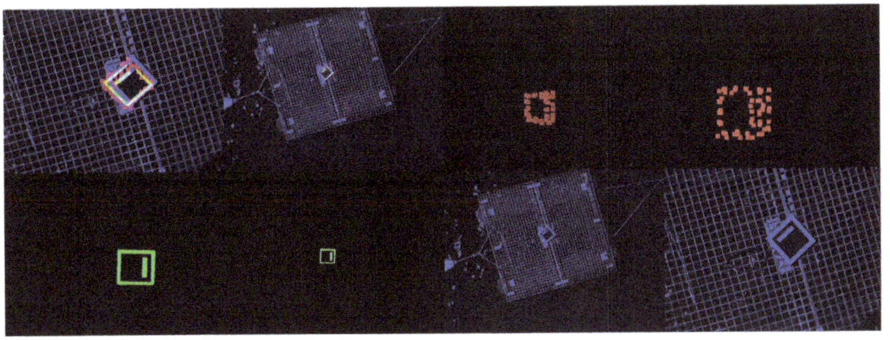

Figure 10. Samples from the Early Fusion dataset.

4.1.5. Annotation

For each sample in the dataset, there is a correspondent annotation. Manual annotations in the form of a bounding box are performed using roboflow framework [38]. For the unimodal datasets, the bounding box fully encloses the unique object present as shown in the Figure 11a–c. In this way, it is noted that there is a care for pixel tightness to drive the model to the best accuracy. As for the case of the fused sample, the annotation considers the stream with the best resolution (shape and colour), as shown in Figure 11d. Additionally, negative samples are present in the dataset to teach the model when an object is not present as shown in the Figure 11e or partially present as depicted by the Figure 11f. For these negative cases, the annotation is not performed, instead it is marked as *null*.

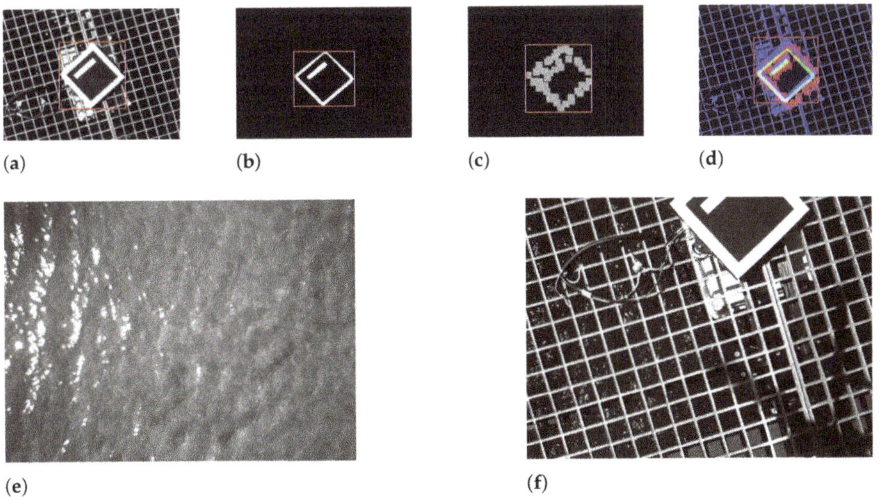

Figure 11. Non-null annotations on the top and null annotations on the bottom. (**a**) A visual sample annotation. (**b**) A thermal sample annotation. (**c**) A LiDAR sample annotation. (**d**) A fused sample annotation. (**e**) A null example (no object present). (**f**) A null example (partially present object).

5. Results

This section exposes (i) the training settings and results, (ii) an ablation test to evaluate the model's resilience during landing for all cases of sensor failure, and (iii) a resilience test against challenging weather conditions for fog, dark, and glary scenarios.

5.1. Training

5.1.1. Training Settings

The training hyperparameters were tuned according to useful guidelines towards reliable training performance provided by the official documentation of the YOLO framework (https://github.com/ultralytics/yolov5/wiki/Train-Custom-Data—Accessed on 19 February 2023) . The training procedure is executed for 400 epochs. The selected batch size is 32. The software training platform resorts to Google Colab's servers (https://colab.research.google.com/—Accessed on 19 February 2023), having access to its free of charge NVIDIA Tesla T4 GPUs. An IoU threshold of 0.2 is chosen. A learning rate cosine scheduling [39] is applied with an initial learning rate of 0.01. An Adam optimizer is used because over SGD and RMS Prop it had the best mAP@0.5:0.95 performance. To prevent overfitting, data augmentation techniques inherently exist in the dataset, and a weight decay of 5×10^{-4} is applied. The training hyperparameters are presented in Table 1.

Table 1. The main training hyperparameters.

Parameter	Quantity
Epochs	400
Batch size	32
IoU threshold	0.2
Momentum	0.937
Learning rate	1×10^{-2}
Weight decay	5×10^{-4}

5.1.2. Training Results

The training results for all the detectors are presented in the Figure 12. Commonly, a clear training convergence is observed for all models because the train regression loss (`train/box_loss`) and the objectness loss (`train/obj_loss`) monotonically decrease. The *box* loss describes the decrease in the regression error between the predicted and ground-truth bounding boxes (x, y, w, h) values. The *obj* loss describes the error between the confidence the model has on the object presence and the true presence. A *cls* loss has a particular behaviour because this is a single-class problem: when there is a class, it always the class 0. Therefore the problem reduces to predicting the presence of an object and the better enclosing bounding box. In addition, the model does not overfit because the validation losses monotonically decrease and eventually settle into a final value.

Regarding accuracy, all models achieved high levels of accuracy (above 95%) both on the The mAP@0.5 and The mAP@0.5:0.95 metrics (above 70%). The mAP@0.5:0.95 is significantly lower because it is more demanding. The early fusion model has the lower accuracy on the mAP@0.5:0.95 metric because of the redundancy of the predictions. Due to being a multimodal detector there is always some uncertainty about the true location of the object because the modalities are not perfectly aligned. This misalignment between modalities is caused by the temporal asynchronism between sensors. Hence, on average, the accuracy is always lower when compared to the accuracy achieved by unambiguous unimodal detection. The precision and recall metrics also demonstrate the accuracy convergence across all models. The precision describes the validity of true positive prediction on the universe of the selected true positive labels. For this metric, there is a general convergence towards 1, despite the thermal detector. On the thermal samples, due to the non-linearity of the binarization operation, some detail is occasionally lost which negatively influences the prediction.

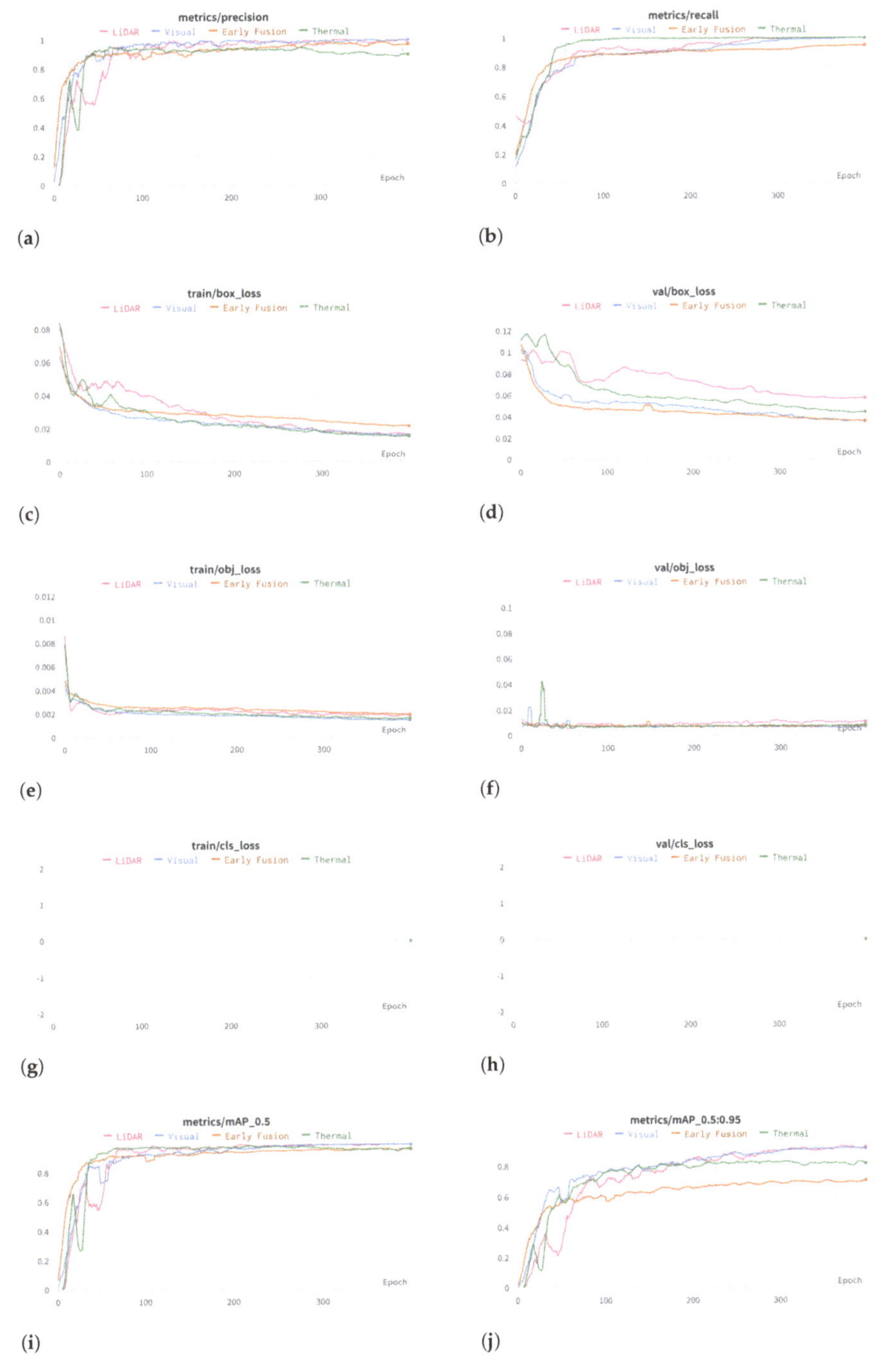

Figure 12. Training results. (**a**) The precision. (**b**) The recall. (**c**) The train regression box loss. (**d**) The validation regression box loss. (**e**) The train regression object loss. (**f**) The validation regression object loss. (**g**) The train regression class loss. (**h**) The validation regression class loss. (**i**) The Mean Average Precision 0.5 accuracy. (**j**) The Mean Average Precision 0.5:0.95 accuracy.

Similarly, for the `recall` metric, it is clear to observe a convergence towards 1, demonstrating that ultimately the predictions are complete: on the universe of all the images in an epoch containing a marker, the model accurately recognized a marker. The early fusion model slightly loses on the recall. This could be caused by the misalignment of some samples. When this misalignment is pronounced, the redundancy is lost and the model produces false negative predictions. This limitation could be avoided by including more unimodal samples on the early fusion dataset, such that the model trains more on unimodal situations (when the redundancy is not present).

Finally, to evaluate the performance of the detectors outside the training domain we compare the mAP score obtained by inferring the correspondent test datasets. Table 2 exposes the mean average precision for each detector.

Table 2. The performance of each detector on its own test set.

Detector	Modality	mAP@.5	mAP@.5:.95	Precision	Recall
Visual	Unimodal	0.999	0.999	1	1
Thermal	Unimodal	1	1	1	1
LiDAR	Unimodal	0.999	0.999	1	1
Early Fusion	Multimodal	0.989	0.989	1	0.985

All detectors achieve outstanding generalizable behaviour. Comparatively, the thermal detector stands out slightly. The thermal information, specifically, discriminates the object better which facilitates detection. Apart from that, thermographic information alone could suffer from particular extreme heat weather conditions (e.g., hot and sunny days). Thus, visual and point cloud information can complement this limitation. Decently, the early fusion detector nearly matches the fine-tuned detectors.

Lastly, the generalization capability of the detector is evaluated by inferring an external dataset called TNO Image Fusion [40] comprised by fused multispectral images as depicted by the Figure 13. Since this dataset does not contain the ArTuga marker, the aim is to examine the resilience of the detector against false positive predictions. Table 3 exposes the number of false positive predictions for five different levels of confidence thresholds across 127 images.

Table 3. False positive evaluation on an external dataset.

Confidence Threshold (%)	False Positives
60	19
70	9
75	4
80	1
90	0

From Table 3, it is clear that below 75% the model starts to become sensitive. However, in a real application if the threshold is set above 80%, the model can be considered reliable. In conclusion, the results from this evaluation are substantial since they demonstrate the resilience of the detector against novel and noisy information.

Figure 13. Some images from the TNO dataset [40].

5.2. Testing

The testing phase is comprised by an ablation and a weather tests to evaluate the model's resilience against sensor failure and challenging weather conditions, respectively. Glary, dark, and foggy weather conditions are addressed. Particularly, it is not feasible to operate current UAV technology in rainy conditions, hence rain settings are not addressed. Additionally, it is inopportune to apply LiDAR technology for rainy conditions [41]. The hardware setup for testing comprises an Intel® Core™ i7-10700F CPU @ 2.90GHz × 16 processor and a NVIDIA GeForce RTX 3060 GPU.

5.2.1. Ablation Test

To evaluate the resilience of the proposed early-fusion detector, we perform an ablation study where sensor failure is simulated by intentionally deactivating several combinations of specific modalities during a UAV landing (from an high to a low altitude). Several cases of unable (deactivated) modalities are simulated as follows: only LiDAR, only thermal, only visual, LiDAR and thermal, LiDAR and visual, thermal and visual. As a reference all cases are compared to the baseline (None) where all streams are activated. For each case, 18 images, acquired during landing, are inferred to the model. Recall and inference time results are exposed in the Table 4. True positive predictions are considered for confidences above 80%. For evaluating the accuracy only the recall metric is considered since all the images have a marker present. Moreover, Figure 14 depicts the cases and the predictions.

Table 4. Performance over 18 images during a landing operation for different unable signals.

Unable Signals	True Positives (Conf > 0.8)	Recall	Inference Time (ms)
None	100%	1	5.6
Visual and Thermal	100%	1	5.5
Visual and LiDAR	100%	1	4.2
Thermal and LiDAR	100%	1	4.7
LiDAR	100%	1	4.2
Visual	100%	1	4.1
Thermal	100%	1	4.8

The so needed redundancy against sensor failure is clearly concluded from the results exposed. Solid performance is demonstrated by the Recall results equaling 1 across all unable signal cases. This robustness is expected considering the presence of several unimodal samples in the dataset. Training with unimodal samples prepares the model against unable signals. Summing up, it can be concluded the reliability of the early fusion approach while operating under a sensor failure situation.

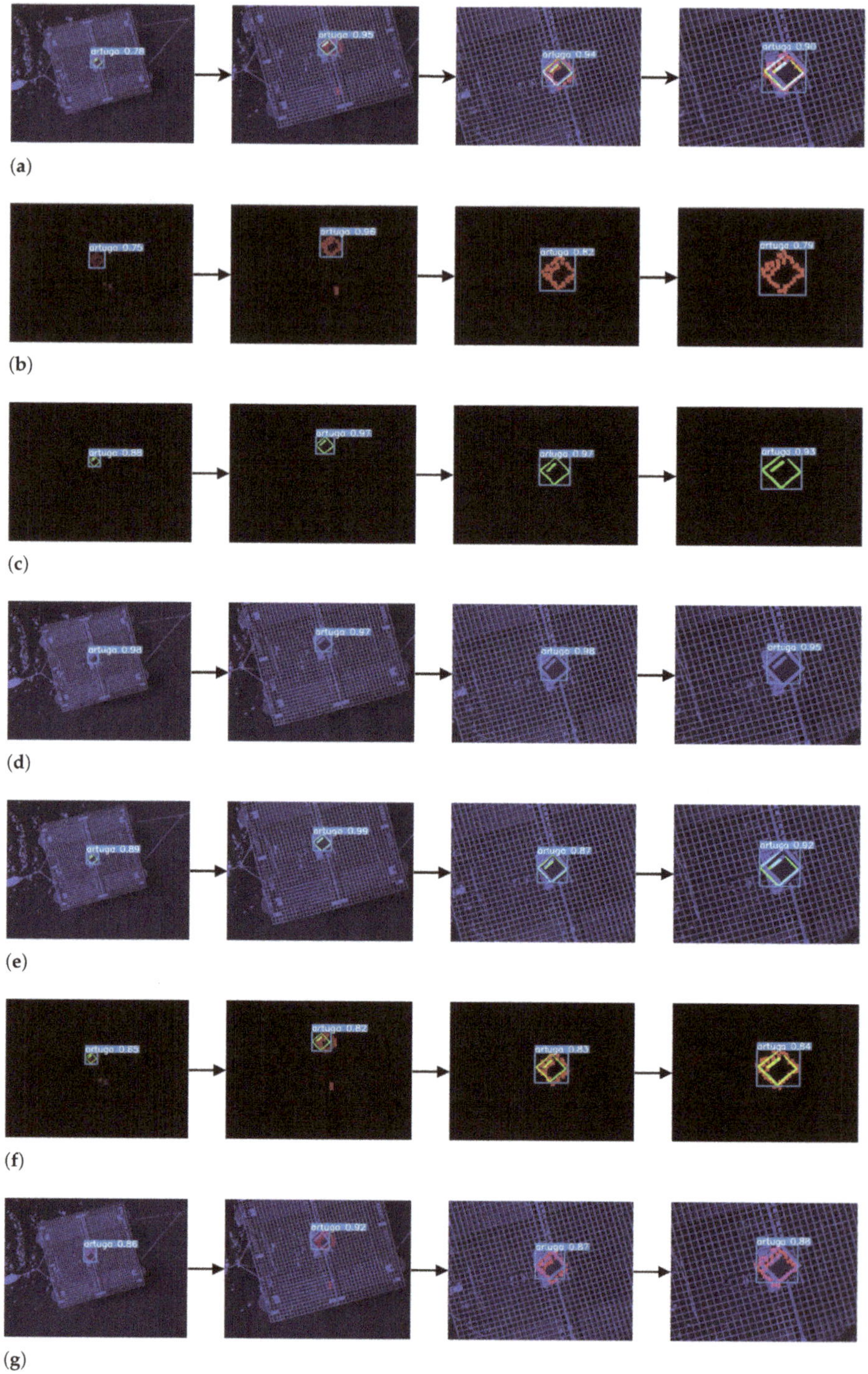

Figure 14. Examples of ablation test predictions for all sensor failure cases. (**a**) Early-fusion (baseline). All sources are activated. (**b**) Relying on LiDAR source only. (**c**) Relying on thermal source only. (**d**) Relying on visual source only. (**e**) LiDAR source failure. (**f**) Visual source failure. (**g**) Thermal source failure.

5.2.2. Weather test

To motivate the use of an early-fusion based detector for challenging weather conditions a stress test for extreme simulated weather conditions is conducted. Glary and dark environments are simulated for increasing and decreasing, respectively, variations in brightness of 10%, 50% and 90% of the original image. Fog environments are simulated by applying stochastic fog augmentation using the Image Augmentation library (https://imgaug.readthedocs.io/en/latest/source/api_augmenters_weather.html—Accessed on 19 February 2023).

Furthermore, every prediction above a 0.5 confidence threshold is considered a true positive. Otherwise, it is considered a false negative. An amount of 100 images are inferred for testing. For every image, there is a marker present, thus, at best, the model should produce 100 true positive predictions. For evaluating the accuracy only the recall metric is considered since all the images have a marker present. The results achieved are presented in the Table 5. Examples of the extreme scenarios are depicted by the Figure 15.

(**a**) (**b**) (**c**)

Figure 15. Examples of predictions under extreme weather conditions. (**a**) A 90% glary (sunny) scenario example. (**b**) A 90% dark (night) scenario example. (**c**) A fog scenario example.

Table 5. Performance over 100 images described by challenging weather conditions.

Weather Condition	True Positives (Conf > 0.6)	False Negatives	Recall	Inference Time (ms)
Bright (10%)	99	1	0.99	2.2
Bright (50%)	98	2	0.98	2.1
Bright (90%)	87	13	0.87	2.0
Dark (10%)	99	1	0.99	2.6
Dark (50%)	100	0	1	2.1
Dark (90%)	96	4	0.96	2.0
Fog (stochastic)	96	4	0.96	2.0

It can be concluded that the model is more sensitive under extreme bright conditions. Conversely, it is resilient against dark and fog conditions. Considering the rare occurrence of extreme brightness situations on the dataset, the model naturally performs poorly against it. Resilience is expected against dark conditions because the dataset includes unimodal samples that mimic sensor failure. Concretely, the model is trained against unable visual signals or, in other words, extreme dark environments. The worthy performance (96% recall) against the stochasticity of fog conditions is the most surprisingly result, which demonstrates the robustness against a more noisy and random signal. Hence, it can be said that the model is presumably robust against real noisy environments.

6. Conclusions

This research introduced a novel methodology that proposes an early-fusion module capable of introducing the required reliability to a lightweight state-of-the-art object detector under sensor failure and extreme weather conditions.

The proposed early fusion approach provided an expressive input that clearly facilitated the detection of a multimodal fiducial marker during a UAV offshore operation. Together, the early fusion detector and the multimodal marker operate in a robust and transparent fashion against challenging weather and sensor failure conditions. In addition, if there is a GPU onboard a robotic solution, it should be emphasized the assurance of a fast system for real-time operating conditions, as demonstrated by the inference time results of less than 6 ms. Consequently, it can be said that assuring robust performance empowered by an early fusion and end-to-end approach for a lightweight detector is one of the major contributions of this work.

Author Contributions: Conceptualization, F.S.N. and A.M.P.; methodology, F.S.N. and A.M.P.; software, F.S.N.; validation, F.S.N., R.M.C. and A.M.P.; investigation, F.S.N.; resources, A.M.P.; data curation, F.S.N. and R.M.C.; writing—original draft preparation, F.S.N.; writing—review and editing, F.S.N., R.M.C. and A.M.P.; visualization, F.S.N., R.M.C. and A.M.P.; supervision, A.M.P.; project administration, A.M.P.; funding acquisition, A.M.P. All authors have read and agreed to the published version of the manuscript.

Funding: This work is funded by the European Union's Horizon 2020 research and innovation programme under grant agreement No. 871571 (ATLANTIS), and by FLY.PT-P2020 Mobilizador within project POCI-01-0247-FEDER046079. An european patent for the ArTuga multimodal fiducial marker with the reference number EP22212945.4 has been submitted.

Institutional Review Board Statement: Not applicable.

Informed Consent Statement: Not applicable.

Data Availability Statement: The data produced by this research is described by four datasets designed to train Early Fusion Multimodal fiducial marker detectors for UAV landing operations. It is available at [37].

Conflicts of Interest: The authors declare no conflict of interest.

References

1. Pendleton, S.D.; Andersen, H.; Du, X.; Shen, X.; Meghjani, M.; Eng, Y.H.; Rus, D.; Ang, M.H. Perception, planning, control, and coordination for autonomous vehicles. *Machines* **2017**, *5*, 6. [CrossRef]
2. Lim, T.Y.; Ansari, A.; Major, B.; Fontijne, D.; Hamilton, M.; Gowaikar, R.; Subramanian, S. Radar and camera early fusion for vehicle detection in advanced driver assistance systems. In Proceedings of the Machine Learning for Autonomous Driving Workshop at the 33rd Conference on Neural Information Processing Systems, Vancouver, BC, Canada, 14 December 2019; Volume 2.
3. Yeong, D.J.; Velasco-Hernandez, G.; Barry, J.; Walsh, J. Sensor and Sensor Fusion Technology in Autonomous Vehicles: A Review. *Sensors* **2021**, *21*, 2140. [CrossRef] [PubMed]
4. Kim, J.; Koh, J.; Kim, Y.; Choi, J.; Hwang, Y.; Choi, J.W. Robust deep multi-modal learning based on gated information fusion network. In Proceedings of the Asian Conference on Computer Vision, Perth, Australia, 2–6 December 2018; pp. 90–106.
5. Boulahia, S.Y.; Amamra, A.; Madi, M.R.; Daikh, S. Early, intermediate and late fusion strategies for robust deep learning-based multimodal action recognition. *Mach. Vis. Appl.* **2021**, *32*, 1–18. [CrossRef]
6. Ngiam, J.; Khosla, A.; Kim, M.; Nam, J.; Lee, H.; Ng, A.Y. Multimodal deep learning. In Proceedings of the ICML, Bellevue, WA, USA, 28 June–2 July 2011.
7. Yi, S.; Jiang, G.; Liu, X.; Li, J.; Chen, L. TCPMFNet: An infrared and visible image fusion network with composite auto encoder and transformer–convolutional parallel mixed fusion strategy. *Infrared Phys. Technol.* **2022**, *127*, 104405. [CrossRef]
8. Panigrahy, C.; Seal, A.; Mahato, N.K. Parameter adaptive unit-linking dual-channel PCNN based infrared and visible image fusion. *Neurocomputing* **2022**, *514*, 21–38. [CrossRef]
9. Shopovska, I.; Jovanov, L.; Philips, W. Deep visible and thermal image fusion for enhanced pedestrian visibility. *Sensors* **2019**, *19*, 3727. [CrossRef]
10. Farahnakian, F.; Heikkonen, J. Deep Learning Based Multi-Modal Fusion Architectures for Maritime Vessel Detection. *Remote Sens.* **2020**, *12*, 2509. [CrossRef]
11. Choi, J.D.; Kim, M.Y. *A Sensor Fusion System with Thermal Infrared Camera and LiDAR for Autonomous Vehicles and Deep Learning Based Object Detection*; ICT Express: Jeju, Republic of Korea, 2022.
12. Liu, L.; He, J.; Ren, K.; Xiao, Z.; Hou, Y. A LiDAR and Camera Fusion 3D Object Detection Algorithm. *Information* **2022**, *13*, 169. [CrossRef]
13. Claro, R.; Silva, D.; Pinto, A. ArTuga: A Novel Multimodal Fiducial Marker for Aerial Robotics. *Robot. Auton. Syst.* 2022, *in press*.

14. Soviany, P.; Ionescu, R.T. Optimizing the trade-off between single-stage and two-stage deep object detectors using image difficulty prediction. In Proceedings of the 2018 20th International Symposium on Symbolic and Numeric Algorithms for Scientific Computing (SYNASC), Timisoara, Romania, 20–23 September 2018; pp. 209–214.
15. Girshick, R.; Donahue, J.; Darrell, T.; Malik, J. Rich feature hierarchies for accurate object detection and semantic segmentation. In Proceedings of the IEEE Conference on Computer Vision and Pattern Recognition, Columbus, OH, USA, 23–28 June 2014; pp. 580–587.
16. Girshick, R. Fast r-cnn. In Proceedings of the IEEE International Conference on Computer Vision, Washington, DC, USA, 7–13 December 2015; pp. 1440–1448.
17. Ren, S.; He, K.; Girshick, R.; Sun, J. Faster r-cnn: Towards real-time object detection with region proposal networks. *arXiv* **2015**, arXiv:1506.01497.
18. Redmon, J.; Divvala, S.; Girshick, R.; Farhadi, A. You only look once: Unified, real-time object detection. *arXiv* **2015**, arXiv:1506.02640.
19. Liu, W.; Dragomir Anguelov, D.E.; Szegedy, C.; Reed, S.E.; Fu, C.Y.; Berg, A.C. SSD: Single shot multibox detector. *arXiv* **2015**, arXiv:1512.02325.
20. Akkus, C.; Chu, L.; Djakovic, V.; Jauch-Walser, S.; Koch, P.; Loss, G.; Marquardt, C.; Moldovan, M.; Sauter, N.; Schneider, M.; et al. Multimodal Deep Learning. *arXiv* **2023**, arXiv:2301.04856.
21. Katsaggelos, A.K.; Bahaadini, S.; Molina, R. Audiovisual Fusion: Challenges and New Approaches. *Proc. IEEE* **2015**, *103*, 1635–1653. [CrossRef]
22. Gadzicki, K.; Khamsehashari, R.; Zetzsche, C. Early vs. Late Fusion in Multimodal Convolutional Neural Networks. In Proceedings of the 2020 IEEE 23rd International Conference on Information Fusion (FUSION), Rustenburg, South Africa, 6–9 July 2020; pp. 1–6. [CrossRef]
23. Azam, S.; Munir, F.; Sheri, A.M.; Ko, Y.; Hussain, I.; Jeon, M. Data fusion of lidar and thermal camera for autonomous driving. In Proceedings of the Applied Industrial Optics: Spectroscopy, Imaging and Metrology, Optical Society of America, Washington, DC, USA, 8–10 July 2019; pp. 2–5.
24. Bhanushali, D.; Relyea, R.; Manghi, K.; Vashist, A.; Hochgraf, C.; Ganguly, A.; Kwasinski, A.; Kuhl, M.; Ptucha, R. LiDAR-Camera Fusion for 3D Object Detection. *Electron. Imaging* **2020**, *2020*, 257. [CrossRef]
25. Geiger, A.; Lenz, P.; Urtasun, R. Are we ready for autonomous driving? The kitti vision benchmark suite. In Proceedings of the 2012 IEEE Conference on Computer Vision and Pattern Recognition, Providence, RI, USA, 16–21 June 2012; pp. 3354–3361.
26. Zhang, Z. A flexible new technique for camera calibration. *IEEE Trans. Pattern Anal. Mach. Intell.* **2000**, *22*, 1330–1334. [CrossRef]
27. Benjumea, A.; Teeti, I.; Cuzzolin, F.; Bradley, A. YOLO-Z: Improving small object detection in YOLOv5 for autonomous vehicles. *arXiv* **2021**, arXiv:2112.11798.
28. Murthy, J.S.; Siddesh, G.; Lai, W.C.; Parameshachari, B.; Patil, S.N.; Hemalatha, K. ObjectDetect: A Real-Time Object Detection Framework for Advanced Driver Assistant Systems Using YOLOv5. *Wirel. Commun. Mob. Comput.* **2022**, *2022*, 9444360. [CrossRef]
29. Al-Qubaydhi, N.; Alenezi, A.; Alanazi, T.; Senyor, A.; Alanezi, N.; Alotaibi, B.; Alotaibi, M.; Razaque, A.; Abdelhamid, A.A.; Alotaibi, A. Detection of Unauthorized Unmanned Aerial Vehicles Using YOLOv5 and Transfer Learning. *Electronics* **2022**, *11*, 2669. [CrossRef]
30. Kasper-Eulaers, M.; Hahn, N.; Berger, S.; Sebulonsen, T.; Myrland, Ø.; Kummervold, P.E. Detecting heavy goods vehicles in rest areas in winter conditions using YOLOv5. *Algorithms* **2021**, *14*, 114. [CrossRef]
31. Jung, H.K.; Choi, G.S. Improved YOLOv5: Efficient Object Detection Using Drone Images under Various Conditions. *Appl. Sci.* **2022**, *12*, 7255. [CrossRef]
32. Mseddi, W.S.; Sedrine, M.A.; Attia, R. YOLOv5 Based Visual Localization for Autonomous Vehicles. In Proceedings of the 2021 29th European Signal Processing Conference (EUSIPCO), Dublin, Ireland, 23–27 August 2021; pp. 746–750.
33. Liu, W.; Wang, Z.; Zhou, B.; Yang, S.; Gong, Z. Real-time signal light detection based on yolov5 for railway. *IOP Conf. Ser. Earth Environ. Sci.* **2021**, *769*, 042069.
34. Ultralytics. YoloV5. Available online: https://ultralytics.com/yolov5 (accessed on 19 February 2023).
35. Li, C.; Li, L.; Jiang, H.; Weng, K.; Geng, Y.; Li, L.; Ke, Z.; Li, Q.; Cheng, M.; Nie, W.; et al. YOLOv6: A single-stage object detection framework for industrial applications. *arXiv* **2022**, arXiv:2209.02976.
36. Wang, C.Y.; Bochkovskiy, A.; Liao, H.Y.M. YOLOv7: Trainable bag-of-freebies sets new state-of-the-art for real-time object detectors. *arXiv* **2022**, arXiv:2207.02696.
37. (2023): Early Fusion Datasets for UAV Landing Operations Using Multimodal Fiducial Marker. Figshare. Dataset. Available online:. (accessed on 19 February 2023). [CrossRef]
38. Roboflow. 2022. Available online: https://roboflow.com/ (accessed on 21 January 2023).
39. He, T.; Zhang, Z.; Zhang, H.; Zhang, Z.; Xie, J.; Li, M. Bag of tricks for image classification with convolutional neural networks. In Proceedings of the IEEE/CVF Conference on Computer Vision and Pattern Recognition, Long Beach, CA, USA, 15–20 June 2019; pp. 558–567.

40. Toet, A. *TNO Image Fusion Dataset*; Springer Data in Brief: Soesterberg, The Netherlands, 2022.
41. Heinzler, R.; Piewak, F.; Schindler, P.; Stork, W. Cnn-based lidar point cloud de-noising in adverse weather. *IEEE Robot. Autom. Lett.* **2020**, *5*, 2514–2521. [CrossRef]

sensors

MDPI

Article

Improving Odometric Model Performance Based on LSTM Networks

Bibiana Fariña *, Daniel Acosta, Jonay Toledo and Leopoldo Acosta

Computer Science and System Department, Universidad de La Laguna, 38200 San Cristobal de La Laguna, Spain
* Correspondence: bfarinaj@ull.edu.es

Abstract: This paper presents a localization system for an autonomous wheelchair that includes several sensors, such as odometers, LIDARs, and an IMU. It focuses on improving the odometric localization accuracy using an LSTM neural network. Improved odometry will improve the result of the localization algorithm, obtaining a more accurate pose. The localization system is composed by a neural network designed to estimate the current pose using the odometric encoder information as input. The training is carried out by analyzing multiple random paths and defining the velodyne sensor data as training ground truth. During wheelchair navigation, the localization system retrains the network in real time to adjust any change or systematic error that occurs with respect to the initial conditions. Furthermore, another network manages to avoid certain random errors by using the relationship between the power consumed by the motors and the actual wheel speeds. The experimental results show several examples that demonstrate the ability to self-correct against variations over time, and to detect non-systematic errors in different situations using this relation. The final robot localization is improved with the designed odometric model compared to the classic robot localization based on sensor fusion using a static covariance.

Keywords: mobile robot; self-localization; odometry; sensor fusion; long short-term memory

Citation: Fariña, B.; Acosta, D.; Toledo, J.; Acosta, L. Improving Odometric Model Performance Based on LSTM Networks. *Sensors* **2023**, *23*, 961. https://doi.org/10.3390/s23020961

Academic Editor: Natividad Duro Carralero

Received: 22 December 2022
Revised: 8 January 2023
Accepted: 9 January 2023
Published: 14 January 2023

1. Introduction

Auto-localization is an essential component of autonomous robots. It enables the robot to navigate safely and accurately by providing real-time information about its pose and surroundings. To achieve this, auto-localization systems typically use fusion algorithms to combine data from multiple sensors, such as cameras, LIDARs, and GPS, providing a robust and accurate localization module even in dynamic environments. The fact of receiving several data based on different operating principles, allows the system to become resilient to sensor failure, ensuring that the robot can continue to navigate effectively.

There are many fusion techniques used in auto-localization, including the Kalman filter. This last one requires each sensor to provide a noise covariance, which describes its measurement accuracy and determines its influence on the final localization. For non-linear systems, there are variants of the Kalman filter, such as the Extended Kalman Filter (EKF) and the Unscented Kalman Filter (UKF). The EKF linearizes the system using the Taylor series, but it can be less effective in highly non-linear systems [1]. The UKF, on the other hand, uses the unscented transform to fit the probability density distribution of non-linear equations, which allows it to avoid the loss of higher-order terms that can occur with linearization [2]. This makes the UKF a simple, fast, and precise option for non-linear systems [3–5].

In order to achieve accurate state estimation for auto-localization systems, it is not enough to simply use the right fusion techniques. The reliability of the sensor data is also crucial. Sensors can be easily affected by noise, and overcoming this uncertainty is essential for improving the efficiency of the auto-localization.

One of the key sensors used in wheeled mobile robots is the odometric sensor, which is a precise, cheap, and easy-to-process sensor. However, the pose calculated by this sensor is incremental, which can lead to the accumulation of small deviations over time. This can result in significant errors in the pose estimation. Additionally, the odometric sensor is sensitive to random errors, such as drifting, which can make it difficult to track and potentially cause inaccurate poses.

Considering these limitations and how artificial intelligence is one of the fastest-growing technologies in mobile robotics, this paper presents the application of neural networks to odometry, improving the data accuracy and also reducing the sensor errors. In this way, the encoder data are processed by a neural network to include effects that a static model does not represent.

This method is implemented in an autonomous wheelchair that has been designed to transport people with severe disabilities [6]. The wheelchair is controlled by the user indicating the destination, while the navigation module will automatically execute the optimal path, using the sensors for the reconstruction of the environment and the obstacles detection [7]. In this way, to achieve safe autonomous navigation, the localization is an essential part of the chair movement, since its failure would affect all other subsystems.

The wheelchair has the following set of sensors:

- Odometry: it estimates the wheel robot movement by optical encoders coupled to the motor. In the prototype, the resolution of the encoder is 8800 pulses per revolution with a resolution of 0.04 degrees. The pose is calculated by the wheel movement knowing the radii and distance between them.
- Light Detection and Ranging (LIDAR): they are two Sick TiM 551 sensors, with a maximum distance of 10 m, an angular resolution of 1 degree and a viewing angle of 270 degrees. The relative movement of the robot can be calculated using two consecutive laser scans. The measurement accuracy is estimated by the ICP (Iterative Closest Point) algorithm [8,9]. It will return a low covariance value if two consecutive sweeps have many singular points that ensure their precise matching. However, if they have few singular points, the covariance will be high.
- Inertial Measurement Unit (IMU): an MPU9250 sensor which consists of an accelerometer, a magnetometer and a gyroscope. Our tests indicate that the magnetometer an the accelerometer are too imprecise to consider its measurements so only the gyroscope will be used. Likewise, the sensor covariance is estimated by setting a static value.

The robot state $\tilde{x}_k(V, W)$ is estimated by a UKF. In this case, the algorithm estimates the state at the next instant (k) by using the uniform rectilinear motion model and considering the sensor data $z_k(V, W)$ and their covariances $(COV_k^{v/w})_z$ as measurements. The output of the filter not only includes the estimated robot state but also its covariance $(COV_k^{v/w})_{filter}$, which measures the precision of the filter at that moment. Moreover, the pose (X, Y, θ) is calculated by integrating the output state of the filter. Figure 1 shows the wheelchair and a schema of the implemented localization where our technique is applied.

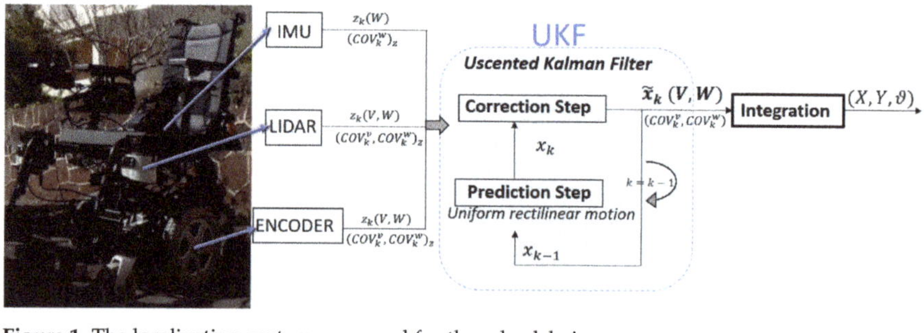

Figure 1. The localization system proposed for the wheelchair.

Accordingly, our paper focuses on the following key points: the robot self-localization problem applied to autonomous wheelchair and the LSTM (long short-term memory) neural network technique to improve the odometric sensor. The paper is structured as follows. Section 2 includes a review of the studies that have been completed prior to this work. Section 3 describes the wheelchair kinematics model. Likewise, Section 4 explains the methodology carried out to improve the accuracy of the odometric sensor. The experiments using the designed model are shown in Section 5, and the conclusions are summarized in Section 6.

2. Previous Work

The study of assistive robots is gaining attention for the possible social, economic, and scientific applications. There are investigations that collect the design of different types of autonomous wheelchair. Ref. [10] designs an autonomous wheelchair capable of segmenting passable areas and anomalies on the roads through deep learning. Refs. [11,12] show the characteristics of two wheelchairs for the transport of people in predefined hospital. Ref. [13] describes an adaptive neural control system for governing the movements of a robotic wheelchair. Likewise, [14,15] use an ROS (Robotic Operative System) in the wheelchair for the architecture of their modules.

Regarding novel techniques in the current literature on self-localization, there have been significant advances in the application of artificial intelligence (AI) to mobile robotics. This has led to the development of more intelligent and autonomous robots that can navigate complex environments and complete tasks more efficiently and accurately. Ref. [16] describes a survey of the different matching learning techniques for motion planing and control for mobile robots. In [17,18], a track fusion algorithm based on the LSTM method are proposed achieving better results in the fusion effect. Likewise, Refs. [19,20] present a deep reinforcement learning approach for the motion planning of autonomous robots in dynamic surroundings.

Machine learning is also applied in the odometry pose estimation to improve the results of using the traditional method by geometric equations. In [21], a feedforward neural network model is used in order to learn the odometric model from data. However, these sensors contain errors that need to be corrected in order to obtain a reliable pose. Real-time calibration of the structure's parameters is important to adapt to any changes in the model over time. There are various techniques in the literature that attempt to adjust these parameters. Refs. [22,23] solves the car odometric systematic errors by using the final pose difference in a predefined path. Other works try to correct them in real time by the design of an AKF (Augmented Kalman Filter) [24] or by estimating the deviation of the radius nominal value by using a marginalized particle filter (MPF) [25].

With respect to non-systematic errors, such as slip, there are some methods that attempt to reduce them. Ref. [7] uses a Doppler speed sensor in order to measurement the odometric noise depending on the data quality and obtain a dynamic odometric covariance. Likewise, Refs. [26,27] try to correct them by reducing their influence in the output filter. They design a real-time outlier detection in the observations, applying a saturation function to the residual. Moreover, Refs. [28,29] try to avoid them by using the dynamic structure model. This last one proposes an slip ratio estimator based on the motion equations, the input torque and encoders data.

Taking into account the AI advances in the field of self-localization and the previously mentioned limitations of the odometric sensor, this paper presents a new approach based on LSTM networks for predicting odometry poses. The proposed system is able to self-calibrate in real-time through network retraining and another LSTM network is used to learn and compensate for non-systematic errors in the final localization.

3. Robot with Differential Kinematics

The wheelchair has differential kinematics. It consist of two drive wheels on a common axis. The robot can move and turn thanks to the independent wheel driving forwards or backwards. Figure 2 shows the scheme of a differential mobile robot with radii $R_{r/l}$ and a distance D between them. The robot movement between two time instants would correspond to a circle trajectory with radius ρ around the point ICC (Instantaneous Centre of Curvature). From the wheel speeds $V_{r/l}$, it is possible to estimate the linear (V) and angular (W) robot velocities, as well as its pose (X, Y) and its orientation (θ).

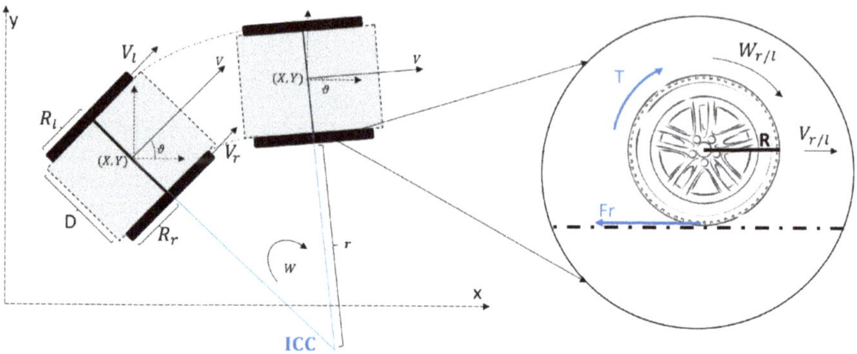

Figure 2. Differential kinematics schema for a mobile robot.

Figure 2 shows the robot wheel dynamics, where T corresponds to the motor torque that generates a linear ($V_{r/l}$) and angular ($W_{r/l}$) velocities. A friction force (F_r) opposes to the movement, it can have a significant impact on the behavior and movement of the robot. These forces depend on the environment in which the robot operates, and small variations in their values can affect the robot's movement. For example, certain types of robots, such as wheeled mobile robots, are very sensitive to small changes in the speed of each wheel. Even minor errors in the relative speeds between the wheels can affect the robot's path. These robots are also sensitive to small variations in the ground plane, and they may need additional wheels (swivel casters) for support.

3.1. Wheel-Encoder-Based Traditional Odometry

The wheelchair's pose at each instant can be obtained from the wheel speeds following the scheme of the Figure 2 by using odometric equations. In this case, we can estimate the respective wheel speeds by using the counts provided from the encoder sensor as follow,

$$V^l_{ok} = \frac{2\pi R_l \triangle ct^l_k}{encRes\triangle T}; \qquad V^r_{ok} = \frac{2\pi R_r count^r_k}{encRes\triangle T} \tag{1}$$

where $V^{l/r}_{ok}$ corresponds to the left and right wheel velocities. $\triangle ct^{l/r}_k$ are the number of each encoder ticks that the electronics receive in time period ($\triangle T$), and $encRes$ is the encoder resolution in one turn. The wheels parameters are the corresponding radii R_r and R_l and the distance between the wheels D.

Considering the circle trajectory with radius ρ around the ICC point, we can estimate the linear (V_k) and angular (W_k) robot velocities (Equation (2)). This calculation is obtained by geometric equations knowing that the linear velocity of a circumference multiplied by the radius is equal to the angular velocity.

$$\begin{gathered} \frac{V^r_{ok}}{r} = W_k; \quad \frac{V^l_{ok}}{\rho+D} = W_k; \quad \frac{V_k}{\rho+D/2} = W_k \\ W_k = \frac{(V^l_{ok}-V^r_{ok})}{D}; \quad V_k = \frac{(V^l_{ok}+V^r_{ok})}{2} \end{gathered} \tag{2}$$

Once the current robot velocities $(V, W)_k$ has been estimated, it is necessary to integrate, considering the time interval $\triangle T$, to obtain the estimated robot pose $(X, Y, \theta)_{k+1}$ at each moment,

$$\begin{aligned} \theta_{k+1} &= \theta_k + W_k \triangle T \\ X_{k+1} &= X_k + V_k cos(\theta_k) \triangle T \\ Y_{k+1} &= Y_k + V_k sin(\theta_k) \triangle T \end{aligned} \quad (3)$$

3.2. Wheel-Motion Pattern Effect

The behavior of wheeled mobile robots depends on the environment characteristics. The contact type between the wheel and the ground and its friction force value significantly influence the robot's movement.

In ideal conditions, regardless of the terrain type, the robot speed depends on the motor torque and its power. An increase in the motor power will lead to an increase in angular wheel speed that will be translated into linear velocity. The opposite case is given to reduce the robot speed. However, there are situations where, depending on the current robot speed and the adhesion degree with the terrain, the robot can slip or skid, producing two situations. The first is due to the motor's dead zone that does not produce enough torque to move the wheel, causing the wheel to stall instead of turning with the robot in motion. The second occurs when the wheel rotates freely, slipping and without any robot movement.

From this assumption, we anticipate that there is a logical correlation between the angular wheel speed and the motor power consumed. This relation disappears when non-systematic errors occur due to the type contact with the ground.

Figure 3 shows graphically an example of the relationship between the motor power and its angular velocity, given by the encoder sensor counts. In Figure 3a, there is no significant slip, following wheel speed and motor power, a similar pattern, increasing and decreasing in a correlated way. However, in Figure 3b there is a slip, highlighted by the green lines, where, although the power increases, the real angular velocity decreases, contrarily to the expected effect. Using this property, the motor power can be used to detect non-systematic errors in the odometric system. There is not a clear mathematical relation between motor power and speed in order to detect non systematic errors, so a Neural Network is used to learn this relation from data.

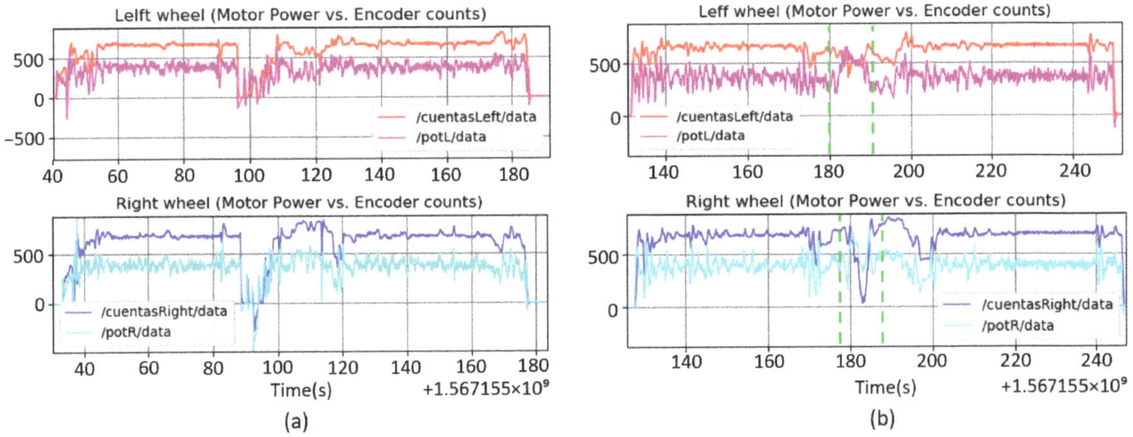

Figure 3. Motor power vs. encoder counts for both wheels.(**a**) Without slipping and (**b**) with slipping.

4. Odometry System Proposed

In this localization system, the data accuracy provided by the odometric sensor is improved using neural networks that learn from the robot behavior. It avoids using manually estimated parameters, improving and correcting some errors, which in the mathematical model are not appreciable and, therefore, difficult to adjust.

The proposed model consists of two neural networks that have, as input, the incremental encoder counts ($\triangle ct_k^{r/l}$) of the right and left wheels, and provides, as output, the robot state, that are the linear (V_k) and angular (W_k) speeds for the respective network. We decided to use two neural networks for each output instead of a single one in order to allow each network to specialize in predicting a specific output and converge more quickly and easily. This is because in the following sections we will have to retrain them online and we need the network to quickly adapt and converge to any changes that may occur in the model. Using a single network with two outputs makes retraining much more difficult and it does not adequately adjust to our desired requirements. This has also been tested experimentally, and we found that the architecture proposed resulted in more precise execution.

The training has been carried out offline by creating a dataset from the information of multiples paths with different characteristics. During the training, the speed data provided by a velodyne HDL 32 sensor at each moment have been collected as the desired output of the robot state. It has been installed in the upper part of the chair and consists of 32 LIDARs with a precision of 2 cm that enables the calculation of the state with high reliability using the LOAM slam algorithm [30,31].

We decided to use recurrent networks, specifically, the LSTM-type [32] was chosen over other models, such as MLP, because of its ability to learn and remember patterns over time. This is particularly useful as it can take into account long-term patterns in the encoder data sequence. Moreover, during the neural network design, we compared the use of other models using the wheelchair dataset and it demonstrates a better effectiveness during training and validation. For the LSTM network, it has been considered a time window of 20 samples that are provided by the sensor at a frequency of 10 Hz. Figure 4 shows the structure carried out to obtain the linear (V_k) and angular (W_k) velocity at each time k. In order to train the designed model, pre-processing data are necessary. First, the raw data were resized into a tensor whose dimension was (samples, time step, features), which in our case was (10,115,20,3). The inputs at one moment correspond to the encoder count increments and the speed at the previous time $(\triangle ct^r, \triangle ct^l, V/W)_{k-1}$. The outputs are the lineal V_k and angular W_k speeds of the robot. Moreover, the input must be normalized to the same scale by eliminating the mean and scaling to a standard deviation equal to 1. So, the samples of an input type are pre-processed using:

$$z = \frac{s-\mu}{\sigma} \qquad with \qquad \mu = \frac{1}{N}\sum_{i=1}^{N} s_i \qquad \sigma = \sqrt{\frac{1}{N}\sum_{i=1}^{N}(s_i-\mu)^2} \tag{4}$$

where z is the new normalized value and s is the raw sample. μ is the mean and σ is the standard deviation of the input, and N correspond to the samples dimension.

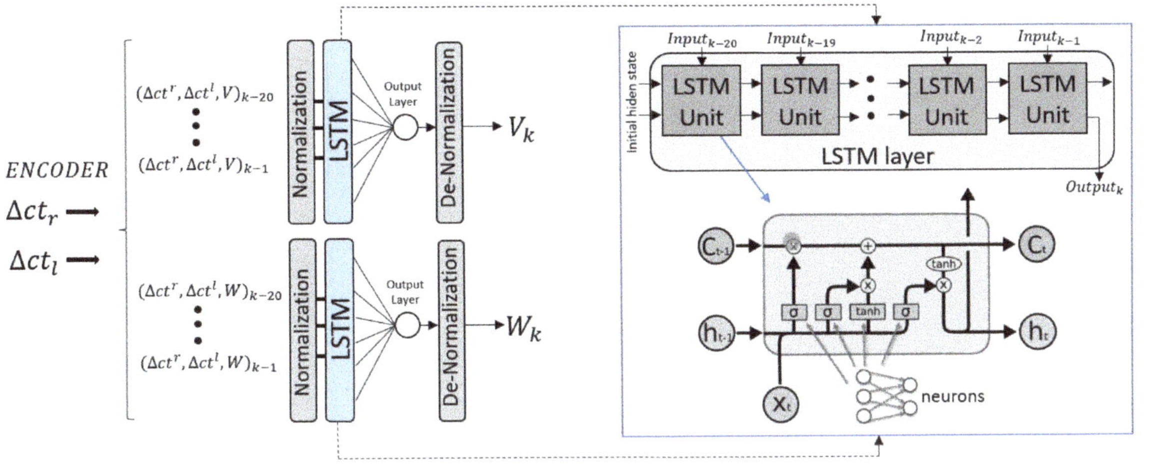

Figure 4. Schema of the odometric model proposed using LSTM networks.

Regarding the implemented LSTM layer, it analyzes an input data stream and produces a prediction. In this structure, data are introduced sequentially in an LSTM unit (cell) that can be divided into different parts. In each cell, x_t corresponds to the input at the current time, h_{t-1} is the previous cell output and C_{t-1} is the cell state. First, the forget gate controls how much information about C_{t-1} is relevant from the inputs h_{t-1} and x_t, by using a sigmoid layer (σ),

$$f_t = \sigma(\omega_f[h_{t-1}, x_t] + b_f) \tag{5}$$

where ω_f represents the weight matrix and b_f the bias. f_t value in each dimension is in the range (0, 1). The information will be forgotten when f_t is close to 0, and the information will be retained when is close to 1.

Subsequently, the gateway is responsible for processing the current input and updating the relevant information. It has two parts, the first (i_t) controls how much of the input is stored using a sigmoid layer and the second ($\tilde{C}_t$) generates a new current state candidate by using the tanh function,

$$i_t = \sigma(\omega_i[h_{t-1}, x_t] + b_i) \qquad \tilde{C}_t = tanh(\omega_c[h_{t-1}, x_t] + b_c) \tag{6}$$

where ω_i and ω_c represent weight matrix and b_i and b_c represent the bias. From this information, the cell state can be updated from the previous one (C_{t-1}) and from the new generated candidate ($\tilde{C}_t$),

$$\tilde{C}_t = f_t * C_{t-1} + i_t * \tilde{C}_t \tag{7}$$

Finally, the output gate generates the cell output (h_t) from the selection of the relevant information in the current cell state,

$$h_t = \sigma(\omega_o[h_{t-1}, x_t] + b_o) * tanh(C_t) \tag{8}$$

where ω_o is the weight matrix and b_o its bias.

Therefore, considering the Figure 4, the data are processed and normalized, then divided into a 70% of training and a 30% as test data. In this case, the LSTM is 1 layer with 5 neurons in each cell, chosen ad hoc, providing the best training and validation results for different tests. This state is passed to a full connected output layer with 1 neuron for a single output and linear activation. Regarding the training, the ADAM optimizer algorithm and the MAE (mean absolute error) metric has been used with a learning rate of 0.001, a batch size of 10, 150, and 200 epochs depending on the model. Figure 5 shows the loss in both training and validation, using the MAE errors obtained with respect to the epochs for each model, V and W.

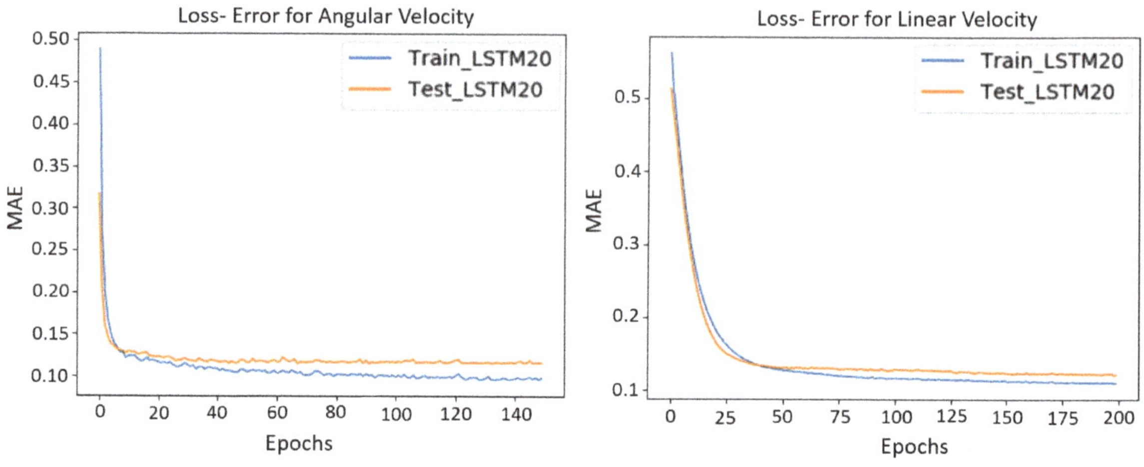

Figure 5. Loss function per epochs using the MAE for the angular and linear velocities.

4.1. Odometry Error Covariance Modeling

The odometric system can be improved using an LSTM network to detect non-systematic errors that occur in unpredictable environments. For example, local bumps, holes, rocks, and non-level terrain can cause non-systematic errors. Using gyro sensors to detect these errors has shown good performance, as demonstrated in previous research [33,34]. However, there are other cases that are more difficult to detect, such as wheel slippage on level terrain. In these cases, a non-systematic detection method is necessary.

As explained in the first section, the fusion algorithm estimates the robot localization from different sensors, where each measurement is defined with a covariance that characterizes its confidence. In the case of odometry, it takes the speed covariances $(COV_k^V, COV_k^W)_{odom}$. Then, the filter prioritizes the measurements with a small covariance and avoids those with greater values. So, it is important to dynamically adjust the covariance of the odometric sensor to assign a high number in the presence of non-systematic errors and, thus, reduce its influence on the final localization result, compared to other sources. This variable covariance is designed based on the assumptions in Section 3.2, where we demonstrate how the wheel speeds depend on the motor torque and its power, and how the adhesion degree with the terrain causes the wheel to slip, affecting this dependence. Therefore, we need a system that is able to learn the relationship between the power of each wheel and its corresponding angular velocity in order to detect when this relationship is not fulfilled and is in an error state. In this case, the system will act accordingly by assigning a very large static value to all elements in the covariance matrix, considering that they are stochastically independent random variables.

Figure 6 shows the method implemented on the wheelchair for this purpose. It consists of an LSTM neural network that takes as input the powers of each wheel (P^r, P^l), the encoder increment counts $(\triangle ct^r, \triangle ct^l)$, and the sensor covariance in the previous measurements (COV_{k-n}^V), (COV_{k-n}^W). As output, it is able to estimate the next step covariance (COV_k^V), (COV_k^W). The data have been preprocessed using the same method discussed earlier, organized into time windows with memory of 20 samples, and normalized to have a standard deviation of one and a mean equal to zero. In our case, several samples have been collected with different slip types and whose input dimension is (12442,20,5), which have been divided into 70% for training and 30% for validation. The training has been carried out offline by using the velodyne sensor data. The network structure is a LSTM layer consisting of 10 neurons and a full connected output layer with 1 neuron for a single output and ReLU activation. We considered the ADAM optimizer algorithm and the MAE metric with a learning rate of 0.001, 120 epochs, and a batch size of 10. A de-normalization process is required in the final phase to obtain the estimated covariance data.

Figure 6 also shows the variance of the angular odometry velocity estimated by our designed network with respect to time, where a slippage happens (non-systematic error) in the period 400–405 s. As we can see, the covariance remains at a low value when the sensor operates correctly and increases considerably when it is in an error state. This information allows the UKF filter to weight the input data in real time. If the LSTM Network detects a failure condition, the covariance will increase, so the data coming from odometry will be considered by the filter with a small weight. The odometry data together with this variable covariance will be used in a subsequent filter to be fused with other sensors and its influence on the final localization result will depend on that estimated covariance.

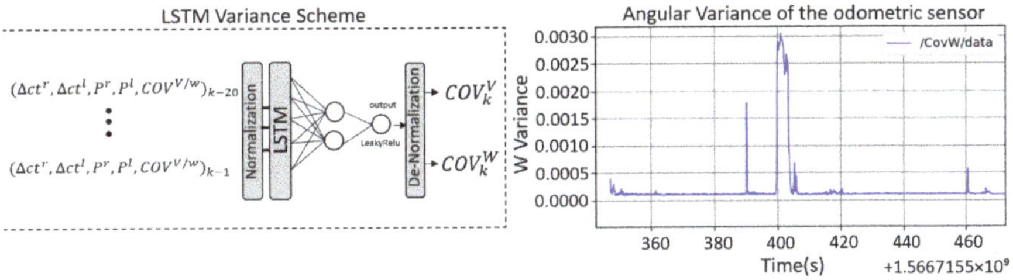

Figure 6. Scheme implemented to avoid non-systematic odometry errors, and the evolution of the angular velocity variance when a slippage happens in an example path.

4.2. Real Time Auto-Calibration

Real-time calibration of the model is crucial when designing an odometry system. It must be able to detect and correct any systematic error that may arise in relation to the offline-trained model. For example, a sudden change in the model may occur due to a shift in the weight of passenger, which can significantly impact the model accuracy. Over time, other situations, such as a change in tire pressure, can also affect the model performance, causing errors in the estimated localization. In order to ensure high accuracy, the system should be able to adapt to these changes in real-time by fine-tuning its model.

The scheme used for this calibration is shown in Figure 7. In this case, the neural network models for both angular and linear velocity estimation are retrained in real-time using input data from encoders $(\triangle ct^r, \triangle ct^l)$ and the output from the UKF filter speeds $(V, W)_k$. It is an iterative process where the data are stored in real-time, and when 300 samples are collected, the network weights are adjusted accordingly. These samples are only stored whenever the covariance of the filter (COV_{Filter}) and the variable covariance of the odometry (COV_{Odom}) are small, which ensures that the data used for re-training are reliable and accurate. In this way, the model is changed only when the robot is well localized, avoiding introducing errors in the retraining process.

Figure 7 also shows the error in angular velocity over time with respect to the ground truth (velodyne) when there is a change in the model, causing the error to increase. These graphs compare the behavior of the odometric system with and without re-training. Without considering the calibration, the model does not detect any change and the error grows and remains constant. In case of the real-time calibration, the system is able to detect these changes and adjust the model by retraining, reducing the relative error with respect to the velodyne sensor.

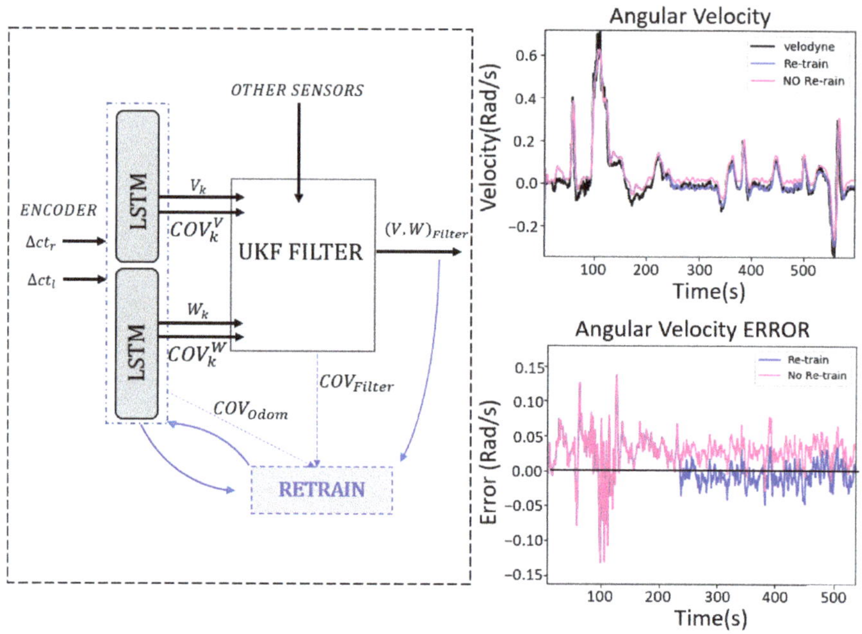

Figure 7. Retraining scheme for odometric model calibration. The error using the model with real-time retraining is smaller than using the model without it.

5. Results

The validation of the implemented wheelchair localization system was demonstrated through the study and analysis of several paths with different characteristics. These experiments were carried out in the corridor of the robotics laboratory at the University of La Laguna. In addition, to evaluate the proposed odometric model, different parameters and speeds were used on the paths to demonstrate the functionality of its variable covariance and the online calibration, as well as to ensure that a wide range of circumstances were covered. The effectiveness of the calibration was tested by varying the passengers and altering the diameters of the wheels through inflation. Furthermore, in the experiments to detect errors from the environment, the wheelchair speed has been increased in various sections and we have used other wheels with greater wear to generate slipping. We have also created these situations by applying pressure on one side, preventing the rotation of a wheel while allowing the other to turn freely.

On the one hand, Figure 8a,b shows two different example paths where the estimated pose of the odometry using LSTM networks (green line) and the pose obtained using traditional odometry (blue line) have been compared with the reference data from the velodyne (black line). As can be observed, the use of our model significantly improves the estimated pose accuracy, providing results that are closer to the ground truth.

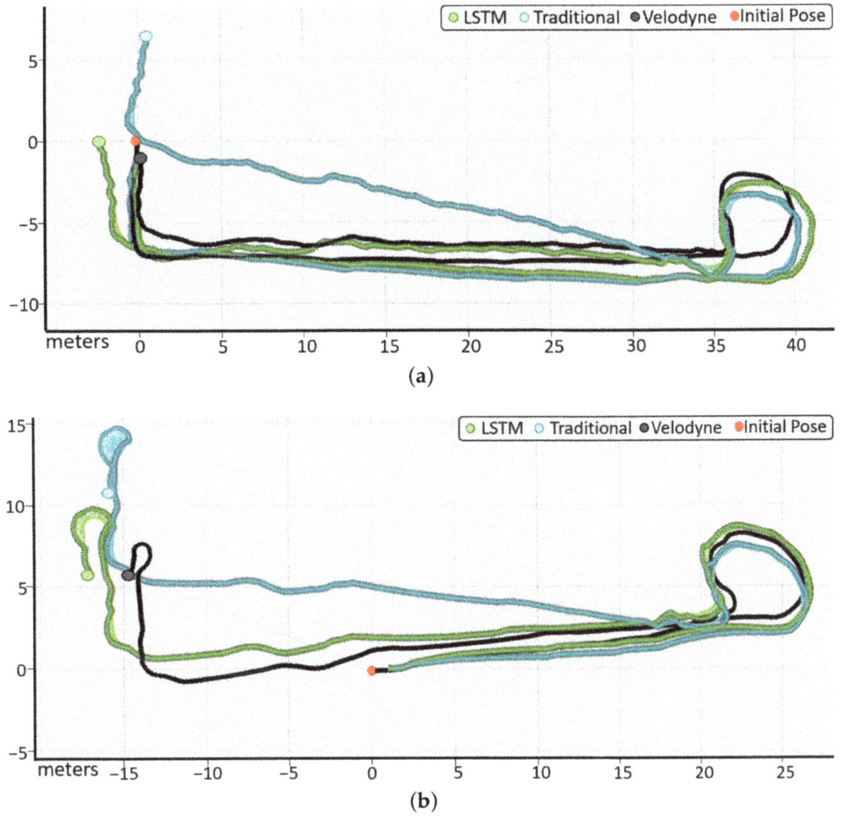

Figure 8. Paths to validate the odometry model proposed. Odometry with LSTM networks (green), traditional odometry (blue), and ground truth (black).

This fact is also demonstrated numerically in Table 1. In this case, multitude of trajectories have been executed and the odometry error with respect to the velodyne has been collected for both cases, using our LSTM network and using traditional model. We have analyzed the errors at each time in the velocities (V, W) and in the robot pose (X, Y, θ). The table shows the mean squared error (RMSE), the mean absolute error (MAE), and the R squared, which are defined by the following formulae:

$$RMSE = \sqrt{\frac{1}{n}\sum_{i=1}^{n}(\hat{y}_i - y_i)^2} \quad MAE = \frac{1}{n}\sum_{i=1}^{n}|\hat{y}_i - y_i| \quad R^2 = 1 - \frac{\sum_{i=1}^{n}(\hat{y}i - y_i)^2}{\sum_{i=1}^{n}(y_i - \bar{y})^2} \tag{9}$$

where $\hat{y}_i$ is the predicted value, y_i is the true value, n is the number of samples, and $\bar{y}$ is the mean value of the true values. Based on the obtained values, the errors of our system are smaller than the traditional model. Therefore, despite the small difference between the errors, it demonstrates the effectiveness of our model in improving the odometry localization of the wheelchair.

Table 1. Error values obtained by comparing the LSTM odometry model and the traditional odometry with respect to the velodyne data.

		V (m/s)	W (rad/s)	Pose	
				X,Y (m)	θ (rad)
LSTM odometry	MAE	0.04646	0.08103	1.56692	0.07131
	RMSE	0.10282	0.1791	2.076499	0.0864
	R_{score}	0.799122	0.84129	0.9569	-
	Accuracy %	79.91%	84.13%	95.69%	-
	Global Model Accuracy (%)			86.57%	
Traditional odometry	MAE	0.062744	0.102723	1.83725	0.1720
	RMSE	0.1237	0.200329	2.48464	0.2151
	R_{score}	0.6734	0.7953	0.9389	-
	Accuracy %	67.43%	79.53%	93.89%	-
	Global Model Accuracy (%)			80.2%	

On the other hand, the proposed real-time calibration model has been studied in several paths and with different changes in the system that have occurred both initially and during it. Figure 9 shows two examples where the characteristics of one wheel have been intentionally modified from the initial model to represent the odometry behavior when a systematic error arises. It causes the pose to deviate, generating an error in the velocities that, when integrated, accumulates over time in the pose. We compared the behavior of three different odometry models: the neural network-based odometry with real-time retraining (green line), odometry with neural networks but without retraining (yellow line), and traditional odometry (blue line). As reference, we consider the velodyne(black line).

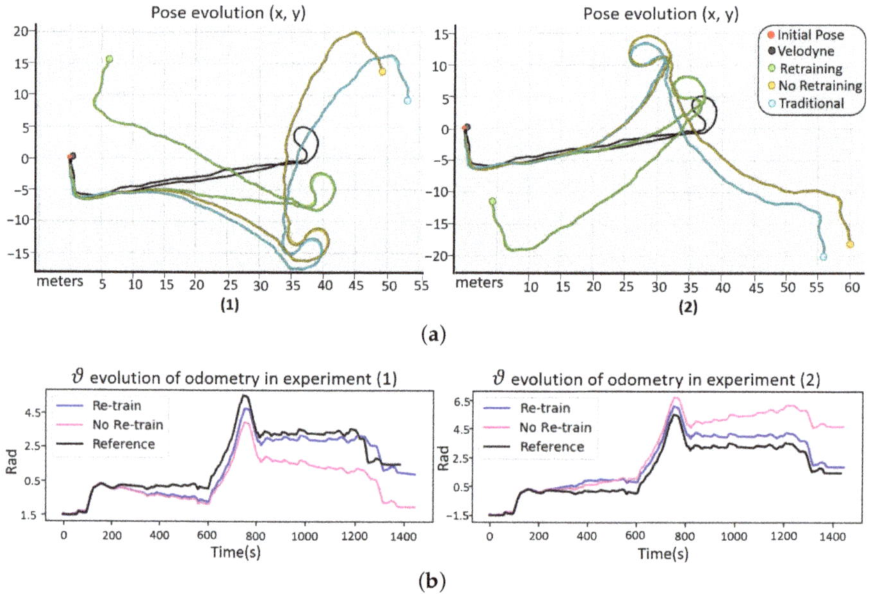

Figure 9. Example paths to validate the calibration model by retraining in real-time. (**a**) Shows the position "X, Y" and (**b**) shows the orientation "θ".

This figure shows how, thanks to retraining, the model is adjusted to new changes and corrects the pose increments of the trajectory, approaching the velodyne. Table 2 collects the errors obtained from several experiments comparing the proposed odometry model with and without retraining. Based on these results, the neural network-based odometric model and traditional odometry are unable to perceive these changes, increasing the error indefinitely. If we consider the calibration, the error in the velocities is reduced. However, it does not directly correct the error pose, but rather gradually improves the pose increments at each time interval. Moreover, we are interested in correcting the velocities, which will be the input measurement data for the filter in estimating the final pose of the robot.

Table 2. Error values obtained by comparing the LSTM-odometry model with and without retraining considering the velodyne data as reference.

		V (m/s)	W (rad/s)	Pose	
				X,Y (m)	θ (rad)
LSTM Retraining	MAE	0.0384577	0.06484	4.27604	0.56772
	RMSE	0.069028	0.115653	8.34219	0.629522
	R_{score}	0.848956	0.71317	0.25404	-
	Accuracy %	84.9%	71.32%	25.4%	-
	Global Model Accuracy (%)			60.54%	
LSTM No retraining	MAE	0.05496	0.07919	5.156277	1.5677
	RMSE	0.081812	0.22197	8.90407	1.7094
	R_{score}	0.67216	0.57198	0.03743	-
	Accuracy %	67.22%	57.2%	3.74%	-
	Global Model Accuracy (%)			42.7%	

Regarding the process implemented to reduce the influence of non-systematic errors, the output of the UKF filter has been studied in various experiments with different slip types. Figure 10 shows a path example where a considerable non-systematic error occurs. The green line represents the pose estimated by the odometry, the blue one the LIDAR, the red is the UKF localization, and the black line defines the ground truth. This error type causes the odometry pose to deviate significantly, and since it is a random error, it is impossible to correct its pose. The proposed solution is to modify the odometry covariance. During the corridor, the LIDAR sensor has a high covariance due to the low number of possible matching points and its estimated pose could be erroneous, however the odometry provides accurate results. In turn, due to the slip, the error in the odometric sensor grows and the LIDAR is more accurate.

Figure 10a shows the UKF-filter output when the traditional odometry is used with a static covariance manually adjusted. The result is worse since it either relies too much on odometry, underestimating the LIDAR or vice versa. Figure 10b shows the localization result of the UKF filter with the proposed LSTM neural network with non-systematic error detection and dynamic covariance estimation. During the slip, the power of the motors and wheel speed is not correlated, the error classifier detects a non-systematic error during the turn, and increases the covariance, reducing the odometry influence on the final localization during that error state. The final pose of the filter is close to ground truth, and clearly better than without the detector.

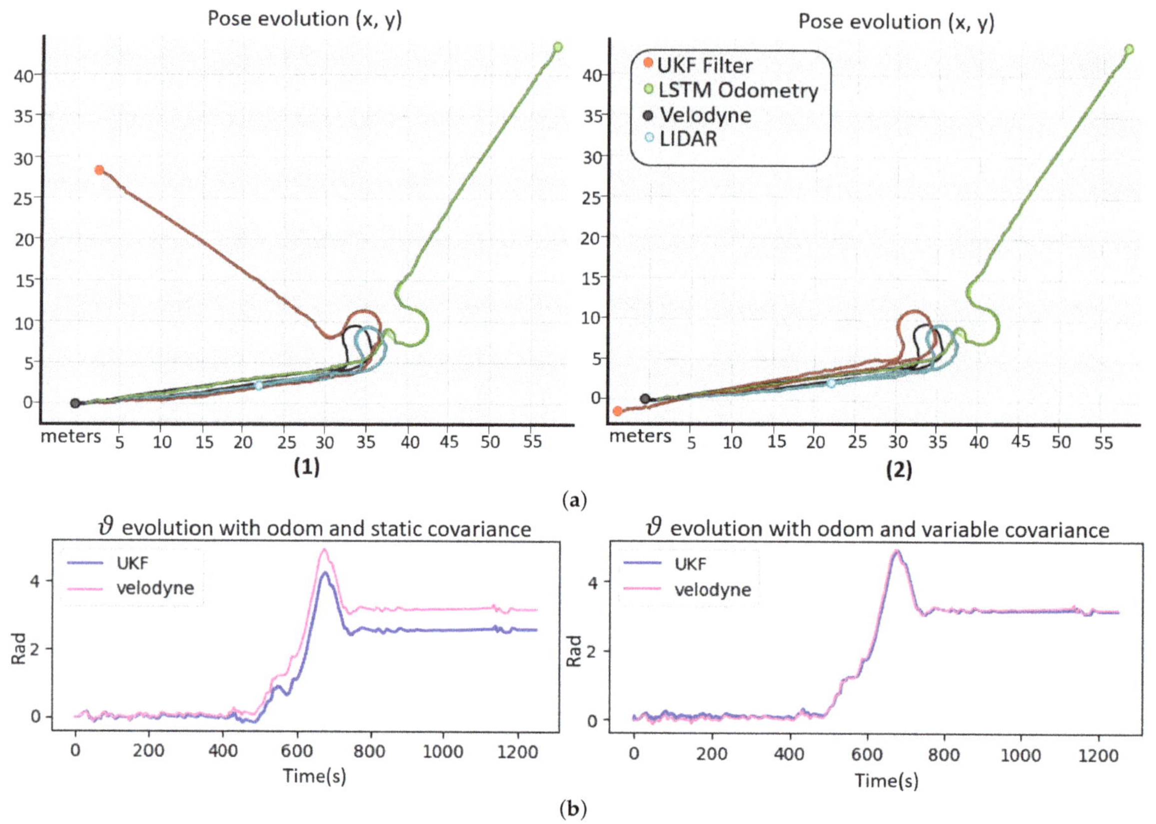

Figure 10. Localization system for a path where the red line is the filter output, the green is the odometry, the blue the LIDAR, and the black line for the ground truth. (**a**) Shows the position "X, Y" and (**b**) shows the orientation "θ". (**1**) Odometry with static covariance and (**2**) odometry with variable covariance.

In addition to the graphical study, the errors of the UKF filter, with respect to the velodyne, have been collected in different experiments for both cases, using traditional odometry with static covariance and the proposed odometry with variable covariance. Table 3 shows the metric values used for its comparison that are MAE, RMSE, R squared, and we have also considered the normalized estimation error squared (NEES) using the filter results and the estimated odometry poses against the ground truth. This metric takes into account the state covariance and is defined as:

$$NEES = \frac{1}{n}\sum_{i=1}^{n}(x_i - \hat{x}_i)^T P_i^{-1}(x_i - \hat{x}_i) \quad (10)$$

where n is the number of samples, x_i is the true state, $\hat{x}_i$ is the estimated state, and P_i is the covariance matrix for the ith sample.

Therefore, the proposed odometry model with a variable covariance has been shown to improve the accuracy of the UKF filter in the presence of non-systematic errors. This has been demonstrated through graphical analysis and by comparing the errors of the UKF filter using the proposed odometry model and the traditional odometry model with static covariance. The NEES metric has also been used to evaluate the accuracy of the filter, with a lower NEES value indicating better performance.

Table 3. Error values obtained by comparing the UKF filter output with respect to the velodyne data in two cases. Using the odometry model with variable covariance and with static covariance.

		V (m/s)	W (rad/s)	Pose X,Y (m)	Pose θ (rad)
LSTM odometry	MAE	0.04125	0.03149	1.4021	0.0547
	RMSE	0.03804	0.05464	2.8699	0.07037
	R_{score}	0.83329	0.80617	0.9279	-
	$NEES_{Odom}$	1087	1982	-	-
	$NEES_{Filter}$	1614	2963	-	-
	Accuracy %	83.33%	80.62%	92.8%	-
	Global Model Accuracy (%)			85.58%	
Traditional odometry	MAE	0.05333	0.0452	2.0592	0.38356
	RMSE	0.0661	0.071619	5.3323	0.4759
	R_{score}	0.7697	0.67634	0.678483	-
	$NEES_{Odom}$	6439	7650	-	-
	$NEES_{Filter}$	5986	4779	-	-
	Accuracy %	76.97%	67.63%	67.85%	-
	Global Model Accuracy (%)			70.82%	

6. Conclusions

This work presents a new odometry system implemented in an autonomous wheelchair. It consists of LSTM neural networks that are able to estimate the robot speed using the data from the encoder sensor. The real-time retraining allows the system to self-calibrate and adapt to changes in the defined model, further improving its performance. Likewise, it is able to reduce the influence of some non-systematic errors by training LSTM networks that learn the relationship between the power of the wheels and their angular velocity to design a variable covariance.

In conclusion, this system significantly improves the accuracy of the wheelchair estimated pose by the UKF filter, compared to the use of classic odometry methods. This is demonstrated through graphical analysis and numerical comparisons of the errors between the different methods. The results show how our system errors are smaller and closer to the velodyne data, providing more accurate final robot localization, particularly in presence of systematic and non-systematic errors. Overall, the proposed wheelchair localization provides a more robust solution for state estimation in challenging environments.

Author Contributions: Conceptualization, B.F., D.A., J.T. and L.A.; methodology, B.F., and D.A.; software, B.F., D.A., J.T. and L.A.; validation, B.F. and J.T.; formal analysis, D.A. and L.A.; investigation, B.F., D.A., J.T. and L.A.; data curation, D.A. and J.T.; writing—original draft preparation, B.F. and D.A.; writing—review and editing, B.F., J.T. and L.A.; visualization, J.T. and L.A.; supervision, L.A. All authors have read and agreed to the published version of the manuscript.

Funding: This research received no external funding.

Data Availability Statement: Not applicable.

Acknowledgments: The authors gratefully acknowledge the contribution from the Spanish Ministry of Science and Technology under the SIRTAPE project DPI2017-90002-R. Author B.F. is partially supported by the grant FPU (Formación de Personal Universitario) of Spanish Ministry of Education.

Conflicts of Interest: The authors declare no conflict of interest.

References

1. Chadaporn, K.; Baber, J.; Bakhtyar, M. Simple Example of Applying Extended Kalman Filter. In Proceedings of the 1st International Electrical Engineering Congress, Chiangmai, Thailand, 13–15 March 2013.
2. Wan, E.A.; Van Der Merwe, R. The unscented Kalman filter for nonlinear estimation. In Proceedings of the IEEE 2000 Adaptive Systems for Signal Processing, Communications, and Control Symposium (Cat. No.00EX373), Lake Louise, AB, Canada, 4 October 2000; pp. 153–158. [CrossRef]

3. Liu, H.; Hu, F.; Su, J.; Wei, X.; Qin, R. Comparisons on Kalman-Filter-Based Dynamic State Estimation Algorithms of Power Systems. *IEEE Access* **2020**, *8*, 51035–51043. [CrossRef]
4. Konatowski, S.; Kaniewski, P.; Matuszewski, J. Comparison of Estimation Accuracy of EKF, UKF and PF Filters. *Annu. Navig.* **2016**, *23*. [CrossRef]
5. Fariña, B.; Toledo, J.; Acosta, L. Sensor fusion algorithm selection for an autonomous wheelchair based on EKF/UKF comparison. In Proceedings of the 2022 10th International Conference on Control, Mechatronics and Automation (ICCMA), Luxembourg, 9–12 November 2022.
6. Arnay, R.; Hernández-Aceituno, J.; Toledo, J.; Acosta, L. Laser and Optical Flow Fusion for a Non-Intrusive Obstacle Detection System on an Intelligent Wheelchair. *IEEE Sens. J.* **2018**, *18*, 3799–3805. [CrossRef]
7. Fariña, B.; Toledo, J.; Estevez, J.; Acosta, L. Improving Robot Localization Using Doppler-Based Variable Sensor Covariance Calculation. *Sensors* **2020**, *20*, 2287. [CrossRef] [PubMed]
8. Bonnabel, S.; Barczyk, M.; Goulette, F. On the covariance of ICP-based scan-matching techniques. In Proceedings of the 2016 American Control Conference (ACC), Boston, MA, USA, 6–8 July 2016; pp. 5498–5503.
9. Brossard, M.; Bonnabel, S.; Barrau, A. A New Approach to 3D ICP Covariance Estimation. *IEEE Robot. Autom. Lett.* **2020**, *5*, 744–751. [CrossRef]
10. Wang, H.; Sun, Y.; Liu, M. Self-Supervised Drivable Area and Road Anomaly Segmentation Using RGB-D Data For Robotic Wheelchairs. *IEEE Robot. Autom. Lett.* **2019**, *4*, 4386–4393. [CrossRef]
11. Alkhatib, R.; Swaidan, A.; Marzouk, J.; Sabbah, M.; Berjaoui, S.; O.Diab, M. Smart Autonomous Wheelchair. In Proceedings of the 2019 3rd International Conference on Bio-engineering for Smart Technologies (BioSMART), Paris, France, 24–26 April 2019; pp. 1–5. [CrossRef]
12. Sinyukov, D.; Desmond, R.; Dickerman, M.; Fleming, J.; Schaufeld, J.; Padir, T. Multi-modal control framework for a semi-autonomous wheelchair using modular sensor designs. *Intell. Serv. Robot.* **2014**, *7*, 145–155. [CrossRef]
13. Boquete, L.; Barea, R.; García, R.; Mazo, M.; Sotelo, M.A. Control of a Robotic Wheelchair Using Recurrent Networks. *Auton. Robot.* **2005**, *18*, 1573–7527. [CrossRef]
14. Baltazar, A.; Petry, M.R.; Silva, M.F.; Moreira, A.P. Driverless Wheelchair for Patient's On-Demand Transportation in Hospital Environment. Ph.D. Thesis, University Oporto, Porto, Portugal, 2020. [CrossRef]
15. Grewal, H.; Matthews, A.; Tea, R.; George, K. LIDAR-based autonomous wheelchair. In Proceedings of the 2017 IEEE Sensors Applications Symposium (SAS), Glassboro, NJ, USA, 13–15 March 2017; pp. 1–6. [CrossRef]
16. Xiao, X.; Liu, B.; Warnell, G.; Stone, P. Motion Planning and Control for Mobile Robot Navigation Using Machine Learning: A Survey. *arXiv* **2020**, arXiv:2011.13112. [CrossRef]
17. Zhang, C.; Hu, L.; Xu, Y.; Rasol, J.; Zhou, Q.; Zhang, Z. Multiple Sensor Track Fusion Algorithm Based on LSTM Network. In Proceedings of the 2022 International Conference on Image Processing and Media Computing (ICIPMC), Xi'an, China, 27–29 May 2022; pp. 154–158. [CrossRef]
18. Wang, M.; Xu, C.; Zhou, C.; Gong, Y.; Baochun, Q. Study on Underwater Target Tracking Technology Based on an LSTM–Kalman Filtering Method. *Appl. Sci.* **2022**, *12*, 5233. [CrossRef]
19. Everett, M.; Chen, Y.F.; How, J.P. Motion Planning Among Dynamic, Decision-Making Agents with Deep Reinforcement Learning. *arXiv* **2018**, arXiv:1805.01956.
20. Kahn, G.; Abbeel, P.; Levine, S. *BADGR: An Autonomous Self-Supervised Learning-Based Navigation System*; IEEE: Piscataway, NJ, USA, 2020. [CrossRef]
21. Toledo, J.; Piñeiro, J.D.; Arnay, R.; Acosta, D.; Acosta, L. Improving Odometric Accuracy for an Autonomous Electric Cart. *Sensors* **2018**, *18*, 200. [CrossRef] [PubMed]
22. Lee, K.; Chung, W. Calibration of kinematic parameters of a Car-Like Mobile Robot to improve odometry accuracy. In Proceedings of the 2008 IEEE International Conference on Robotics and Automation, Pasadena, CA, USA, 19–23 May 2008; pp. 2546–2551. [CrossRef]
23. Borenstein, J. Experimental results from internal odometry error correction with the OmniMate mobile robot. *IEEE Trans. Robot. Autom.* **1998**, *14*, 963–969. [CrossRef]
24. Hassanzadeh, I.; Fallah, M. Design of Augmented Extended and Unscented Kalman Filters for Differential-Drive Mobile Robots. *J. Appl. Sci.* **2008**, *8*. [CrossRef]
25. Lundquist, C.; Karlsson, R.; Ozkan, E.; Gustafsson, F. Tire Radii Estimation Using a Marginalized Particle Filter. *IEEE Trans. Intell. Transp. Syst.* **2014**, *15*, 663–672. [CrossRef]
26. Haile, M.; Wang, Y. *Robust Extended Kalman Filtering for Systems With Measurement Outliers*; IEEE: Piscataway, NJ, USA, 2021. [CrossRef]
27. Ting, J.A.; Theodorou, E.; Schaal, S. A Kalman filter for robust outlier detection. In Proceedings of the 2007 IEEE/RSJ International Conference on Intelligent Robots and Systems, San Diego, CA, USA, 29 October–2 November 2007; pp. 1514–1519. [CrossRef]
28. Kim, J.; Kim, B.K. Cornering Trajectory Planning Avoiding Slip for Differential-Wheeled Mobile Robots. *IEEE Trans. Ind. Electron.* **2020**, *67*, 6698–6708. [CrossRef]
29. Kono, U.; Fujimoto, H.; Hori, Y. Localization of Wheeled Mobile Robots from Slip Ratio Estimation with Simple Model. In Proceedings of the 2021 IEEE International Conference on Mechatronics (ICM), Takamatsu, Japan, 8–11 August 2021; pp. 1–6. [CrossRef]

30. Zhang, J.; Singh, S. LOAM: Lidar Odometry and Mapping in Real-time. In Proceedings of the Robotics: Science and Systems, Berkeley, CA, USA, 12–26 July 2014. [CrossRef]
31. Gonzalez, C.; Adams, M. An improved feature extractor for the Lidar Odometry and Mapping (LOAM) algorithm. In Proceedings of the 2019 International Conference on Control, Automation and Information Sciences (ICCAIS), Chengdu, China, 23–26 October 2019; pp. 1–7. [CrossRef]
32. Hochreiter, S.; Schmidhuber, J. Long Short-Term Memory. *Neural. Comput.* **1997**, *9*, 1735–1780. [CrossRef] [PubMed]
33. Nagatani, K.; Endo, D.; Yoshida, K. Improvement of the Odometry Accuracy of a Crawler Vehicle with Consideration of Slippage. In Proceedings of the 2007 IEEE International Conference on Robotics and Automation, Rome, Italy, 10–14 April 2007; pp. 2752–2757. [CrossRef]
34. Borenstein, J.; Feng, L. Gyrodometry: A new method for combining data from gyros and odometry in mobile robots. In Proceedings of the IEEE International Conference on Robotics and Automation, Minneapolis, MN, USA, 22–28 April 1996; Volume 1; pp. 423–428. [CrossRef]

sensors

Article

Active Actions in the Extraction of Urban Objects for Information Quality and Knowledge Recommendation with Machine Learning

Luis Augusto Silva [1,*], André Sales Mendes [1], Héctor Sánchez San Blas [1], Lia Caetano Bastos [2], Alexandre Leopoldo Gonçalves [2] and André Fabiano de Moraes [2,3,*]

[1] Expert Systems and Applications Lab (ESALAB), Faculty of Science, University of Salamanca, 37008 Salamanca, Spain
[2] Department of Knowledge Engineering and Management, Federal University of Santa Catarina, Florianopolis 88040-900, Brazil
[3] Department Information Technology, IT Institute Federal of Science Technology IFC, Camboriú 88340-055, Brazil
* Correspondence: luisaugustos@usal.es (L.A.S.); andre.moraes@ifc.edu.br (A.F.d.M.)

Citation: Silva, L.A.; Sales Mendes, A.; Sánchez San Blas, H.; Caetano Bastos, L.; Leopoldno Gonçalves, A.; Fabiano de Moraes, A. Active Actions in the Extraction of Urban Objects for Information Quality and Knowledge Recommendation with Machine Learning. *Sensors* **2023**, *23*, 138. https://doi.org/10.3390/s23010138

Academic Editor: Natividad Duro Carralero

Received: 30 October 2022
Revised: 16 December 2022
Accepted: 20 December 2022
Published: 23 December 2022

Abstract: Due to the increasing urban development, it has become important for municipalities to permanently understand land use and ecological processes, and make cities smart and sustainable by implementing technological tools for land monitoring. An important problem is the absence of technologies that certify the quality of information for the creation of strategies. In this context, expressive volumes of data are used, requiring great effort to understand their structures, and then access information with the desired quality. This study are designed to provide an initial response to the need for mapping zones in the city of Itajaí (SC), Brazil. The solution proposes to aid object recognition employing object-based classifiers OneR, NaiveBayes, J48, IBk, and Hoeffding Tree algorithms used together with GeoDMA, and a first approach in the use of Region-based Convolutional Neural Network (R-CNN) and the YOLO algorithm. All this is to characterize vegetation zones, exposed soil zones, asphalt, and buildings within an urban and rural area. Through the implemented model for active identification of geospatial objects with similarity levels, it was possible to apply the data crossover after detecting the best classifier with accuracy (85%) and the kappa agreement coefficient (76%). The case study presents the dynamics of urban and rural expansion, where expressive volumes of data are obtained and submitted to different methods of cataloging and preparation to subsidize rapid control actions. Finally, the research describes a practical and systematic approach, evaluating the extraction of information to the recommendation of knowledge with greater scientific relevance. Allowing the methods presented to apply the calibration of values for each object, to achieve results with greater accuracy, which is proposed to help improve conservation and management decisions related to the zones within the city, leaving as a legacy the construction of a minimum technological infrastructure to support the decision.

Keywords: machine learning; information extraction; object spatial; smart cities; gis detection

1. Introduction

Constant changes occur due to technological evolution in various areas of knowledge, generating new evolutionary cycles, bringing demands, and presenting a large number of still fragmented solutions. For example, the laws of fiscal responsibility, access to information, the civil mark of the internet, and the general law of data protection, guarantee the right to be well informed about what is produced in the public sector. They regulate the implementation of general concepts on data protection, rules for active transparency, and operational procedures. However, the provision of public data in an open format aims to ensure the transparency of stored information that is not under secrecy or access restriction to strengthen the quality of the services offered.

There has been less previous evidence in some studies from the context, focused on analyzing the evolution of data quality [1], basic probability [2], information quality evaluation method [3], text mining techniques [4], research on data and information quality [5], evaluation methods for information quality criteria [6], and mainly higher accuracy for data quality [7]. Innovative and important aspects are highlighted for the main models applied in object learning, and are still being adopted in conjunction with numerous solutions and methodologies. However, different forms and implementation strategies were observed, and among these strategies that enable the monitoring of urban and rural areas, it is essentially necessary to collect and update large volumes of data, consequently of the improvement in the quality of information for the delineation of territorial and social expansion policies.

Until a few years ago, the processes of cartographic revision and, particularly, those aimed at calculating the fiscal area have always been carried out manually. Specifically, these processes required large investments in airplanes or helicopters, making the processes more expensive. Because of this, municipalities are unable to perform mapping surveys frequently. Currently, one of the most current fields of research is to investigate technological capabilities for local authorities to perform detailed surveys of the territory of municipalities at a reasonable cost.

Remote sensing studies based on very high-resolution images have increased in the last few years (e.g., [8–12]), partly because of the availability of satellite images worldwide and the popularization of unmanned aerial vehicles (UAV). The images acquired in all these processes differ in scale, resolution, sensor type, orientation, quality, and ambient illumination conditions. In addition to these difficulties, buildings, cities, neighborhoods, rivers and vegetation may have complicated structures and could be hidden by each other. Both structural and deterministic clues must be taken into account when constructing the solution. Up-to-date and accurate data are essential for municipalities, applied at the Smart Cities concept, that use disruptive technology to solve some specifics problems, in this case, to solve this problem, the solution is the use of satellite and UAV imagery [13] in combination with machine learning techniques [14,15].

However, it is necessary to treat large volumes of data with caution [16], adopting computational and technological resources in conjunction with appropriate use of machine learning methods and techniques. This reflection is based on the proposed [17], in which it had identified that machine learning in several cases had lost connection with problems for other issues due to its complexity. From this perspective, limitations are noted in many existing datasets, metrics employed for evaluation, and the degree to which results express the domain of a problem. As [18–20] state, changes are needed in the way research is conducted to increase the impact of ML, and six impact challenges are highlighted to focus explicitly on problems. Aiming to inspire further discussions and focus on the implementation of ML is the main contribution of this paper, highlighting: (1) regulatory framework for use and implementation, (2) cost reduction with rules for informed decision making, (3) avoiding conflicts of interest between nations, (4) extending cyber security through defenses, (5) saving human lives with diagnostics or recommendations, and (6) improving the Human Development Index (HDI) with at least 10% fair taxation in the country. In the following, this study describes evaluation of the main recognition and extraction methods for active identification of geospatial objects, their characteristics, processes, relationships, and integration for rule generation.

The contributions of this paper are summarized as follows: (a) acquisition of spatial data and satellite images; (b) image processing and machine learning; (c) use of GeoDMA (Geographic Data Mining Analyst) and TerraView technologies for remote sensing; (d) use of Weka software applied to the spatial and geographic context; and (e) use deep learning techniques for object detection in GIS images and GIS detection.

To be able to make all the contributions, a methodology of standardization of the steps is applied with the DSR (Design Science Research) proposed by [21], to add new practices and build a literature review in parallel to the technological steps that help in the implementation of the proposal. To this end, the processes of (a) classification in the context of remote sensing (RS) are also reviewed; (b) text mining techniques with the software Weka (workbench for machine learning) performing text analysis, quantifying the words and extracting statistics with the TF-IDF method (Frequency-Inverse Document Frequency). Through a case study carried out in the municipality of Itajaí in the state of Santa Catarina, Brazil. The results collected for the urban and regional planning evaluation period are presented, considering the crossing of environmental and social data, referring to territorial occupation.

For a better understanding of the research and results of this work, the paper is organized as follows: Section 2 presents the development; Section 3 describes the methodology, in Section 4 the materials and methods used to apply the case study are detailed; in Section 5 we present the discussion of the results, and finally, in Section 6 are the final conclusions and future work.

2. Development

To develop the project, actions were initially planned for the acquisition of spatial data and satellite images. In the second stage, cataloging and standardization processes are carried out for data interoperability. The third stage presents the technological solutions adopted from GeoDMA and TerraView for remote sensing (RS). In the fourth step occurs the Weka implementation applied to the spatial context. To orchestrate the processes, the main methodology of the work aims from the experiments to define the best practices for the classification process focused on the SR, enabling the integration of all processes towards the active learning of objects. At the end, the case study is presented with specific results and discussions.

2.1. Spatial Data Acquisition and Satellite Imagery

In particular for matrix data, [22] defines spatial data acquisition from images from a few approaches, those being: input data definition, segmentation, detection cycles, cell space (matrix) creation, and preliminary extraction features. From the proposals [23–25] implemented through the GeoDMA framework (GEOBIA), in synthesis provides the realization of the steps of segmentation of satellite images, extraction of attributes, creation of classification rules, hierarchical classification and visualization of results. Additionally, the works [19,26–29] describe in detail the precautions to be taken in image acquisition and processing. In particular, according to [27,30], the monitoring of the interactions with the terrestrial surfaces is very important, where each intensity of the solar radiation must be observed. That is, the intensity of solar radiation received by the surface depends on the solar zenith angle that is calculated, considering [27,31] the incident solar radiation, the solar radiation intensity and the solar zenith angle. After acquiring the digital data from the sources provided according to [32], new processes for cataloging emerge. However, the extraction characteristics, initially present the need for specific module of resources extraction and storage in a database. From this, it is possible to extract several features, depending on the size of the scanning process and the amount of objects this task can be very time consuming to complete. Therefore, cataloging guarantees that the features will be extracted without losing original characteristics. Experiments conducted using the images collected by the CBERS-4 (China-Brazil Earth Resources Satellite) and the CBERS-4A satellite, located in http://www.dgi.inpe.br/documentacao/dgi/documentacao/satelites/cbers/capa-cbers (accessed on 1 June 2021), considering different periods.

2.2. Machine Learning and Processing

The availability of images from satellites and aerial platforms over the Earth's surface in the most diverse resolutions has been enabling an unprecedented approach between technology and society, as [28] the processing of large volumes of data and geolocation for the use of mobile devices in most different devices makes the insertion of various technologies flexible. However, large volumes of data are generated, and for analysis, new challenges arise involving interoperability, from those related to data collection and storage, through ethics and privacy [33–35], to the development of efficient and robust algorithms to extract the most unimaginable information.

However, processing large volumes of data requires technical expertise in remote sensing, raw data processing, information extraction, a transformation of orthogonal models and spectral mixing, calculation of physical indices, arithmetic operations, frequencies, and statistical classification of data. However, all of these resources seek to assist in classifying image pixels associated with the presence of observed spatial object characteristics. To enable this, only with the use of consolidated technologies as [36,37], various classification algorithms have been developed, as there is a growth in the data obtained by images.

Each classifier has its strengths and weaknesses. Hybridizing classifiers with each other have the potential to combine the strengths and overcome the weaknesses by analyzing level by level as per Figure 1.

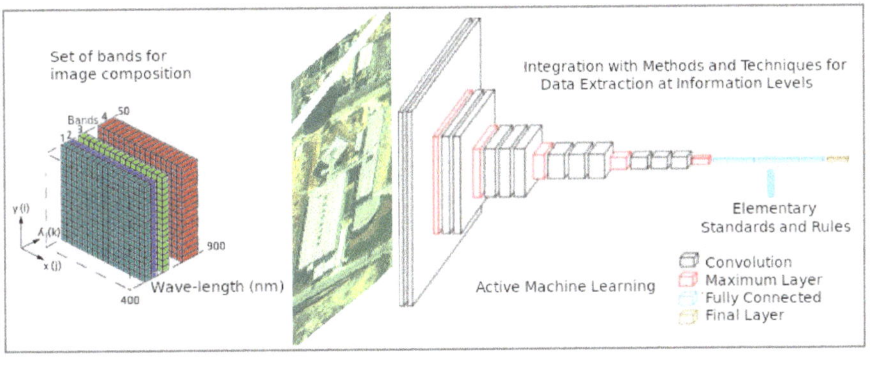

Figure 1. Architecture of the ecosystem adopted for active detection, extraction, and learning of geospatial objects.

It is necessary to advance with new studies on the behavior of different algorithms for hybrid classification as KNN (K-Nearest Neighbors) and SVM (Support Vector Machine) addressed by [38], with genetic programming, decision trees with an artificial neural network, Naive Bayes with decision trees and decision trees with K-means. Additionally, in hybrid image processing, specifically for the segmentation process, basic procedures are established according to the works of [22,39], and there are still procedures to be evaluated.

For the processing of images according to [29], together with the extraction of attributes from the regions and their spectral characteristics, they must be previously identified, validated, and calculated. The attributes can be determinant in interpretations and classification processes that involve many classes and some with little separability.

From this, the challenge arises to increase the amount of coherent information to facilitate the discrimination of spectrally similar classes. To this end, determining procedures that help increase identification with greater accuracy provides a set of rules that can be instrumental in identifying distributed objects.

For image processing according to [29], besides the extraction of attributes from the regions and their spectral characteristics, they must be previously identified, validated and calculated. The attributes can be determinant in interpretations and classification processes that involve many classes and some with little separability. From this, the challenge arises to increase the amount of coherent information to facilitate the discrimination of spectrally similar classes.

To this end, determining procedures that help increase identification with greater accuracy provides a set of rules that can be instrumental in identifying distributed objects.

However, new attributes can be associated with the spectral attributes from new dimensions attached to the original image. The spectral attributes refer to the color of the pixels and are calculated separately for each band of the input image based on the pixels belonging to the segment.

As per categorical data analysis [40], three standards of texture verification are still evaluated: structural, statistical, and spectral approaches. A statistical approach is the most widely used and considers the texture of an image as a quantitative measure of the arrangement of intensities in a given region. In this context, the concurrency matrix contemplates the numerical characteristics of the texture using similar shades of gray between a pixel and its adjacencies determined by the $N \times N$ pixel quadrant. The main formulations adopted to quantify the texture in digital images, in this case, obtained by (the CBERS-4 satellite), are mean, variance and entropy. The mean, according to (1), corresponds to the value of the arithmetic mean of the gray levels of a region in each band of the image. Where, $R(i)$ equals each element (i) segment R and N the total number of pixels.

$$Average = \left(\sum_{i=1}^{N} .Ri \right) N \quad (1)$$

$$Variance = \left(\sum_{i=1}^{N} .i - M^2 \right) \quad (2)$$

$$Entropy = \left(\sum_{i=1}^{N} .Pi.InPI \right) \quad (3)$$

The variance (2) is a measure of the dispersion of the gray level values of the pixels in the region around the mean, and M is the mean of the gray levels of the segment. Entropy (3) is calculated based on the distribution of pixel values in the region and is a measure equivalent to the "distortion" of the values in the region. Where $P(i)$ contains a normalized histogram of the segment elements. The geometric attributes [29] are calculated based on the polygon that defines the segment boundary, being: area, perimeter, compactness, convexity, and elongation. In other words, the geometric metrics of a segment are defined in advance for the behavior of the processing algorithm. The main considerations about image segmentation refer to the selection of the optimal parameters for each application. However, new active and adaptive processes have presented important results with machine learning, such as GeoDMA.

2.3. GeoDMA and TerraVIEW for Remote Sensing

To analyze altered patterns [22,23], in large remote sensing datasets, GeoDMA was created. Implemented in TerraView software, a tool to integrate the most essential image analysis algorithms, ecology metrics, a scheme for multitemporal analysis [41] and data mining techniques to automate the analysis of large databases. Addressing only implementation aspects of active extraction features, it seeks to provide new perspectives for generating automaton functions for data collection, management, analysis, and representation, both for basic functionalities, and the integration extraction, and transformation of geospatial objects [19].

2.4. Weka Applied to the Spatial and Geographic Context

Through the work environment for machine learning Weka, it is possible to perform various analyses on a specific data set, or on several sets, provided that these sets have the format in which it can perform the analyses. In this regard, several discussions about GDPM (Geographic Data Preprocessing Module) [23,42,43] arise, extending the Weka Data Mining Toolkit to support geographic data. Additionally, [22] presents discussions of geographic data integration techniques, for example, the ID3, C4.5, and C5.0 algorithms for rule generation. Improvements with Weka-3.9.3 (2019), operating through MOA (Massive

Online Analysis) different types of datasets available at http://moa.cms.waikato.ac.nz/downloads (accessed on 20 January 2022).

That is, types such as JSON, XML, SHP, DAT, TXT, CSV, PostgreSQL/PostGIS tables, MySQL/MyGIS, ARFF and XRRF, among others, must necessarily be evaluated and formatted for machine learning processes. The discretization process of Weka is another interesting way to enhance the processes and information extraction, considering the geospatial representation. However, it is necessary to improve experiments for the automation of many of the data transformation tasks for the generation of information with higher quality [44,45].

2.5. Deep Learning and Object Detection

The main problem that arises in the processes of acquiring knowledge from images is that of relating the images collected by satellites or drones to object detection systems and the corresponding verification of the same within the databases of local systems. One approach may be to follow the advances in the machine learning algorithm literature, with a focus on using Deep Learning (DL), which is a class of Machine Learning algorithms. This type of algorithm uses multiple layers to progressively extract features from the input images [46].

DL-based approaches are efficient when large datasets are available. The word deep specifies more layers and deep neural networks. DL uses nonlinear functions. Thanks to deep learning, Intelligent Document Processing (IDP) is able to combine various AI technologies not only to automatically classify photographs, but also to describe the different elements of images. Deep learning models, with their multi-level structures, are very useful for extracting complicated information from input images. Convolutional Neural Networks (CNN) are also able to dramatically reduce computational time by leveraging the GPU for computation, something that many networks do not utilize. In the field of object identification in images, two methods stand out: regional proposal algorithms and regression object detection algorithms.

The first method is to discover in advance the possible target locations to be detected in the picture. This can ensure that the highest retrieval rate is maintained when fewer windows are selected. Suppose an image is input and, after a series of convolutions and backbone clustering, a feature map of size M × N is obtained, which corresponds to the division of the original image into areas M × N. The center of each area of the original image is represented by the coordinates of a pixel in this feature map.

Region Proposition Algorithms are used to determine whether the k anchor boxes corresponding to each pixel contain a target. The network must learn to classify the anchor boxes as background or foreground. From this, it must calculate regression coefficients to modify the position, width and height of the foreground anchor box. Within these classifiers, we find algorithms such as R-CNN [47], Fast R-CNN [48], Faster R-CNN [49] and MASK-CNN [50]. Of the algorithms mentioned above, Mask R-CNN stands out. This algorithm is an extension of Faster R-CNN and works by adding a branch to predict an object mask in parallel with the existing branch for bounding box recognition. The key element of R-CNN Mask is pixel-to-pixel alignment, which is the main missing piece in Fast/Faster R-CNN. The R-CNN mask adopts the same two-phase procedure with an identical first phase (which is RPN). In the second phase, in parallel with class prediction and box clearing, Mask R-CNN also produces a binary mask for each RoI. This is in contrast to more recent systems, where classification depends on mask predictions. Furthermore, Mask R-CNN is simple to implement and train thanks to the faster R-CNN framework, which facilitates a wide range of flexible architecture designs. Furthermore, the mask branch only adds a small computational overhead, allowing for a fast system and rapid experimentation. Figure 2 shows a visual example of the segmentation performed by the algorithm.

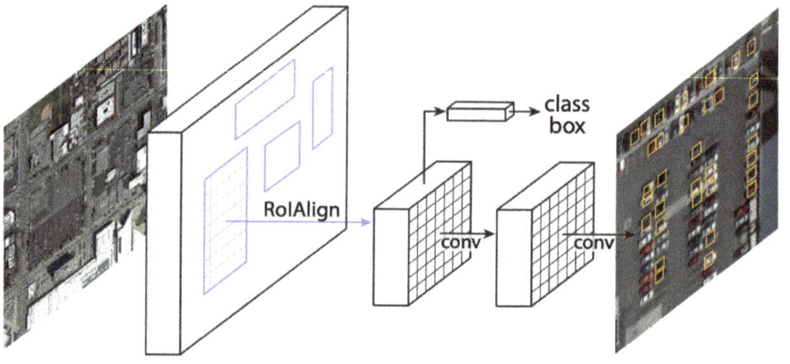

Figure 2. Mask R-CNN framework.

The above algorithms use detection as a classification problem, i.e., object proposals are first generated and then these proposals are sent to the classification/regression regions. However, some methods approach detection as a regression problem based on a similar operation. Within this field, the YOLO (You Only Look Once) and SSD (Single Shot Detector) algorithms stand out. The SSD algorithm [51] strikes a good balance between speed and accuracy. SSD runs a convolutional network on the input image only once and computes a feature map. It then runs a small 3 × 3 convolutional kernel on this feature map to predict bounding boxes and classification probability. SSD also uses anchor boxes in various aspect ratios, similar to Faster-RCNN, and learns the offset instead of learning the box. To handle scale, SSD predicts bounding boxes after multiple convolutional layers. Since each convolutional layer operates at a different scale, it is able to detect objects of various scales. An example of how the SSD algorithm works can be seen in Figure 3.

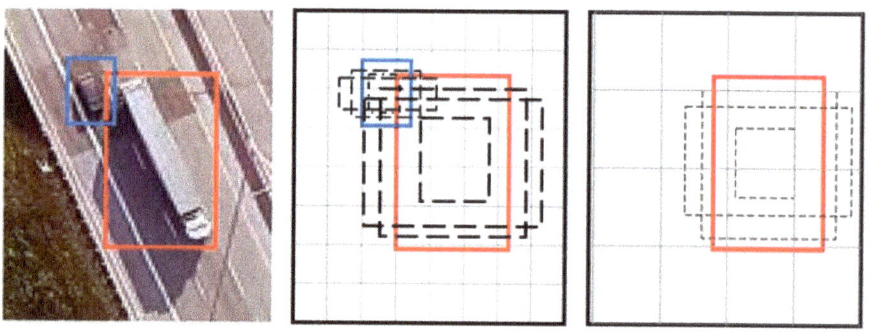

Figure 3. SSD framework.

For YOLO [52], detection is a simple regression problem that takes an input image and learns the class probabilities along with the coordinates of the bounding box. YOLO divides each image into an S × S grid, and each grid predicts N bounding boxes along with their confidence. The confidence reflects the accuracy of the bounding box and whether the bounding box actually contains an object, regardless of the class. YOLO also predicts the classification score of each bounding box for each class in the training. It can combine both classes to calculate the probability that each class is present in a predicted box. Thus, a total of SxSxN bounding boxes are predicted. However, most of these boxes have low confidence scores, so if we set a threshold, for example of 30% confidence, we can eliminate most of them, as shown in Figure 4.

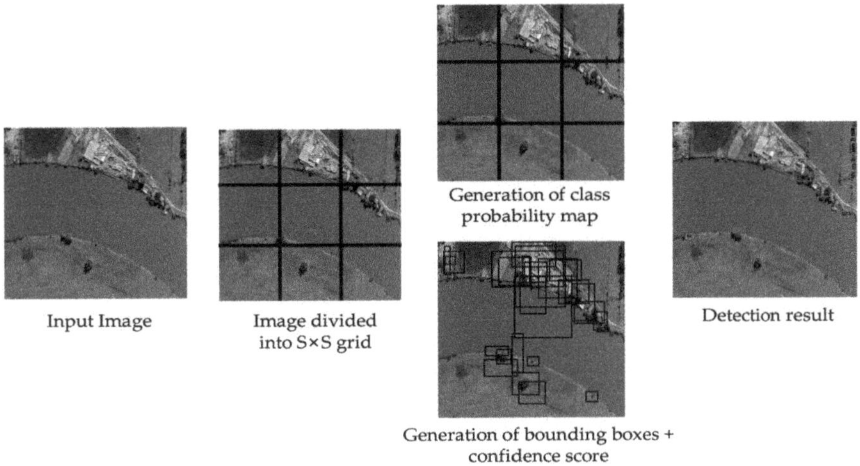

Figure 4. Yolo framework example.

YOLO is a faster algorithm than all other detection algorithms, allowing it to be run in real time. Another key difference is that YOLO sees the entire image at once, rather than looking only at the proposals of a region generated in previous methods. Thus, this contextual information helps to avoid false positives. However, one of the limitations of YOLO is that it only predicts one type of class in a grid, so it has difficulties with very small objects. There are several versions of YOLO such as YOLOv2 [53], YOLOv3 [54], YOLOv4 [55], YOLOv4-tiny [56,57], YOLO-Fine [58] and recently YOLOv7 [59]. There are also available versions of YOLO applied to Satellite Imagery, such as YOLT [60] and MRFF-YOLO [61].

3. Methodology

From the concepts of DSR (Design Science Research) proposed by [21], whose study considers it essential to also deepen the area of management. In this context, according to [62], hierarchies are applied for knowledge-intensive tasks on each identified problem. Added to the discussions of [24,63–66] allied to the classification methods being divided according to the processing, into visual or digital, known as supervised, unsupervised and hybrid as per [36].

Additionally, observing the metrics, in parametric or non-parametric and according to the approach by pixel or by regions (objects), the methodology proposed in the work aims, from the survey of satellite images and/or images obtained by RPAS and also by crossing previously shared textual data, to identify through active learning the recognition of geospatial objects with the generation of elementary rules. For this, an architecture for systematic detection and extraction supported by machine learning is proposed, see Figure 5.

In stage 1 meetings, interviews, surveys for questionnaires implementation, documentation for support, and a survey of the prerequisites of the required project are planned. The installation, testing, and homologation processes are also planned. This is where different work platforms are made available for individual or collaborative use (groupware) for integration and standardization. In stage 2, the requirements engineering processes are carried out with the construction of artifacts using UML (Unified Modeling Language). The important delimitation of the coverage area also takes place with the objective of project execution. Data acquisition processes, images, and related documents. The cataloging of data with centralized and shared storage. The processes of treatment, qualification, and homologation of the collected data with due certification. In step 3, acquisitions are made, such as contracting satellite image collection services with specific parameters. Scene processing for example (CBERS) and (LANDSAT-8). Definition of scenes imaged by the satellites through date parameters, bands, and other relevant details, and also the survey and integration of demands with the definition of the goals.

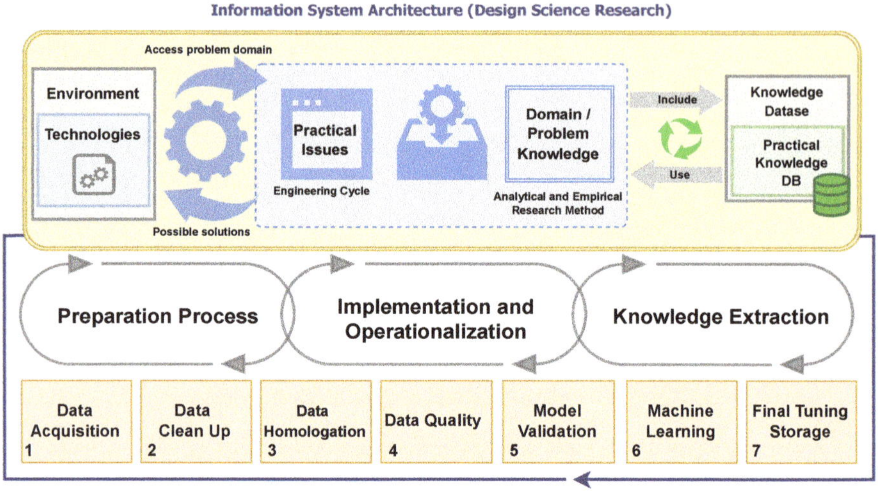

Figure 5. Architecture of the ecosystem adopted for active detection, extraction and learning of geospatial objects.

Step 4 is homologation and image processing, with the choice of the contrast method chosen for the visualization of the objects to be evaluated. Optionally, the methods can vary; for example, the linear method, histogram equalization, square contrast, square root contrast, log contrast, mean and standard deviation, decorrelation enhancement, cumulative 2% enhancement, composition and decomposition method, arithmetic operations of image bands with NDBI (Normalized Difference Built-in Index), fusion method, and image segmentation method.

In step 5, the homologation of each processing generated by the choice and application of the methods is subsequently performed, the indexes are prepared, and the methods for detecting and extracting spatial or textual objects are made specifically available. The data structures generated in the previous step are necessarily reused, according to cataloging by date, time, function, data sources, and coverage regions, enabling the import and centralized integration for sharing, through specific infrastructure for networks and sensors.

In step 6, specifically different algorithms for adaptive rule generation are evaluated. Adaptive rules are statistical patterns detected for representing the analyzed dataset. This made it possible to combine them with the intersection of new attributes already stored in the repository. In step 7, the optimization of the data structures for the repository and subsequent reverse engineering is a priority. From known rules, it is possible to actively generate learning about the experiences with the availability of large volumes of data to support the other decision-support processes. To synchronize each step and process, an infrastructure [67–69] computer network for remote communication between various devices, data collectors, cameras, and sensors is implemented.

Furthermore, through the fruit of several research discussions comes the development of the model for active knowledge extraction, presented by [45], intensifying the interoperability of the data and the advancement of the implementation of the concepts in this work. It also aims at improving the functions through the prototype as presented in Figure 6. From the application architecture idealized by [22], the same also provides the opportunity for the derivation of new experiments for machine learning, since the collection of data, cataloging, discretization of data and application of algorithms is of great importance for the detailing of each process and recording of operations for possible control.

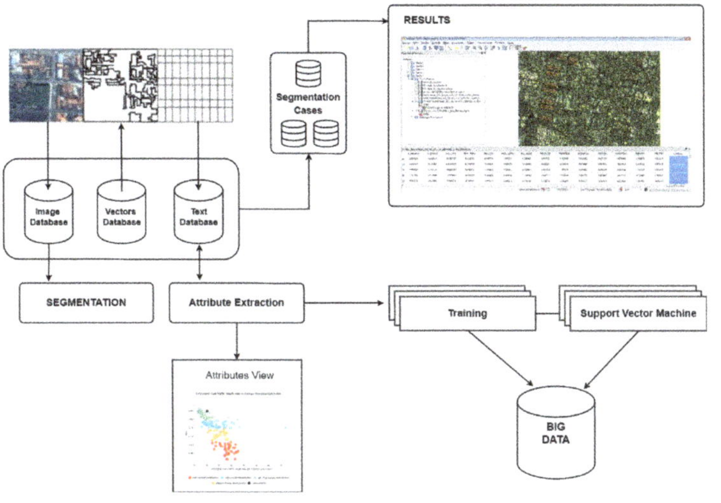

Figure 6. Adapted architecture for systematization and generation of recommendation cases.

This defines a process for detecting geographic patterns, urban and rural devices through segmentation and other records for automatic observation and evaluation of territorial expansion from the extraction of shared knowledge.

One of the main reasons is the difficulty of constantly processing a large volume of data, due to satellite images collected periodically and which can present files with expressive sizes (megabytes or gigabytes), fundamentally important for pattern recognition.

3.1. Remote Sensing Classification

In applications that require image classification, the availability of labeled samples (training data) is closely associated with the choice the analyst will make for extracting information from the images. Two families of techniques are distinguished, called supervised and unsupervised, according to the presence or absence of labeled samples, respectively.

In the context of SR, classification is the process that seeks to assign a label to certain data described by a set of attributes. In digital terrestrial remote sensing imagery, this process is equivalent to determining, for each pixel, which category is present on the surface, such as water, soil, and forest, which is usually done by spectral attributes, such as the gray level in each band.

They are commonly used over the radiometric indices as [70], arithmetic contrast operations with NDBI, NDVI (Normalized Difference Vegetation Index), and NDWI (Normalized Difference Water Index). Considering the processes of unsupervised classification in (SR) and supervised classification, from data collected by (satellites) or unmanned aircraft, these images can be analyzed in different scenarios, whether knowledgeable or not about the observed area. The discussions of [29] describe in detail the implementation of each process that can be adapted to different experiments.

3.2. Active Training and Machine Learning

We also consider using parametric classifiers that model the decision boundaries between training classes with a fixed number of parameters, regardless of the number of samples [71,72].

It is the simplest classifier in existence and therefore ends up having a more didactic than operational role. The decision boundaries are positioned on lines equidistant between midpoints of the various classes present.

The classification process by Euclidean minimum distance is performed by Figure 7A the distribution of the sample elements of each class in two bands of a generic image and

Figure 7B the averages calculated for each sample and respective distances to a pixel to be classified.

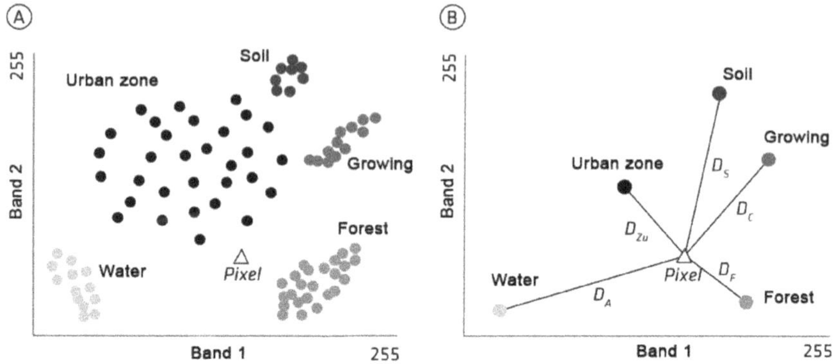

Figure 7. Representation of Euclidean distances.

Especially with the evolution of text mining techniques according to [36], the development of the StringToWordVector function can optionally assist in scanning large amounts of text using TF-IDF concepts in Weka treated IDFTransform and TFTransform.

Additionally, when searching with textual data, a statistical measure is adopted that is intended to indicate the importance of a word in a document relative to a collection of documents or in a linguistic corpus. According to [73], it is possible to distinguish the importance between different word features, and it is necessary to calculate the weights of the prominent words. For this, the TF-IDF method is implemented and used to calculate the weight according to (4).

$$TF - IDF = tfxid = \left(\frac{a}{t}\right)Xlog\left(\frac{b}{c+1}\right) \tag{4}$$

In the formula, a is the frequency of the resource in the document set, this is the total number of times of all resources in the document set, b is the document number in the document set, and t is the number of documents that contain the resource. Then, with the use of the TF-IDF method, it is possible to select n resources with the maximum value of TF-IDF as per the candidate resource set. Using IDFTransform and TFTransform, scans and learning are performed on the datasets prepared for the textual data matching, as detailed in the case study of Figure 8.

This enabled the cataloging of each geographic object properly identified from the satellite image with the items found with IDF-TFTranform. In parallel, from the discussions of [74], another viable strategy to relate and spatially represent different information can be through a geographical matrix of spatial queries, being a two-dimensional representation of intrinsic relationships between locations. To exemplify, implement the forest code and limits of permanent preservation areas in each municipality. Many impasses arise, and although this is not a complex task, it requires great human effort and skilled labor for permanent monitoring. In this sense, the municipalities that make up the basin of the Itajaí-Açú River were mapped to generate the geographical matrix. Through the implementation, the integration of data was carried out, resulting in a large volume of distinct information, providing important relationships for the evolution of monitoring through the recognition of spatial objects as explored in the case study.

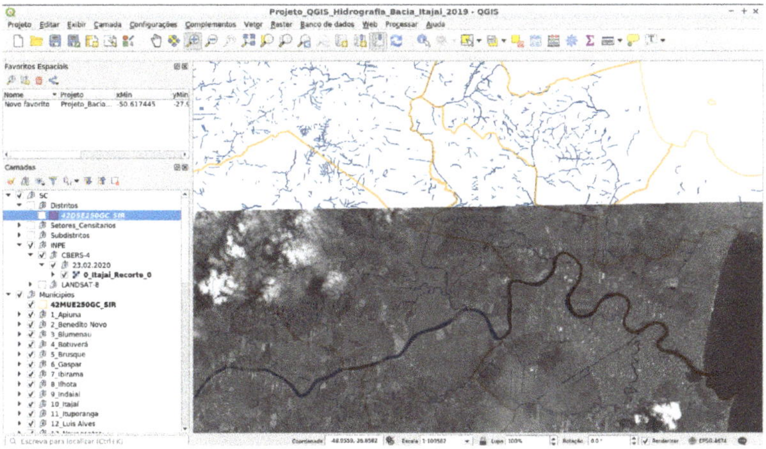

Figure 8. Integration in QGIS with CBERS4 satellite images and CTM data (Itajaí-SC) for monitoring the expansion on riverbanks.

4. Materials and Methods

4.1. Study Area

The study sites were located in the south of Brazil, in the Itajaí Municipality along the Itajaí-Açu basin river, in Santa Catarina (SC) state, Brazil (Figure 9).

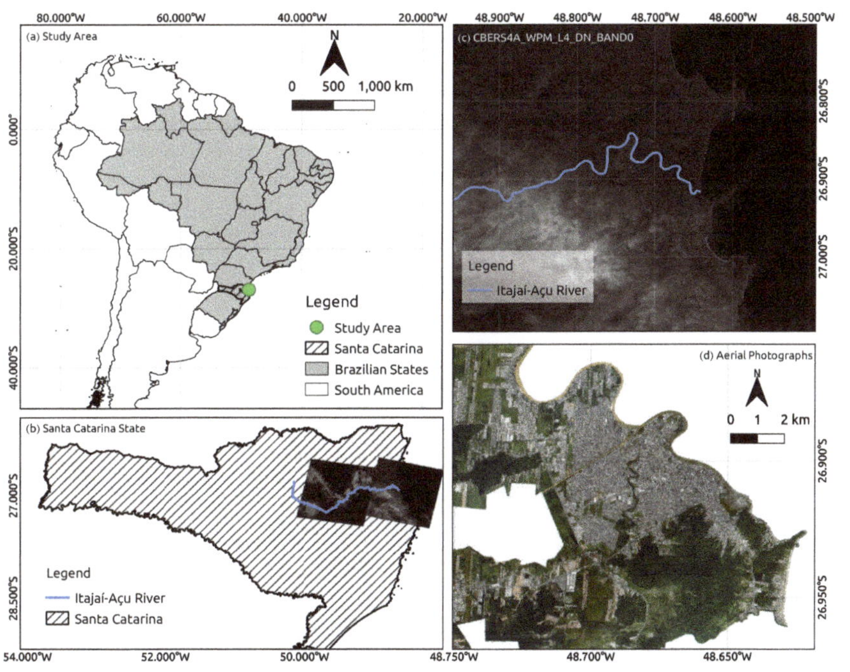

Figure 9. Location of study sites: (**a**) in South America, Brazil, Santa Catarina state. (**b**) State of Santa Catarina with the CBERS4A Panchromatic 2 m Image, including the Itajaí-Açu River layer. (**c**) Satellite image from the CBERS4A of the mouth of the Itajaí-Açu River (**d**) Airborne from the Itajaí—SC Municipality.

4.2. Materials

New challenges arise with updating a municipality's land registry. According to [73], a good cadastre contributes to the equitable distribution of tax resources, promotes property security and creates bases for urban and regional planning. For this, bases for urban and regional planning are created, using the Multifinality Technical Cadastre (*Cadastro Técnico Multifinalitário*—CTM), available at https://geoitajai.github.io/geo/plantacadastral.html (accessed on 20 October 2022). And from the collection of images through the CBERS-4 and CBERS-4A satellite with spatial resolution of 5 and 2 m, respectively http://www.dgi.inpe.br/catalogo (accessed on 20 October 2022). Additionally, the aerial photogrammetry are obtained from the portal https://arcgis.itajai.sc.gov.br/geoitajai/plantacadastral/plantacadastral.html (accessed on 20 October 2022), they are used to identify spatial objects in the municipality of Itajaí, state of Santa Catarina—Brazil. The map layouts were generated using QGIS software [75].

4.3. Methods

With the advancement of technologies and the emergence of large volumes of data, it is possible to cross data through automated processes and solutions with greater reliability and precision. The policies are evaluated with the crossing of environmental and social data, referring to territorial occupation, and considering traditional economic, physical, and legal aspects, among others.

However, one of the limiting factors for updating base registers is still the high cost of developing the entire cartographic framework. However, with the advancement of collaborative technologies, more affordable alternatives should be considered, especially for small and medium-sized cities. In this sense, we highlight the possibility of using photogrammetric techniques, since the logistics involved in the operation of these systems are more flexible and economical when compared to cartography by topographic or geodetic techniques, or even conventional photogrammetry with UAV [76].

In the case study, image acquisition processes were carried out, and subsequently the prior selection of the delimited perimeter for application of the present study. Figure 10 presents the result of this process and the application of the linear contrast method. During the investigation, the pixel transformation functions were evaluated from the image contrast, and in the application of active object learning, the contrast presented a more suitable visual result. From the transformation function (T) for a single pixel (r = original pixel value), the resulting pixel value (S) is generated through different techniques to obtain better processing and visualization of objects. Where, (s) = T(r). Table 1 shows the results with the contrast method.

Figure 10. Image of the municipality of Itajaí SC—Brazil, after the contrast method in the area chosen for object detection and active learning.

Table 1. Preparation for active learning experiments.

Satellite		Segmentation		Generated Records
CBERS Sensor Banda [a]	Contrast Method [b]	CPU Time i7 6.5 GB GNU/Linux	Similarity	Object Detection
PAN5M—Band Espec = 1	Linear	327 s 70	0.050	1,066,788
		390 s 46	0.045	1,345,492
		435 s 84	0.040	1,639,031
		464 s 94	0.035	2,158,357
		516 s 02	0.030	2,770,752
		572 s 11	0.025	3,760,484

[a] Image acquisition criteria, e.g., (date, location, quality and others). [b] Software used for image processing, such as TerraLib/TerraView (INPE) and Qgis.

The lower the degree of similarity threshold, the higher the generation of objects for analysis processing and active learning. In other words, higher demand and availability of hardware resources to support higher processing volumes are essential.

Additionally, the availability of a relational database management system, such as PostgreSQL/PostGIS, is another important implementation due to the need of storage for manipulation. The case study was designed to integrate public data with the main objective of evaluating the existence of relationships, direct or indirect crossover. Different data sources were considered for experiments of the algorithms.

After two comparative implementations, over the data set (49,325 occurrences) obtained from the CTM. Where, initially in the first implementation was prioritized application of the classification method to elaborate the decision tree with ID3 algorithm, using the Weka tool (version 2). It generated 78.83% (38,885) correctly classified instances and 21.15% (10,436) incorrectly classified instances.

In the second implementation on the same dataset, a new tree structure was generated with J48 algorithm using Weka tool (version 3.9.3). After visualizing the tree, it was possible to detect the levels and the class (Conservation), with more information gain. For correctly classified instances, it was obtained 85.2002% (42025) of success and 14.79% (7300) not classified. The final kappa statistic of 76.11% also determined the classifier that obtained the best learning. After the implementations with the software Weka, for the same area investigated, another study was performed using the software TerraView/GeoDMA, where the methods of segmentation, image vectorization and later extraction of the set of attributes were applied, see Figure 10.

Subsequently, with GeoDMA for object classification, 212 attributes were generated. From the identification of the characteristics of each pixel was performed the conference of the generation of each polygon and definition of relations for characterization of basic rules of each object. For a better understanding of the objects detected in the study, it was important to evaluate the different types of scales.

5. Discussion

From the expected results are presented some reflections and discussions about the development of the work, initially idealized and later obtained. That is, the results achieved by the research and their implementation are presented.

Reflection 1: Through the studies presented, was it possible to detect the applicability of knowledge extraction in CTM in conjunction with other areas? Clearly, and as is proven through the case study presented in Figures 5 and 6, along with a comparison of the Weka software in version 3.2 (2001) and version 3.9.3 (2019), both versions allowed for obtaining models, enabling machine learning, and expansion of analysis with new processes of data integration and information extraction.

Reflection 2: Specifically in this paper, new performance tests were presented with the Weka classifiers: OneR, IBk, NaiveBayes, and J48. All classifiers use the same CTM data and under the same conditions. That is, at this point it is worth mentioning the inclusion of analysis with the Weka "Hoeffding Tree" classifier, allowing the generation of a tree with less criticality.

Reflection 3: Was it possible to use new satellite images and run GeoDMA for automatic learning of new objects? Yes, in this regard, it is worth highlighting the specifications adopted according to Table 1, enabling the correct acquisition of images through standards for the next stage of segmentation, then the vectorization of objects. In particular, several prodecures are performed for image segmentation, which not part of the scope of this work, but will be detailed with new experiments.

Reflection 4: Did the development and discussion of the experiments occur with other CTM databases integrated for the recommendation? Partially, some experiments use static datasets successfully, but have not been evaluated by mining continuous stream data using Weka-MOA.

Reflection 5: Were performance evaluations of the Weka software performed? Yes, exhaustive performance experiments were conducted, as shown in Table 2 and made available for online access. In all classifiers, the CTM dataset was analyzed with the same parametrizations and specifications, such as the 10-fold cross-validation over the "property conservation" class. The Weka classifiers performed well on a dataset containing 18 attributes and 49,325 instances. The OneR, NaiveBayes, J48, IBk, and Hoeffding Tree classifiers showed satisfactory results. The results of the classifiers are explained one by one below:

For Weka → OneR: OneR processing took 0.09 s to build and run the model. Additionally, with the application of OneR, a correct classification of the instances of (38,404) records was obtained totaling an accuracy of 77.8591% by the algorithm. It also presented satisfactory learning with 63.75% evaluated by the kappa statistic.

For Weka → NaiveBayes: Initially, the processing took 0.13 s to build and run the model. Additionally, with the application of NaiveBayes, a correct classification of the instances of (37,445) records was obtained, totaling an accuracy of 75.9149% by the algorithm. It also presented satisfactory learning with 62.48% evaluated by the kappa statistic.

For Weka → J48: The processing took 2.45 s to build and run the model. Additionally, with an application of J48, a correct classification of the instances of (42,025) records was obtained, totaling an accuracy of 85.2002% by the algorithm. It also showed satisfactory learning with 76.11% evaluated by the kappa statistic.

For Weka → IBk: Processing took 3 min 34 s to build and run the model. Additionally, with an application of IBk, a correct classification of the instances of (40,042) records was obtained, totaling an accuracy of 81.1799% by the algorithm. It also showed satisfactory learning with 69.97% evaluated by the kappa statistic.

For Weka → Hoeffding Tree: Processing took 0.69 s to build and run the model. Additionally, with an application of the Hoeffding Tree, a correct classification of the instances of (39,203) records was obtained, totaling an accuracy of 79.48% by the algorithm. It also presented satisfactory learning with 66.54% evaluated by the kappa statistic.

Experiments with Weka → IBk from the vector with 25 attributes and containing (1,000,000) instances, generated after image segmentation Weka with the IBk classifier built the classification model quickly, but presented a very large slowness (9 h) to measure the distances of all instances. However, the alternative found to speed up the processing was to retain in memory only a "window" of instances, instead of the complete dataset. In Weka, the default parameter "window size = 0" allows you to set the maximum number of instances allowed in the training pool, and adding additional instances simply removes the old ones, freeing up memory to improve performance.

In addition, for a better understanding, analysis was performed on the datasets below, being separated into three different sets to initially compare the training with 100 (objects), 1000 (objects), and 15,000 respective training results, it was possible to decide which algorithm to use first to be prioritized and adopted for further processing, as shown in Table 2, containing the results obtained from training the different datasets and compared with the performance of processing the entire set of objects. The final test was evaluated with a total of 49,325 instances, and these instances are isolated from the training dataset.

Table 2. Training datasets (objects).

Algorithm	Training (100 Objects) %	Training (1000 Objects) %	Training (15,000 Objects) %	Performance Evaluation (Final) %
J48	52	77.9	80.4067	85.2002
IBk	78	78.2	80.4467	81.1799
Hoeffding Tree	76	77.1	80.3733	79.4800
OneR	75	77.9	80.3600	77.8591
NaiveBayes	76	77.1	79.4133	75.9149

Training datasets and evaluation/tests are available in: http://sadpreaigeo.org/ufsc-egc/mtec2022/ (accessed on 15 December 2022).

With Weka → Hoeffding Tree, a Hoeffding tree (VFDT—Very Fast Decision Trees) is a very fast decision tree algorithm for incremental decision tree induction at any time, capable of learning from massive data streams, assuming the distribution generation instances do not change over time. Hoeffding trees exploit the fact that a small sample can be sufficient to choose an optimal splitting attribute. This idea is supported mathematically by the Hoeffding limit, which quantifies the number of observations (examples) needed to estimate some statistics within a prescribed precision (according to the goodness of an attribute). A theoretically attractive feature of Hoeffding Trees not shared by other additional decision tree learners is that it has good performance guarantees. Using the Hoeffding boundary, one can show that its output is asymptotically nearly identical to that of a non-incremental study using infinite examples proposed by [41]. This classifier is a successful reference in dealing with large spatial representation datasets, for example, the evaluated dataset (Weka→ ConvtypNom), regarding spatial coverage of forests with quadrants defined in 30 × 30 m, 581,012 instances, and 54 attributes, elaborated the model in 41 s and completed the evaluation processing with cross-validation 10 times in 6 min and 49 s. All the results of Weka 3.9.3 performance are presented in Table 3.

Table 3. Performance comparison of Weka 3.9.3.

Nr	Weka 3.9.3 Classifiers	Time	Correct Instances	% Hits	% Kappa
1°	J48	0.30 s	42.0250	85.2002%	76.11%
2°	IBk	3 m 34 s	40.0420	81.1799%	69.97%
3°	Hoeffding Tree	0.69 s	39.2030	79.4800%	66.54%
4°	OneR	0.03 s	38.4040	77.8591%	63.75%
5°	NaiveBayes	0.04 s	37.4450	75.9149%	62.48%

Results available in: http://sadpreaigeo.org/ufsc-egc/mtec2019/ (accessed on 20 January 2022).

In all classifiers, the CTM dataset was analyzed with equal parameterizations and with the same specifications, such as the "k = 10 cross-validation" on the "property conservation" class. For property conservation in the "good condition" category, from the confusion matrix generated by the first classifier established, the true positives (TP) with 22,279 units, and the true negatives (VN) with 19,746, totaling 42,025 units to be certified, were first identified. Afterward, the false positives (FP) with 4002 units enabled the separation for re-evaluation of each occurrence. However, unlike the IBk classifier, the other evaluated classifiers had good performance regarding processing time using 18 attributes and 49,325 instances, over the same computational infrastructure provided.

With the satisfactory results obtained with OneR, NaiveBayes, J48, IBk, and Hoeffding Tree, allows the use of GeoDMA for automatic learning of new objects to be positively proven by Figure 10. Additionally, the fusion between textual classifiers and geospatial classifiers made possible through this work, the verification of an innovative form of knowledge extraction engineering.

From the use of GeoGMA [22], to perform the extraction of attributes, after exhaustive performance tests on the hardware used Table 1, adjustments were applied opting for the selection of all statistical methods, except "Percent of each class by area", because this method increases the consumption of processing and memory. Still in the process of extraction of attributes, it was possible from this procedure to obtain a better response time with a duration of up to 15 min of processing load.

Specifically, in Figure 11, it is shown how this made it possible to start the elaboration of queries through the filter on the attribute "B0Mean" > 0.4, especially to obtain the selection of objects with the highest "vegetation index" in the image.

In the second query, it was possible through the filter on the attribute "B7Mean" > 300.0, to obtain the selection of objects with the "shadow" characterization on the image. In this particular case, the query changed the return color for the objects. However, it was not rendered after processing, changing the color parameter set as "Yellow" to "Green", but remaining the best identification of the color "yellow" for the recognition of the object "shadow" in the image.

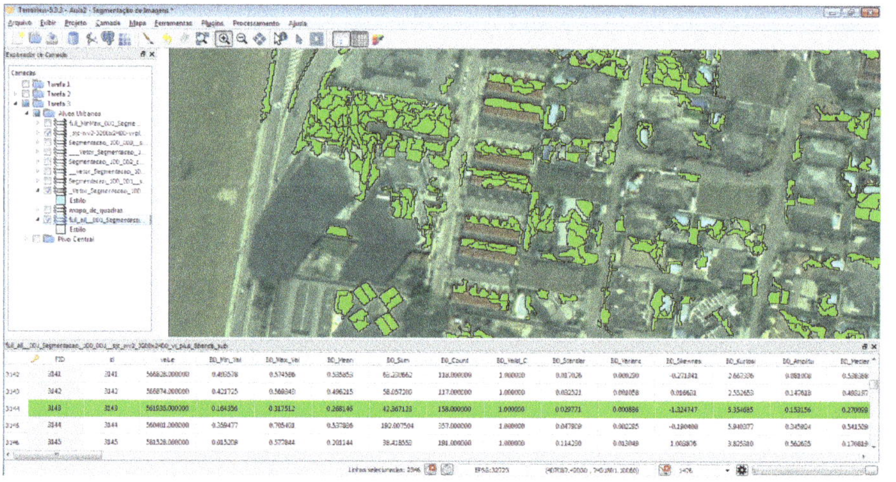

Figure 11. Results after calibration of values and recognition of objects.

Still, for the definition and characterization of the objects, some queries were performed, allowing, through the filter on the attribute "Band 5-Mean" using the GeoDMA [22], us to obtain different values for the selection of objects and characterization, such as "ceramic roofs", as shown in Figure 12. Thus, for the tested objects, the following values were obtained and are available in Table 4.

Table 4. Obtained values.

ID	Min	Max	Mean
2128	386.000000	511.000000	426.769231
2135	241.000000	724.000000	425.036585
2240	235.000000	631.000000	532.342541
2302	326.000000	692.000000	531.554622
2398	27.000000	556.000000	355.121339
2999	310.000000	664.000000	551.215962
3056	385.000000	645.000000	553.728000
3075	460.000000	650.000000	596.877160
3116	260.000000	526.000000	426.401042
3144	372.000000	529.000000	436.227848
3732	218.000000	638.000000	425.622951
3767	270.000000	521.000000	391.483974
3768	416.000000	584.000000	529.207207
3867	229.000000	555.000000	398.095023

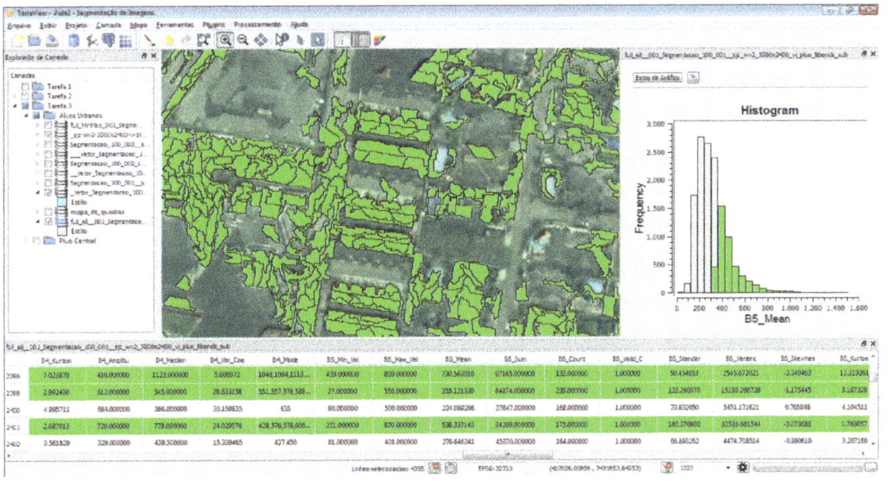

Figure 12. Results after calibration of values and recognition of objects with the results in the historigram.

The choice of samples was made randomly within the dataset images used for training, aiming initially to understand how the learning was performed, especially with the use of GeoDMA, as in Figures 13 and 14. This provided an important experience in the choice of classes. Specifically, the classes of investigation were: (a) asphalt, (b) roofs—light, dark or ceramic, (c) swimming pools, (d) shadows, (e) exposed soil, and (f) vegetation. Figure 13 describes the calibration process to obtain the final result, detailed in Figure 14.

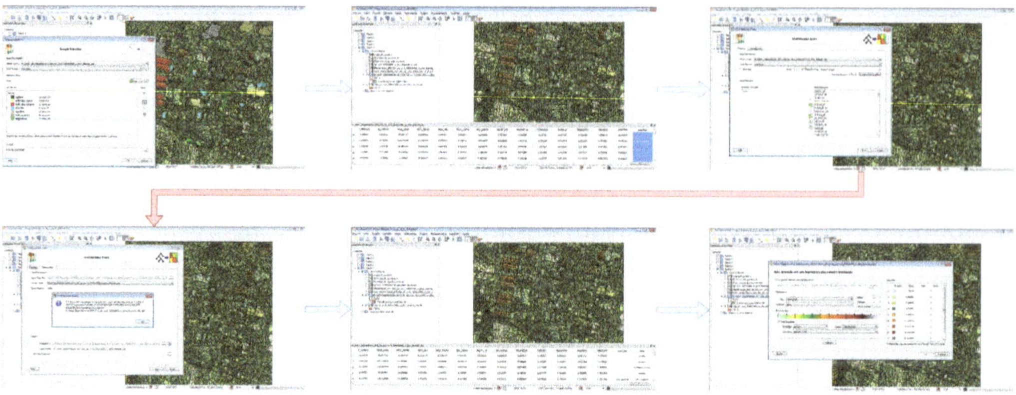

Figure 13. Calibration for segmentation process.

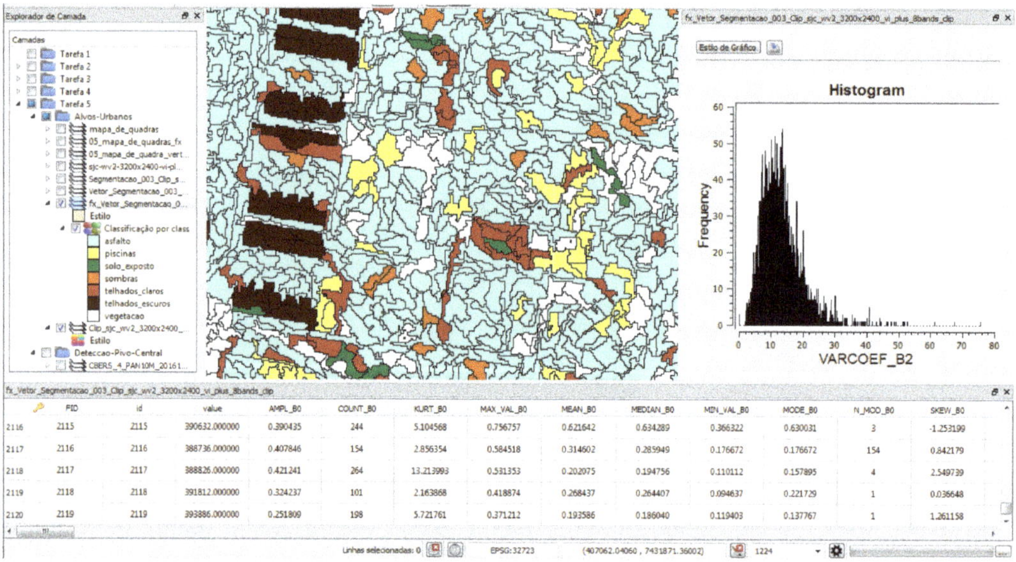

Figure 14. Segmentation results of the recognized objects.

From this study, it was possible to carry out new experiments to acquire a larger number of samples and improve the rule generation. For example, with respect to the class "asphalt", the rule was very broad, in which a greater number of representations were obtained, as illustrated in Figure 14. On the other hand, for "dark ceramic roofs", the rule was generated with better results. However, for future work, new implementations will be developed, aiming to train new classes.

Certainly, much still has to evolve computationally, especially with regard to constant active machine learning. In particular, urban object recognition and monitoring is noted. All documents related to the Weka training, datasets and other complementary documents are available at http://sadpreaigeo.org/ufsc-egc/mtec2019 and http://sadpreaigeo.org/ufsc-egc/mtec2022 (accessed on 15 December 2022).

6. Conclusions

Finally, according to the initial objectives of this work, through extensive research, it was possible to prove the applicability of the extraction of knowledge with the integration of data collected from the Cadastro Técnico Multifinalitário (CTM). Additionally, as prioritized, the research contemplated through investigation of the publications made in the last 5 years. During the study, it was noted that there is a great involvement of the academic and scientific community in the development of technologies that understand the geospatial and earth phenomena.

Growth and strong trends were observed in the use of the SVM (Support Vector Machine) method for evaluating large volumes of textual and geospatial data, as was the use of data discretization to enhance the execution (performance) of classifiers (algorithms). In this sense, the possibility of implementation and integration of the software Weka with TerraView/GeoDMA was proven, and they were compatible and operationalized because both complement each other from the collection to the structuring of textual and geospatial data for the evaluation of datasets, as presented in the Section 5.

An important characteristic to highlight with the Weka software was experimention with vectors that presented only numeric values, in which linear regression (Weka 3.9.3) proved to be faster to deal with large volumes of data. Regarding linear regression, the J48 decision tree showed the best results with the best classifier with accuracy (85%) and the kappa agreement coefficient (76%) in an average time of 0.30 s.

For future work, further studies to advance experiments from evolving data streams—those generated by mechanisms that change or fluctuate over time, by implementing the Weka/MOA package, designed specifically for data stream mining including new adaptations with Deep Learning algorithms. Furthermore, we intend to advance the development of a module for a pre-processing face as proposed [77], prioritizing data collection, transformation, and preparation of datasets and images. This is essential for the crossing of data and construction of rules to ensure the quality of the information to the user and decision maker. Another need concerns the improvement of models, being more or less robust and that can be reusable through vectors for TF-IDF x TerraViewGeoDMA application.

Although great demands of work are generated, one must prioritize care for the quality of data and information both for technical and operational issues, as well as for strategic issues that require constant validations, allowing certifications to occur for each step performed during all the extraction processes, generating a reliability indicator for the quality of information. The continuity of actions to intensify the implementation of processes for information quality, in this work, is indispensable so that all stages of knowledge extraction are guaranteed and certified.

Importantly, our results provide evidence for an implementation of an innovative practical, and systematic approach. The extraction of information and recommendation of knowledge shows a greater scientific relevance. Allowing the methods presented to apply calibration parameters for each object and achieve results with greater accuracy.

Author Contributions: Conceptualization, L.A.S. and A.F.d.M.; methodology, L.A.S. and A.F.d.M.; software, L.A.S., A.S.M. and H.S.S.B.; validation, L.A.S. and A.F.d.M.; formal analysis, L.A.S.; investigation, L.A.S. and A.F.d.M.; resources, L.A.S. and A.F.d.M.; data curation, L.A.S. and A.F.d.M.; writing—original draft preparation, L.A.S. and A.F.d.M.; writing—review and editing, L.A.S., A.S.M., H.S.S.B., A.L.G., L.C.B. and A.F.d.M.; visualization, L.A.S. and A.F.d.M.; supervision, A.F.d.M.; project administration, A.F.d.M.; funding acquisition, L.A.S. All authors have read and agreed to the published version of the manuscript.

Funding: The research of Luis Augusto Silva has been funded by the call for predoctoral contracts USAL 2021, co-financed by Banco Santander. Héctor Sánchez San Blas's research was supported by the Spanish Ministry of Universities (FPU Fellowship under Grant FPU20/03014).

Conflicts of Interest: The authors declare no conflict of interest.

References

1. Shankaranarayanan, G.; Blake, R. From Content to Context: The Evolution and Growth of Data Quality Research. *J. Data Inf. Qual.* **2017**, *8*, 9. [CrossRef]
2. Li, D.; Gao, X.; Deng, Y. A Generalized Expression for Information Quality of Basic Probability Assignment. *IEEE Access* **2019**, *7*, 174734–174739. [CrossRef]

3. Lin, G.; Jing, G.; Fang-Fang, D. Evaluation method of enterprise information quality based on QFD. In Proceedings of the 2011 International Conference on Consumer Electronics, Communications and Networks (CECNet), Xianning, China, 16–18 April 2011; pp. 325–328.
4. Tseng, Y.H.; Lin, C.J.; Lin, Y.I. Text mining techniques for patent analysis. *Inf. Process. Manag.* **2007**, *43*, 1216–1247. [CrossRef]
5. Madnick, S.E.; Wang, R.Y.; Lee, Y.W.; Zhu, H. Overview and Framework for Data and Information Quality Research. *J. Data Inf. Qual.* **2009**, *1*, 2. [CrossRef]
6. Naumann, F.; Rolker, C. Assessment Methods for Information Quality Criteria. Mathematisch-Naturwissenschaftliche Fakultät, Humboldt-Universität zu Berlin, Berlin, Germany, 2005. . [CrossRef]
7. Wang, R.Y.; Strong, D.M. Beyond Accuracy: What Data Quality Means to Data Consumers. *J. Manage. Inf. Syst.* **1996**, *12*, 5–33. [CrossRef]
8. Mahdianpari, M.; Granger, J.E.; Mohammadimanesh, F.; Warren, S.; Puestow, T.; Salehi, B.; Brisco, B. Smart solutions for smart cities: Urban wetland mapping using very-high resolution satellite imagery and airborne LiDAR data in the City of St. John's, NL, Canada. *J. Environ. Manag.* **2021**, *280*, 111676. . [CrossRef] [PubMed]
9. Hu, Q.; Wu, W.; Xia, T.; Yu, Q.; Yang, P.; Li, Z.; Song, Q. Exploring the use of Google Earth imagery and object-based methods in land use/cover mapping. *Remote Sens.* **2013**, *5*, 6026–6042. [CrossRef]
10. Fawcett, D.; Bennie, J.; Anderson, K. Monitoring spring phenology of individual tree crowns using drone-acquired NDVI data. *Remote Sens. Ecol. Conserv.* **2021**, *7*, 227–244. [CrossRef]
11. Mahdavi Saeidi, A.; Babaie Kafaky, S.; Mataji, A. Detecting the development stages of natural forests in northern Iran with different algorithms and high-resolution data from GeoEye-1. *Environ. Monit. Assess.* **2020**, *192*, 1–15. [CrossRef]
12. Alphan, H.; Çelik, N. Monitoring changes in landscape pattern: Use of Ikonos and Quickbird images. *Environ. Monit. Assess.* **2016**, *188*, 1–13. [CrossRef]
13. Albuquerque, R.; Vieira, D.; Ferreira, M.; Soares, L.; Olsen, S.; Araujo, L.; Vicente, L.; Tymus, J.; Balieiro, C.; Matsumoto, M.; et al. Mapping Key Indicators of Forest Restoration in the Amazon Using a Low-Cost Drone and Artificial Intelligence. *Remote Sens.* **2022**, *14*, 830. [CrossRef]
14. Habibie, M.I.; Ahamed, T.; Noguchi, R.; Matsushita, S. Deep Learning Algorithms to determine Drought prone Areas Using Remote Sensing and GIS. In Proceedings of the 2020 IEEE Asia-Pacific Conference on Geoscience, Electronics and Remote Sensing Technology (AGERS), Jakarta, Indonesia, 7–8 December 2020; pp. 69–73. [CrossRef]
15. Hosseiny, B.; Mahdianpari, M.; Brisco, B.; Mohammadimanesh, F.; Salehi, B. WetNet: A Spatial–Temporal Ensemble Deep Learning Model for Wetland Classification Using Sentinel-1 and Sentinel-2. *IEEE Trans. Geosci. Remote Sens.* **2022**, *60*, 1–14. [CrossRef]
16. Klein, V.B.; Todesco, J.L. Um Modelo Conceitual para uso de Big Data e Open Data nas Smart Cities. *Intern. J. Knowl. Eng. Manag.* **2017**, *6*, 28–50. [CrossRef]
17. Wagstaff, K.L. Machine Learning that Matters. In Proceedings of the International Conference on Machine Learning, Edinburgh, UK, 26 June–1 July 2012; pp. 529–536.
18. Souza, J.; Francisco, A.; Piekarski, C.; Prado, G.; Oliveira, L. Data Mining and Machine Learning in the Context of Sustainable Evaluation: A Literature Review. *IEEE Lat. Am. Trans.* **2019**, *17*, 372–382. [CrossRef]
19. Mei, J.; Jiang, X.; Cai, J. Learning local feature representation from matching, clustering and spatial transform. *J. Vis. Commun. Image Represent.* **2019**, *63*, 102601. [CrossRef]
20. Gonçalves, V.P.; Ribeiro, E.A.W.; Imai, N.N. Mapping Areas Invaded by Pinus sp. from Geographic Object-Based Image Analysis (GEOBIA) Applied on RPAS (Drone) Color Images. *Remote Sens.* **2020**, *12*, 2805. [CrossRef]
21. Bayazit, N. Investigating Design: A Review of Forty Years of Design Research. *Des. Issues* **2004**, *20*, 16–29. [CrossRef]
22. Korting, T.S. GEODMA: A Toolbox Integrating Data Mining with Object-Based and Multi-Temporal Analysis of Satellite Remotely Sensed Imagery. Ph.D. Thesis, INPE, Sao Jose dos Campos, Brazil, 2012.
23. Korting, T.S.; Fonseca, L.M.G.; Camara, G. GeoDMA-Geographic Data Mining Analyst. *J. Comput. Geosci.* **2013**, *57*, 133–145. [CrossRef]
24. Maretto, R.V.; Korting, T.S.; Fonseca, L.M.G. An Extensible and Easy-to-use Toolbox for Deep Learning Based Analysis of Remote Sensing Images. In Proceedings of the IEEE International Geoscience and Remote Sensing Symposium (IGARSS), Kuala Lumpur, Malaysia, 17–22 July 2019; pp. 9815–9818.
25. Braz, A.M.; Korting, T.S.; Martins, A.P.; Braz, A.M. Geobia E Mineração de Dados para Classificação de Imagens de Altíssima Resolução Espacial. *Estud. Geográficos Rev. Eletrônica Geogr.* **2021**, *19*, 209–224. [CrossRef]
26. Kanjir, U.; Veljanovski, T.; Oštir, K. Multilevel Segmentation of Soil Sealing in Urban Areas. In Proceedings of the Asian Conference on Remote Sensing (ACRS), Taipei, Taiwan, 3–7 October 2011; pp. 1–7.
27. Liu, W.T.H. Aplicacoes de Sensoriamento Remoto. In *Plastics*; Oficina de Textos: Rio de Janeiro, Brazil, 2019; pp. 10–100.
28. Rajbhandari, S.; Aryal, J.; Osborn, J.; Lucieer, A.; Musk, R. Leveraging Machine Learning to Extend Ontology-Driven Geographic Object-Based Image Analysis (O-GEOBIA): A Case Study in Forest-Type Mapping. *Remote Sens.* **2019**, *11*, 503. [CrossRef]
29. Zanotta, D.C.; Ferreira, M.P.; Zortea, M. Processamento de Imagens de Satélite. In *Plastics*; Blucher: São Paulo, Brasil, 2019; pp. 107–301.
30. Barbosa, A.L.; Loureiro, G.; Manea, S.; Duarte, J.M.L.; Garbi, G.P. Ranking of Fault Mitigation Techniques for Spatial Radiation in Commercial Off-the-Shelf Field Programmable Gate Array. *IEEE Lat. Am. Trans.* **2020**, *18*, 736–743. [CrossRef]

31. Obando, E.D.; Carvajal, S.X.; Agudelo, J.P. Solar Radiation Prediction Using Machine Learning Techniques: A Review. *IEEE Lat. Am. Trans.* **2019**, *17*, 684–697. [CrossRef]
32. Pletsch, M.A.; Körting, T. Information mining for automatic search in remote sensing image catalogs. *Rev. Bras. Cartogr.* **2019**, *70*, 1860–1884. [CrossRef]
33. Cesconetto, J.; Augusto Silva, L.; Bortoluzzi, F.; Navarro-Cáceres, M.; Zeferino, C.A.; Leithardt, V.R.Q. PRIPRO—Privacy Profiles: User Profiling Management for Smart Environments. *Electronics* **2020**, *9*, 1519. [CrossRef]
34. de Moraes Rossetto, A.G.; Sega, C.; Leithardt, V.R.Q. An Architecture for Managing Data Privacy in Healthcare with Blockchain. *Sensors* **2022**, *22*, 8292. [CrossRef]
35. Lima, R.; Filippetto, A.S.; Heckler, W.; Barbosa, J.L.V.; Leithardt, V.R.Q. Towards ubiquitous requirements engineering through recommendations based on context histories. *PeerJ Comput. Sci.* **2022**, *8*, 794. [CrossRef]
36. Mehta, R.; Kumar, A. An Analysis of Hybrid Layered Classification Algorithms for Object Recognition. *IOSR J. Comput. Eng.* **2018**, *20*, 57–64. [CrossRef]
37. Yahav, I.; Shehory, O.; Schwartz, D. Comments Mining with TF-IDF: The Inherent Bias and Its Removal. *IEEE Trans. Knowl. Data Eng.* **2019**, *31*, 437–450. [CrossRef]
38. Khorshid, M.M.H.; Abou-El-Enien, T.H.M.; Soliman, G.M.A. A Comparison among Support Vector Machine and other Machine Learning Classification Algorithms. *Int. J. Comput. Sci.* **2015**, *3*, 25–35.
39. Moen, E.; Bannon, D.; Kudo, T.; Graf, W.; Covert, M.; Valen, D.V. Deep learning for cellular image analysis. *Nat. Methods* **2019**, *16*, 1233. [CrossRef]
40. Giolo, S.R. Introdução à análise de dados categóricos com aplicações. In *Plastics*; Blucher: São Paulo, Brasil, 2017; pp. 75–190.
41. Hulten, G.; Spencer, L.; Domingos, P. Mining time-changing data streams. In Proceedings of the ACM International Conference on Knowledge Discovery and Data Mining SIGKDD, San Francisco, CA, USA, 26–29 August 2001; pp. 97–106.
42. Bogorny, V.; Avancini, H.; de Paula, B.C.; Rocha Kuplich, C.; Alvares, L.O. Weka-STPM: A Software Architecture and Prototype for Semantic Trajectory Data Mining. *Trans. GIS* **2011**, *15*, 227–248. [CrossRef]
43. Witten, I.H.; Frank, E.; Hall, M.A.; Pal, C.J. Data Mining: Practical Machine Learning Tools and Techniques. In *Plastics*; Kaufmann Morg.: Cambridge, MA, USA, 2016; pp. 515–621.
44. Damacena, A.P.C. Avaliação da qualidade de dados geoespaciais. In *Plastics*, 2nd ed.; IBGE: Rio de Janeiro, Brasil, 2019; pp. 41–52.
45. Moraes, A.F.; Bastos, L.C.; Koehler, F.S.; Kracik, M.S.; Moraes, N.I.G. Open integration for knowledge quality distributed: Approach to geospatial data collect in sustainable cities. *Braz. J. Dev.* **2019**, *5*, 31219–31236. [CrossRef]
46. Goodfellow, I.; Bengio, Y.; Courville, A. *Deep Learning*; MIT Press: Cambridge, MA, USA, 2016.
47. Girshick, R.; Donahue, J.; Darrell, T.; Malik, J. Rich feature hierarchies for accurate object detection and semantic segmentation. In Proceedings of the Conference on Computer Vision and Pattern Recognition (CVPR), Columbus, OH, USA, 23–28 June 2014; pp. 580–587. [CrossRef]
48. Girshick, R. Fast R-CNN. In Proceedings of the IEEE International Conference on Computer Vision, Santiago, Chile, 7–13 December 2015; pp. 1440–1448. [CrossRef]
49. Rampersad, H. Faster R-CNN: Towards Real-Time Object Detection with Region Proposal Networks. In *Total Performance Scorecard*; Routledge: London, UK, 2020; pp. 159–183. [CrossRef]
50. Doll, P.; Girshick, R.; Ai, F. Mask R-CNN. *arXiv* **2017**, arXiv:1703.06870.
51. Liu, W.; Anguelov, D.; Erhan, D.; Szegedy, C.; Reed, S.; Fu, C.Y.; Berg, A.C. *SSD: Single Shot MultiBox Detector*; Springer: Berlin/Heidelberg, Germany, 2015. [CrossRef]
52. Redmon, J.; Divvala, S.; Girshick, R.; Farhadi, A. You only look once: Unified, real-time object detection. In Proceedings of the IEEE Computer Society Conference on Computer Vision and Pattern Recognition, Las Vegas, NV, USA, 27–30 June 2016; pp. 779–788. [CrossRef]
53. Redmon, J.; Farhadi, A. YOLO9000: Better, faster, stronger. In Proceedings of the 30th IEEE Conference on Computer Vision and Pattern Recognition, CVPR 2017, Honolulu, HI, USA, 21–26 July 2017; pp. 6517–6525. [CrossRef]
54. Redmon, J.; Farhadi, A. YOLOv3: An Incremental Improvement. *arXiv* **2018**, arXiv:1804.02767.
55. Bochkovskiy, A.; Wang, C.Y.; Liao, H.Y.M. YOLOv4: Optimal Speed and Accuracy of Object Detection. *arXiv* **2020**, arXiv:2004.10934.
56. Tang, Y.; Zhou, H.; Wang, H.; Zhang, Y. Fruit detection and positioning technology for a *Camellia oleifera* C. Abel orchard based on improved YOLOv4-tiny model and binocular stereo vision. *Expert Syst. Appl.* **2023**, *211*, 118573. [CrossRef]
57. Silva, L.A.; Sanchez San Blas, H.; Peral García, D.; Sales Mendes, A.; Villarubia González, G. An Architectural Multi-Agent System for a Pavement Monitoring System with Pothole Recognition in UAV Images. *Sensors* **2020**, *20*, 6205. [CrossRef]
58. Pham, M.T.; Courtrai, L.; Friguet, C.; Lefèvre, S.; Baussard, A. YOLO-Fine: One-Stage Detector of Small Objects Under Various Backgrounds in Remote Sensing Images. *Remote Sens.* **2020**, *12*, 2501. [CrossRef]
59. Wang, C.Y.; Bochkovskiy, A.; Liao, H.Y.M. YOLOv7: Trainable bag-of-freebies sets new state-of-the-art for real-time object detectors. *arXiv* **2022**, arXiv:2207.02696.
60. Li, W.; Li, W.; Yang, F.; Wang, P. Multi-Scale Object Detection in Satellite Imagery Based On YOLT. In Proceedings of the IGARSS 2019—2019 IEEE International Geoscience and Remote Sensing Symposium, Yokohama, Japan, 28 July–2 August 2019; pp. 162–165. [CrossRef]

61. Xu, D.; Wu, Y. MRFF-YOLO: A Multi-Receptive Fields Fusion Network for Remote Sensing Target Detection. *Remote Sens.* **2020**, *12*, 3118. [CrossRef]
62. Schreiber, G.; Akkermans, H.; Anjewierden, A.; Hoog, R.; Shadbolt, N.; Velde, W.V.; Wielinga, B. Knowledge Engineering and Management—The CommonKADS Methodology. In *Plastics*; MIT Press: Cambridge, MA, USA, 2000; pp. 69–121.
63. Dresch, A.; Lacerda, D.P.; Júnior, J.A.A. Design Science Research: Método de pesquisa para a engenharia de produção. In *Plastics*; Bookman: Porto Alegre, Brazil, 2015; pp. 22–102.
64. Rajyalakshmi, D.; Raju, K.K.; Varma, G.P.S. Taxonomy of Satellite Image and Validation Using Statistical Inference. In Proceedings of the IEEE 6th Intern Conference on Advanced Computing, Toulouse, France, 18–21 July 2016; pp. 352–361.
65. Li, Y.; Fang, Y.; Cheng, R.; Zhang, W. Spatial Pattern Matching: A New Direction for Finding Spatial Objects. *ACM SIGSPATIAL Spec.* **2019**, *11*, 3–12. [CrossRef]
66. Arganda-Carreras, I.; Kaynig, V.; Rueden, C. Trainable Weka Segmentation: A machine learning tool for microscopy pixel classification. *Bioinformatics* **2017**, *33*, 2424–2426. [CrossRef] [PubMed]
67. Kotikot, S.M.; Kar, B.; Omitaomu, O.A. A Geospatial Framework Using Multicriteria Decision Analysis for Strategic Placement of Reserve Generators in Puerto Rico. *IEEE Trans. Eng. Manag.* **2020**, *67*, 659–669. [CrossRef]
68. Gimenez, F.; Zerbini, C.; Riva, G. Extending SMS Service Coverage in Rural Areas by using LoRa Communication Technology. *IEEE Lat. Am. Trans.* **2020**, *18*, 214–222. [CrossRef]
69. Huai, M.; Miao, C.; Li, Y.; Suo, Q.; Zhang, A. Learning Distance Metrics from Probabilistic Information. *ACM Trans. Knowl. Discov. Data* **2020**, *14*, 53–86. [CrossRef]
70. Rodrigues, M.L.; Körting, T.S.; de Queiroz, G.R.; Sales, C.P.; Silva, L.A.R.d. Detecting Center Pivots in Matopiba Using Hough Transform and Web Time Series Service. In Proceedings of the 2020 IEEE Latin American GRSS ISPRS Remote Sensing Conference (LAGIRS), Santiago, Chile, 22–26 March 2020; pp. 189–194. [CrossRef]
71. Russell, S.; Norvig, P. Artificial Intelligence: A moen approach. In *Plastics*; Prentice Hall: Hoboken, NJ, USA, 2009; pp. 149–297.
72. Zhang, Q.; Zhou, J.; He, J.; Xiaodong, C.; Zhang, Z.S.B. A shell dataset, for shell features extraction and recognition. *Sci. Data* **2019**, *6*, 226. [CrossRef]
73. Erba, D.A.; Loch, C. Cadastro técnico multifinalitário rural e urbano. In *Plastics*; Lincoln: Cambridge, MA, USA, 2007; pp. 50–108.
74. Ferreira, M.C. *Iniciação à Análise Geoespacial: Teoria, téCnicas e Exemplos para Geoprocessamento*; Unesp: Sao Paulo, Brazil, 2014; pp. 297–333.
75. QGIS Development Team. QGIS Geographic Information System, 2022. Open Source Geospatial Foundation Project. Available online: http://qgis.osgeo.org (accessed on 15 December 2022).
76. Awrangjeb, M.; Ravanbakhsh, M.; Fraser, C.S. Automatic detection of residential buildings using LIDAR data and multispectral imagery. *J. Photogramm. Remote Sens.* **2010**, *65*, 457–467. [CrossRef]
77. Chenthamarakshan, V.; Desphande, P.; Krishnapuram, R.; Varadarajan, R.; Stolze, K. *WYSIWYE: An Algebra for Expressing Spatial and Textual Rules for Visual Information Extraction*; Cornell University Computer Sciense: Ithaca, NY, USA, 2016.

sensors

MDPI

Article

Real-Time Implementation of the Prescribed Performance Tracking Control for Magnetic Levitation Systems

Thanh Nguyen Truong, Anh Tuan Vo and Hee-Jun Kang *

Department of Electrical, Electronic and Computer Engineering, University of Ulsan, Ulsan 44610, Republic of Korea
* Correspondence: hjkang@ulsan.ac.kr; Tel.: +82-52-259-2207

Abstract: For magnetic levitation systems subject to dynamical uncertainty and exterior perturbations, we implement a real-time Prescribed Performance Control (PPC). A modified function of Global Fast Terminal Sliding Mode Manifold (GFTSMM) based on the transformed error of the novel PPC is introduced; hence, the error variable quickly converges to the equilibrium point with the prescribed performance, which means that maximum overshoot and steady-state of the controlled errors will be in a knowledge-defined boundary. To enhance the performance of Global Fast Terminal Sliding Mode Control (GFTSMC) and to reduce chattering in the control input, a modified third-order sliding mode observer (MTOSMO) is proposed to estimate the whole uncertainty and external disturbance. The combination of the GFTSMC, PPC, and MTOSMO generates a novel solution ensuring a finite-time stable position of the controlled ball and the possibility of performing different orbit tracking missions with an impressive performance in terms of tracking accuracy, fast convergence, stabilization, and chattering reduction. It also possesses a simple design that is suitable for real-time applications. By using the Lyapunov-based method, the stable evidence of the developed method is fully verified. We implement a simulation and an experiment on the laboratory magnetic levitation model to demonstrate the improved performance of the developed control system.

Keywords: prescribed performance tracking control; terminal sliding mode control; disturbance observer; magnetic levitation systems

Citation: Truong, T.N.; Vo, A.T.; Kang, H.-J. Real-Time Implementation of the Prescribed Performance Tracking Control for Magnetic Levitation Systems. *Sensors* **2022**, *22*, 9132. https://doi.org/10.3390/s22239132

Academic Editor: Natividad Duro Carralero

Received: 7 October 2022
Accepted: 22 November 2022
Published: 24 November 2022

1. Introduction

The potential applications of magnetic levitation systems are huge. They have become quite popular as a testing system in control engineering labs and advanced nonlinear control programs. A major reason for the popularity of this nonlinear system is the fact that it is relatively easy to construct and manage. Their many applications can be found in real systems such as contactless melting, rocket-guiding projects, gyroscopes, high-speed trains, frictionless bearings, vibration isolation systems, etc. The characteristics of Magnetic Levitation Systems (MLSs) tend towards nonlinearity and instability described by nonlinear differential formulas. The proportional-integral-differential (PID) and proportional-derivative (PD) controllers are typically proposed for the effective regulation of systems under the assumption of well-known gain parameters [1]. The tracking control performance of those controllers can rapidly deteriorate as deviations from their nominal operating point increase. There is no doubt that the nonlinearity and inherent instabilities of the system prevent classical PD or PID controllers from being applied to these more complex problems. A trajectory tracking task typically involves the gain of the system no longer being constant, and it is determined by changes in distance from the magnet. In order to maintain operating time with a high steady-state as long as possible, advanced control methods, such as nonlinear controllers [2–6], should be applied to MLSs. Ref. [7] developed a nonlinear controller based on fast online algebraic identification of the input gain for an MLS. Unfortunately, this controller was not considered the effects of external disturbances, and it only guarantees an asymptotically exponentially stable. In addition, variation factors, such as system parameters, resistance, inductance, and suspending mass, should be taken into account. The implemented algorithms in real-time systems about the issue of tracking output trajectory tasks for the suspended metal ball did not use trajectories of certain difficulty to produce controlled motions such as sinusoidal or rest-to-rest reference trajectory positions. Thus, the verification is not general. Most of the methods introduced for MLS achieve only asymptotic stability, and a few achieve finite-time stability. However, achieving prescribed performance has hardly been introduced for MLSs in the literature. Therefore, there is still much interest in improving the performance of MLSs.

In order to handle the influences of uncertainties and exterior disturbances, sliding mode control (SMC) with powerful and immutable properties can be employed effectively. In recent years, SMC has been successfully applied for a wide range of practical applications such as unmanned aerial vehicles (UAVs) [8–10], autonomous underwater vehicles (AUVs) [11–15], robotic manipulators [16–20], and so on. In the approach stage, SMC cannot maintain uniform characteristics due to the existence of unidentified uncertain elements. Moreover, the SMC's control performance depends on the linear sliding mode surface and a discontinuous control law to drive the state variables to that sliding surface. Consequently, this method only has an asymptotically exponentially stable along with a non-smooth control signal that is known as chattering. Whenever higher accuracy and faster convergence are needed under the SMC method, a very large control force is required, which is not possible because the capability of the hardware devices is limited, and chattering behavior also becomes much more serious.

Terminal SMC (TSMC) [21], Fast TSMC [22], GFTSMC [16,23], Non-singular TSMC [24], or Non-singular Fast TSMC (NFTSMC) [25] were proposed to inherit the invariant properties of the SMC and at the same time overcome the SMC's disadvantages. Several attractive properties of these methods include robustness against uncertain terms, finite-time convergence, and high accuracy. Thus far, these methods have also been successfully applied to real-time nonlinear systems in general and MLSs in particular. However, each method also has its disadvantages. For example, slow convergence speed, singularity, and chattering are the main weaknesses of TSMC, while singularity and chattering are the main weaknesses of Fast TSMC. The main weakness of the two remaining control methods is chattering. The issue of chattering attenuation has therefore become a popular one. Moreover, according to the knowledge of the authors, there are currently no studies related to the prescribed performance control based on the aforementioned methods proposed for MLSs. This is also one of the main motivations of this article.

The use of quasi-sliding mode control, neural-network-based SMC methods [6], disturbance observers [26,27], super-twisting SMC methods [28,29], second-order SM observers (SOSMOs) [30], third-order SM observers (TOSMOs) [17,31], and so on have been proposed for resolving the chattering. It is noted that the active control methods have been proven to provide better tracking performance if the uncertain components are approximated correctly when compared to the passive control methods. For example, the technique of applying an observer to reduce chattering behavior can be understood as a disturbance observer will approximate the uncertain components to obtain a precise dynamic model. As a consequence, the approximate error from the observer can only be compensated by using a reasonable sliding gain. Among the observers discussed, TOSMO is superior to the rest when convergence is achieved in finite time and only needs information from the position sensor. However, its convergence speed also needs to be further improved to avoid causing delays in the control system.

In our paper, an MLS suspends a metal ball in the air by electromagnetic force. The ball and MLS do not interact mechanically. It is inherently unstable and highly nonlinear. Adjusting and tracking the reference trajectory of the levitated ball is a highly challenging process. Therefore, the control target is to develop a new real-time prescribed performance tracking controller for MLSs under the effects of uncertainty and exterior disturbance. The important contributions of the proposed methods can be listed below:

- A modified TOSMO is applied to quickly estimate the approximate value of the uncertainty and exterior disturbance;
- The novel Prescribed Performance Function (PPF) does not contain a singularity problem and can flexibly adjust lower and upper bounds. Furthermore, it can extend the operation domain at a steady state compared to that of the conventional PPF. With the proposed PPF, the steady-state error boundaries will be symmetric to zero, so, when the transformed error converges to zero, the tracking error also converges to zero;
- A modified function of GFTSMM based on the transformed errors of the PPC is introduced; hence, the error variables quickly converge to the equilibrium point with the prescribed performance;
- The maximum overshoot, convergence index, and steady-state error can be managed within a predefined domain under the proposed controller;
- A novel solution ensures a finite-time stable position of the controlled ball and the possibility of performing different orbit tracking missions with an impressive performance in the terms of tracking accuracy, fast convergence, stabilization, and chattering reduction;
- The effectiveness of the designed control solution was confirmed by simulation and experiment;
- This controller is presented in a way that can be applied to real-time applications. In addition, it can apply not only to MLSs but also to a class of second-order nonlinear systems.

Our manuscript is structured as follows: The formulation of the problem takes place in Section 2. The control method design is discussed in Section 3. Section 4 is the simulation and experimental results. Section 5 is conclusions.

2. Problem Statements

The structure of the MLS is depicted in Figure 1. In addition, an explanation of this MLS is provided in the article [4].

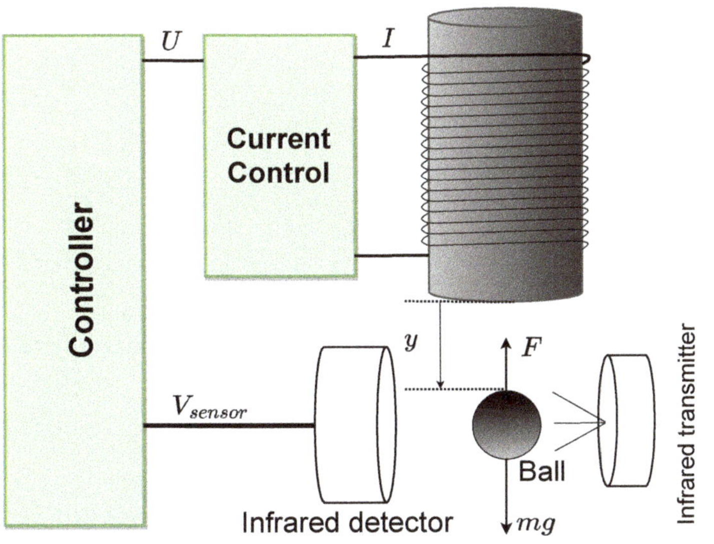

Figure 1. Magnetic levitation system graph.

The dynamic model of the MLS is described as:

$$m\ddot{y} = mg - \lambda\left(\frac{I}{y}\right)^2 \tag{1}$$

$$I = KU \tag{2}$$

where m is the mass of the levitated ball, g denotes the acceleration due to gravity, U is the control input voltage, I is the winding current, y is the position of the levitated ball, K is the constant related to the input voltage and the current through the coil, and λ is a constant related to the mutual inductance of the ball and coupling coefficients.

Substituting Equation (2) into Equation (1) yields:

$$\ddot{y} = g - \frac{\mu U^2}{y^2} \tag{3}$$

where $\mu = \frac{\lambda K^2}{m}$.

The exact value of μ cannot be known, it can be identified by using an identification method such as [7]. In addition, considering the effects of uncertainty and exterior disturbance, Equation (3) can be described as:

$$\ddot{y} = g - \frac{\hat{\mu} U^2}{y^2} + \Delta(y, \delta, t), \tag{4}$$

where $\hat{\mu}$ represents the estimated value of μ, and $\Delta(y, \delta, t) = \frac{\mu - \hat{\mu}}{y^2} U^2 + \delta(t)$ is a function of exterior disturbance and interior uncertainty, $\delta(t)$ is exterior disturbance. Let $y_1 = y$, and $y_2 = \dot{y}$; then, Equation (4) is rewritten as:

$$\begin{cases} \dot{y}_1 = y_2 \\ \dot{y}_2 = g - \frac{\hat{\mu} U^2}{y_1^2} + \Delta(y, \delta, t) \end{cases} \tag{5}$$

Assumption 1. *It is assumed that uncertain terms are bound by the following:*

$$|\Delta(y, \delta, t)| \le \bar{\Delta}, \tag{6}$$

where $\bar{\Delta}$ is a positive constant.

MLS suspends a metal ball in the air by electromagnetic force. The ball and MLS do not interact mechanically. It is inherently unstable and highly nonlinear. Adjusting and tracking the reference trajectory of the levitated ball is a highly challenging process. Therefore, the objective of our paper is to develop a new real-time prescribed performance tracking controller for MLSs under the effects of uncertainty and exterior disturbance ensuring a finite-time stable position of the controlled ball and the possibility of performing different orbit tracking missions with an impressive performance in the terms of tracking accuracy, fast convergence, stabilization, and chattering reduction.

3. Design of the Proposed Control Method

3.1. Design of the Sliding Mode Surface

The position error and velocity error are respectively defined as $\tilde{y} = y_1 - y_r$ and $\dot{\tilde{y}} = \dot{y}_1 - \dot{y}_r$, where y_r and $\dot{y}_r$ represent the reference trajectory and its derivative.

For system (5), the sliding mode surface is constructed by using the position error and velocity error, as follows:

$$s = \dot{\tilde{y}} + \lambda^* sig(\tilde{y})^\alpha + \omega^* sig(\tilde{y})^\beta \tag{7}$$

where $\lambda^* = \frac{2\lambda}{1+e^{-\eta(|\tilde{y}|-\gamma)}}$, $\omega^* = \frac{2\omega}{1+e^{\vartheta(|\tilde{y}|-\gamma)}}$, $\lambda > 0, \eta > 0, \omega > 0, \vartheta > 0$, $\alpha > 1, 0 < \beta < 1$, $\gamma = (\omega/\lambda)^{1/(\alpha-\beta)}$, and the $sig()^k$ function is defined as $sig(x)^k = |x|^k sign(x)$ with $k > 0$.

The finite-time stable evidence of the selected sliding manifold is given in [23]. The convergence time T_s of the sliding mode motion is stated in Equation (27) of [23].

3.2. Design of Global Fast Terminal Sliding Mode Control

Using the selected sliding mode surface, we can calculate its time derivative as follows:

$$\begin{aligned} \dot{s} = \ddot{\tilde{y}} + \frac{2\lambda\alpha}{1+e^{-\eta(|\tilde{y}|-\gamma)}}|\tilde{y}|^{\alpha-1}\dot{\tilde{y}} + \frac{2\lambda\eta e^{-\eta(|\tilde{y}|-\gamma)}}{\left(1+e^{-\eta(|\tilde{y}|-\gamma)}\right)^2}|\tilde{y}|^{\alpha}\dot{\tilde{y}} \\ +\frac{2\omega\beta}{1+e^{\vartheta(|\tilde{y}|-\gamma)}}|\tilde{y}|^{\beta-1}\dot{\tilde{y}} - \frac{2\omega\vartheta e^{\vartheta(|\tilde{y}|-\gamma)}}{\left(1+e^{\vartheta(|\tilde{y}|-\gamma)}\right)^2}|\tilde{y}|^{\beta}\dot{\tilde{y}} \end{aligned} \tag{8}$$

Equation (8) is rewritten as:

$$\dot{s} = \ddot{\tilde{y}} + M_{\tilde{y}} = \dot{y}_2 - \ddot{y}_r + M_{\tilde{y}} \tag{9}$$

where $M_{\tilde{y}} = \frac{2\lambda\alpha}{1+e^{-\eta(|\tilde{y}|-\gamma)}}|\tilde{y}|^{\alpha-1}\dot{\tilde{y}} + \frac{2\lambda\eta e^{-\eta(|\tilde{y}|-\gamma)}}{\left(1+e^{-\eta(|\tilde{y}|-\gamma)}\right)^2}|\tilde{y}|^{\alpha}\dot{\tilde{y}} + \frac{2\omega\beta}{1+e^{\vartheta(|\tilde{y}|-\gamma)}}|\tilde{y}|^{\beta-1}\dot{\tilde{y}} - \frac{2\omega\vartheta e^{\vartheta(|\tilde{y}|-\gamma)}}{\left(1+e^{\vartheta(|\tilde{y}|-\gamma)}\right)^2}|\tilde{y}|^{\beta}\dot{\tilde{y}}$.

Substituting system (5) into Equation (9) yields:

$$\dot{s} = g - \frac{\hat{\mu}}{y_1^2}U^2 + \Delta(y, \delta, t) - \ddot{y}_r + M_{\tilde{y}} \tag{10}$$

From Equation (10), GFTSMC is designed as:

$$U = \sqrt{\frac{y_1^2}{\hat{\mu}}(u_{eq} + u_r)} \tag{11}$$

in which the terms of u_{eq} and u_r are designed as:

$$\begin{cases} u_{eq} = g + M_{\tilde{y}} - \ddot{y}_r \\ u_r = \sigma_1 s + (\bar{\Delta} + \kappa_1)\,\text{sign}(s) \end{cases}$$

where κ_1 and σ_1 are positive constants.

The following evidence will be given to prove the GFTSMC's correctness and stability.

Evidence: Applying the control signal system (11) to Equation (10), we have:

$$\dot{s} = \Delta(y, \delta, t) - (\bar{\Delta} + \kappa_1)\,\text{sign}(s) - \sigma_1 s \tag{12}$$

Considering the Lyapunov function $V_1 = 0.5s^2$, then we see that:

$$\dot{V}_1 = s\dot{s} \tag{13}$$

By substituting Equation (12) into Equation (13), we obtain:

$$\begin{aligned} \dot{V}_1 &= \Delta(y, \delta, t)s - \bar{\Delta}|s| - \kappa_1|s| - \sigma_1 s^2 \\ &\leq -\kappa_1|s| - \sigma_1 s^2 \leq 0 \end{aligned} \tag{14}$$

Obviously, $\dot{V}_1 \leq 0$ and $V_1 \geqslant 0$. It means that the sliding variables $s, \dot{s}$ can reach the sliding manifold ($s \to 0$ and $\dot{s} \to 0$), and $\tilde{y}, \dot{\tilde{y}}$ can reach the equilibrium point.

Generally, it is hard to provide an exact function to the nominal controller by offering a mathematical model of dynamical uncertainties and disturbances. In addition, the designed controller (11) does not provide a prescribed controlled performance. This means that the maximum overshoot, convergence index, and steady-state error are not managed within a predefined domain. To overcome the mentioned challenges, a modified observer is applied to quickly estimate the approximate value of the uncertainty and exterior disturbance while the proposed controller is designed based on PPC to achieve the desired-prescribed control performance.

3.3. Design of a Disturbance Observer for Magnetic Levitation Systems

In this section, a modified TOSMO is designed to estimate the whole interior uncertainty and exterior disturbance. This observer is designed based on the modification of TOSMO [32] to improve the convergence speed of TOSMO as follows:

$$\begin{cases} \dot{\hat{y}}_1 = \pi_1 [\bar{y}_1]^{\frac{2}{3}} + \rho \bar{y}_1 + \hat{y}_2 \\ \dot{\hat{y}}_2 = g - \frac{\beta}{y_1^2} U^2 + \pi_2 [\bar{y}_1]^{\frac{1}{3}} \\ \quad + \rho(\dot{\hat{y}}_1 - \hat{y}_2) + \hat{\Delta} \\ \dot{\hat{\Delta}} = \pi_3 sign(\bar{y}_1) \end{cases} \tag{15}$$

where $\hat{y}_1$, $\hat{y}_2$, and $\hat{\Delta}$ are the estimated value of y_1, y_2, and $\Delta(y, \delta, t)$, respectively. $\pi_i (i = 1, 2, 3)$ represents observer gain which is selected as [33], and ρ is a design positive constant. By selecting the suitable design parameter ρ, the convergence rate of the modified TOSMO (15) can be improved immensely.

We can obtain the estimation error by subtracting Equation (15) from Equation (5) as follows:

$$\begin{cases} \dot{\bar{y}}_1 = -\pi_1 [\bar{y}_1]^{\frac{2}{3}} - \rho \bar{y}_1 + \bar{y}_2 \\ \dot{\bar{y}}_2 = -\pi_2 [\bar{y}_1]^{\frac{1}{3}} - \rho \pi_1 [\bar{y}_1]^{\frac{2}{3}} - \rho^2 \bar{y}_1 \\ \quad + \Delta(y, \delta, t) - \hat{\Delta} \\ \dot{\hat{\Delta}} = \pi_3 sign(\bar{y}_1) \end{cases} \tag{16}$$

where $\bar{y}_1 = y_1 - \hat{y}_1$, $\bar{y}_2 = y_2 - \hat{y}_2$. Define $X = \bar{y}_2 - \rho \bar{y}_1$ and use Equation (16), therefore,

$$\begin{aligned} \dot{X} &= \dot{\bar{y}}_2 - \rho \dot{\bar{y}}_1 \\ &= -\pi_2 [\bar{y}_1]^{\frac{1}{3}} - \rho \pi_1 [\bar{y}_1]^{\frac{2}{3}} - \rho^2 \bar{y}_1 \\ &\quad + \Delta(y, \delta, t) - \hat{\Delta} - \rho(-\pi_1 [\bar{y}_1]^{\frac{2}{3}} - \rho \bar{y}_1 + \bar{y}_2) \\ &= -\pi_2 [\bar{y}_1]^{\frac{1}{3}} + \Delta(y, \delta, t) - \hat{\Delta} - \rho \bar{y}_2 \end{aligned} \tag{17}$$

Using Equation (17), Equation (16) can be rewritten in form of the TOSMO as follows:

$$\begin{cases} \dot{\bar{y}}_1 = -\pi_1 [\bar{y}_1]^{\frac{2}{3}} + X \\ \dot{X} = -\pi_2 [\bar{y}_1]^{\frac{1}{3}} + Z \\ \dot{Z} = -\pi_3 sign(\bar{y}_1) + \dot{\Sigma} \end{cases} \tag{18}$$

where $\Sigma = \Delta(y, \delta, t) - \rho \bar{y}_2$ and $Z = \Sigma - \hat{\Delta}$. Suppose that $\left| \frac{d\Sigma}{dt} \right| \leq \bar{\Sigma}$.

Selecting the positive Lyapunov function and performing the proof process like [33], we concluded that the TOSMO (18) is stable and the differentiators including $\bar{y}_1$, X, and Z reach zero in finite time. Consequently, the proposed observer (15) is finite-time stable, and the estimate errors reach zero in finite time.

3.4. Prescribed Performance Control

Remark 1. *Most of the existing studies of PPC [34–39] use only one prescribed performance function (PPF) to generate specified performance boundaries, for example [34], $p(t)$ is a PPF, the upper boundary can be $p(t)$, and the lower boundary can be $-Lp(t)$ $(0 < L < 1)$. We can clearly see that generating boundaries from a PPF will have disadvantages, such as the lower boundary will be L times smaller than the upper boundary at steady-state, which makes the operating area of specified performance be scaled down over a specified static error value (e.g., p_∞ is the specified error value of tracking error at steady-state, the upper boundary value specified as p_∞ and the lower boundary value specified as $-Lp_\infty$ at steady-state). As a result, the upper and lower boundaries* *will not be symmetrical about zero at the steady state, and this leads to the fact that, although the transformed error equals zero, the tracking error deviates from zero. Using a ratio of a PPF to create the lower boundary makes it more difficult to choose the error transformation function (ETF) (due to the unsymmetrical about zero of the boundaries at steady-state). In addition, some ETFs [34,40–42] have a singularity problem, which seriously affects the operation of the real system.*

In our paper, we propose two separate PPFs to assign each of the upper and lower bounds of the prespecified performance for tracking error, in which one PPF limits the convergence rate and steady-state error, and the other PPF limits the overshoot and steady-state error on the other side. Setting the same allowable range of tracking error at the steady state of two PPFs makes the specified performance space at the steady-state larger than in the traditional way. Moreover, the steady-state error boundaries will be symmetric to zero, so when the transformed error is zero, the tracking error is also zero. ETFs can be designed more easily using the above design. In addition, we design an ETF that does not suffer from singularity problems. Consequently, the stated drawbacks in Remark 1 have been overcome.

From the objective of the proposed controller, the prescribed performance indicates that the tracking error $\tilde{y}$ is confined within a preset region as follows:

$$-p_L(t) < \tilde{y}\,\mathrm{sign}(\tilde{y}_0) < p_U(t) \tag{19}$$

with

$$\begin{cases} p_U(t) = (p_0 - p_\infty)exp(-rt) + p_\infty \\ p_L(t) = (p_1 - p_\infty)exp(-rt) + p_\infty \end{cases},$$

where $\tilde{y}_0$ is the initial tracking error, $p_U(t)$ and $p_L(t)$ are the PPFs. The $p_U(t)$ and $p_L(t)$ are defined as [34]: $p_U(t)$ and $p_L(t) : R_+ \to R_+$ are smoothly, positive, and decreasing functions, respectively satisfying $\lim_{t\to\infty} p_U(t) = p_\infty > 0$, $\lim_{t\to\infty} p_L(t) = p_\infty > 0$, and R_+ is a set of positive real numbers. $p_0 > |\tilde{y}_0| > 0, p_0 \geqslant p_1 \geqslant p_\infty$, r are positive constants used to tune the specified performance region. Figure 2 illustrates the prescribed tracking error behavior definition.

Remark 2. *It is noted that, when the sign of the initial error changes, the position of the lower and upper bounds will be reversed through the signum function. The PPFs $p_U(t)$ and $p_L(t)$ are assigned to the upper and lower bounds of the specified performance domain. $p_U(t)$ limits the convergence rate and the maximum allowable size of the tracking error $\tilde{y}$ at the steady state of the upper boundary. $p_L(t)$ limits the maximum allowable boundary of the overshoot and the maximum allowable size of the tracking error $\tilde{y}$ at the steady state of the lower boundary. The constant p_∞ represents the maximum allowable size of the tracking error $\tilde{y}$ at the steady-state, the maximum overshoot of $\tilde{y}$ is prescribed as less than p_1, and the constraint on the convergence rate of $\tilde{y}$ depends on the decreasing rate of $p_U(t)$ which is adjusted by r. Hence, the appropriate selection of the PPFs ($p_U(t)$, $p_L(t)$), which imposes the output trajectory of the system.*

In order to assure that the prescribed performance is maintained, an ETF is used to convert the constrained error dynamics into the equivalent unconstrained dynamics. More specifically, we define:

$$\tilde{y} = p(t)T(\tilde{z}) \tag{20}$$

where $\tilde{z}$ is a transformed error, and

$$p(t) = \begin{cases} p_U(t) & \text{if } \mathrm{sign}(\tilde{y}.\tilde{y}(0)) > 0 \\ p_L(t) & \text{if } \mathrm{sign}(\tilde{y}.\tilde{y}(0)) < 0 \end{cases},$$

and $T(\tilde{z})$ is an ETF that possesses the following properties:

- $T(\tilde{z})$ is a smooth and strictly increasing function;

- $-1 < T(\tilde{z}) < 1$;
- $T(\tilde{z}) = 0$ if $\tilde{z} = 0$;
- $\begin{cases} \lim\limits_{\tilde{z}\to-\infty} T(\tilde{z}) = -1 \\ \lim\limits_{\tilde{z}\to+\infty} T(\tilde{z}) = 1 \end{cases}$.

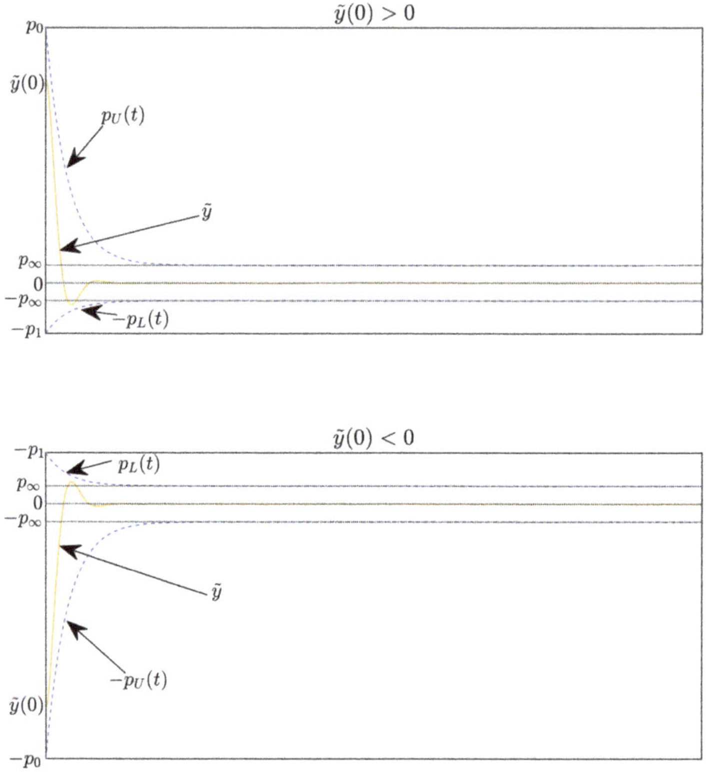

Figure 2. The prescribed tracking error behavior definition.

Owing to the definition of the PPF, when $\tilde{y}(0) > 0$ and $\tilde{y} > 0$, we have $0 < T(\tilde{z}) < 1$ and $p_U(t) > 0$. Hence,

$$0 < p_U(t)T(\tilde{z}) < p_U(t) \tag{21}$$

When $\tilde{y}(0) > 0$ and $\tilde{y} < 0$, we have $-1 < T(\tilde{z}) < 0$ and $p_L(t) > 0$. Hence,

$$-p_L(t) < p_L(t)T(\tilde{z}) < 0 \tag{22}$$

From Equations (20)–(22) whenever $\tilde{y}(0) > 0$, we can obtain

$$-p_L(t) < \tilde{y} < p_U(t) \tag{23}$$

Otherwise, when $\tilde{y}(0) < 0$ and $\tilde{y} < 0$, we have

$$-p_U(t) < p_U(t)T(\tilde{z}) < 0 \tag{24}$$

and, if $\tilde{y}(0) < 0$ and $\tilde{y} > 0$, when we have

$$0 < p_L(t)T(\tilde{z}) < p_L(t) \tag{25}$$

From Equations (20), (24) and (25) whenever $\tilde{y}(0) < 0$, we can obtain

$$-p_U(t) < \tilde{y} < p_L(t) \tag{26}$$

From Equations (23) and (26), we completely can achieve Equation (19), which represents a mathematical interpretation of the control objective to achieve the prescribed tracking error behavior over transient and steady-state scenarios.

The ETF is designed as

$$T(\tilde{z}) = \frac{2}{\pi}\arctan(\tilde{z}) \tag{27}$$

Based on Equations (20) and (27), the transformed error $\tilde{z}$ can be calculated as follows:

$$\tilde{z} = \tan\left(\frac{\pi\tilde{y}}{2p(t)}\right) \tag{28}$$

Taking time derivative of $\arctan(\tilde{z})$, we have

$$(\arctan(\tilde{z}))' = \frac{\dot{\tilde{z}}}{1+\tilde{z}^2} \tag{29}$$

where $\dot{\tilde{z}}$ is the time derivative of $\tilde{z}$.

Taking the time derivative of Equation (20) and using Equations (27) and (29), we have

$$\begin{aligned}\dot{\tilde{y}} &= \dot{p}(t)T(\tilde{z}) + p(t)\dot{T}(\tilde{z}) \\ &= \dot{p}(t)\frac{2}{\pi}\arctan(\tilde{z}) + p(t)\frac{2}{\pi}\frac{\dot{\tilde{z}}}{1+\tilde{z}^2}\end{aligned} \tag{30}$$

where $\dot{p}(t) = \begin{cases}\dot{p}_U(t) \text{ if } \mathrm{sign}(\tilde{y}.\tilde{y}(0)) > 0 \\ \dot{p}_L(t) \text{ if } \mathrm{sign}(\tilde{y}.\tilde{y}(0)) < 0\end{cases}$.

From Equation (30), the time derivative of transformed error $\tilde{z}$ is obtained:

$$\dot{\tilde{z}} = \frac{\pi(1+\tilde{z}^2)}{2p(t)}\left(\dot{\tilde{y}} - \frac{2\dot{p}(t)}{\pi}\arctan(\tilde{z})\right) \tag{31}$$

Taking the second-order time derivative of $\arctan(\tilde{z})$, we have

$$(\arctan(\tilde{z}))'' = \frac{\ddot{\tilde{z}}(1+\tilde{z}^2) - 2\tilde{z}\dot{\tilde{z}}^2}{(1+\tilde{z}^2)^2} \tag{32}$$

where $\ddot{\tilde{z}}$ is the second-order time derivative of $\tilde{z}$.

Taking the second-order time derivative of Equation (20) and using Equations (27), (29) and (32), we have

$$\begin{aligned}\ddot{\tilde{y}} &= \left(\dot{p}(t)T(\tilde{z}) + p(t)\dot{T}(\tilde{z})\right)' \\ &= \ddot{p}(t)T(\tilde{z}) + 2\dot{p}(t)\dot{T}(\tilde{z}) + p(t)\ddot{T}(\tilde{z}) \\ &= \frac{2}{\pi}\left(\ddot{p}(t)\arctan(\tilde{z}) + \frac{2\dot{p}(t)\dot{\tilde{z}}}{1+\tilde{z}^2} - \frac{2p(t)\tilde{z}\dot{\tilde{z}}^2}{(1+\tilde{z}^2)^2}\right) \\ &+ \frac{2p(t)}{\pi}\frac{\ddot{\tilde{z}}}{(1+\tilde{z}^2)}\end{aligned} \tag{33}$$

where $\ddot{p}(t) = \begin{cases}\ddot{p}_U(t) \; if \; \mathrm{sign}(\tilde{y}.\tilde{y}(0)) > 0 \\ \ddot{p}_L(t) \; if \; \mathrm{sign}(\tilde{y}.\tilde{y}(0)) < 0\end{cases}$.

From Equation (33), we can derive the second-order time derivative of transformed error as follows:

$$\ddot{\tilde{z}} = A(\ddot{\tilde{y}} - N) \tag{34}$$

where $A = \frac{\pi(1+\tilde{z}^2)}{2p(t)} > 0$, $N = \frac{2}{\pi}\left(\ddot{p}(t)\arctan(\tilde{z}) + \frac{2\dot{p}(t)\dot{\tilde{z}}}{1+\tilde{z}^2} - \frac{2p(t)\tilde{z}\dot{\tilde{z}}^2}{(1+\tilde{z}^2)^2}\right)$.

3.5. Proposed Controller Design

For system (5), the sliding manifold is constructed by using the transformed position error and the transformed velocity error as follows:

$$s = \dot{\tilde{z}} + \lambda^* sig(\tilde{z})^{\alpha} + \omega^* sig(\tilde{z})^{\beta} \tag{35}$$

where $\lambda^* = \frac{2\lambda}{1+e^{-\eta(|\tilde{z}|-\gamma)}}$, $\omega^* = \frac{2\omega}{1+e^{\vartheta(|\tilde{z}|-\gamma)}}$, $\lambda > 0, \eta > 0, \omega > 0, \sigma > 0$, $\alpha > 1$, $0 < \beta < 1$, $\gamma = (\omega/\lambda)^{1/(\alpha-\beta)}$.

The time derivative of the selected sliding mode surface is:

$$\begin{aligned}\dot{s} = \ddot{\tilde{z}} + \frac{2\lambda_j\alpha}{1+e^{-\eta(|\tilde{z}|-\gamma)}}|\tilde{z}|^{\alpha-1}\dot{\tilde{z}} + \frac{2\lambda\eta e^{-\eta(|\tilde{z}|-\gamma)}}{(1+e^{-\eta(|\tilde{z}|-\gamma)})^2}|\tilde{z}|^{\alpha}\dot{\tilde{z}} \\ + \frac{2\omega\beta}{1+e^{\vartheta(|\tilde{z}|-\gamma)}}|\tilde{z}|^{\beta-1}\dot{\tilde{z}} - \frac{2\omega\vartheta e^{\vartheta(|\tilde{z}|-\gamma)}}{(1+e^{\vartheta(|\tilde{z}|-\gamma)})^2}|\tilde{z}|^{\beta}\dot{\tilde{z}},\end{aligned} \tag{36}$$

Equation (36) is rewritten as:

$$\dot{s} = \ddot{\tilde{z}} + M_{\tilde{z}} \tag{37}$$

where $M_{\tilde{z}} = \frac{2\lambda_j\alpha}{1+e^{-\eta(|\tilde{z}|-\gamma)}}|\tilde{z}|^{\alpha-1}\dot{\tilde{z}} + \frac{2\lambda\eta e^{-\eta(|\tilde{z}|-\gamma)}}{(1+e^{-\eta(|\tilde{z}|-\gamma)})^2}|\tilde{z}|^{\alpha}\dot{\tilde{z}} + \frac{2\omega\beta}{1+e^{\vartheta(|\tilde{z}|-\gamma)}}|\tilde{z}|^{\beta-1}\dot{\tilde{z}} - \frac{2\omega\vartheta e^{\vartheta(|\tilde{z}|-\gamma)}}{(1+e^{\vartheta(|\tilde{z}|-\gamma)})^2}|\tilde{z}|^{\beta}\dot{\tilde{z}}$.

Substituting Equations (5) and (34) into Equation (37) yields:

$$\dot{s} = A\left(g - \frac{\hat{\mu}}{y_1^2}U^2 + \Delta(y,\delta,t) - \ddot{y}_r - N\right) + M_{\tilde{z}} \tag{38}$$

The proposed controller is designed to obtain the control object as follows:

$$U = \sqrt{\frac{y_1^2}{\hat{\mu}}(u_{eq} + u_{ob} + u_r)}. \tag{39}$$

where the elements of U are designed as:

$$\begin{cases} u_{eq} = g - \ddot{y}_r - N + A^{-1}M_{\tilde{z}} \\ u_{ob} = \hat{\Delta} \\ u_r = \sigma_2 s + (\bar{\delta} + \kappa_2)\,\mathrm{sign}(s). \end{cases}$$

where $\bar{\delta} > 0$ is the bounded value of the estimation error of the disturbance observer ($\bar{\delta} \geqslant |\Delta(y,\delta,t) - \hat{\Delta}|$), and κ_2 and σ_2 are positive constants.

The block diagram of the proposed control system is shown in Figure 3.

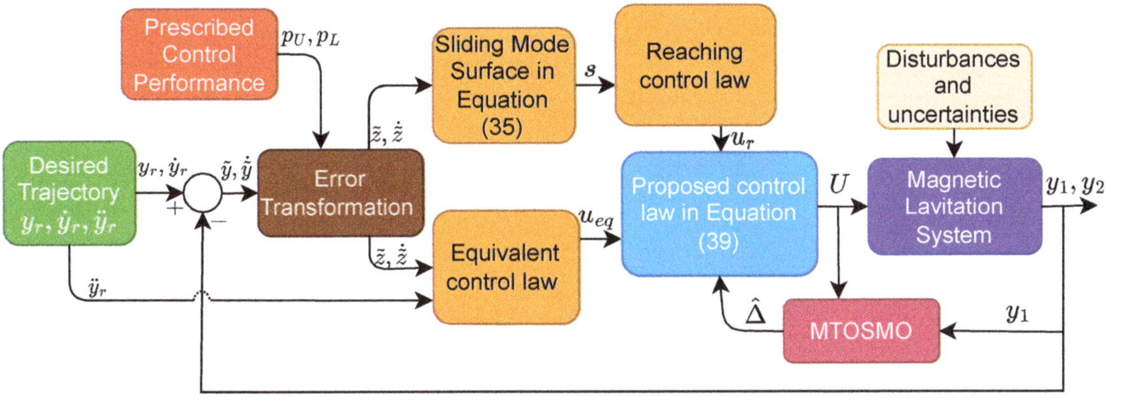

Figure 3. The block diagram of the proposed control system.

The following theorem summarizes the control design procedure.

Theorem 1. *For the magnetic levitation system presented in Equation (5), if the control input signal is designed as (39), then the system (5) is finite-time stable. Furthermore, the maximum overshoot and steady state of position tracking error are guaranteed in prescribed control performance.*

Proof of Theorem 1. Applying the control signal system (39) to (38) gains:

$$\dot{s} = A(\Delta(y,\delta,t) - \hat{\Delta} - (\bar{\delta} + \kappa_2)\,\text{sign}(s) - \sigma_2 s) \tag{40}$$

Considering the following Lyapunov function $V_2 = \frac{p_0}{\pi}s^2$, then we see that:

$$\dot{V}_2 = \frac{2p_0}{\pi}s\dot{s} \tag{41}$$

By substituting Equation (40) into Equation (41), we obtain:

$$\begin{aligned} \dot{V}_2 &= \frac{2p_0}{\pi}sA\begin{pmatrix}\Delta(y,\delta,t) - \hat{\Delta} \\ -(\bar{\delta}+\kappa_2)\,\text{sign}(s) - \sigma_2 s\end{pmatrix} \\ &\leqslant \frac{2p_0}{\pi}sA(\bar{\delta} - (\bar{\delta}+\kappa_2)\,\text{sign}(s) - \sigma_2 s) \\ &\leqslant -\frac{2p_0}{\pi}sA_{\min}(\kappa_2\,\text{sign}(s) + \sigma_2 s) \\ &\leqslant -\kappa_2|s| - \sigma_2 s^2 \\ &\leqslant -\kappa_2\sqrt{\frac{\pi}{p_0}}|V_2|^{1/2} - \frac{\pi\sigma_2}{p_0}V_2 \leqslant 0 \end{aligned} \tag{42}$$

From Equations (41) and (42), we can conclude that system (5) is finite-time stable according to [43]. Thus, the sliding variables $s, \dot{s}$ can reach the sliding manifold in finite-time ($s \to 0$ and $\dot{s} \to 0$) that means $\tilde{z}, \dot{\tilde{z}}$ converge to its equilibrium point in finite time. Consequently, the position tracking error $\tilde{y}$ is converged to the equilibrium point in finite time and guaranteed in prescribed control performance. The proof is completed. □

4. Simulation and Experimental Results

In this section, two examples include a simulation and an experiment on the laboratory magnetic levitation model to demonstrate the improved performance of the developed control system.

4.1. Simulation Results

The MLS is established according to [4]. The considered system parameters are stated in Table 1. The maximum control voltage is $U_{\max} < 5\text{V}$, and the sampling time is 1.10^{-3} s.

Table 1. System parameters.

System Parameters	Value	Unit
g	9.81	m/s^2
m	0.02	kg
λ	$2.48315625 \times 10^{-5}$	Nm^2/A^2
K	1.05	A/V
$\hat{\mu}$	0.00136487	$(\text{N.m}^2/\text{kg.V}^2)$

A magnetic ball started in a certain position and followed a prescribed trajectory at the start of the simulation. The desired trajectory as sinusoidal is planned below:

$$y_r = 15 + 3\sin(0.2\pi t) \quad (\text{mm}) \tag{43}$$

With the assumed disturbance $\delta(t) = 2\sin(t)$, the upper bound of uncertain terms is defined by:

$$|\Delta(y, \delta, t)| \leq \frac{|\mu - \hat{\mu}|}{y_{\min}^2} U_{\max}^2 + \delta_{max}(t) = 2.7 \tag{44}$$

where $y_{r\min} = 12(\text{mm})$, and the initial value of $y_0 = 26(\text{mm})$.

In addition, to check the influence of sensor measurement noise, an external noise in the form of Gaussian random noise with the variance of 0.0001 is added to the velocity sensor. Thus, the simulation case will be similar to the real system where there is always noise from the measuring device.

The root-mean-square tracking error (RMSTE) is calculated as follows:

$$\text{RMSTE} = \sqrt{\frac{1}{N}\sum_{i=1}^{N} (y_{ri} - y_i)^2} \tag{45}$$

in which N is the number of samples to be taken into account in this calculation. y_{ri} and y_i are respectively the desired trajectory and the real trajectory at the time index ith.

To facilitate the evaluation of the accuracy of the controllers, the RMSTE is calculated according to Equation (45), in the time interval after convergence (3rd to 30th second).

In the simulation example, the control parameters of the four controllers were selected as in Table 2.

Table 2. The control parameters.

Controller	Symbol	Value
PID	K_p, K_i, K_d	300, 100, 10
SMC	$c, \sigma, \bar{\Delta} + \kappa$	$20, 50, 3.8$
GFTSMC	$\lambda, \omega, \eta, \vartheta, \alpha, \beta$	$10, 10, 1.1, 2.2, 1.1, 0.8$
	$\sigma_1, \bar{\Delta} + \kappa_1$	$20, 50, 3.8$
Proposed Method	p_0, p_1, p_∞, r	$0.023, 0.006, 0.0015, 3$
	$\lambda, \omega, \eta, \vartheta, \alpha, \beta$	$10, 10, 1.1, 2.2, 1.1, 0.8$
	$\sigma_2, \bar{\delta} + \kappa_2$	$0.13, 0.1$
	$\pi_1, \pi_2, \pi_3, \rho$	$5.45, 3.67, 6.6, 100$

We compared the simulated control performance of the four controllers, PID, SMC, GFTSMC (11), and the proposed controller, to determine which is the most effective.

The PID controller has control voltage as follows:

$$U = K_p\tilde{y} + K_i\int \tilde{y} + K_d\dot{\tilde{y}} \tag{46}$$

where K_p, K_i, K_d are control gains.

The SMC controller has control voltage as follows:

$$U = \sqrt{\frac{y_1^2}{\hat{\mu}}\left(g + M_{\tilde{y}} - \ddot{y}_r + \sigma s + (\bar{\Delta} + \kappa)sign(s)\right)} \tag{47}$$

where $s = \dot{\tilde{y}} + c\tilde{y}$ is a linear sliding surface, $c > 0$. κ and σ are positive constants.

The simulation process and the result evaluation are carried out in four steps:

- Step 1: simulates and evaluates the approximation ability of the proposed observer through a comparison between its approximation ability and the conventional TOSMO;
- Step 2: investigates the management of the terms of the proposed PPC including maximum overshoot and steady-state of the controlled errors;
- Step 3: compares the tracking accuracy, maximum overshoot and steady-state of the controlled errors among the four control methods through figures plotted from MATLAB and RMS methods;
- Step 4: considers the chattering behavior that appeared in the control signal of the four methods.

To compare the estimation performance of the TOSMO and MTOSMO, we use the SMC (47) to control the MLS, while the two observers are run in parallel with SMC to compare their outputs as shown in Figure 4. The TOSMO is designed as [17]:

$$\begin{cases} \dot{\hat{y}}_1 = \pi_1[\tilde{y}_1]^{\frac{2}{3}} + \hat{y}_2 \\ \dot{\hat{y}}_2 = g - \frac{\hat{\mu}}{y_1^2}U^2 + \pi_2[\tilde{y}_1]^{\frac{1}{3}} + \hat{\Delta} \\ \dot{\hat{\Delta}} = \pi_3 sign(\tilde{y}_1) \end{cases} \tag{48}$$

where $\hat{y}_1$, $\hat{y}_2$, and $\hat{\Delta}$ are the estimated value of y_1, y_2, $\Delta(y,\delta,t)$, respectively. $\pi_i(i = 1,2,3)$ represents observer gains. The parameters of SMC and MTOSMO are selected in Table 2, and the parameters of TOSMO are set the same as MTOSMO.

From Figure 5, the uncertain components can be approximated accurately by both observers. However, the proposed TOSMO has a faster convergence speed than the traditional TOSMO. This is absolutely necessary to avoid causing a delay to the control system. Therefore, it plays a role in improving control performance.

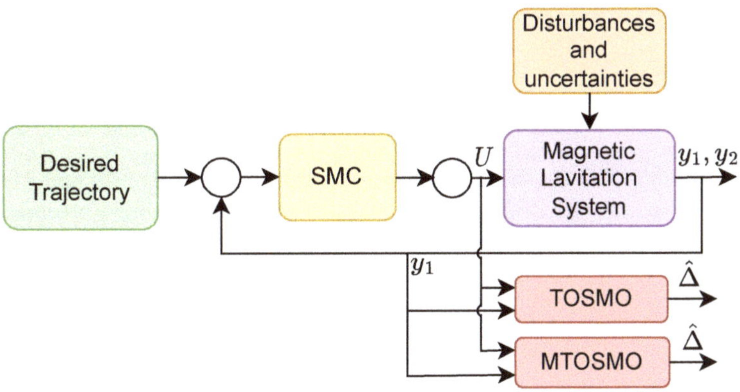

Figure 4. The setting for comparing two observers.

Looking at Figures 6 and 7, it is clear that the proposed controller has the smallest maximum overshoot and the smallest controlled errors at a steady state. Both of these metrics can be predefined by adjusting the design parameters of the PPFs in Equation (19). In contrast, we cannot manage the maximum overshoot as well as the tracking error at the steady state of the controlled magnetic ball through the remaining three controllers. Specifically, the PID controller can not satisfy both stated terms. In addition, comparing the boundary values of the proposed PPF with the boundary values of the existing PPFs [36,38–40], we found that the boundary values of conventional PPFs are designed largely up to 10^{-2}. It is noted that the existing controllers can almost obtain this accuracy level. Therefore, it does not see clearly the management of the prescribed performance of the PPCs, while the boundary values of the proposed PPF are designed smaller than that of conventional PPFs, and its value is 10^{-3}.

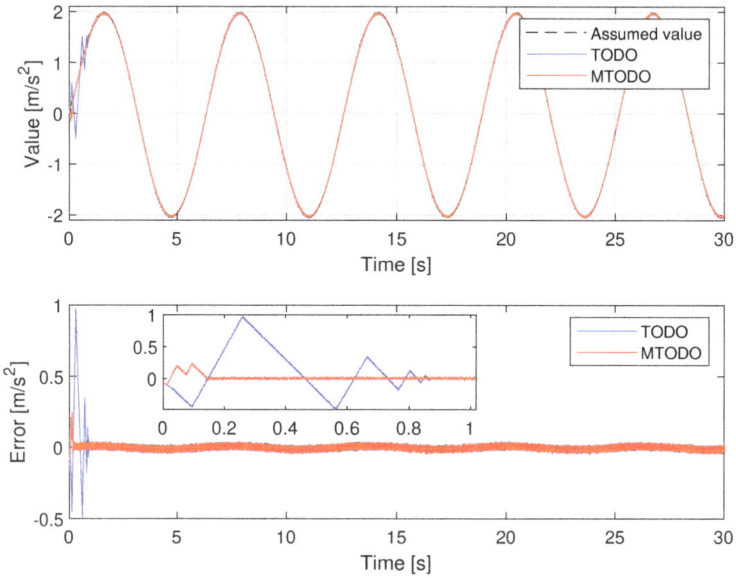

Figure 5. Time evolution of observer output.

The simulation tracking performance of the controlled ball from four different control methods is illustrated in Figures 6 and 7. It can be seen clearly in Figure 6 that the trajectory deviation of the magnetic ball controlled by the PID versus the desired trajectory is the largest, and the GFTSMC provides a smaller trajectory deviation of the ball than that of the PID and the SMC, while the proposed controller provides the smallest trajectory deviation of the ball among the four methods. We investigate in detail the tracking accuracy illustrated in Figures 7 and Table 3 and recognize that the proposed controller controls the ball with the smallest steady-state errors, RMSTE $= 4.2119 \times 10^{-5}$, and the smallest maximum overshoot that satisfied the predefined prescribed performance; the SMC and the GFTSMC control the ball with seemingly equivalent performance, the control errors of these two methods are RMSTE $= 3.1097 \times 10^{-4}$ and RMSTE $= 1.979 \times 10^{-4}$, respectively; the PID controls the ball with the largest control error, RMSTE $= 1.0034 \times 10^{-3}$ and the largest maximum overshoot that do not satisfy the predefined operation domain.

Table 3. RMSTE using the four separate controllers.

Controller	RMSTE
PID	1.0034×10^{-3}
SMC	3.1097×10^{-4}
GFTSMC	1.979×10^{-4}
Proposed Method	4.2119×10^{-5}

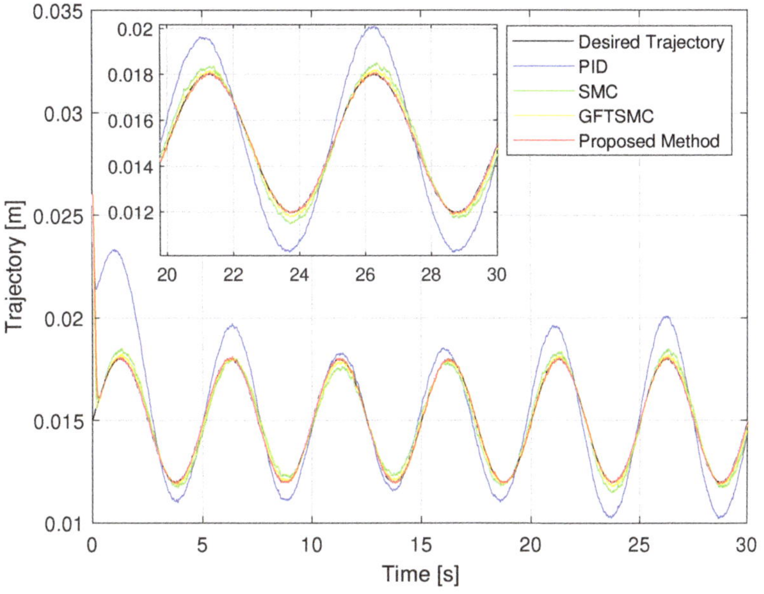

Figure 6. The controlled ball trajectory under four controllers: planned trajectory and actual trajectory.

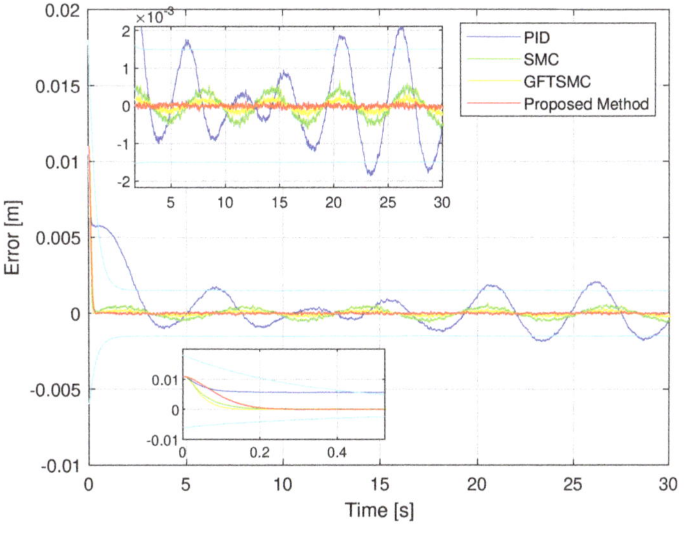

Figure 7. Time evolution of the trajectory errors of the controlled ball using four control methods.

Regarding the problem of oscillations in the control voltage, all four control methods appear oscillation phenomena. Due to the influence of sensor measurement noise, even a linear controller like a PID also has oscillations. The fluctuation amplitude of the control voltage of the SMC and the GFTSMC is almost equivalent because both methods apply the same sliding value, which is chosen to be greater than or equal to the upper bound of the uncertain components, $\bar{\Delta} + \kappa = 3.8$, to compensate for the effects of the uncertain elements. As shown in Figure 8, the suggested TOSMO can estimate accurately and quickly the assumed uncertain terms.

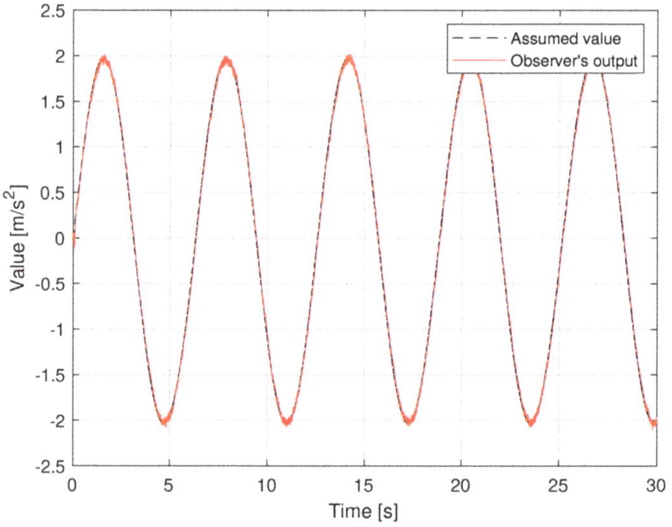

Figure 8. Time evolution of observer output versus assumed disturbance.

Thanks to the TOSMO's accurate information, the developed controller applied a smaller sliding value than that of the SMC and the GFTSMC. Therefore, the fluctuation amplitude of its control voltage is smaller than that of the two remaining methods as shown in Figure 9.

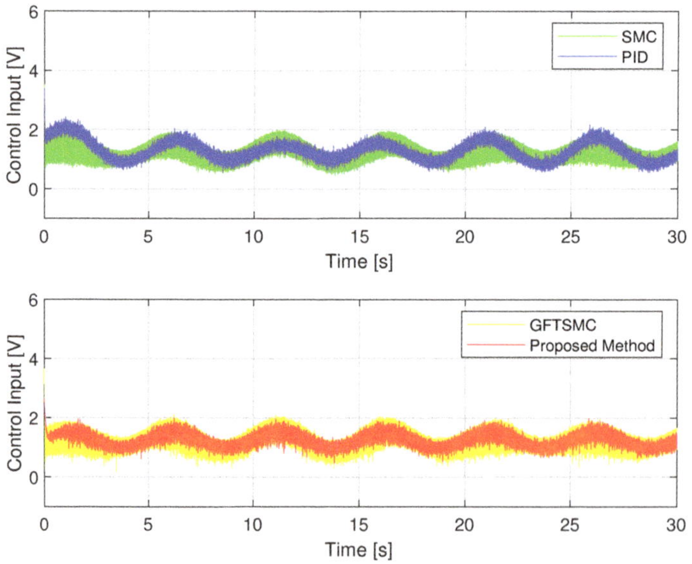

Figure 9. Time evolution of control voltage from four control methods.

Simulation results show that the developed controller has the best control performance among the compared control algorithms.

4.2. Experimental Results

In order to test the effectiveness of the proposed control algorithm in the practical control system, an experiment on MLS was implemented.

The experimental MLS was produced by Feedback Instrument, as configured in Figure 10. The experimental system includes a mechanical unit (model Feedback 33-210) and an analog control interface (Feedback model 33-301). A PCI1711 I/O card was inserted into a PCI computer slot, and then it was connected to the feedback SCSI adapter box by the SCSI cable. The control programming was implemented by using MATLAB/Simulink, Real-Time Workshop (RTW), Microsoft Visual C++ Professional, Control Toolbox, and Real-Time Windows Target.

For more information on installing this experimental system, readers can refer to it in [4].

The maximum control voltage is $U_{\max} < 5$ V, and the sampling time is 1.10^{-3} s.

The magnetic ball started in a specific position and followed a prescribed trajectory when the experiment began. There were two desired trajectories: a sinusoidal as Equation (43) and a rest-to-rest line with an upper value of 17 mm and lower value 12 mm. There was an assumption that the disturbance $\delta(t) = 2\sin(t)$ would affect the system.

The control parameters of the four separate controllers selected in Table 2 are also used for the experimental example.

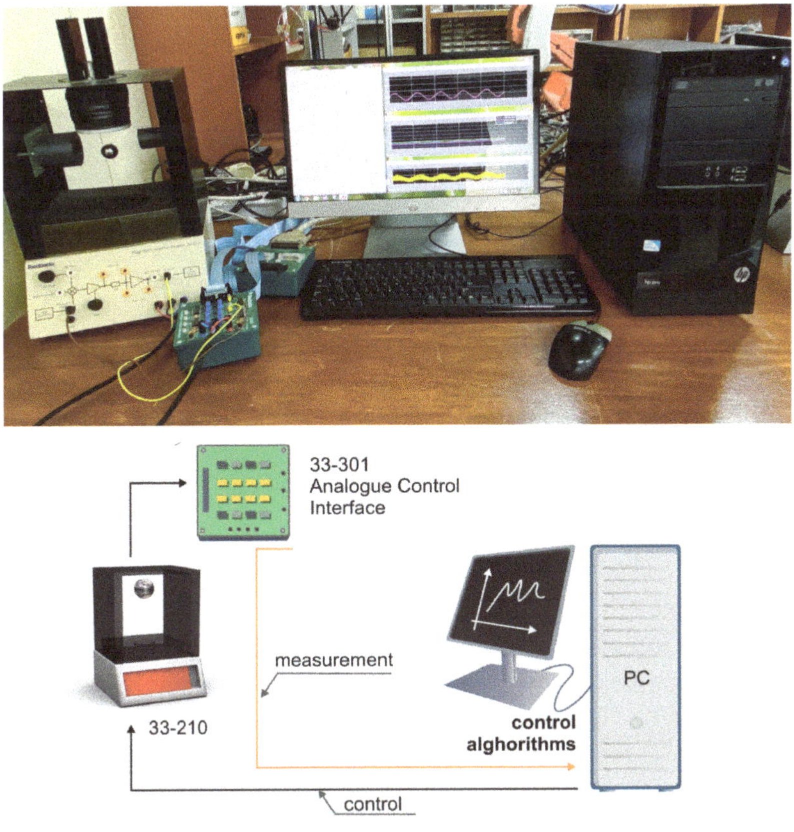

Figure 10. Configuration of experimental system.

The experimental performance provided by four control methods for a magnetic ball by tracking a sinusoidal and a rest-to-rest line is exhibited in Figures 11–13. The experiment process and results evaluation are carried out in three steps from step 2 to step 4 as in the simulation example.

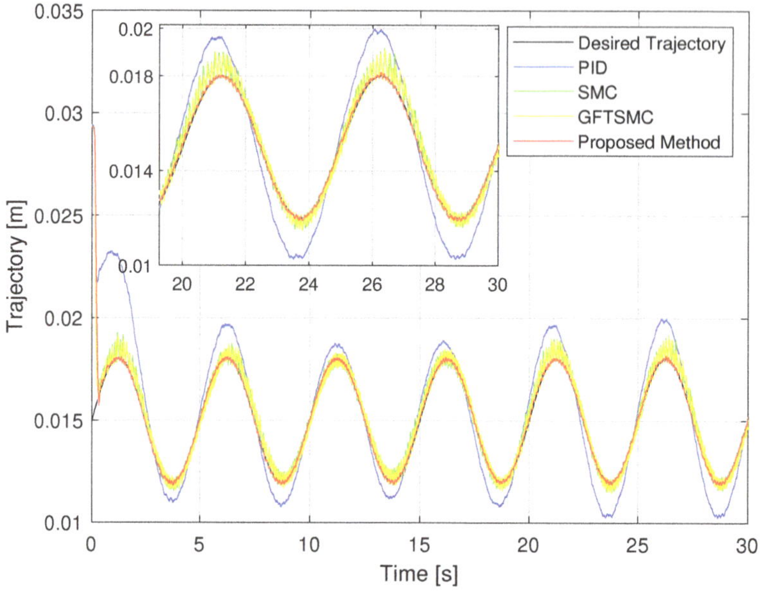

(**a**)

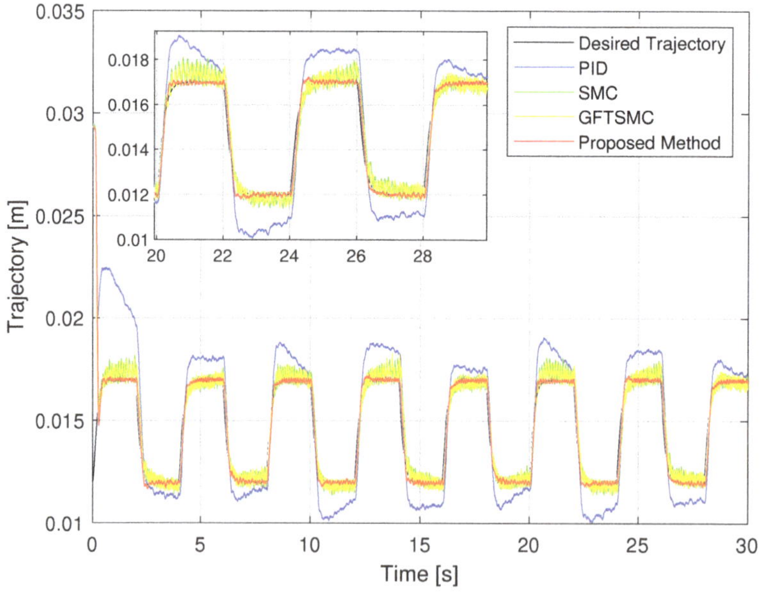

(**b**)

Figure 11. Comparison of the controlled ball trajectory with the desired trajectory using four separate controllers; (**a**) in case of sinusoidal orbit; (**b**) in case of a rest-to-rest line.

Figures 11 and 12 show that the proposed controller has the smallest maximum overshoot and the smallest controlled errors at a steady state, similar to the simulation results. Neither of these conditions can be satisfied by PID, hence the ball controlled by the PID can not track the reference trajectory well. The SMC and the GFTSMC cannot also obtain the controlled errors with small oscillations. It is seen that, at the 14th second, both SMC and GFTSMC do not satisfy a prescribed performance. Consequently, the ball controlled by the SMC and the GFTSMC will be vibrated around its operation point.

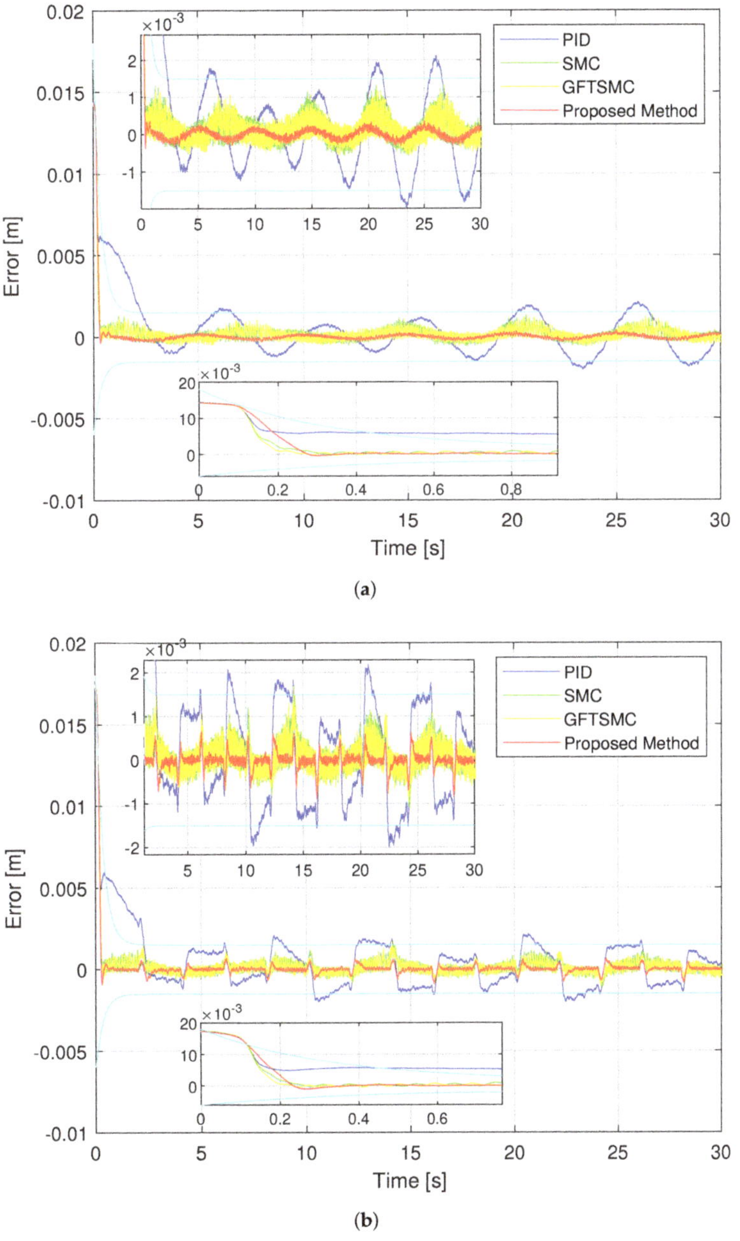

Figure 12. The trajectory errors of the controlled ball under four controllers; (**a**) in case of sinusoidal orbit; (**b**) in case of a rest-to-rest line.

As shown in Figure 11, the proposed controller provides the smallest trajectory disparity of the magnetic ball among the four methods, the GFTSMC provides a smaller trajectory disparity of the ball than the PID or the SMC, and the PID provides the largest trajectory disparity of the ball. From Figure 12 and Table 4, the proposed controller also achieves the smallest steady-state errors with two different types of orbits. Apparently, it provided a control efficiency that was within a pre-specified boundary. The SMC and the GFTSMC achieve control efficiency with equivalent accuracy. With two different types of orbits, the maximum overshoot of the PID does not satisfy the predefined operation domain. The PID achieves the worst accuracy with both trajectories.

The noise in sensor measurement appears to be influencing all four control methods, which cause oscillations in control voltage. With two different types of orbits, the fluctuation amplitude of the control voltage of each controller is similar to the simulation results, as shown in Figure 13. Figure 14 shows the time evolution of the proposed TOSMO's output.

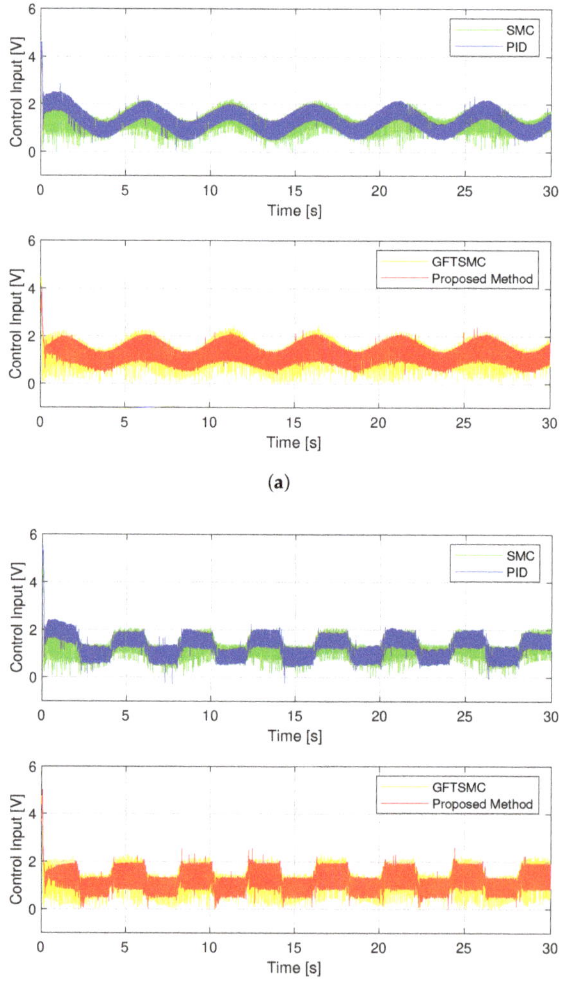

Figure 13. Control voltage of four controllers; (**a**) in case of sinusoidal orbit; (**b**) in case of a rest-to-rest line.

Table 4. RMSTE of four controllers.

Controller	RMSTE in Case of Sinusoidal Orbit	RMSTE in Case of a Rest-to-Rest Line
PID	1.0411×10^{-3}	1.1342×10^{-3}
SMC	3.6039×10^{-4}	3.7681×10^{-4}
GFTSMC	3.3052×10^{-4}	3.4805×10^{-4}
Proposed Method	1.2203×10^{-4}	1.9601×10^{-4}

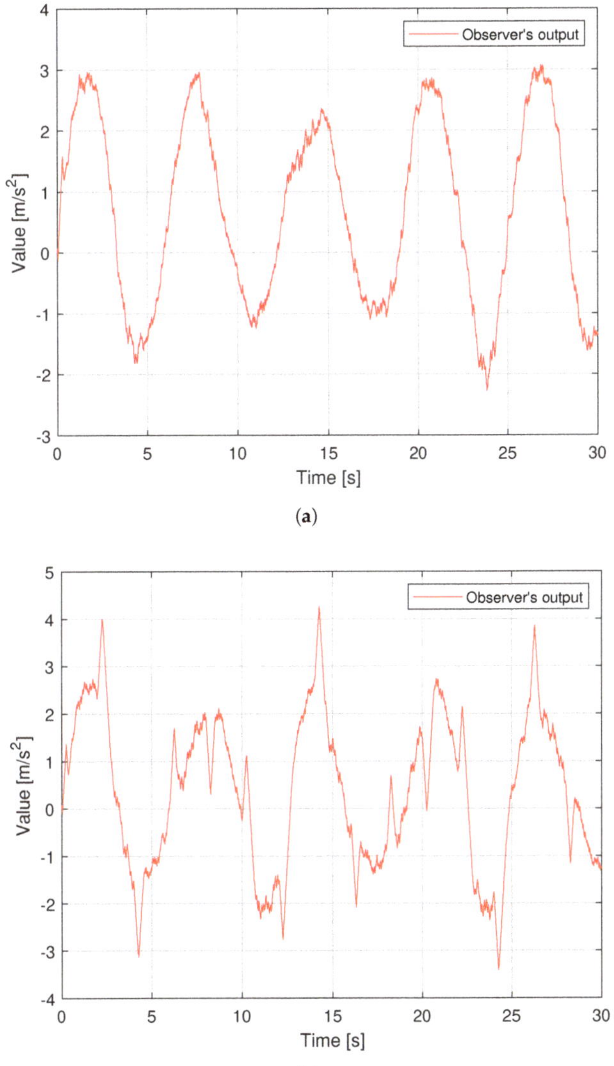

Figure 14. Time evolution of the MTOSMO's output; (**a**) in case of sinusoidal orbit; (**b**) in case of a rest-to-rest line.

In a comparison of the experimental performance, the developed controller also stands out as the best controller.

Remark 3. *From the simulation and experimental results, we can see that the tracking performance of the proposed controller is guaranteed within a prescribed performance in both cases. In the experiment, the tracking accuracy of the proposed controller has slightly reduced compared to the simulation case; however, it is still at a high tracking accuracy. The chattering behavior in the control signal of the proposed controller in the experimental case is not significantly increased compared to the simulation case. It is concluded that the proposed control method is effective to control MLSs.*

5. Some Remarkable Conclusions

We implemented a real-time PPC for MLSs subject to dynamical uncertainty and exterior perturbations. A modified function of GFTSMM based on the transformed errors of the proposed PPC was developed; hence, the tracking error variables quickly converge to the equilibrium point with the prescribed performance. Maximum overshoot and steady-state of the controlled errors have been prescribed in a predefined boundary. By using the designed observer, it is possible to know the approximate value of the entire uncertainty, which contributes to reducing chattering and improving control performance. The combination of the GFTSMC, the PPC, and the MTOSMO generated a novel PPC strategy ensuring a finite-time stable position of the controlled ball and the possibility of performing different orbit tracking missions with an impressive performance in the terms of tracking accuracy, fast convergence, stabilization, and chattering reduction in real time. With a simple design, the proposed strategy is suitable for real-time applications of MLSs. Mathematical proof using Lyapunov theory, a simulation, and an experimental example on a laboratory magnetic levitation model have both been used to demonstrate the proposed controller's stability and effectiveness.

Author Contributions: Conceptualization, methodology, validation, writing—original draft preparation, and writing—review and editing, T.N.T.; software, visualization, and resources, A.T.V.; supervision, funding acquisition, and project administration, H.-J.K.; formal analysis, investigation, and data curation, T.N.T., A.T.V., and H.-J.K. All authors have read and agreed to the published version of the manuscript.

Funding: This research was funded by the Ministry of Education (NRF-2019R1D1A3A03103528).

Institutional Review Board Statement: Not applicable.

Informed Consent Statement: Not applicable.

Data Availability Statement: The data sets generated and/or analyzed during the current study are available from the corresponding author on reasonable request.

Acknowledgments: This research was supported by the Basic Science Research Program through the National Research Foundation of Korea (NRF) funded by the Ministry of Education (NRF-2019R1D1A3A03103528).

Conflicts of Interest: The authors declare no conflict of interest.

References

1. Golob, M.; Tovornik, B. Modeling and control of the magnetic suspension system. *ISA Trans.* **2003**, *42*, 89–100. [CrossRef] [PubMed]
2. Wang, J.; Rong, J.; Yang, J. Adaptive Fixed-Time Position Precision Control for Magnetic Levitation Systems. *IEEE Trans. Autom. Sci. Eng.* **2022**, 1–12. doi: 10.1109/TASE.2022.3156943. [CrossRef]
3. Wang, J.; Zhao, L.; Yu, L. Adaptive terminal sliding mode control for magnetic levitation systems with enhanced disturbance compensation. *IEEE Trans. Ind. Electron.* **2020**, *68*, 756–766. [CrossRef]
4. Truong, T.N.; Vo, A.T.; Kang, H.J. Implementation of an Adaptive Neural Terminal Sliding Mode for Tracking Control of Magnetic Levitation Systems. *IEEE Access* **2020**, *8*, 206931–206941. [CrossRef]
5. Vo, A.T.; Truong, T.N.; Kang, H.J. A Novel Fixed-Time Control Algorithm for Trajectory Tracking Control of Uncertain Magnetic Levitation Systems. *IEEE Access* **2021**, *9*, 47698–47712. [CrossRef]
6. Truong, T.N.; Vo, A.T.; Kang, H.J. An Adaptive Terminal Sliding Mode Control Scheme via Neural Network Approach for Path-following Control of Uncertain Nonlinear Systems. *Int. J. Control Autom. Syst.* **2022**, *20*, 2081–2096. [CrossRef]
7. Morales, R.; Feliu, V.; Sira-Ramirez, H. Nonlinear control for magnetic levitation systems based on fast online algebraic identification of the input gain. *IEEE Trans. Control Syst. Technol.* **2010**, *19*, 757–771. [CrossRef]

8. El-Sousy, F.F.; Alattas, K.A.; Mofid, O.; Mobayen, S.; Asad, J.H.; Skruch, P.; Assawinchaichote, W. Non-Singular Finite Time Tracking Control Approach Based on Disturbance Observers for Perturbed Quadrotor Unmanned Aerial Vehicles. *Sensors* **2022**, *22*, 2785. [CrossRef]
9. Alattas, K.A.; Mofid, O.; Alanazi, A.K.; Abo-Dief, H.M.; Bartoszewicz, A.; Bakouri, M.; Mobayen, S. Barrier Function Adaptive Nonsingular Terminal Sliding Mode Control Approach for Quad-Rotor Unmanned Aerial Vehicles. *Sensors* **2022**, *22*, 909. [CrossRef]
10. Fan, Y.; Liu, B.; Wang, G.; Mu, D. Adaptive Fast Non-Singular Terminal Sliding Mode Path Following Control for an Underactuated Unmanned Surface Vehicle with Uncertainties and Unknown Disturbances. *Sensors* **2021**, *21*, 7454. [CrossRef]
11. Wang, Y.; Wang, Z.; Chen, M.; Kong, L. Predefined-time sliding mode formation control for multiple autonomous underwater vehicles with uncertainties. *Chaos Solitons Fractals* **2021**, *144*, 110680. [CrossRef]
12. Manzanilla, A.; Ibarra, E.; Salazar, S.; Zamora, Á.E.; Lozano, R.; Munoz, F. Super-twisting integral sliding mode control for trajectory tracking of an Unmanned Underwater Vehicle. *Ocean Eng.* **2021**, *234*, 109164. [CrossRef]
13. González-García, J.; Gómez-Espinosa, A.; García-Valdovinos, L.G.; Salgado-Jiménez, T.; Cuan-Urquizo, E.; Escobedo Cabello, J.A. Experimental Validation of a Model-Free High-Order Sliding Mode Controller with Finite-Time Convergence for Trajectory Tracking of Autonomous Underwater Vehicles. *Sensors* **2022**, *22*, 488. [CrossRef]
14. Zheng, J.; Song, L.; Liu, L.; Yu, W.; Wang, Y.; Chen, C. Fixed-time sliding mode tracking control for autonomous underwater vehicles. *Appl. Ocean Res.* **2021**, *117*, 102928. [CrossRef]
15. González-García, J.; Gómez-Espinosa, A.; García-Valdovinos, L.G.; Salgado-Jiménez, T.; Cuan-Urquizo, E.; Cabello, J.A.E. Model-Free High-Order Sliding Mode Controller for Station-Keeping of an Autonomous Underwater Vehicle in Manipulation Task: Simulations and Experimental Validation. *Sensors* **2022**, *22*, 4347. [CrossRef]
16. Truong, T.N.; Vo, A.T.; Kang, H.J. A backstepping global fast terminal sliding mode control for trajectory tracking control of industrial robotic manipulators. *IEEE Access* **2021**, *9*, 31921–31931. [CrossRef]
17. Nguyen, V.C.; Vo, A.T.; Kang, H.J. A Non-Singular Fast Terminal Sliding Mode Control Based on Third-Order Sliding Mode Observer for a Class of Second-Order Uncertain Nonlinear Systems and its Application to Robot Manipulators. *IEEE Access* **2020**, *8*, 78109–78120. [CrossRef]
18. Vo, A.T.; Truong, T.N.; Kang, H.J.; Van, M. A Robust Observer-Based Control Strategy for n-DOF Uncertain Robot Manipulators with Fixed-Time Stability. *Sensors* **2021**, *21*, 7084. [CrossRef]
19. Zhao, Z.Y.; Jin, X.Z.; Wu, X.M.; Wang, H.; Chi, J. Neural network-based fixed-time sliding mode control for a class of nonlinear Euler-Lagrange systems. *Appl. Math. Comput.* **2022**, *415*, 126718. [CrossRef]
20. Nguyen, V.C.; Vo, A.T.; Kang, H.J. A finite-time fault-tolerant control using non-singular fast terminal sliding mode control and third-order sliding mode observer for robotic manipulators. *IEEE Access* **2021**, *9*, 31225–31235. [CrossRef]
21. Mu, C.; He, H. Dynamic behavior of terminal sliding mode control. *IEEE Trans. Ind. Electron.* **2017**, *65*, 3480–3490. [CrossRef]
22. Solis, C.U.; Clempner, J.B.; Poznyak, A.S. Fast terminal sliding-mode control with an integral filter applied to a van der pol oscillator. *IEEE Trans. Ind. Electron.* **2017**, *64*, 5622–5628. [CrossRef]
23. Truong, T.N.; Vo, A.T.; Kang, H.J.; Van, M. A Novel Active Fault-Tolerant Tracking Control for Robot Manipulators with Finite-Time Stability. *Sensors* **2021**, *21*, 8101. [CrossRef] [PubMed]
24. Zhai, J.; Xu, G. A novel non-singular terminal sliding mode trajectory tracking control for robotic manipulators. *IEEE Trans. Circuits Syst. II Express Briefs* **2020**, *68*, 391–395. [CrossRef]
25. Xu, S.S.D.; Chen, C.C.; Wu, Z.L. Study of nonsingular fast terminal sliding-mode fault-tolerant control. *IEEE Trans. Ind. Electron.* **2015**, *62*, 3906–3913. [CrossRef]
26. Vo, A.T.; Truong, T.N.; Kang, H.J. A novel tracking control algorithm with finite-time disturbance observer for a class of second-order nonlinear systems and its applications. *IEEE Access* **2021**, *9*, 31373–31389. [CrossRef]
27. Truong, T.N.; Kang, H.J.; Vo, A.T. An active disturbance rejection control method for robot manipulators. In *Proceedings of the International Conference on Intelligent Computing*; Springer: Berlin/Heidelberg, Germany, 2020; pp. 190–201.
28. Vo, A.T.; Truong, T.N.; Kang, H.J. A Novel Prescribed-Performance-Tracking Control System with Finite-Time Convergence Stability for Uncertain Robotic Manipulators. *Sensors* **2022**, *22*, 2615. [CrossRef]
29. Vo, A.T.; Truong, T.N.; Kang, H.J.; Le, T.D. An Advanced Terminal Sliding Mode Controller for Robot Manipulators in Position Tracking Problem. In *Proceedings of the International Conference on Intelligent Computing*; Springer: Berlin/Heidelberg, Germany, 2022; pp. 518–528.
30. Truong, T.N.; Vo, A.T.; Kang, H.J.; Le, T.D. An Observer-Based Fixed Time Sliding Mode Controller for a Class of Second-Order Nonlinear Systems and Its Application to Robot Manipulators. In *Proceedings of the International Conference on Intelligent Computing*; Springer: Berlin/Heidelberg, Germany, 2022; pp. 529–543.
31. Huang, J.; Zhang, M.; Ri, S.; Xiong, C.; Li, Z.; Kang, Y. High-order disturbance-observer-based sliding mode control for mobile wheeled inverted pendulum systems. *IEEE Trans. Ind. Electron.* **2019**, *67*, 2030–2041. [CrossRef]
32. Nguyen, V.C.; Tran, X.T.; Kang, H.J. A Novel High-Speed Third-Order Sliding Mode Observer for Fault-Tolerant Control Problem of Robot Manipulators. *Actuators* **2022**, *11*, 259. [CrossRef]
33. Levant, A. Higher-order sliding modes, differentiation and output-feedback control. *Int. J. Control* **2003**, *76*, 924–941. [CrossRef]
34. Bechlioulis, C.P.; Rovithakis, G.A. Robust adaptive control of feedback linearizable MIMO nonlinear systems with prescribed performance. *IEEE Trans. Autom. Control* **2008**, *53*, 2090–2099. [CrossRef]

35. Yu, Z.; Zhang, Y.; Liu, Z.; Qu, Y.; Su, C.Y.; Jiang, B. Decentralized finite-time adaptive fault-tolerant synchronization tracking control for multiple UAVs with prescribed performance. *J. Frankl. Inst.-Eng. Appl. Math.* **2020**, *357*, 11830–11862. [CrossRef]
36. Koksal, N.; An, H.; Fidan, B. Backstepping-based adaptive control of a quadrotor UAV with guaranteed tracking performance. *ISA Trans.* **2020**, *105*, 98–110. [CrossRef]
37. Lu, X.; Jia, Y. Adaptive coordinated control of uncertain free-floating space manipulators with prescribed control performance. *Nonlinear Dyn.* **2019**, *97*, 1541–1566. [CrossRef]
38. Wang, X.; Wang, Q.; Sun, C. Prescribed Performance Fault-Tolerant Control for Uncertain Nonlinear MIMO System Using Actor-Critic Learning Structure. *IEEE Trans. Neural Netw. Learn. Syst.* **2021**, *33*, 4479–4490. [CrossRef]
39. Huang, H.; He, W.; Li, J.; Xu, B.; Yang, C.; Zhang, W. Disturbance Observer-Based Fault-Tolerant Control for Robotic Systems With Guaranteed Prescribed Performance. *IEEE Trans. Cybern.* **2022**, *52*, 772–783. doi: 10.1109/TCYB.2019.2921254. [CrossRef]
40. Wang, S.; Na, J.; Chen, Q. Adaptive Predefined Performance Sliding Mode Control of Motor Driving Systems With Disturbances. *IEEE Trans. Energy Convers.* **2021**, *36*, 1931–1939. [CrossRef]
41. Yang, P.; Su, Y. Proximate Fixed-Time Prescribed Performance Tracking Control of Uncertain Robot Manipulators. *IEEE/ASME Trans. Mechatron.* **2021**, *27*, 3275–3285. [CrossRef]
42. Ding, Y.; Wang, Y.; Chen, B. Fault-tolerant control of an aerial manipulator with guaranteed tracking performance. *Int. J. Robust Nonlinear Control* **2022**, *32*, 960–986. [CrossRef]
43. Yu, S.; Yu, X.; Shirinzadeh, B.; Man, Z. Continuous finite-time control for robotic manipulators with terminal sliding mode. *Automatica* **2005**, *41*, 1957–1964. [CrossRef]

sensors

MDPI

Article

Health, Security and Fire Safety Process Optimisation Using Intelligence at the Edge

Ollencio D'Souza [1], Subhas Chandra Mukhopadhyay [1,*] and Michael Sheng [2]

[1] School of Engineering, Faculty of Science and Engineering, North Ryde Campus, Macquarie University, Sydney, NSW 2109, Australia
[2] Department of Computing, Macquarie University, Sydney, NSW 2109, Australia
* Correspondence: subhas.mukhopadhyay@mq.edu.au; Tel.: +61-401-108-244

Abstract: The proliferation of sensors to capture parametric measures or event data over a myriad of networking topologies is growing exponentially to improve our daily lives. Large amounts of data must be shared on constrained network infrastructure, increasing delays and loss of valuable real-time information. Our research presents a solution for the health, security, safety, and fire domains to obtain temporally synchronous, credible and high-resolution data from sensors to maintain the temporal hierarchy of reported events. We developed a multisensor fusion framework with energy conservation via domain-specific "wake up" triggers that turn on low-power model-driven microcontrollers using machine learning (TinyML) models. We investigated optimisation techniques using anomaly detection modes to deliver real-time insights in demanding life-saving situations. Using energy-efficient methods to analyse sensor data at the point of creation, we facilitated a pathway to provide sensor customisation at the "edge", where and when it is most needed. We present the application and generalised results in a real-life health care scenario and explain its application and benefits in other named researched domains.

Keywords: TinyML; machine learning; edge analytics; energy harvesting; health care; security; safety; fire safety

Citation: D'Souza, O.; Mukhopadhyay, S.C.; Sheng, M. Health, Security and Fire Safety Process Optimisation Using Intelligence at the Edge. *Sensors* **2022**, 22, 8143. https://doi.org/10.3390/s22218143

Academic Editor: Natividad Duro Carralero

Received: 9 September 2022
Accepted: 19 October 2022
Published: 24 October 2022

1. Introduction

The prediction for sensor usage indicates that [1] several billion devices will be installed by 2025, leading to an exponential growth in sensor data [2]. Individual systems in the wild compete to transfer sensor data to centralised repositories, such as cloud infrastructure, for analysis. Communication networks such as LPWAN, 5G, NB/10T or even the WiFi 2.4G or 5Gig (Free To Air) options will find it challenging to maintain the synchronous nature of time-sensitive data with sufficient temporal accuracy that life-saving domains require. The volume of disparate data trying to share network resources progressively reduces the bandwidth available to each user. Analysts also have to manage massive amounts of error-prone sensor data reaching them, making processing and analysis a resource-hungry task that must be conducted efficiently and effectively [3].

Real-time domain sensors are always active [4], monitoring parameters continuously to expose problems [5]. The high detection throughput needs advanced techniques to conserve energy and optimise real-time operational reliability over long, sustained periods. Evaluating the contribution of each sensor type is necessary to determine the right mix of data required in each operating domain. A tabular listing of domain features, Table 1, presents the commonality of needs in the health, safety, security and fire safety domains.

Table 1. The features and performance of a sensor fusion operational environment.

SNo	Operational Requirements	Desired Performance (Output)
1.	Embedded device for sensor fusion, real-time multiple sensor data analysis and results in storage	TinyML edge analytics ML model optimised for sensor data fusion and edge analytics. Only temporally synchronous application data will be stored so "prediction" and "detection" are achieved reliably by analysis.
2.	Power supply optimised for extended operational periods	"Wake" triggers conserve power. Renewable energy source, reliability monitoring and energy harvesting at the edge.
3.	Operating firmware reliability	Error checked for OTA (over the air) model updates.
4.	Reliable results transfer (upload)	Error checked, the optimised payload for transmission of analysed and computed results.

1.1. Key Issues to Be Considered

Reports suggest that only snapshots of the generated real-time data are sent successfully due to network constraints and reduced sampling rates. This truncated data, Figure 1, affects the accuracy of research and analysis in real-time operational domains [5]. The raw source data are not sent over networks to reduce privacy risk, communication bottlenecks, costs, processing time, the expense of handling, storing and maintaining infrastructure and data manipulation required by large, stored data sets before analysis.

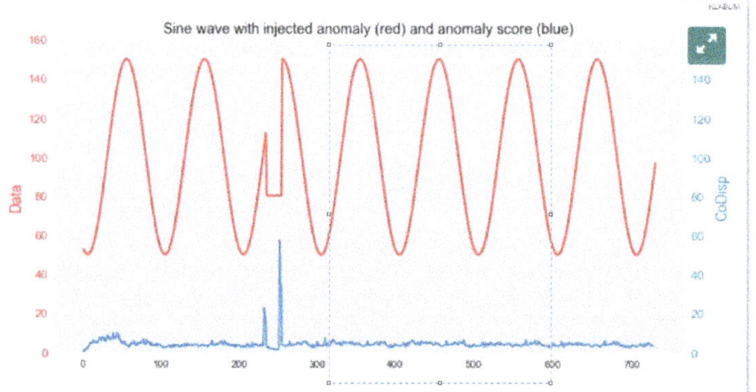

Figure 1. "When time is of the essence" [6] waveform sampling study of truncated data.

1.2. Research Considerations

We identified, measured, corrected and compensated for issues that maintain data integrity and operational consistency [7]. Attention to the selection of sensors, the authenticity of real-time data [8], and the fusion of sensor data close to where the data are created is a crucial process adopted. For a pre-defined "wake up" period, constrained low-power computing devices are needed to analyse all the data in real-time to deliver the results to a localised repository. This technique ensures that all essential data are captured and analysed in real time. Machine learning models were created by training them to meet domain operational needs. The post-fusion analysis models developed in the cloud were distilled down to versions using TinyML (a compressed version of the larger version) to fit onto constrained microcontroller platforms, as presented in later sections. Results are sent as an optimised "payload" to a "first responder" [9] or a central control facility [10], effectively reducing the requirement for temporally synchronous computing resources [11] at the management end.

Conserving power is essential in field-deployed devices, so domain-specific anomaly detection "wake up" triggers turn on power-hungry analytic modes to optimise the power consumption of the microcontroller devices.

In health care, data from sensors have an essential role in maintaining operational care compliance and improving the response to those that need urgent care [12,13]. These deliverables are stipulated in legislation; hence, measuring and sustaining performance in real-time is a serious requirement.

In the security, safety and fire safety domains, the need is to generate an alert based on a multifactor assessment of complex sensor data, to keep "unwanted event" triggers to a minimum. Industry statistics for false or "unwanted" alarms are close to 90% [14], resulting in a waste of resources, time to attend these events and severe economic loss in some cases. In all these situations, constant changes occur, and sensor management systems that send ineffective post-event advice continue to disrupt normal processes [15]. Our method improves detection quality and ensures event predictability to improve outcomes [16]. Flow charts, tables and a framework based on machine learning techniques at the "edge" explain the model's detection, prediction processes and performance outcomes.

Highlights and original ideas over existing state of the art, presented in several sections of the paper, are summarised for clarity. The basic idea of using machine learning is not new, but the creation of a model to overcome specific "real-time" industry domain application issues is one of the key highlights. Our development overcomes issues with the management of real-time operational process flows and optimises situational awareness by delivering event information directly to the "first responder". Several examples demonstrated in different sections show how "event detection" is more than just the result of a single sensor trigger but the result of a "cluster" of sensors of different types delivering "intelligence" using "collaborative, corroborative and reinforcement" techniques. This, we believe, is our unique contribution using machine learning in health, safety, security and fire safety where time is of the essence and critical to life-saving processes.

2. Materials and Methods

Sensor data from monitored parameters at the site [17] are part of the organisation's established risk management strategy. Identifying and managing data loss, data "cleaning" and data storage in the repository from where the analytic engine ingests, corrects and analyses data to comply with the requirement of the data-driven environment is a core responsibility [7]. The two key issues that reduce the efficacy of many data-harvesting operations are the quality of data arriving and data analysis in temporal sync. The first issue tackled is the quality of data. Multiple sensors deliver multiple data streams, so data fusion techniques reduce a sensor cluster's results payload. The results are sent directly to the "first responder" dashboards to improve real-time situational awareness (see simplified explanation in Appendix A).

2.1. Related Work—A Two-Year Quest to Meet Real-Time Smart City Objectives

A prior two-year design and implementation of a smart city research system funded by local government and a university research endowment provided adequate experience in data management. The knowledge and understanding of how vital temporal sync is to support health and well-being concerns are documented in system configuration drawings (graphs in Figure 2) [7], data management records, published papers and reports. The research explained how seventy-one in-house developed multisensor IoT nodes connected to two city-wide gateways delivered vital data over public LoRaWAN infrastructure. Over two years, the experience demonstrated that reliability, data throughput and temporal sync were vital to the usability, effectiveness and efficiency of city-wide services. Other instances also reinforce this conclusion [18].

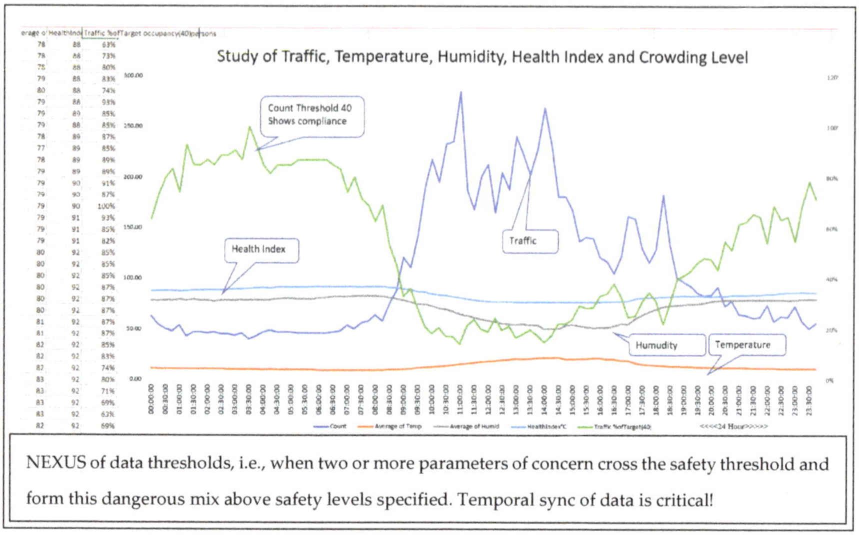

Figure 2. Related prior work building a city-wide IoT infrastructure for health and safety.

2.2. *"Intelligence"—Training Machine Learning Resources to Produce Efficient and Effective Outcomes*

"Intelligence" describes the outcome of a fusion process used to securely package the results of the data analysis at the "edge". This process no longer requires expensive data handling to produce accurate results or to maintain the integrity of the data at the point of collection where the data generation and analysis will occur. Results from analytics at the "edge" are fed directly to a live operational dashboard [19].

The process uses techniques that include the use of "situational knowledge" or "features" (Table 1) essential to train the machine learning system to improve decision-making and prediction capability. Machine learning (ML) [20] requires a training data set to extract associations and insights from disparate data. Hardware-constrained microcontroller "edge" devices use (TinyML) [11] models created in the cloud to function at the "edge" [21]. These models enable sensor fusion analysis to improve "classification" and detect anomalies, leading to an improved detection performance and prediction capability.

2.3. *"Feature" Logic Designs for Targeted Real-Time Domains*

The sensor grouping enables the mapping of multiple sensor outputs to "collaborate, corroborate and reinforce" to confirm an event. The concept developed in this research is explained diagrammatically in Figure 3. The human algorithm builder can only do so much based on the associations seen in the data. Machines can do this many times faster and build many more unbiased associations. The core learning is that allowing ML to extract relationships, associations and relevance produces better insights but requires plenty of relevant data to be used in training, as per the logic flow example described in Figure 3.

2.4. *"Event Reinforcement" Techniques Using "Collateral" Sensor Data Example*

The process workflow of a "reinforced" data point is explained in the flow chart Figure 4 and Table 1. The accompanying "pressure wave" is detected if a properly shut door is opened. This "collateral" sense confirms the door has been opened from a shut position rather than from an open (ajar) state.

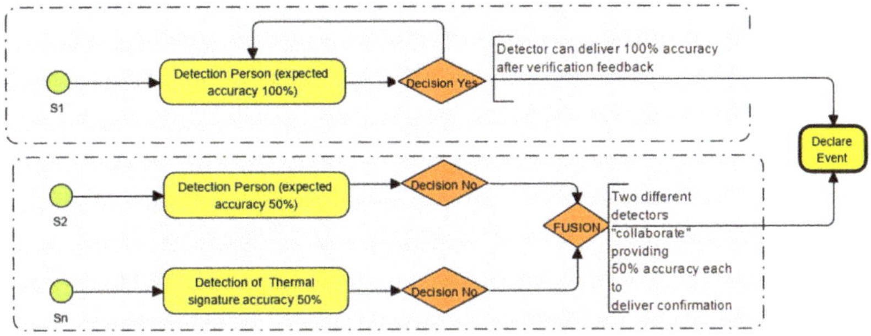

Figure 3. Reinforcing and collateral data workflows demonstrate how data from different sensors "collaborate, corroborate and confirm" outcomes. (Sensors S1 to Sn, Probability & accuracy in yellow, decision points in mauve diamond and outcome from "fusion" in yellow).

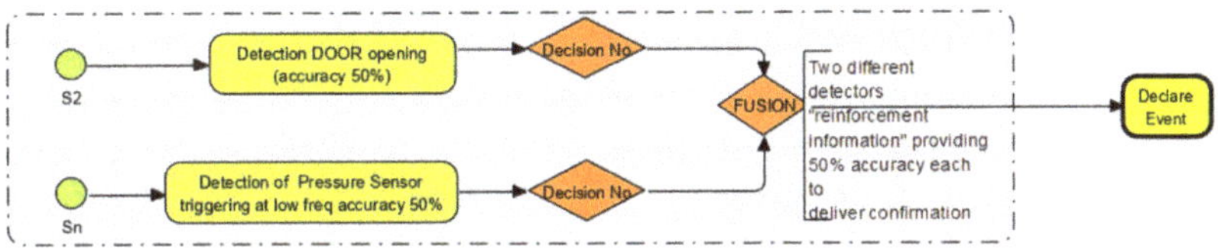

Figure 4. How one sensor data point and another sensor data point "collaborate" and "corroborate" to confirm an event.

A similar phenomenon is used to "corroborate" activity such as using "gesture" recognition. Just the motion sensor triggering might not help define the event, but a side-to-side motion (directional detection) would generate a unique motion signature as temporally synchronised "collateral" data from the motion sensor on board. This event could be assigned to a different response or output—such as a "nurse call" to care staff.

2.5. Visual Colour Sense "Camera"-Like "Features" (No Privacy Broken—Single-Pixel Resolution)

The "reinforcement" techniques Figure 5 are essential because the ability of the sensor to detect colour, albeit just a few pixels, "reinforces" sensor data that point towards a specific characteristic such as the colour yellow in fire or red on a caregiver uniform.

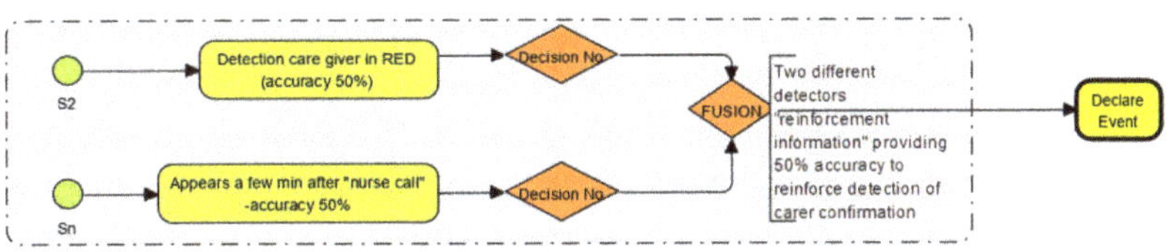

Figure 5. Using colour information to "reinforce" the arrival of a "nurse" in red clothing just after a "nurse-call" trigger is initiated. (sensor data in green detected but not accurate enough at decision point in mauve diamond. However at a fusion decision point in mauve diamond they reinforce and confirm the event in yellow).

Humans build unique algorithms with the capability to build associations using a few sensors. Machine learning can build associations using many or all sensor inputs of different types and in temporal sync. This widens the opportunity to find similarities, correlations, dependencies, links, etc., to enrich the data and detection quality, accuracy, depth, reach, etc., and perform it in seconds to ensure that nothing is missed.

In the case shown in Figure 5, a single pixel colour detection sensor is already on board. Therefore, associating this colour detection with the colour of the nurse uniform will reinforce the association of detecting caregiver movement near patients. In addition, in the case of a fire, the yellow colour reinforces the existence of a flame.

The few associations developed by humans are helpful but not as useful as the thousands that are possible across the entire operation, which machines (MCU and ML) can perform very fast in human time.

2.6. The Feature Set and Performance Specification

Key system operational "features" and performance requirements are shown in Table 1.

2.7. Machine Learning Training Requirements and Operational Examples

The targeted domain-sensitive training outcomes are shown in Table 2. The domains considered are a whole class of environments that require real-time event detection and immediate response. The physical environment is not so critical, for example, indoor, outdoor, small room, large hall, bedroom, lift, etc. The key issue is "what is to be detected?" (i.e., the objective). Enough examples of the type of event are required in the training of the ML model. ML will build associations based on triggers from various sensors in the systems that are in temporal sync.

Table 2. A tabular listing of typical "wake" trigger options in each activity domain (truncated to fit).

Standard Multi-Domain "Wake" Trigger Options	Typical Risk Factor—Health	Security	Typical Risk Factor—Safety	Typical Risk Factor—Fire
Detecting a human or humans entering and leaving the sensor area in daylight	Several "sterile" areas could be "compromised" if people without PPE are allowed to roam unchallenged.	Designated "no go" (sterile) security areas are typical risk triggers set up to keep "intruders" away.	Building areas with cranes, etc., must be isolated so no one gets hurt. Thus, a man or machine "wake triggers" applied.	Often a "roll call" is taken on a fire alert to see who is on the premises. Any human or animal presence triggers "wake triggers".
Detecting a human or humans entering and leaving the sensor area in "darkness" (no light)	Dark areas have severe risk implications. Special sensors with IR illumination or high gain deliver similar situational awareness as in the day.	Dark areas have severe risk implications, but illumination triggered by human movement helps.	Dark areas have severe risk implications, and illumination is essential for safety. Lights could be automatically turned on.	Dark areas have severe risk/rescue implications. Smoke and carbon monoxide are the most significant threats—smoke/CO sensors are used to trigger the siren and lights.
Establishing the time of day	All incoming and outgoing personnel must be timed in and out.	All incoming and outgoing personnel must be timed in and out,	All incoming and outgoing personnel must be timed in and out.	All incoming and outgoing personnel must be timed in and out.
Establishing the route	Movement history	Movement history	Movement history	Movement history
Establishing the count for 24 h	Counts of people through to waiting rooms could be used to control congestion.	Counts determine traffic flow into service areas.	Counts are used to restrict people's traffic to stop overloading.	Counts are used during the evacuation of people to safety.
Data generation failure	Fused sensor detectors deliver continuity.	Fused sensor detectors offer continuity.	Fused sensor detectors offer continuity.	Fused sensor detectors provide continuity.
Establishing "door open/closed/ajar	Essential to establish the status of the door to stop the spread of infections.	Door status is vital to isolating the threat.	Door status is essential to isolating threats by sealing the area.	Fire door status is very critical to establishing or maintaining safety.
Establishing bed movement and positioning	Specific to "aged care". Location of crucial caregiving assets.	Unusual activity of assets in working space related to security.	Assets left unattended block vital thoroughfare.	Legislated fire safety requirements supervised autonomously.
Establishing staff attendance	Specific to "aged care". Colour of staff clothing (see image Figure 6, blue).	Use of HiVisibility vests or uniforms (see image Figure 6, yellow).	Use of high-visibility and safety gear (see image Figure 6, yellow).	All fire kits are a specific colour (yellow/red), and hence can be easily recognised.

Figure 6. Staff in uniform.

2.8. Examples to Explain How ML Would Build Associations, Collate, Collaborate, Corroborate, Reinforce Better Than Humans by Using Multiple Sensors on Board the MCU

A motion is detected along with a "blue uniform" colour recognition trigger; it does not matter where the incident occurs. The association of two key known criteria, person movement and the colour of the uniform worn by the responder, would flag a caregiving episode. Similarly, a vehicle travelling outdoors and a group of people detected in the path could be an accident waiting to happen. In a room, if a door was opened and the pressure sensor was also triggered, then the association of the opening door and pressure wave indicates the opening of a closed door, not a door left ajar. Some other sensors will indicate human movement, which could mean intrusion, the arrival of help, etc. ML thrives on building associations of such value that detection of incidents (with recognition of event type) or "anomalies" (unusual activity) is no longer conducted by a single sensor but by a group of sensor types because of the speed and capability to create associations by a trained ML model. This heightened situational awareness is due to the creation of many associations made by the ML model during training. This becomes an efficient and effective tool to support a human operator's "inference" capability. These benefits are explained in the decision tree tables and workflow diagrams provided later in the description of the trials.

2.9. Implementation of an Experimental Environment to Test the Hypothesis

At the "edge", each domain will have a "wake" trigger to activate the "intelligent edge analysis" on the data generated by a sensor or a group of sensors (fusion) [22]. We put together the target "behaviour" outcomes for our training set, as listed in the domain segmentation Table 2.

A small system combines location- or task-specific sensors [7]. A cluster of small systems in a larger configuration (i.e., multiple small systems tied together) require a federated learning approach to using "intelligence" at the "edge". We propose scalability using a fusion sensor approach and BLE (Bluetooth Low Energy) communications.

2.10. Applied Traditional and Machine Learning Workflows

We use value engineering [23] to design the model to ensure it is scalable. To this end, all logic (including design, programming and simulation) goes into delivering these benefits Figure 7.

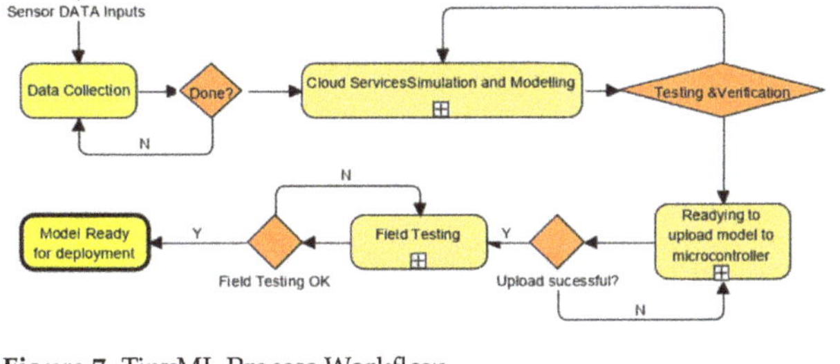

Figure 7. TinyML Process Workflow.

Live simulations, from the existing system, with as many examples of match and non-match as possible, were used to meet machine learning (ML) ingestion workflow standards (Figure 8). The accuracy of the model generated by the machine learning build process depends very much on the variations fed during training. After this training, the model serves as the "selector", looking through all the new data to explore all associations, relationships and relevance to produce the best match determination that improves situational awareness.

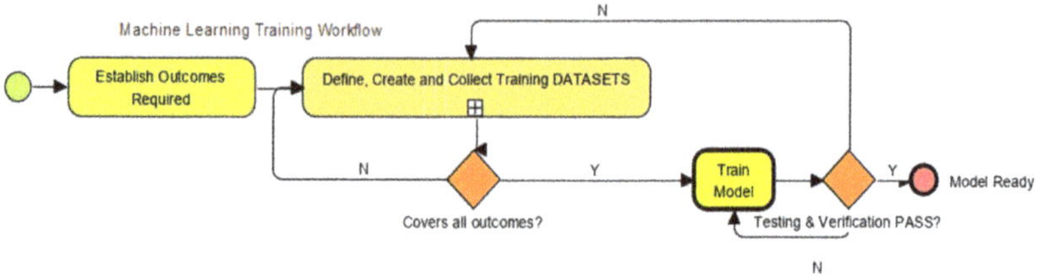

Figure 8. Machine Learning Training Workflow.

2.11. Embedded Systems Evolution and Deployment in This Research

Embedded systems with I/O and processing power deployed at "the edge" is enhanced using BLE (low-energy mesh communications). Typical system configuration [24] is shown in Figure 9. Tiny Machine Learning (TinyML) reduces the size of the generated models to fit into resource-constrained hardware with BLE capability enabling an ecosystem to emerge from this work. Each "cluster" of sensors on board an MCU device forms an intelligent node. As we expand connectivity between "clusters" over BLE, the network grows in capability to share information [25], and "intelligence at the edge" is created.

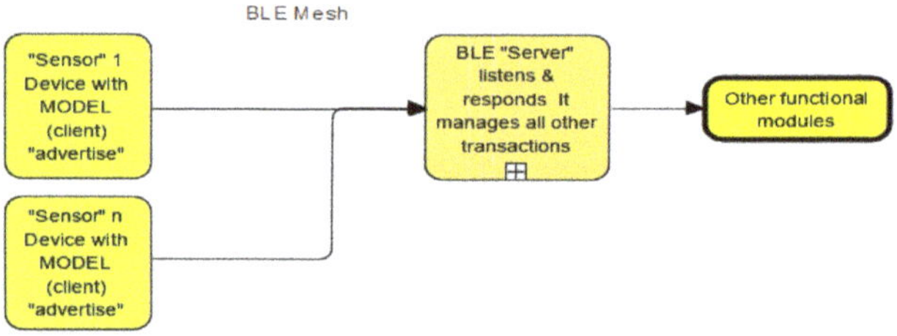

Figure 9. Typical machine learning BLE mesh architecture delivering "intelligence at the edge".

2.12. Architecture Deployed and Lessons Learned

Multisensor microcontroller (MCU) boards currently have physical, chemical, environmental, video, audio and optical sensors built on the single embedded board. These multidimensional data sensor architectures provide machine learning systems access to rich data to unearth observations that contribute to a greater understanding of each dimension's inputs to the system's overall operational performance.

As demonstrated earlier in logic diagrams, ML-based event detection is no longer a single sensor endeavour. ML uses thousands of associations to collaborate, corroborate and reinforce new scenario settings and unique event detection episodes.

The MCU used is the Arduino Nano BLE Sense with eleven (11) independent sensory capabilities in five (5) sensor groups. These MEMS-based sensors are very reliable, capable devices on the MCU as a cluster of sensors of different types—ideal for machine learning to generate complex associations (Figure 10).

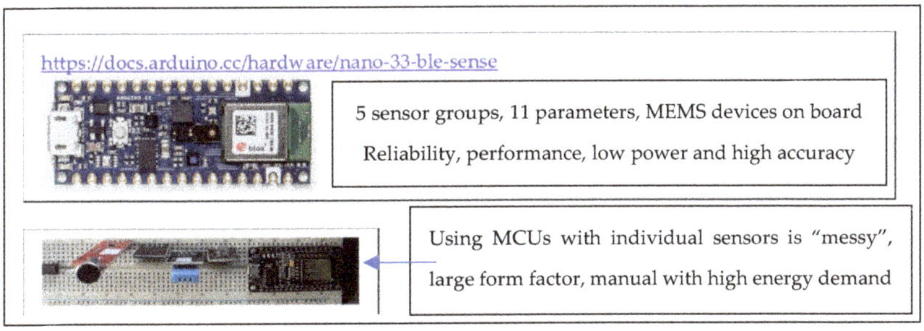

Figure 10. MCU features and choices.

Our research uses off-the-shelf MCUs. The choice, capability and sensor density vary; hence, as a first step, sensors relevant to the domains were identified. In the health, safety, security and fire safety domains, person detection, parameter detection (such as temperature, humidity, air quality, airflow, hygrometric levels wetness), vibration, movement, colour, pressure and sound thresholds (unique sounds and keywords) were prime requirements.

Applications in health care are explained with bed-mounted and wall-mounted configurations shown later in graphed results. Each MCU with sensors on board act as a "sensor cluster". They have an ML model uploaded to it and perform as per the training provided during the model creation. Layouts and images of actual test training procedures and set ups are explained with relevant test frame grabs.

There is no single sensor per area because each MCU is a sensor cluster connected to other clusters via BLE. Each cluster could have eleven or more parametric sensory outputs to an ML model, which would then proceed to extract events, anomalies, etc., based on the model uploaded to it. By using associations, it generates from multiple sensors as explained in the collaborations, corroboration and reinforcement framework workflows; ML conducts them many times faster to present a range of associations and result options to a live dashboard.

TinyML models are uploaded to different form factor MCU configurations. Tiny compute hardware, with constrained resources and different onboard sensors, as shown in Figure 11, provide a rich source of functional performance measurements required in health, safety, security and fire safety environments. Thermal sensor add-ons for predictive fire safety also follow the same principles but use an extended spectral dimension to predict the onset of fire risk.

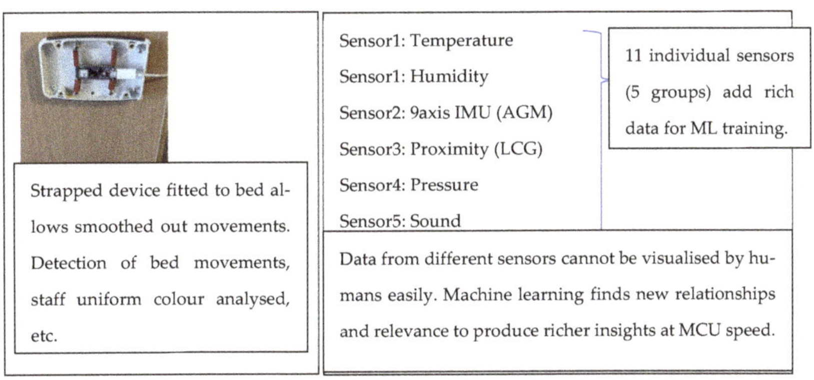

Figure 11. Example of an installed multisensor microcontroller device.

A systems approach establishes steady-state operating parameters. This process is shown in graphs produced by device simulations in fixed and flexible configurations. It measures quiescent operating conditions and real-time sensing variations for different sensor inputs, which provides an understanding of how the device (and ML model) would function in a fixed or tailored configuration. Most of these simulations use cloud services connected live to the "cluster" or use data uploaded to it, which is usually the process followed to "fine tune" the system.

2.13. Machine Learning (TinyML) Sensor Fusion Process

The fusion of all relevant detection input data is performed by uploading all relevant data to the service provider cloud repository. Once all data types are uploaded to the cloud after all preliminary sensor tests are finalised, the cloud service creates the model in the cloud using the software learning block apps provided. Subsequently, the model is reduced to run on a standard low-cost microcontroller (MCU) [22] device installed, for example, on a wall (Figure 11), with the TinyML model on it. The MCU and model run at less than 1 mW in "listening mode" (awaiting the "wake word"). When triggered, they take up to a maximum of 10 mW to process an event and then drop back into "listen" mode (Figure 12).

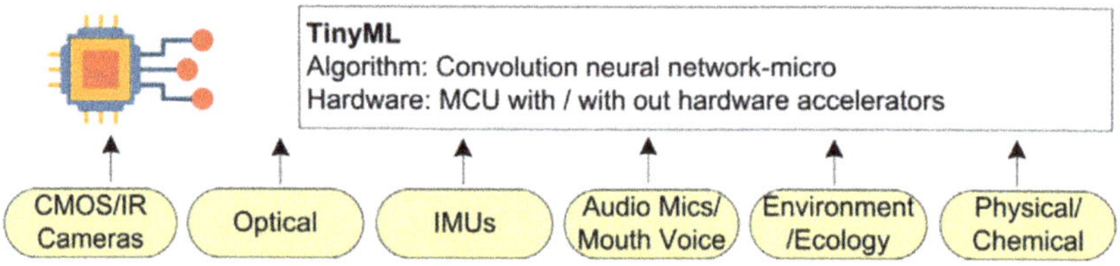

Figure 12. Industry-issued figure of a device ecosystem.

One of the many industry-developed cloud-based tools was used to understand the process and work on the key steps in producing an "intelligent edge device". Current device technology uses "digitised" (MEMS [26]) sensor technology, so a combination of sensors are built on board (Figure 12), allowing the "edge intelligence"application (Figure 13) that performs all existing functions to be implemented.

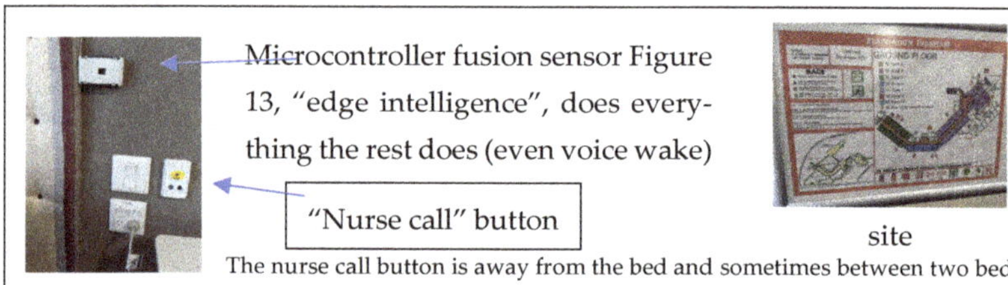

The nurse call button is away from the bed and sometimes between two beds, making it difficult to reach. Individual sensing and analysis will provide the clarity required to optimise time and effort to meet the response required. The "wall fixed" configuration does not deliver "bed movement" data, which are useful to understand restfulness, movement patterns, length of time between moves and other issues such as bed moved out of place. The facility is large and operationally staff require all the help they can get to "know" what is happening, when it happens, so "decisive" and timely responses can be provided to meet legislated performance requirements.

Figure 13. Typical bedside "resident" support assets.

Site layout for BLE mesh communications.

2.14. Deploying a Machine Learning Process

The "tooling" is a term used to describe the set of instructions, methods, support software and infrastructure provided by some providers of cloud-based ecosystems to build TinyML models. Support infrastructure allows uploading sensor data and external sources to the cloud for analysis and preparation [27,28], before the model is created.

Figure 14 describes the workflow followed when developing the solution from live data [5], which are collected based on training requirements to ensure the trained model delivers effective outcomes. Sometimes the results are below expectations, so retraining is needed, or the results are very encouraging and can be installed confidently.

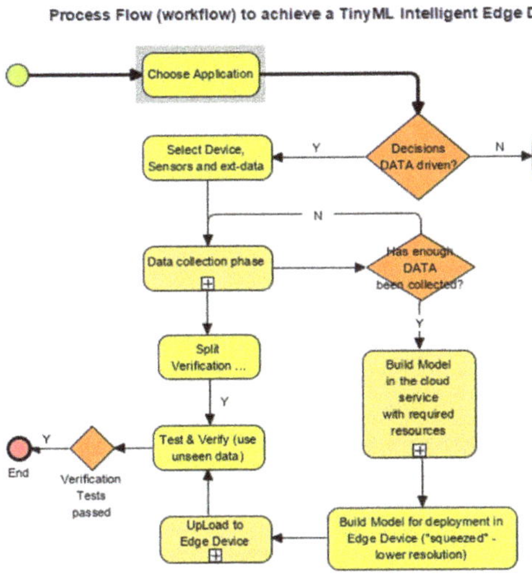

Figure 14. Typical workflow to deploy TinyML.

2.15. Training the ML Model (Figure 14)

The key steps to follow in planning the application in the selected domains are:

1. Choosing the right MCU "appliance" (sometimes referred to as the microcontroller or "silicon").
2. Collecting the data (achieved using the device itself connected to the cloud service provider). This data can also be collected locally and uploaded to the cloud service storage for analysis.

The service provider also provides the testing and training tools that help create a working model in the cloud, which one can test, verify and test again on unseen data. After the successful first passes based on the domain requirements, the next step is to "squeeze" the model down to the device requirements and upload it to the firmware. At this point, a fully functioning "intelligent edge device" emerges with a model loaded to its firmware customised to suit our application. The Actual Working Steps (soft considerations in the application such as the "Help Me!" call) are expanded below. The steps to successfully create a model in the cloud and upload it to a microcontroller device vary from provider to provider. Details on how we achieved this follow.

2.16. The Application Objective

Choosing the ecosystem described earlier, we needed an "intelligent edge" device to meet the detection requirements of four distinct domains with very similar detection demands (see Tables 1 and 2).

2.17. Data Collection—Designing What and How

The microcontroller device with multiple sensors on board is connected to the service provider server via a custom CLI or a WEBLink. Figure 15 demonstrates the process of collecting data for the "wake word" "Help Me!".

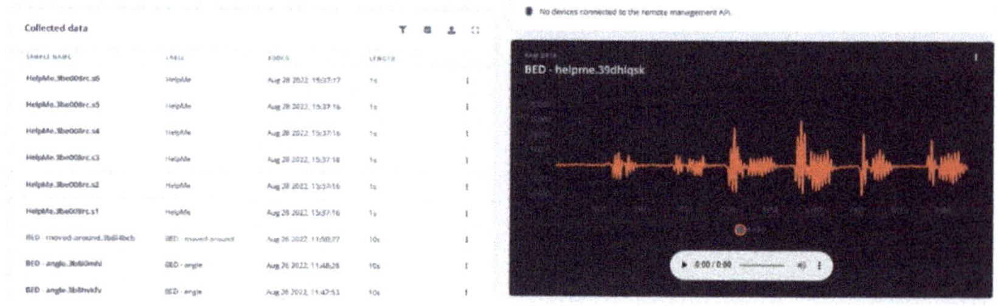

Figure 15. Collecting data for "Help Me!" "wake word".

The system designer makes a distinct choice on what the "wake word" will be. "Help Me!" is a common but distinct two-syllable phrase. The training outcomes in Figure 16 show that 95.4% accuracy was achieved.

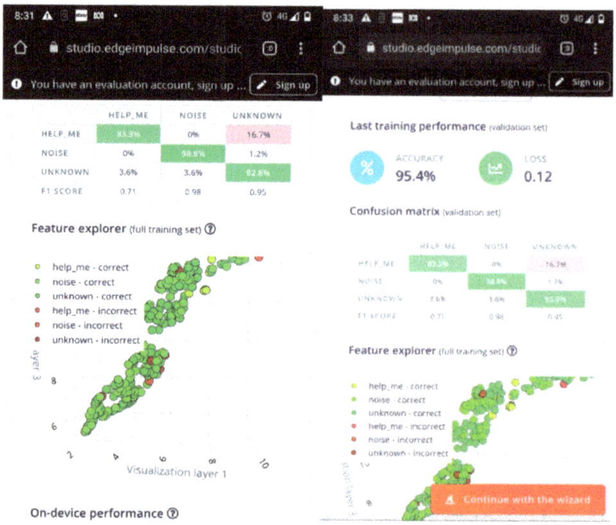

Over three epochs from 83.3%, model filters were adjusted until 95.4% accuracy was achieved. Note: three streams were used; the first was to get a pure unadulterated call in many pronunciations and pitches. The second stream was noise, "radio crackle", and the third stream was lots of general talking. From these inputs the cloud-based service provider developed a model that could be loaded onto the microcontroller to act as the wake word. (Step-1) The next steps were to upload the rest of the access control, "nurse call" and noise data (non-data to teach the ML system what not to select).

Figure 16. "Wake word" planning, configuration, tests and results.

2.18. Collecting and Obtaining the Right Data

A comprehensive data set was collected from multiple sensors to help the ML process find many more associations between sensor triggers. The "wake word" will trigger the analytic system to respond to a higher-level risk analysis triggered by the microphone sensor, thereby conserving energy to process the other "feature detection" requirements designed into the ML application. We found this to be a powerful framework that allows the ML analysis tool to optimise the associations using the "collaborative, corroborative and reinforcement" techniques designed into the framework. The ML process will find many more associations in temporal sync, depending on the level of training data used, which is just impossible using human observation alone, as shown in Figure 17 below.

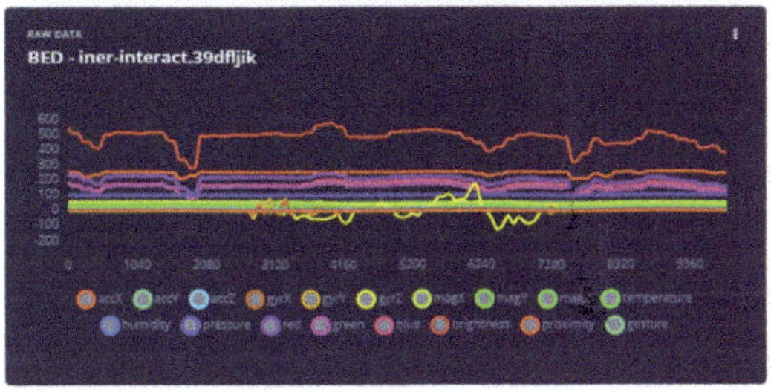

Figure 17. Multiple sensors produce rich data for the ML algorithm to learn and build unique associations within the synchronous time frame (i.e., in temporal sync).

Process flows presented earlier show "collateral", "collaborative" and "reinforcement" data analysis techniques confirmed the occurrence of "events", reducing the "FALSE ALARM" rate that beleaguers the industry. Machine learning analysis produces many more associations when using the framework's "collaborative, corroborative and reinforcement" techniques. Learning to develop the model in the cloud with adequate computing resources is a very effective method to develop the "wake word" mechanism from scratch.

2.19. Summary of Materials and Methods

Our research exposed several areas of interest in the quest to build a universal sensor for the chosen domains. We focused on health care and its immediate needs, i.e., to respond to issues of care.

To call for help assistance/support, one "wake up" call, "Help Me!", was developed. Once this "wake word" is invoked the system begins to analyse issues based on the training objectives of the TinyML model.

The bed vibrations with sensor fixed to the wall Figure 18 and bed vibrations with strapped sensor Figure 19 directly triggers a call for help (as just another "wake up" trigger) with better precision than a human who might pick up only one aspect of the problem. Most occurrences can be "taught" to the ML system using pre-recorded event history. Sensor data play an essential part during collection to provide input on the temporally synchronous activity that can precisely identify the problem.

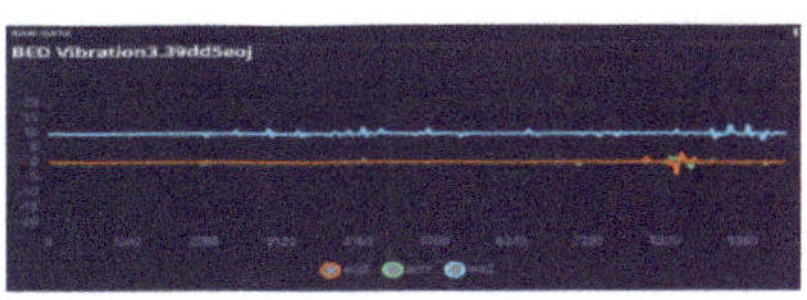

Figure 18. Bed vibration with fixed sensor.

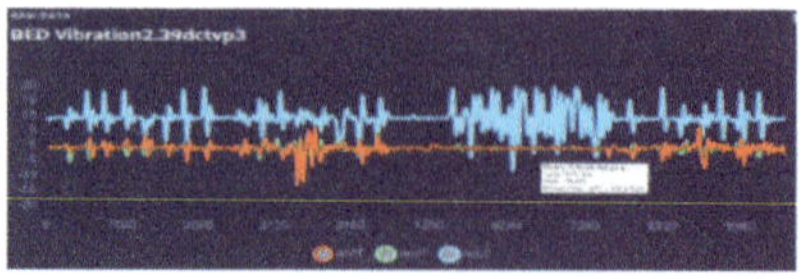

Figure 19. Bed vibrations with strapping suspended sensor.

Audio is used as a "wake word" to provide the best opportunity for the patient to be heard. Training ML on how to ignore irrelevant sounds was conducted as explained in the following graphs. The strapping proved to help "listening" because all sympathetic noise generation is smoothed out with the strapping (Figures 20 and 21).

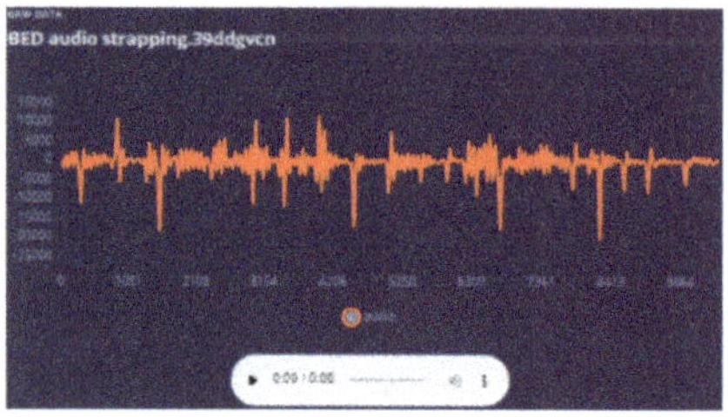

Figure 20. Audio sensor is strapped—waveform smoother.

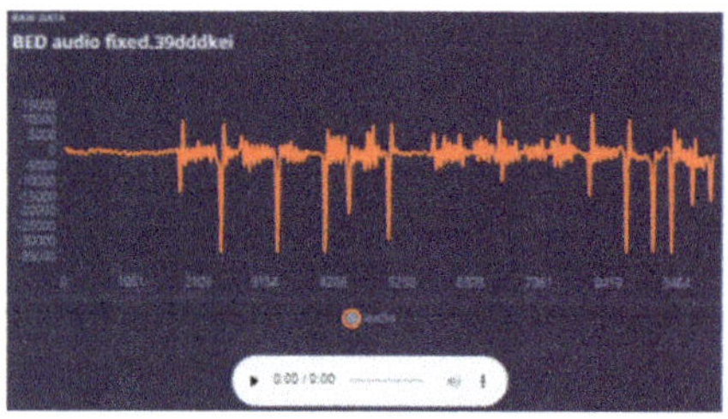

Figure 21. Response from audio when sensor fixed to a hard surface.

The analysis of the "anomaly" triggers from multiple sensors determines much more than bed shaking. The multidimensional data from, for example, the magnetometer, can provide information on the bed itself being moved, misaligned, etc., leading caregivers to investigate why (Figure 22). In all domains the scope of this technology is very encouraging.

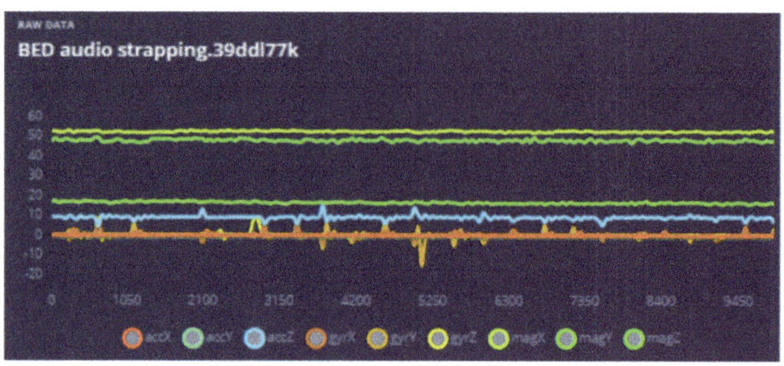

Figure 22. The accelerometer and magnetometer provide positional information instantaneously when disturbed.

Failure cost is a serious consideration because operational performance depends on reliable sensors with long battery life. We considered low energy management strategies such as the "wake word" and BLE (Bluetooth Low Energy) connectivity and are investigating energy harvesting options such as infrared wireless power to extend the system's operational life.

2.20. Application Overview

Health, security, safety and fire safety are the four domains targeted in this research (Table 2). The configurations were developed to use a multidimensional approach using MCUs with multiple "digital" sensors on board to maximise the sensor fusion outcomes. Fusion techniques enabled ML to optimise the benefits of having many data points to collaborate, corroborate and reinforce outcomes.

The BLE [29] device uses several modes to communicate with other devices or the outside world (even via the internet). These mechanisms follow the Generic Access Profile (GAP) guidelines. GAP defines how BLE-enabled devices can make themselves available and how two devices can communicate directly (Figure 23). The technology is flexible, reliable and has a maximum range of up to 1 Km with a suitable antenna.

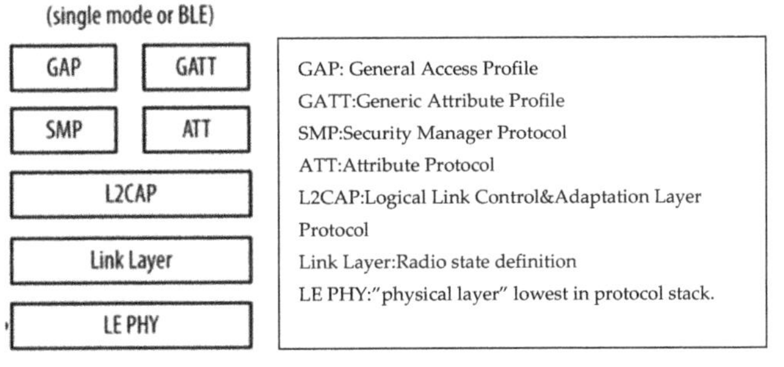

Figure 23. BLE protocol stack that allows meshed and long-range links.

2.21. Data Used in Training the Machine Learning Model

Data were generated from two sources:

(a) Simulations (because site access was not possible for some tests).
(b) Anonymised data provided from actual live sites.
(c) The simulated data collection described in later sections.

The anonymised data were generously provided by a trusted external source and contain the following event information:

1. Access control data describing entry and egress events in temporal sync.
2. "Nurse call" data describe health care residents' activity in calling for help using the push button on the bed.

2.22. Using Data Provided along with Simulated Data to Train Model

The process of using different data structures is quite complex but made easy using tools provided by the TinyML service provider (Figure 24).

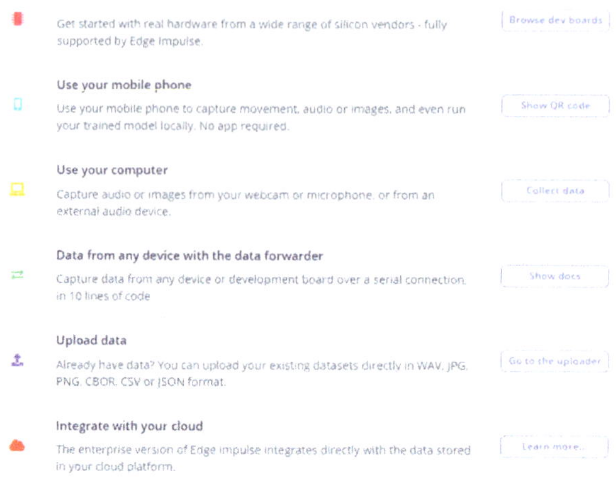

Figure 24. Data upload options provided by the service provider.

The data samples are shown in Figures 25 and 26.

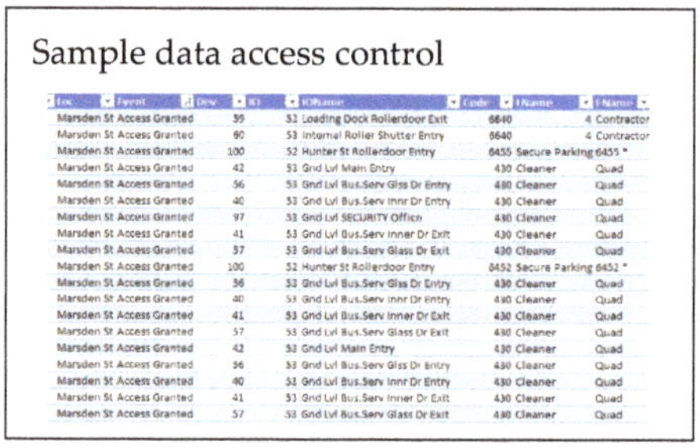

Figure 25. Access control data sample.

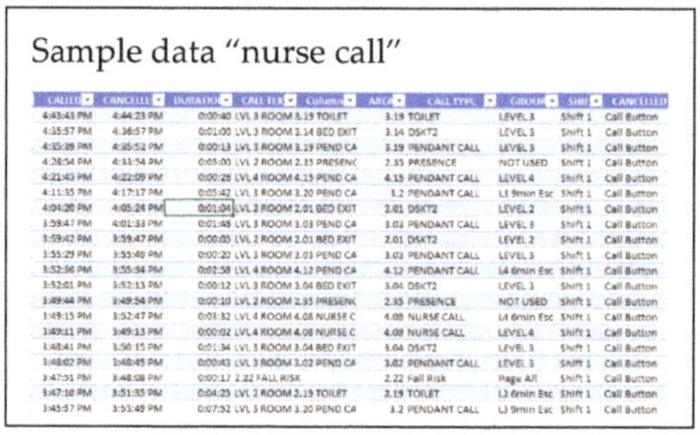

Figure 26. "Nurse call" data sample.

This data had to be converted into a format that is time stamped and follows the prescribed format: timestamp, Data-1, Data-1, Data-3, ...

The data (Figures 25 and 26) were uploaded using the interface provided (Figure 24). The ML model creation was initiated in the cloud service. Advanced "learning blocks" (sophisticated processing steps Figure 27) such as Transfer Learning and Regression filters, NN, etc., were provided to improve model performance.

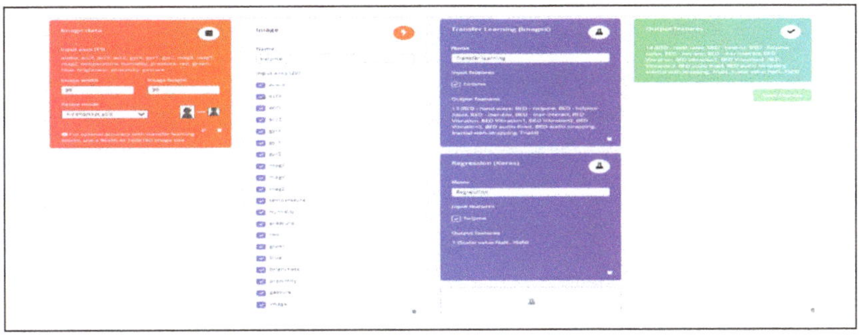

Figure 27. Additional "learning blocks" (TL, KERAS, NN) provided to improve outcomes.

The TinyML ecosphere is still developing, allowing many opportunities to discover new techniques to improve the performance of models. The key outcome in this research is understanding the stepped approach to developing models that make it possible to have "intelligence at the edge" embedded into constrained computing devices such as microcontrollers.

3. Results

3.1. "Feature"/Function Table

The results table explains how different aspects of the project objectives were met with system flexibility and performance in focus.

"Feature"/Function	"Edge Intelligence" Using Sensor "Clusters" and Low-Energy Techniques
Quality of sensors and functionality	Many unique MEMS low-power reliable sensors on computing device board (five parametric sensing groups, eleven sensor types).
Sensor fusion in real-time	Sensors are in close proximity, close to where the data are captured, so optimised, temporally. Synchronised data fusion analysis is successfully performed.
Data collection	Relevant data in many forms can be added to the fusion step.
Power optimisation	New techniques such as "wake words" conserve power for those times that require computing resources.
Advanced feature sets	Data are analysed in real-time—only results stored or sent to storage or dashboards. The result "payload" is in Kilobytes. Machine learning extracts associations from the data to improve situational awareness. It can achieve this many times faster than humans can after appropriate model training.
"Intelligence"	The power of associations generated by ML enabled "collaboration, corroboration and reinforcement" avoiding "unwanted" "event detection" based on just a single sensor trigger.

3.2. The "Wake Word" for Energy Harvesting

Activating and testing the "wake word" simulation returned a 95.4% accuracy rate (Figure 28). Tests were conducted on the cloud service of EdgeImpule.com (personal demo account).

The wake word "Help Me!" trigger is set up as a two-syllable sound to avoid false detection.
A "wake" trigger is also chosen from the event detection list.
High risk maneuvers are tracked by multiple sensors. "Collaborative, corroborative and reinforcement" techniques, as presented in workflow drawings, are used to confirm the event.
Machine learning models unearth many more associations quicker and depending on tools used (such as CNN filters), producing "actionable intelligence".
https://docs.edgeimpulse.com/docs/edge-impulse-studio/learning-blocks

Figure 28. The wake word "Help Me!" turns on event recognition computing resources.

3.3. *The Confusion Matrix*

A table of the results using the confusion matrix for event detection using machine learning is recorded in Excel. The number of tests and trials required automatically calculating the results of the "model testing" in an automated spreadsheet, made it easier to record the consistency and to test different scenarios providing the flexibility to try out many different operational modes: voice, time sync data, fusion techniques, etc.

The trials (Figure 29) were carried out to experience the machine learning process and to experience its benefits, applicability and how it relates to existing processes. We also felt that the real power of TinyML is its ability to deliver a model which can be improved at will with training and is uploaded to a microcontroller device to perform application tailored analysis.

Performance Parameters	Trial-1	Trial-2	Trial-3
Accuracy	0.4571	0.6250	0.7143
Precision	0.7895	0.5000	0.5000
Recall	0.5000	0.5000	0.5000
False-Positive Rate (FPR),	0.8000	0.3000	0.2000
False-Negative Rate (FNR), and	0.5000	0.5000	0.5000
Specificity (Sp).	0.2000	0.7000	0.8000

These results showed consistency for different types of "detection" outcomes. The focus was model functionality across the four domains of interest because if the objectives change, retraining helps to tailor performance.

Figure 29. Test results—"event detection" trials.

Automation to Test Multiple Scenarios

A spreadsheet (Figure 30) was developed to automatically take in data and produce confusion matrix results so many tests and trials could be conducted for comparison over multiple tests.

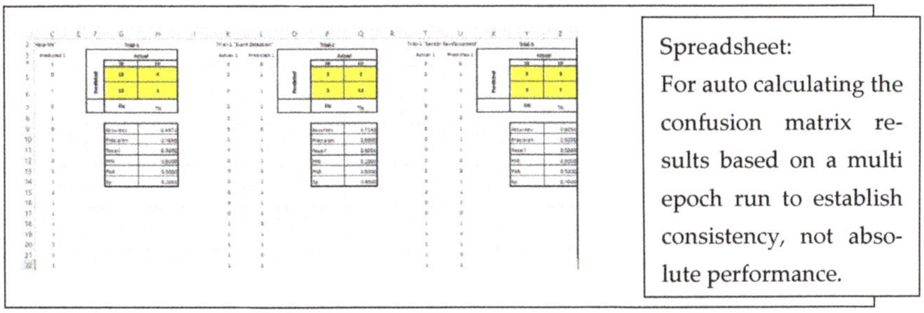

Figure 30. Spreadsheet to perform multiple trials to test for consistency using the confusion matrix.

4. Discussion

The most effective application of model-based sensor fusion would be in a smart city project. We juxtapose two scenarios and represent them in drawings and graphs.

Scenario-1 is a smart city project with 71 sensor nodes sending data over a LoRaWAN network to a central repository. In the design requirements there were no real-time expectations. Therefore, the system performed as per the resources provided to it. Independent sensors were used, which shared a LoRaWAN communication channel and produced Max1 to 100 Kbits/sec of data to a repository. The data were analysed for insights, then transitioned via an analysis package to a dashboard several minutes later depending on network traffic delays and retries or loss and "cleaning" functions to replace lost data. The time to collect, clean and have data ready for analysis for a real-time threat response was in the order of ten to twenty minutes (Figure 31).

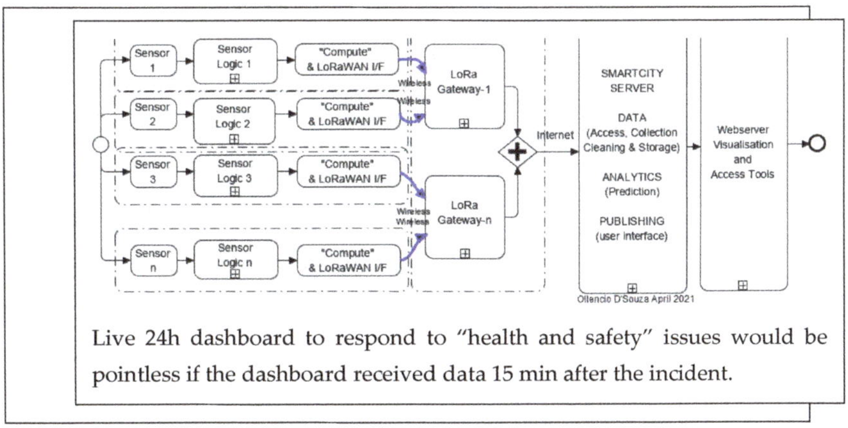

Figure 31. Schematic drawing shows the path of sensor data through infrastructure which could result in loss of temporal sync.

The second test scenario was created using the Arduino Nano BLE Sense with many more sensors on board. After ML training, presented earlier, the system detected and performed instantly over a sub-giga-hertz BLE (Bluetooth Low Energy) channel providing "actionable intelligence", i.e., human deliverable responses.

From the two studies, the following scenarios are described:

(a) Individual sensor data are sent over severely under-resourced networks to a central repository for processing, where issues such as network availability, speed and reliability are not guaranteed. Background analysis and "cleaning" (i.e., replacing missing data) at the receiving end is required before the results are sent to a dashboard as events for operators to respond to (Figure 31).

(b) Another scenario where "intelligence at the edge" generates analysed results using a customised machine learning model where the data are fused and analysed at the point where the data are collected, and "actionable intelligence" is sent out to the first responders (Figure 32).

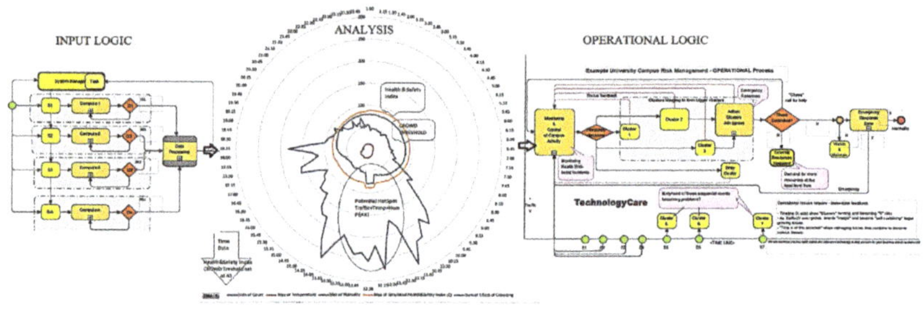

Multiple sensors, fused at the edge—delivering instant results to the dashboard for operations to respond to. No collection, storing, cleaning results sent directly to the first responder in the control room or in the field. On the left is the multiple sensor board managing sensor inputs as defined by the MLModel—uploaded to it, including the audio "Help Me!" call. This progressed to a dashboard (live 24 h event display spider web graph dashboard) which translated into an operational process model as "actionable intelligence".

Figure 32. The fusion dashboard operational mode.

These two scenarios present two different ways of delivering "actionable intelligence", which needs to be "on time" or it is of no benefit to the end user.

Limitations

The "need for speed" to deliver solutions that work forces humans to take short cuts, which does not sit well with machine learning because it expects us to know what kind of an output we want. "Garbage in–Garbage out" is so very true of the machine learning process that works incredibly fast and accurately. The process workflow was explained in depth, but if it is "compromised" by poor input or "reasoning" it is likely to magnify the lack of rigour in the assumptions made and outcomes delivered.

5. Conclusions

Our research explored, experienced, documented and presented several levels of discipline required to obtain reliable results when configuring machine learning systems. We researched and reported preliminary results from the domains where our TinyML model's customised detection capability was trialled, documenting the "features" and model functionality outcomes required, as shown in Tables 1 and 2.

An energy-efficient single-board microcontroller system with MEMS sensors on board (Figure 10) and BLE communications was set up in a test scenario. As the MCUs (embedded microprocessor control units) develop in sensor density and computing power, the opportunity to use low-power, high-throughput "meshed" (BLE inter-device communication) devices can only grow to improve "edge intelligence". This area is highlighted for future work.

Our research explored and deployed these sophisticated customised sensor models using data fusion and techniques that continue to be researched in the industry. The process of training, identifying features and functions that need to be used in training, is an expertise that must be practised. We tested and delivered customisable TinyML model designs to meet real-time operational domain requirements using "edge intelligence". These developed models can be used with many different MCUs with similar sensors and in other domains with similar detection requirements. This solves a significant issue for industries with severely restricted resources. We explored all avenues to ensure reliable data analysis at the "edge" delivers credible reliable results in temporal sync. Therefore, the results are received by "first responder" hand-held or control room dashboards so that they can be responded to immediately reducing "unwanted" or false alarms that are a drain on resources.

This research also explored and presented the possible sophisticated framework that uses data points to collaborate, corroborate and reinforce results in a data-driven operational environment using multiple sensor inputs, using four real domain requirements as examples.

Failure cost is a serious consideration because sensors contribute to operational performance and must be highly reliable. Poor reliability will require a human to physically attend to a system failure to find out the problem. Nevertheless, poor-quality products tend to add costs exponentially to the operational cost of a data-driven ecosystem. It is essential, therefore, to use value-engineered design principles to test and deploy with reliability in mind.

Further research into this area of "edge intelligence" is necessary along with key topics covered in this research paper, such as energy harvesting and wireless power, to extend reliability and satisfy value engineering criteria. TinyML is a formidable development and a useful tool for using sophisticated machine learning training methods to improve the performance of the customised models in real-time health, safety, security and fire safety situations.

Author Contributions: Conceptualization, O.D. and M.S.; Data curation, O.D.; Formal analysis, O.D.; Investigation, O.D. and S.C.M.; Methodology, O.D.; Project administration, S.C.M.; Resources, S.C.M.; Software, O.D.; Supervision, S.C.M. and M.S.; Writing—original draft, O.D.; Writing—review & editing, S.C.M. and M.S. All authors have read and agreed to the published version of the manuscript.

Funding: This research received no direct funding and is part of the work done by a student on a scholarship.

Institutional Review Board Statement: Not applicable.

Informed Consent Statement: The study did not involve humans in the trial which was conducted on the hardware only simulating potential situations observed in the application of the technology.

Data Availability Statement: Data both real and simulated was used in the trial and results presented in the paper. The study did not generate any extensive report data. The results were presented in a "Confusion Matrix Table".

Conflicts of Interest: The authors declare no conflict of interest.

Appendix A Non-Technical Explanation of the Objectives of the Research

The paper highlights key issues because of our reliance on data to function:

1. Data might not be in real time and often become corrupted by constrained infrastructure, before they are analysed. The paper raises these issues and proposes a solution to solve it. The solution uses machine learning to fuse data from different sensors, analyse it and deliver it live to where it is required most. The results are instant and sent directly to the "first responder", saving lives.
2. The solution also solves problems for overloaded infrastructure by reducing payload and ensuring only analysed results are sent.
3. Humans are intelligent and have risen up the evolution tree because they constantly invent tools that help humans do things better and more quickly. What has been developed is an intelligent tool that helps humans sieve through a lot of data to build a solution customised for the chosen real-time domains.
4. If someone is ill, would it be appropriate to obtain a "diagnosis" based on single parameter, or would you consider all factors causing the problem before treatment?

5. The TinyML model is developed from human logic, because we train the model by providing it with many examples of what is required and not required. It ingests our logic and then works incredibly fast to build associations and relevance (edge intelligence) by collating, collaborating, corroborating and reinforcing different data points to deliver a customised solution we designed.
6. We demonstrate how "intelligence at the edge" (close to where the data are collected) is used effectively to improve the reliability of service outcomes.

References

1. Merenda, M.; Porcaro, C.; Iero, D. Edge Machine Learning for AI-Enabled IoT Devices: A Review. *Sensors* **2020**, *20*, 2533. [CrossRef] [PubMed]
2. Ray, P.P.; Dash, D.; De, D. Real-time event-driven sensor data analytics at the edge-Internet of Things for smart personal healthcare. *J. Supercomput.* **2019**, *76*, 6648–6668. [CrossRef]
3. Das, H. *Real-Time Data Analytics for Large Scale Sensor Data*, 1st ed.; Academic Press: London, UK, 2020.
4. Teh, H.Y.; Kempa-Liehr, A.W.; Wang, K.I.-K. Sensor data quality: A systematic review. *J. Big Data* **2020**, *7*, 1–49. [CrossRef]
5. Soro, S. TinyML for ubiquitous edge AI. *arXiv* **2021**, arXiv:2102.01255.
6. IoT Data Collection: When Time is of The Essence, IoT Agenda. Available online: https://www.techtarget.com/iotagenda/post/IoT-data-collection-When-time-is-of-the-essence (accessed on 28 August 2022).
7. D'Souza, O.; Mukhopadhyay, S.; Akhter, F.; Khadivizand, S.; Memar, E. Extracting Operational Insights from Everyday IoT Data, Generated by IoT Sensors Over LoRaWAN. In *Proceedings of the International Conference on Computational Intelligence and Computing*; Springer: Singapore, 2022; pp. 241–249. [CrossRef]
8. Dai, Y.; Zhang, K.; Maharjan, S.; Zhang, Y. Edge Intelligence for Energy-Efficient Computation Offloading and Resource Allocation in 5G Beyond. *IEEE Trans. Veh. Technol.* **2020**, *69*, 12175–12186. [CrossRef]
9. "First Responder," Wikipedia. 30 June 2022. Available online: https://en.wikipedia.org/w/index.php?title=First_responder&oldid=1095789858 (accessed on 8 September 2022).
10. Aguileta, A.A.; Brena, R.F.; Mayora, O.; Molino-Minero-Re, E.; Trejo, L.A. Multi-Sensor Fusion for Activity Recognition—A Survey. *Sensors* **2019**, *19*, 3808. [CrossRef]
11. Reddi, V.J.; Plancher, B.; Kennedy, S.; Moroney, L.; Warden, P.; Suzuki, L.; Agarwal, A.; Banbury, C.; Banzi, M.; Bennett, M.; et al. Widening Access to Applied Machine Learning with TinyML. *arXiv* **2021**, arXiv:2106.04008.
12. Campbell, M.; Jovanovic, M. Conversational Artificial Intelligence: Changing Tomorrow's Health Care Today. *Computer* **2021**, *54*, 89–93. [CrossRef]
13. Mukhopadhyay, S.C.; Postolache, O.; Suryadevara, N.K. Special Issue on Smart Environments and Healthcare. *Appl. Sci.* **2019**, *9*, 1307. [CrossRef]
14. Gladstone, N. False Fire Cost Taxpayers $100 Million a Year. The Sydney Morning Herald. 22 February 2020. Available online: https://www.smh.com.au/national/nsw/false-fire-cost-taxpayers-100-million-a-year-20200220-p542j2.html (accessed on 1 July 2021).
15. Gupta, G.S.; Mukhopadhyay, S.C. Design Issues of Microcontroller Interfacing. In *Embedded Microcontroller Interfacing*; Springer: Berlin/Heidelberg, Germany, 2010; pp. 67–82. [CrossRef]
16. Chaari, L. *Digital Health Approach for Predictive, Preventive, Personalised and Participatory Medicine*, 1st ed.; Springer International Publishing: Cham, Switzerland, 2019. [CrossRef]
17. Azimirad, E.; Haddadnia, J.; Izadipour, A.L.I. A comprehensive review of the multi-sensor data fusion architectures. *J. Theor. Appl. Inf. Technol.* **2015**, 71.
18. Christou, I.T.; Kefalakis, N.; Zalonis, A.; Soldatos, J.; Bröchler, R. End-to-End Industrial IoT Platform for Actionable Predictive Maintenance. *IFAC-PapersOnLine* **2020**, *53*, 173–178. [CrossRef]
19. Laurent, G.; Moussa, M.D.; Cirenei, C.; Tavernier, B.; Marcilly, R.; Lamer, A. Development, implementation and preliminary evaluation of clinical dashboards in a department of anesthesia. *Int. J. Clin. Monit. Comput.* **2020**, *35*, 617–626. [CrossRef]
20. Phatak, A.A.; Wieland, F.-G.; Vempala, K.; Volkmar, F.; Memmert, D. Artificial Intelligence Based Body Sensor Network Framework—Narrative Review: Proposing an End-to-End Framework using Wearable Sensors, Real-Time Location Systems and Artificial Intelligence/Machine Learning Algorithms for Data Collection, Data Mining and Knowledge Discovery in Sports and Healthcare. *Sports Med.—Open* **2021**, *7*, 79. [CrossRef]
21. Ray, P.P. A review on TinyML: State-of-the-art and prospects. *J. King Saud Univ.—Comput. Inf. Sci.* **2021**, *34*, 1595–1623. [CrossRef]
22. Brena, R.F.; Aguileta, A.A.; Trejo, L.A.; Molino-Minero-Re, E.; Mayora, O. Choosing the Best Sensor Fusion Method: A Machine-Learning Approach. *Sensors* **2020**, *20*, 2350. [CrossRef]
23. Hermes, A.; Mandelbaum, J.; Parker, D.; Williams, H. *Value Engineering Synergies with Lean Six Sigma*; Productivity Press: New York, NY, USA, 2017.
24. Bao, W.; Wu, C.; Guleng, S.; Zhang, J.; Yau, K.-L.A.; Ji, Y. Edge computing-based joint client selection and networking scheme for federated learning in vehicular IoT. *China Commun.* **2021**, *18*, 39–52. [CrossRef]

25. Luo, B.; Sun, Z.; Pang, Y.; Ahmad, A.; Lin, J.; Wu, J.; Zhang, H. Neighbor Discovery for IPv6 over BLE Mesh Networks. *Appl. Sci.* **2020**, *10*, 1844. [CrossRef]
26. Agarwal, T. MEMS Sensor: Working Principle, Types, Advantages & Its Applications, ElProCus—Electronic Projects for Engineering Students. 2019. Available online: https://www.elprocus.com/mems-sensor-working-and-its-applications/ (accessed on 21 August 2022).
27. Igwegbe, G. Gigwegbe/Tinyml-Papers-and-Projects. 19 August 2022. Available online: https://github.com/gigwegbe/tinyml-papers-and-projects (accessed on 21 August 2022).
28. Banbury, C.R.; Reddi, V.J.; Lam, M.; Fu, W.; Fazel, A.; Holleman, J.; Yadav, P. Benchmarking TinyML systems: Challenges and direction. *arXiv* **2020**, arXiv:2003.04821.
29. Demrozi, F.; Turetta, C.; Chiarani, F.; Kindt, P.H.; Pravadelli, G. Estimating indoor occupancy through low-cost BLE devices. *IEEE Sens. J.* **2021**, *21*, 17053–17063. [CrossRef]

sensors

MDPI

Article

Trainable Quaternion Extended Kalman Filter with Multi-Head Attention for Dead Reckoning in Autonomous Ground Vehicles

Gary Milam [1,†], Baijun Xie [1,†], Runnan Liu [1,†], Xiaoheng Zhu [1,†], Juyoun Park [2], Gonwoo Kim [3] and Chung Hyuk Park [1,*]

1 Department of Biomedical Engineering, George Washington University, Washington, DC 20052, USA
2 Korea Institute of Science and Technology, Seoul 02792, Korea
3 Department of Control and Robot Engineering, ChunBuk National University, Chungbuk 28644, Korea
* Correspondence: chpark@gwu.edu
† These authors contributed equally to this work.

Abstract: Extended Kalman filter (EKF) is one of the most widely used Bayesian estimation methods in the optimal control area. Recent works on mobile robot control and transportation systems have applied various EKF methods, especially for localization. However, it is difficult to obtain adequate and reliable process-noise and measurement-noise models due to the complex and dynamic surrounding environments and sensor uncertainty. Generally, the default noise values of the sensors are provided by the manufacturer, but the values may frequently change depending on the environment. Thus, this paper mainly focuses on designing a highly accurate trainable EKF-based localization framework using inertial measurement units (IMUs) for the autonomous ground vehicle (AGV) with dead reckoning, with the goal of fusing it with a laser imaging, detection, and ranging (LiDAR) sensor-based simultaneous localization and mapping (SLAM) estimation for enhancing the performance. Convolution neural networks (CNNs), backward propagation algorithms, and gradient descent methods are implemented in the system to optimize the parameters in our framework. Furthermore, we develop a unique cost function for training the models to improve EKF accuracy. The proposed work is general and applicable to diverse IMU-aided robot localization models.

Keywords: localization; inertial navigation system; extended Kalman filter; mobile robot; autonomous ground vehicle

Citation: Milam, G.; Xie, B.; Liu, R.; Zhu, X.; Park, J.; Kim, G.; Park, C.H. Trainable Quaternion Extended Kalman Filter with Multi-Head Attention for Dead Reckoning in Autonomous Ground Vehicles. *Sensors* **2022**, *22*, 7701. https://doi.org/10.3390/s22207701

Academic Editor: Natividad Duro Carralero

Received: 20 September 2022
Accepted: 7 October 2022
Published: 11 October 2022

Publisher's Note: MDPI stays neutral with regard to jurisdictional claims in published maps and institutional affiliations.

1. Introduction

Over the past years, localization has become one of the challenging issues for autonomous ground vehicles (AGVs). In particular, the most difficult issue in navigation is estimating the accurate and stable position and orientation of the robot through data obtained from sensors and other navigation systems [1]. Recently, various technologies have been designed to solve robot localization problems, such as visual-odometry-aided camera localization [2] and Global Positioning System (GPS)-based localization using reinforcement learning [3].

However, the simultaneous localization and mapping (SLAM) methods may fail to function correctly in certain complicated situations due to the physical characteristics of sensors. The laser imaging, detection, and ranging (LiDAR) sensor, for example, is a sensor system that measures distance by transmitting light into spaces and receiving reflected signals from a target [4]. However, LiDAR can lose its signal in situations such as foggy and rainy conditions. In addition to that, the strength of the signal can be affected by the reflectivity of the objects. This could lead a mobile robot or an AGV into a target-blind zone, endangering safety and maneuverability. Therefore, a reliable contingency plan needs to be considered that can compensate for the performance degradation caused by these limitations of the sensors.

The inertial measurement unit (IMU), a combination of a gyroscope, accelerometers, and sometimes magnetometers, could provide an efficient approach to solve this problem. Specifically, the IMU is one of the solutions that can independently measure the state of the body without any external feedback. The accelerometer and gyroscope components measure linear acceleration and angular velocity, which represent the movement of the body to which the sensor is attached. Additionally, the magnetometer component measures orientation based on the Earth's magnetic field, which is available almost everywhere [5].

Nevertheless, there is one drawback that could degrade the performance of an IMU-based localization, which is the accumulated drift. To reliably utilize the IMU, it is imperative to eliminate the accumulated drift [1]. In the field of probabilistic robotics, various methods for correcting errors, such as Bayes filters, Gaussian filters (e.g., information filters), and nonparametric filters (e.g., particle filters) [6], have been proposed. Among them, the extended Kalman filter (EKF) [7] is one of the most widely used methods to reduce the accumulated drift. The EKF is a nonlinear Kalman filter (KF) that linearizes a current mean and covariance estimate. Since EKF can solve nonlinear problems, it has been applied to IMU-aided localization systems [8–10]. The process-noise covariance matrix, Q, and the measurement-noise covariance matrix, R, are constructed with a priori constant values determined by the characteristics of sensors and environments in traditional KF systems, which assume that they remain constant throughout the whole navigation operation. EKF can achieve optimal results if the process noise is well defined. However, depending on external factors, such as complex environments or sensor limitations (e.g., occlusions), sensor noise values can change, and it is difficult to recognize the exact error and in situ information of when and how the change occurs [11].

The following is a list of the major contributions of this study:

1. In this work, we propose a novel approach to improve the accuracy of EKF-based IMU localization with a convolutional neural network (CNN) architecture. Specifically, we design a stable training method that can find the optimal parameters of the system and the observation-noise covariance in real time by reducing the error in each iteration. Furthermore, the system is designed and tested for online training, unlike many other approaches, such as [12], where the algorithm is trained offline using batch and multiple epochs. The intention behind this is that the algorithm is to be trained continuously while SLAM is functioning online, in which case a sequence of IMU data points is observed and acquired.
2. Our proposed CNN module consists of multi-head attention (MHA) layers to model the cross-modal fusion of different sources of modalities (e.g., multiple IMUs, lidars, etc.). The MHA was initially proposed to address the problems of natural language processing (NLP) [13], and it was later discovered to be effective in modeling cross-modal interactions between different modalities [14]. These previous works inspired us to model cross-modal interactions that combine different sensor information sources via the attention mechanism.
3. We conducted extensive experiments using an actual robotic platform to assess the effectiveness of our proposed method in the real world (a factory environment in our case). We designed real-world scenarios for the online training, where the SLAM might fail in some cases and only the IMU(s) can provide sensory information for the EKF-based localization module. The algorithm is also trained continuously while the robot is online and navigating.

2. Related Work

Studies have been conducted recently on Kalman filter (KF)-based localization technology with adaptive noise-covariance estimation. One previous study proposed by Akhlaghi et al. [15] introduced innovation-based and residual-based methods to adaptively adjust the covariance matrices Q and R at each step of the EKF process to improve the state estimation accuracy. In addition, Hu et al. [16] proposed an adaptive unscented Kalman filter (UKF), another variant of KF for the nonlinear system, with process-noise covariance estimation to improve the UKF performance. However, these approaches might not be able to fully characterize the nonlinear stochastic noises that arise in real-world situations.

It is known that the artificial neural network has the capability to approximate nonlinear functions [17]. Haarnoja et al. [18] demonstrated that a backpropagation-based Kalman filter, consisting of a KF and a CNN, was capable of predicting the measurement-noise covariance matrix, where the CNN was trained via minimizing position errors. Brossard et al. [12] proposed an approach for dead-reckoning for wheeled vehicles with the IMU only. Deep neural networks were used to update the parameters of an invariant EKF dynamically. A recent study also explored the use of long short-term memory (LSTM), a type of recurrent neural network (RNN), to model the nonlinear noises for KF [19] to address target tracking problems. Another approach that uses reinforcement learning to adaptively estimate the process-noise covariance matrix was proposed by Gao et al. [20], in which their algorithm used the deep deterministic policy gradient (DDPG) to extract the optimal process-noise covariance matrix estimation from the continuous action space, using an integrated navigation system as the environment and the reverse of the current positioning error as the reward. Wu et al. [21] also proposed a deep learning framework combining a denoising autoencoder and a multitask temporal CNN. Multitask learning was used to optimize the loss for both the process-noise covariance and measurement-noise covariance matrices from KF simultaneously.

3. Quaternion-Based Extended Kalman Filter

3.1. IMU Inclination Calculation

The rotation matrix $\mathbf{R}_n^b$, mapping the navigation frame n to the body frame b, can be represented by ϕ (rotation angle along the x-axis), θ (rotation angle along the y-axis), and ψ (rotation angle along the z-axis), as follows (trigonometric functions *sin* and *cos* are denoted as s and c, respectively):

$$\mathbf{R}_n^b = \begin{bmatrix} c\theta c\psi & c\theta s\psi & -s\theta \\ c\psi s\theta s\phi - c\phi s\psi & c\phi c\psi + s\theta s\phi s\psi & c\theta s\phi \\ c\phi c\psi s\theta + s\phi s\psi & c\phi s\theta s\psi - c\psi s\psi & c\theta c\phi \end{bmatrix}. \tag{1}$$

When the IMU is stationary or moving at a constant speed, the acceleration in the navigation frame should be equal to the gravity constant g, so the inclination angles θ and ϕ can be calculated by Equation (2) as in [22]:

$$\begin{bmatrix} a_x^b \\ a_y^b \\ a_z^b \end{bmatrix} = \mathbf{R}_n^b \begin{bmatrix} 0 \\ 0 \\ g \end{bmatrix} \Rightarrow \begin{cases} \theta = \arcsin(-\frac{a_x^b}{g}) \\ \phi = \arctan(\frac{a_y^b}{a_z^b}) \end{cases}. \tag{2}$$

The IMU coordinate frame is assumed as shown in Figure 1, where the z-axis is upward, so gravity has a positive value.

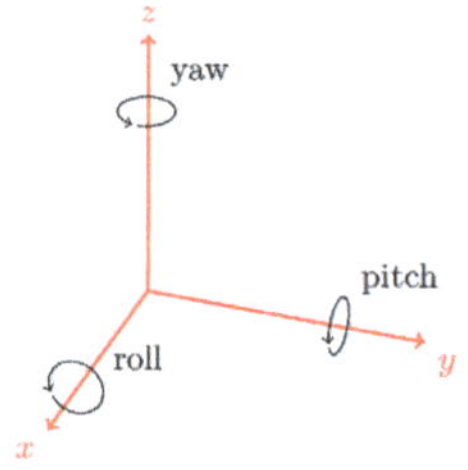

Figure 1. The IMU coordinate frame.

3.2. IMU Integration Model

The continuous-time relationships among position $\mathbf{p}^n$, velocity $\mathbf{v}^n$, and acceleration $\mathbf{a}^n$ are defined as follows:

$$\frac{\partial \mathbf{p}_t^n}{\partial t} = \mathbf{v}_t^n, \ \frac{\partial \mathbf{v}_t^n}{\partial t} = \mathbf{a}_t^n, \tag{3}$$

where $\mathbf{a}_t^n$ represents the acceleration in the navigation frame at time t, which can be calculated using $\mathbf{a}_t^b$, which is the acceleration obtained from the IMU sensor as follows:

$$\mathbf{a}_t^n = \mathbf{R}(\mathbf{q}_b^n)\mathbf{a}_t^b - \mathbf{g}^n, \tag{4}$$

where $\mathbf{R}(\mathbf{q}_b^n)$ is the rotation matrix represented by quaternion $\mathbf{q}_b^n$. The orientation $\mathbf{q}_b^n$ and the angular velocity ω^b are related as

$$\frac{\partial \mathbf{q}_b^n}{\partial t} = \mathbf{q}_b^n \odot \frac{1}{2}\omega^b. \tag{5}$$

From the continuous-time model, the dynamics of position, velocity, and orientation in discrete time are given by Equations (6)–(8), as explained in [23], as follows:

$$\mathbf{p}_t^n = \mathbf{p}_{t-1}^n + \mathbf{v}_{t-1}^n \cdot \delta t + \frac{1}{2}(\mathbf{a}_{t-1}^n + \mathbf{e}_{a,t}) \cdot \delta t^2 \tag{6}$$

$$\mathbf{v}_t^n = \mathbf{v}_{t-1}^n + (\mathbf{a}_{t-1}^n + \mathbf{e}_{a,t}) \cdot \delta t \tag{7}$$

$$\mathbf{q}_{b,t}^n = \mathbf{q}_{b,t-1}^n \odot exp_q(\frac{1}{2}(\omega_{t-1}^b - \mathbf{e}_{\omega,t}) \cdot \delta t), \tag{8}$$

where $\mathbf{e}_{a,t}, \mathbf{e}_{\omega,t}$ are the noise terms of the dynamics model which are assumed to follow the normal distribution, and the distribution axes are independent of each other, as follows:

$$\mathbf{e}_{a,t} \sim \mathcal{N}(0, \sigma_a \mathcal{I}_3) \tag{9}$$

$$\mathbf{e}_{\omega,t} \sim \mathcal{N}(0, \sigma_\omega \mathcal{I}_3). \tag{10}$$

The state variable $\mathbf{x}_t$ is a 10×1 vector consisting of the current position $\mathbf{p}_t^n$, velocity $\mathbf{v}_t^n$, and orientation $\mathbf{q}_{b,t}^n$ which represents the mapping of the body frame onto the navigation frame, as follows:

$$\mathbf{x}_t = \begin{bmatrix} \mathbf{p}_t^n, & \mathbf{v}_t^n, & \mathbf{q}_{b,t}^n, \end{bmatrix}_{10\times 1}^T \tag{11}$$

so the state transition function can be written as

$$\hat{\mathbf{x}}_{t|t-1} = f(\hat{\mathbf{x}}_{t-1|t-1}, \mathbf{u}_t, \mathbf{e}_t), \tag{12}$$

where $\mathbf{u}_t = [\mathbf{a}_t^b, \omega_t^b]^T$ is the control input modeled by the accelerometer and gyroscope measurements, and the noise term $\mathbf{e}_t = [\mathbf{e}_{a,t}, \mathbf{e}_{\omega,t}]^T$.

We linearize Equation (12) at the current estimate and propagate the covariance forward to predict the system covariance

$$\mathbf{P}_{t|t-1} = \mathbf{F}_t \mathbf{P}_{t-1|t-1} \mathbf{F}_t^T + \mathbf{G}_t \mathbf{Q} \mathbf{G}_t^T, \tag{13}$$

where $\mathbf{F}_t$, $\mathbf{G}_t$ are Jacobian matrices of the transition function (12) with respect to $\mathbf{x}_t$ and $\mathbf{u}_t$, as shown below:

$$\mathbf{F}_t = \left.\frac{\partial f(\mathbf{x}_t, \mathbf{u}_t, \mathbf{e}_t)}{\partial \mathbf{x}_t}\right|_{\substack{\mathbf{e}_t=0 \\ \mathbf{x}_t=\hat{\mathbf{x}}_{t-1|t-1}}} \tag{14}$$

$$\mathbf{G}_t = \left.\frac{\partial f(\mathbf{x}_t, \mathbf{u}_t, \mathbf{e}_t)}{\partial \mathbf{u}_t}\right|_{\substack{\mathbf{e}_t=0 \\ \mathbf{x}_t=\hat{\mathbf{x}}_{t-1|t-1}}}, \tag{15}$$

and the process covariance $\mathbf{Q} = diag(\Sigma_a, \Sigma_a, \Sigma_\omega)$, where Σ_a and Σ_ω represent the covariances of acceleration and angular velocity, respectively.

3.3. EKF Correction with Measurements

Since the accelerometer measures the local gravity vector when an AGV is moving at a constant speed or is stationary, it can provide information about the inclination of the sensor [23]. Then, the robot control center can provide the velocity information $\mathbf{v}^b$ along the x-axis and y-axis at current times. In addition, we consider the vertical velocity, which is roughly null in the robot frame, as a pseudo-velocity measurement v^b_{pseudo}, so the total measurement vector can be written as

$$\mathbf{z}_t = \begin{bmatrix} \mathbf{a}^b_t, & \mathbf{v}^b_{cmd,t}, & v^b_{pseudo,t} = 0 \end{bmatrix}^T_{6\times 1}. \tag{16}$$

Thus, we can obtain the measurement function mapping the state space to the measurement space as follows:

$$h(\hat{\mathbf{x}}_{t|t-1}) = \begin{bmatrix} \mathbf{R}(\hat{\mathbf{q}}^n_{b,t|t-1})^T \mathbf{g}^n \\ \mathbf{R}(\hat{\mathbf{q}}^n_{b,t|t-1})^T \hat{\mathbf{v}}^n_{t|t-1} \end{bmatrix}, \tag{17}$$

Then, we obtain the measurement matrix $\mathbf{H}_t$ by linearizing the measurement function:

$$\mathbf{H}_t = \left.\frac{\partial h(\mathbf{x}_t)}{\partial \mathbf{x}_t}\right|_{\mathbf{x}_t = \hat{\mathbf{x}}_{t|t-1}}. \tag{18}$$

Therefore, the measurement residual $\mathbf{y}_t$ and the Kalman gain $\mathbf{K}_t$ are calculated by Equations (19) and (20), as detailed in [24]:

$$\mathbf{y}_t = \mathbf{z}_t - h(\hat{\mathbf{x}}_{t|t-1}) \tag{19}$$

$$\mathbf{K}_t = \mathbf{P}_{t|t-1}\mathbf{H}_t^T(\mathbf{H}_t\mathbf{P}_{t|t-1}\mathbf{H}_t^T + \mathbf{R})^{-1}, \tag{20}$$

where the measurement covariance matrix $\mathbf{R}$ can be defined as $diag(\Sigma_a, \Sigma_v, \Sigma_{v_{pseudo}})$. Finally, the predicted state and covariance are updated as follows:

$$\hat{\mathbf{x}}_{t|t} = \hat{\mathbf{x}}_{t|t-1} + \mathbf{K}_t\mathbf{y}_t, \tag{21}$$

$$\mathbf{P}_{t|t} = (\mathbf{I} - \mathbf{K}_t\mathbf{H}_t)\mathbf{P}_{t|t-1}. \tag{22}$$

4. Covariance Optimization

4.1. Adjustable Covariance

It is well known that the EKF is a model-based optimal filter, which requires exact knowledge of process and measurement models as well as process- and measurement-noise statistics. However, it is difficult to model the dynamic noise changes over time [25]. Thus, we redesign the covariance Σ of the covariance matrices $\mathbf{Q}$ and $\mathbf{R}$ as follows:

$$\Sigma_a = \sigma_a^2 \cdot 10^{\mu \tanh(s_a)} \tag{23}$$

$$\Sigma_\omega = \sigma_\omega^2 \cdot 10^{\mu \tanh(s_\omega)} \tag{24}$$

$$\Sigma_{v_{cmd}} = \sigma_{v_{cmd}}^2 \cdot 10^{\mu \tanh(s_{v_{cmd}})} \tag{25}$$

$$\Sigma_{v_{pseudo}} = \sigma_{v_{pseudo}}^2 \cdot 10^{\mu \tanh(\mathbf{s}_{v_{pseudo}})}, \tag{26}$$

where σ_a, σ_ω, $\sigma_{v_{cmd}}$, and $\sigma_{v_{pseudo}}$ correspond to our initial estimate of the noise parameters and $\mu > 0$. Thus, the covariance can be limited between a factor $10^{-\mu}$ and a factor 10^{μ} with respect to its original value because of the function $\tanh(\cdot)$, which makes the covariance optimized within a reasonable interval set heuristically. By adjusting the value of the parameter s, the covariance matrices could be changed indirectly.

Parameters s_a and s_ω are adjusted during training by backpropagation based on the loss function described by Equation (30). Once the training stops, the parameters are considered fixed for the algorithm.

Although the noise components of the velocity v and pseudo-velocity v_{pseudo} are unknown, the deviation can be assumed to be dynamic rather than stationary in the real world. In other words, the measurement covariance from velocity can be treated as loose strict null instead of strict null, which means that the uncertainty can be encoded in the covariance [12]. A CNN layer is applied to dynamically compute the parameters s_v and $s_{v_{pseudo}}$, taking as input a window size of N IMU data points.

$$\begin{bmatrix} \mathbf{s}_v \\ s_{v_{pseudo}} \end{bmatrix} = \begin{bmatrix} \begin{bmatrix} s_{v_x} \\ s_{v_y} \end{bmatrix} \\ s_{v_{pseudo}} \end{bmatrix} = CNN([u_{t-N}, ...u_t]). \tag{27}$$

Figure 2 shows the CNN architecture used to predict parameters s_v and $s_{v_{pseudo}}$. A window size of 20 IMU data points was used for the input. Each IMU data point consists of acceleration and angular velocity data for the three axes (x,y,z). Using a single channel, where the overall input dimension becomes $1 \times 6 \times 20$, the input is initially split into its individual acceleration and angular velocity matrices, which are processed separately by their respective 3×3 convolution layers followed by a leaky ReLU activation (ConvLR block). Then, a multi-head attention (MHA) layer is introduced to model the feature-fusion data. The output of the individual paths is then concatenated and processed by two more ConvLR blocks and a global average pooling layer. The final result is the three parameters s_{v_x}, s_{v_y}, and $s_{v_{pseudo}}$.

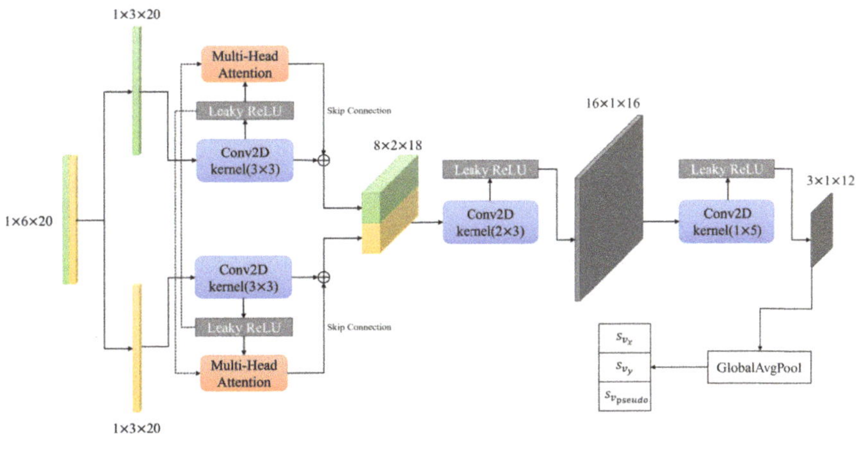

Figure 2. The proposed CNN architecture for predicting the parameters s_{v_x}, s_{v_y}, and $s_{v_{pseudo}}$, which are used to calculate the covariances as per Equations (25) and (26). The input consists of a window size of 20 IMU data points, each containing the acceleration and angular velocity data for all three axes.

The MHA mechanism was initially proposed in the field of natural language processing (NLP) [13]. Later, Tsai et al. [14] explored leveraging MHA mechanism to reinforce a target modality with features from another data modality via learning the cross-modal attention. The following is the formulation of the attention output:

$$\mathbf{Attention}(Q_A, K_B, V_B) = \mathbf{softmax}(\frac{Q_A K_B^\top}{\sqrt{d_k}}) V_B. \tag{28}$$

Following the definition of [14], $Q_A \in \mathbb{R}^{d \times d_A}$ denotes the queries from the modality A, $K_B \in \mathbb{R}^{d \times d_B}$ denotes the set of keys, and $V_B \in \mathbb{R}^{d \times d_B}$ denotes the set of values from the modality B. Then, the information from modality B is passed to modality A by calculating the attention function in Equation (28). In this study, we leverage the MHA to model feature fusion via introducing one from another modality. As shown in Figure 2, taking acceleration data as an example, we reinforce its features via modeling the cross attention with angular velocity data by calculating attention output $\mathbf{Attention}(Q_{acc}, K_{ang}, V_{ang})$. A skip connection is also implemented, to sum up the output from the first 3×3 convolution layers and attention output.

4.2. Online Training Method

SLAM works well unless a sensor fails [26], which is a well-known problem statement. Therefore, our training system initially considers the SLAM outputs under ideal (reliable) conditions as ground truths to calculate the loss function between the output states of SLAM and EKF.

Furthermore, the iterative EKF estimation process in consecutive time steps is a kind of Markov decision process, which means that the current state is only related to the previous one, so our method focuses on optimizing each EKF estimation process. When the estimation performance of each EKF iteration is high, it will show high estimation accuracy in the entire iterative process. In order to evaluate the performance of each EKF estimation, the loss function is designed as follows:

$$loss = \text{MSE}\,(\,\hat{\mathbf{x}}_{t+1|t+1}\,,\,\mathbf{x}_{t+1}^{slam}), \tag{29}$$

where MSE is the mean squared error (MSE) function that expresses the bias of the estimated state by EKF compared to the state of SLAM at timestamp $t+1$. However, the performance of AGV localization depends on the two-dimensional (2D) position errors (p_x, p_y) and the heading angle (ψ) errors, so the loss function only needs to compute the mean squared error of p_x, p_y, and ψ calculated from the orientation state $\mathbf{q}$. Thus, the loss function can be rewritten as

$$loss = \text{MSE}\,(\,[\,\hat{p_x}, \hat{p_y}, \hat{\psi}\,]_{t+1}^T\,,\,[\,p_x^{slam}, p_y^{slam}, \psi^{slam}\,]_{t+1}^T\,). \tag{30}$$

The initial state of each EKF iteration should be the same as the state of SLAM at the previous timestamp, so that the loss function can effectively express the error generated by each EKF iteration. Therefore, the input of trainable EKF that estimates the state at the next timestamp should be the state from SLAM at the current timestamp during training as follows:

$$\hat{\mathbf{x}}_{t+1|t+1} = \text{EKF}(\mathbf{x}_t^{slam}\,,\,\mathbf{u}_t\,,\,\mathbf{v}_{t+1}^{cmd}, \mathbf{P}_t^{slam}), \tag{31}$$

which also considers the initial covariance $\mathbf{P}_t$ of EKF as the current estimation covariance from SLAM.

However, the frequency of SLAM is different from the frequency of IMU in real-time, so we cannot guarantee that the estimated state of each EKF iteration corresponds to the output state of each SLAM at the same time, which means that it is unable to regard the ground truth of EKF as SLAM at that time. Additionally, the frequency of IMU is usually higher than the frequency of SLAM, so to synchronize the output of EKF and SLAM, multiple EKF iterations can be trained by performing backpropagation [27] at once when the output state of SLAM is provided. The training structure is shown in Figure 3. The structure allows the system to optimize the covariance in real-time continuously. The training process begins with the MSE loss function (30), and then calculates the gradient of the loss function corresponding to each $\mathbf{Q}$ and $\mathbf{R}$ based on the derivative chain rule.

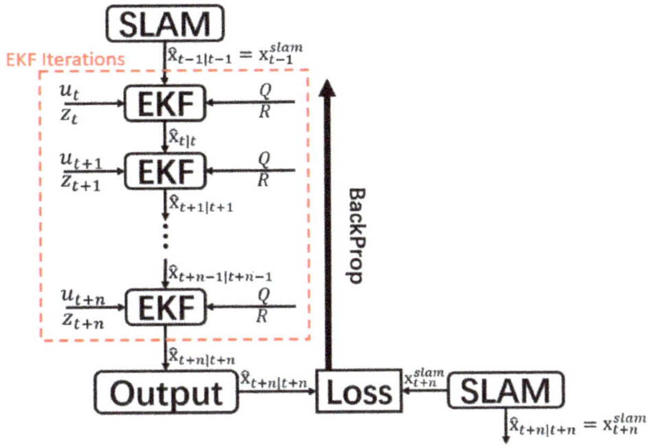

Figure 3. Structure and training flow of our network. The initial state of the EKF iterations is based on the previous SLAM state update at timestamp $t-1$. N number of EKF iterations will be performed until the next update of the SLAM state at $t+n$. The loss is calculated between the SLAM state update and the final predicted output state of the N EKF updates. Then, backpropagation is performed on the basis of the calculated loss.

4.3. Implementation Details

This section introduces the settings and the implementation details of our algorithm. The whole self-adjusting method is implemented in Python with the PyTorch library [28] for training and inference, and the Robot Operating System (ROS) [29] was utilized for collecting sensor data.

The initial parameters of the EKF were set as follows prior to training. The initial system error covariance $\mathbf{P}_0 = \mathbf{I}_{10}$, which is the identity matrix 10×10. We set $\sigma_a = 0.01$ m/s^2 in (23), $\sigma_\omega = 0.01$ rad/s^2 in (24), $\sigma_v = 0.25$ m/s in (25), and $\sigma_{v_{pseudo}} = 0.0225$ m/s in (26). The adjustable parameters s are defined as $\mathbf{s}_a = \mathbf{s}_\omega = \mathbf{s}_{v_{cmd}} = s_{v_{pseudo}} = 0.01$ in order to make initial covariance $\Sigma_a \approx \sigma_a^2$, $\Sigma_\omega \approx \sigma_\omega^2$, $\Sigma_{v_{cmd}} \approx \sigma_{v_{cmd}}^2$, and $\Sigma_{v_{pseudo}} \approx \sigma_{v_{pseudo}}^2$. We defined $\mu = 3$ from (23) to (26), which allows for each covariance element to be 10^3 times higher or smaller than its original values [12].

The Adam optimizer [30] was applied to update parameters with learning rate 10^{-3} and iterated once for each tandem EKF.

5. Experiments and Results

In this section, we present the robotic platform that was used for collecting data and testing a real-time SLAM failure scenario. Three evaluation metrics were used for evaluating the performance. The performance of our proposed localization architecture, the trainable quaternion-based EKF, was evaluated by comparing it with the EKF model without training. Additionally, various deep learning model architectures were compared to determine the most effective model.

5.1. Dataset

The data collected to evaluate the proposed algorithm are based on the usage of a proprietary omniwheel robot platform of dimensions of 2.481×1.595 meters. The robotic system provides the current velocity of the robot, which is used as an input to the EK. The robot contains four 2D LiDARs (two in the front and two in the rear), two stereo cameras (one in the front and one in the rear), and an IMU. Figure 4 shows the robot setup.

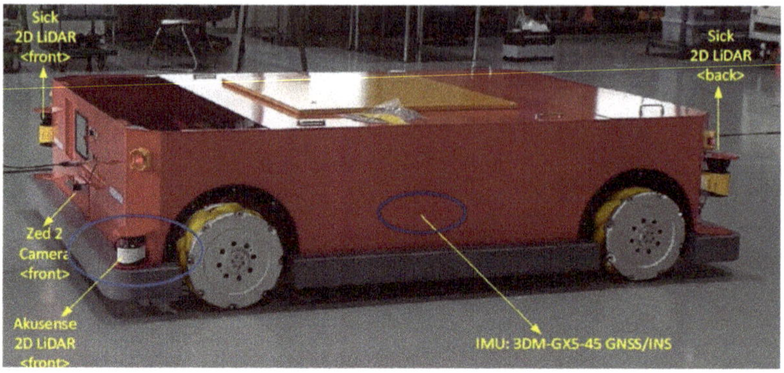

Figure 4. Robot setup used for data collection of the experiment.

Experiments were conducted across four different trajectories with total lengths of 66.46, 145.39, 103.42, and 78.62 meters, respectively. The trajectory path collected as a result of SLAM was used as the ground truth. For this work, the visual–LiDAR–inertial SLAM algorithm in [31] was used, as it provides a very accurate position and orientation. The dataset collected contained the SLAM output (ground truth position and orientation), the acceleration and angular velocity information, and IMU data for the EKF state estimation calculations.

Since the EKF state estimation was calculated at the origin of the IMU, all data points (SLAM and velocity) were transformed to the IMU's origin. Furthermore, all data had to be transformed to the body frame coordinates of the omniwheel robot, where the x-axis runs along the length of the robot in the forward direction, the y-axis is 90 degrees anticlockwise, and the z-axis runs upward.

5.2. Experimental Setup

To evaluate the improvement of the proposed algorithm, the localization output of a traditional EKF was used as the experimental control data.

Our proposed system architecture was designed with the intention that the algorithm would be trained continuously while the robot is online; therefore, the data are collected in real time, and each data point is unique and independent. Taking this into account, we trained the algorithm on three sequence runs and performed the inference test on the fourth unseen sequence, with the data of each sequence only seen once during training and inference. This process was performed four times to run the inference test on all available sequences.

The EKF state estimation was then fused with the SLAM state during the inference mode while the system considered the SLAM positional data reliable, which is determined by the covariance values provided by the SLAM module. The time range when the SLAM positional data were unreliable was considered the SLAM failure period, at which point the SLAM state was no longer fused with the EKF estimation, and the output was purely relying on EKF.

To simulate a real-time SLAM failure scenario, 100 s time intervals within each individual trajectory were selected as the SLAM failure period. Both approaches, the traditional EKF and the proposed trainable EKF, had the same start and end times for the SLAM failure.

In Figures 5 and 6, the period from when SLAM failed ("Start Point") to when SLAM recovered ("End Point") were regarded as the SLAM failure period. The green and orange paths represent the estimated position from the untrained EKF and EKF CNN with MHA, respectively, while the blue path (ground truth) is from the SLAM outputs.

To evaluate the performance of our proposed algorithms, we considered the following three evaluation metrics for our experiments:

1. *Position Error* (P_e) :
   The ratio of the position error to the total path length when SLAM fails.

$$P_e = (Last\ Position\ Error / Total\ Distance) * 100 \tag{32}$$

2. *Rotation Error* (A_e) :
   Azimuth angle error relative to the total path length when SLAM fails.

$$A_e = Last\ Azimuth\ Angle\ Error / Total\ Distance \tag{33}$$

3. *Average Mean Squared Error* (MSE_{avg}):
   The overall average of the squares of the errors between ground truths (SLAM) and object to be assessed (estimated position and orientation) when SLAM fails.

$$MSE_{avg} = Mean(\ MSE\ (\ [\hat{p_x}, \hat{p_y}, \hat{\psi}]^T_{t+1}, [p_x^{slam}, p_y^{slam}, \psi^{slam}]^T_{t+1}\ )\ ) \tag{34}$$

5.3. Result and Analysis

We present the comparison results of our proposed algorithm with those of the traditional method (i.e., EKF). From the trajectory comparisons in Figures 5 and 6, we can see that the proposed EKF CNN with MHA (orange path) was more in line overall with less noise than the traditional EKF (green path). Furthermore, from Figures 7 and 8, which provide the mean squared errors (MSE) during the failure period, we can see that the proposed method (green line) could achieve a lower MSE than the traditional EKF (blue line). From Figures 7 and 8, the maximum MSE for the proposed method was 0.0297 and 0.0323, respectively, while the traditional EKF method reached 0.0690 and 0.1108 for the same trajectories.

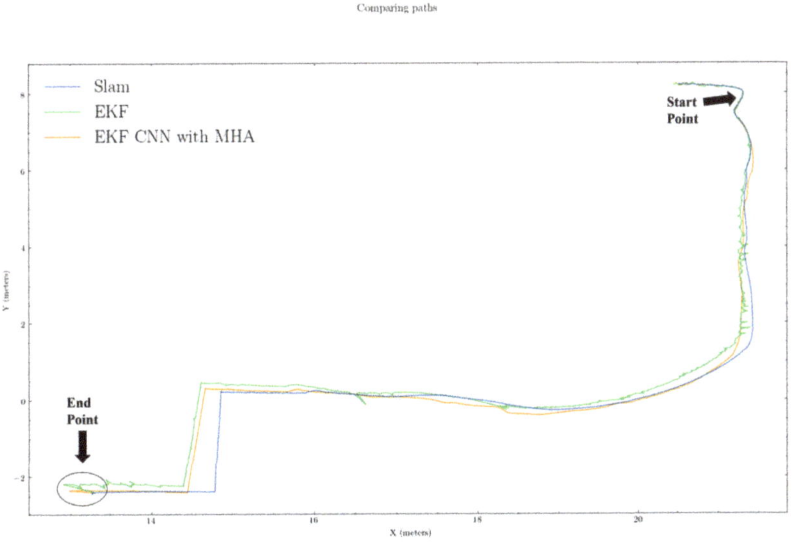

Figure 5. Estimated path comparison results for trajectory 1. Green, orange, and blue lines represent EKF, EKF+CNN with MHA, and ground truths, respectively.

Table 1 shows the full test results we collected and analyzed. As mentioned for each trajectory, an arbitrary interval of 100 s was chosen as the failure period, which could result in different failure lengths. Based on the provided results, it is clear that our proposed method overall outperformed the traditional EKF. In all cases, the position error P_e and the rotation error A_e were lower with the proposed method. In fact, the proposed method achieved 1.24765% for overall P_e and 0.02785 deg/m for A_e. Similarly, the proposed method performed much better on the MSE metric.

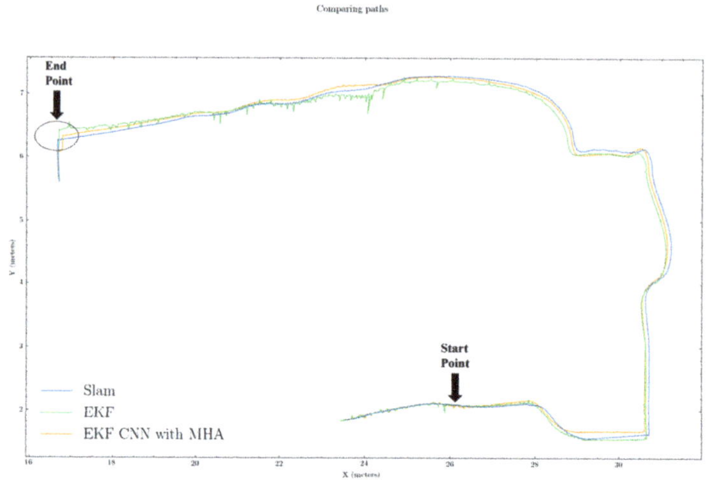

Figure 6. Estimated path comparison results for trajectory 4. Green, orange, and blue lines represent EKF, EKF+CNN with MHA, and ground truths, respectively.

Table 1. Results for the four trajectory paths followed by evaluation metrics. Failure length: moving track length of robot during SLAM failure period; P_e: position error; A_e: rotation error; MSE_{avg}: average mean squared error.

Trajectory	Failure Length (m)	EKF			Proposed		
		P_e (%)	A_e (deg/m)	MSE_{avg}	P_e (%)	A_e (deg/m)	MSE_{avg}
1	18.19	1.64180	0.01320	0.02055	0.97901	0.00639	0.01580
2	23.92	3.56478	0.10087	0.11934	2.79777	0.04045	0.07454
3	22.94	1.00380	0.22513	0.04406	0.84346	0.01948	0.01294
4	24.81	0.38022	0.02335	0.00504	0.37035	0.04509	0.00523
Overall		1.64765	0.09064	0.04725	1.24765	0.02785	0.02713

Our trainable EKF CNN with MHA showed improved results over the traditional EKF estimation due to optimization of process and measurement errors through a combination of CNN inference, backpropagation, and gradient descent.

In Table 2, we also compared the performance of using different model architectures. For the implementation of EKF+CNN_1, we removed the MHA layer with skip connection and kept all the other components, and the EKF+CNN_2 was implemented by removing the last convolution layer of EKF+CNN_1. EKF+LSTM was also introduced to attempt to learn the temporal features from the data. As can be seen, eliminating the final convolution layer reduced Pe and Ae performance overall while somewhat lowering MSE, which EKF+LSTM also achieved. Nevertheless, the proposed architecture combining CNN and MHA showed considerable improvements in MSE and lowered P_e a bit. The A_e was still higher than that of EKF+CNN_1, but in the acceptable range.

Table 2. Results for the four trajectory paths followed by evaluation metrics. Failure length: moving track length of robot during SLAM failure period; P_e: position error; A_e: rotation error; MSE_o: overall mean squared error. CNN_1 is the proposed model architecture without the multi-head attention layer. CNN_2 is the same as CNN_1 but removes the last convolution layer. LSTM is implemented with a number of 2 layers and a hidden size of 256.

Model	P_e (%)	A_e (deg/m)	MSE_o
EKF	1.64765	0.09064	0.04725
EKF+CNN_1	1.25114	0.02111	0.04084
EKF+CNN_2	1.46609	0.03109	0.03696
EKF+LSTM	1.35288	0.03138	0.03573
Proposed	1.24765	0.02785	0.02713

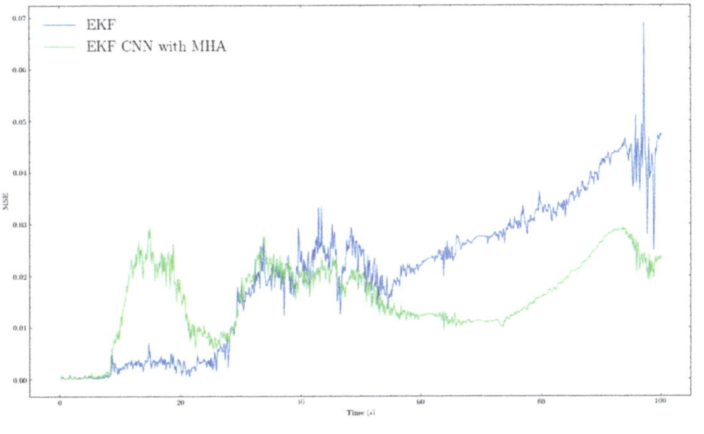

Figure 7. The Mean Squared Error (MSE) comparison between EKF and EKF+CNN with MHA for trajectory 1. Blue and green lines represent EKF and proposed method, respectively

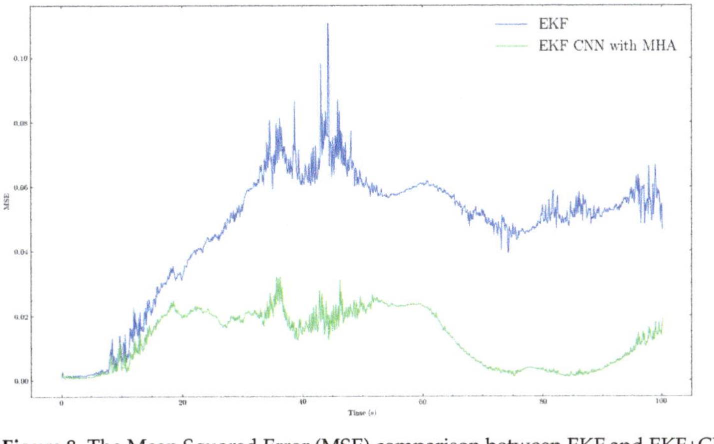

Figure 8. The Mean Squared Error (MSE) comparison between EKF and EKF+CNN with MHA for trajectory 3. Blue and green lines represent EKF and proposed method, respectively.

5.4. Discussion

We demonstrated the advantages of our proposed trainable quaternion-based EKF with CNN and MHA based on the conducted experiments and analysis. Our proposed method could achieve 1.24765% in P_e, 0.02785 in A_e, and it significantly outperforms other model architectures in terms of MSE with 0.02713. The results also suggest that a trainable EKF, which can dynamically adjust process and measurement noise covariance matrices, can improve localization performance. Moreover, our proposed model architecture provides scope for fusing the inputs via reinforcing one modality by introducing features from another modality for the inputs of estimating covariance matrices' parameters for EKF.

However, the online training cost of the proposed method can limit the overall performance. Furthermore, the time lag introduced by the online model training can be accumulated and cause deviations at the endpoints because of the limitations of the embedded computing system. Nonetheless, this issue can be solved by transmitting data to a more powerful server for online training. Furthermore, the future development of the AI-embedded platform will provide more power to achieve better online training performance. In conclusion, our results demonstrate that the deep learning model can be trained and provide predictions in an online training manner in a local integrated system.

6. Conclusions

In this paper, we designed a trainable and adjustable quaternion-based EKF algorithm with CNN and MHA for the sensor fusion of IMU-based localization and visual–LiDAR–inertial SLAM. Specifically, we developed an approach where the EKF-based localization system can provide a more accurate position estimation when SLAM failure occurs during a short time period. This was performed by tuning the process- and measurement-covariance matrices trained by CNN through backpropagation and further adjusting the velocity measurement covariances according to real-time IMU data through network inference. The approach leveraged the SLAM data as ground truths to compute the mean squared error of position and orientation estimated by the EKF while training.

For training and estimation, we designed a tandem EKF structure to adapt to the situation where real-time data from different sources were fused at different frequencies. Our proposed trainable EKF will be effective in dead-reckoning as a complementary process of SLAM when SLAM fails, which will enhance accuracy and stability in localization in complex and dynamic environments. Our future steps will focus on enhancing the structure of our proposed EKF CNN with MHA by performing multi-IMU fusion with multiple EKF modules, where each EKF leverages a unique IMU source.

Author Contributions: Conceptualization, G.M., B.X., R.L., X.Z., J.P., and C.H.P.; methodology, G.M., B.X., R.L., X.Z., and C.H.P.; software, G.M., R.L., X.Z., and J.P.; validation, G.M., B.X., R.L., and X.Z.; formal analysis, G.M. and B.X.; investigation, G.M., B.X., R.L., and X.Z.; resources, G.K. and C.H.P.; data curation, G.M. and B.X.; writing—original draft preparation, R.L., X.Z., and C.H.P.; writing—second draft with additional results and editing, G.M., B.X., and C.H.P.; visualization, J.P., G.K., and C.H.P.; supervision, J.P. and C.H.P.; project administration, G.K. and C.H.P.; funding acquisition, G.K. and C.H.P. All authors have read and agreed to the published version of the manuscript.

Funding: This research was financially supported by the Ministry of Trade, Industry and Energy (MOTIE) and the Korea Institute of Advancement of Technology (KIAT) through the International Cooperative R&D program (Project No. P0004631).

Institutional Review Board Statement: Not applicable.

Informed Consent Statement: Not applicable.

Data Availability Statement: Not applicable.

Conflicts of Interest: The authors declare no conflict of interest.

References

1. Alatise, M.B.; Hancke, G.P. A Review on Challenges of Autonomous Mobile Robot and Sensor Fusion Methods. *IEEE Access* **2020**, *8*, 39830–39846. [CrossRef]
2. Ott, F.; Feigl, T.; Loffler, C.; Mutschler, C. ViPR: Visual-Odometry-aided Pose Regression for 6DoF Camera Localization. In Proceedings of the IEEE/CVF Conference on Computer Vision and Pattern Recognition Workshops, Virtual, 14–19 June 2020; pp. 42–43.
3. Zhang, E.; Masoud, N. Increasing GPS Localization Accuracy With Reinforcement Learning. *IEEE Trans. Intell. Transp. Syst.* **2020**, *22*, 2615–2626. [CrossRef]
4. Lidow, A.; De Rooij, M.; Strydom, J.; Reusch, D.; Glaser, J. *GaN Transistors for Efficient Power Conversion*; John Wiley & Sons: Hoboken, NJ, USA, 2019.
5. Zhao, J. A Review of Wearable IMU (Inertial-Measurement-Unit)-based Pose Estimation and Drift Reduction Technologies. *J. Physics: Conf. Ser.* **2018**, *1087*, 9. [CrossRef]
6. Thrun, S.; Burgard, W.; Fox, D. *Probability Robotics*; MIT Press: Cambridge, MA, USA, 2005.
7. Kalman, R.E. A New Approach to Linear Filtering and Prediction Problems. *J. Basic Eng.* **1960**, *82*, 35–45. [CrossRef]
8. Malyavej, V.; Kumkeaw, W.; Aorpimai, M. Indoor robot localization by RSSI/IMU sensor fusion. In Proceedings of the 2013 10th International Conference on Electrical Engineering/Electronics, Computer, Telecommunications and Information Technology, Krabi, Thailand, 15–17 May 2013; pp. 1–6.
9. Brossard, M.; Bonnabel, S. Learning wheel odometry and IMU errors for localization. In Proceedings of the 2019 International Conference on Robotics and Automation (ICRA), Montreal, QC, Canada, 20–24 May 2019; pp. 291–297.
10. Yan, Y.; Zhang, B.; Zhou, J.; Zhang, Y.; Liu, X. Real-Time Localization and Mapping Utilizing Multi-Sensor Fusion and Visual–IMU–Wheel Odometry for Agricultural Robots in Unstructured, Dynamic and GPS-Denied Greenhouse Environments. *Agronomy* **2022**, *12*, 1740. [CrossRef]
11. Jurado, J.; Kabban, C.M.S.; Raquet, J. A regression-based methodology to improve estimation of inertial sensor errors using Allan variance data. *Navig. J. Inst. Navig.* **2019**, *66*, 251–263. [CrossRef]
12. Brossard, M.; Barrau, A.; Bonnabel, S. AI-IMU dead-reckoning. *IEEE Trans. Intell. Veh.* **2020**, *5*, 585–595. [CrossRef]
13. Vaswani, A.; Shazeer, N.; Parmar, N.; Uszkoreit, J.; Jones, L.; Gomez, A.N.; Kaiser, Ł.; Polosukhin, I. Attention is all you need. *Adv. Neural Inf. Process. Syst.* **2017**, *30*, 5998–6008.
14. Tsai, Y.H.H.; Bai, S.; Liang, P.P.; Kolter, J.Z.; Morency, L.P.; Salakhutdinov, R. Multimodal transformer for unaligned multimodal language sequences. In Proceedings of the 57th Annual Meeting of the Association for Computational Linguistics, Florence, Italy, 28 July–2 August 2019; Volume 2019, p. 6558.
15. Akhlaghi, S.; Zhou, N.; Huang, Z. Adaptive adjustment of noise covariance in Kalman filter for dynamic state estimation. In Proceedings of the 2017 IEEE Power & Energy Society General Meeting, Chicago, IL, USA, 16–20 July 2017; pp. 1–5.
16. Hu, G.; Gao, B.; Zhong, Y.; Gu, C. Unscented kalman filter with process noise covariance estimation for vehicular ins/gps integration system. *Inf. Fusion* **2020**, *64*, 194–204. [CrossRef]
17. Hornik, K.; Stinchcombe, M.; White, H. Multilayer feedforward networks are universal approximators. *Neural Netw.* **1989**, *2*, 359–366. [CrossRef]
18. Haarnoja, T.; Ajay, A.; Levine, S.; Abbeel, P. Backprop kf: Learning discriminative deterministic state estimators. *Adv. Neural Inf. Process. Syst.* **2016**, *29*, 4376–4384.
19. Song, F.; Li, Y.; Cheng, W.; Dong, L.; Li, M.; Li, J. An Improved Kalman Filter Based on Long Short-Memory Recurrent Neural Network for Nonlinear Radar Target Tracking. *Wirel. Commun. Mob. Comput.* **2022**, *2022*, 8280428. [CrossRef]
20. Gao, X.; Luo, H.; Ning, B.; Zhao, F.; Bao, L.; Gong, Y.; Xiao, Y.; Jiang, J. RL-AKF: An adaptive kalman filter navigation algorithm based on reinforcement learning for ground vehicles. *Remote Sens.* **2020**, *12*, 1704. [CrossRef]
21. Wu, F.; Luo, H.; Jia, H.; Zhao, F.; Xiao, Y.; Gao, X. Predicting the noise covariance with a multitask learning model for Kalman filter-based GNSS/INS integrated navigation. *IEEE Trans. Instrum. Meas.* **2020**, *70*, 1–13. [CrossRef]
22. Feng, K.; Li, J.; Zhang, X.; Shen, C.; Bi, Y.; Zheng, T.; Liu, J. A new quaternion-based Kalman filter for real-time attitude estimation using the two-step geometrically-intuitive correction algorithm. *Sensors* **2017**, *17*, 2146. [CrossRef] [PubMed]
23. Kok, M.; Hol, J.D.; Schön, T.B. Using inertial sensors for position and orientation estimation. *arXiv* **2017**, arXiv:1704.06053.
24. Mochnac, J.; Marchevsky, S.; Kocan, P. Bayesian filtering techniques: Kalman and extended Kalman filter basics. In Proceedings of the 2009 19th International Conference Radioelektronika, Bratislava, Slovakia, 22–23 April 2009; pp. 119–122.
25. NGOC, T.T.; KHENCHAF, A.; COMBLET, F. Evaluating Process and Measurement Noise in Extended Kalman Filter for GNSS Position Accuracy. In Proceedings of the 2019 13th European Conference on Antennas and Propagation (EuCAP), Krakow, Poland, 31 March–5 April 2019; pp. 1–5.
26. Khairuddin, A.R.; Talib, M.S.; Haron, H. Review on simultaneous localization and mapping (SLAM). In Proceedings of the 2015 IEEE International Conference on Control System, Computing and Engineering (ICCSCE), Penang, Malaysia, 27–29 November 2015; pp. 85–90.
27. LeCun, Y.; Touresky, D.; Hinton, G.; Sejnowski, T. A theoretical framework for back-propagation. In Proceedings of the 1988 Connectionist Models Summer School, San Mateo, CA, USA, 17–26 June 1988; Volume 1, pp. 21–28.
28. Paszke, A.; Gross, S.; Massa, F.; Lerer, A.; Bradbury, J.; Chanan, G.; Killeen, T.; Lin, Z.; Gimelshein, N.; Antiga, L.; et al. Pytorch: An imperative style, high-performance deep learning library. *Adv. Neural Inf. Process. Syst.* **2019**, 32, 8026–8037.

29. Stanford Artificial Intelligence Laboratory. Robotic Operating System. Available online: https://www.ros.org (accessed on 1 September 2020).
30. Kingma, D.P.; Ba, J. Adam: A method for stochastic optimization. *arXiv* **2014**, arXiv:1412.6980
31. Nam, D.V. Robust Multi-Sensor Fusion-based SLAM using State Estimation by Learning Observation Model. Ph.D. Thesis, Chungbuk National University, Cheongju, Korea, 2022.

sensors

MDPI

Article

Detecting Faults at the Edge via Sensor Data Fusion Echo State Networks

Dario Bruneo *,† and Fabrizio De Vita †

Department of Engineering, University of Messina, 98166 Messina, Italy; fdevita@unime.it
* Correspondence: dbruneo@unime.it
† These authors contributed equally to this work.

Abstract: The pervasive use of sensors and actuators in the Industry 4.0 paradigm has changed the way we interact with industrial systems. In such a context, modern frameworks are not only limited to the system telemetry but also include the detection of potentially harmful conditions. However, when the number of signals generated by a system is large, it becomes challenging to properly correlate the information for an effective diagnosis. The combination of Artificial Intelligence and sensor data fusion techniques is a valid solution to address this problem, implementing models capable of extracting information from a set of heterogeneous sources. On the other hand, the constrained resources of Edge devices, where these algorithms are usually executed, pose strict limitations in terms of memory occupation and models complexity. To overcome this problem, in this paper we propose an Echo State Network architecture which exploits sensor data fusion to detect the faults on a scale replica industrial plant. Thanks to its sparse weights structure, Echo State Networks are Recurrent Neural Networks models, which exhibit a low complexity and memory footprint, which makes them suitable to be deployed on an Edge device. Through the analysis of vibration and current signals, the proposed model is able to correctly detect the majority of the faults occurring in the industrial plant. Experimental results demonstrate the feasibility of the proposed approach and present a comparison with other approaches, where we show that our methodology is the best trade-off in terms of precision, recall, F1-score and inference time.

Keywords: ESN; recurrent neural networks; sensor data fusion; edge computing; industry 4.0; fault detection; deep learning

Citation: Bruneo, D.; De Vita, F. Detecting Faults at the Edge via Sensor Data Fusion Echo State Networks. *Sensors* **2022**, *22*, 2858. https://doi.org/10.3390/s22082858

Academic Editor: Natividad Duro Carralero

Received: 19 March 2022
Accepted: 5 April 2022
Published: 8 April 2022

1. Introduction

The emergence of the Industry 4.0 paradigm has made modern plants involve several set of sensors to enable monitoring under many different aspects. The pervasive presence of sensors and actuators in industrial systems totally changed the interactions with these systems and paved the way for the realization of more precise control and telemetry frameworks [1]. On the other hand, when the number of sensors is large, it becomes difficult to correlate and analyze the information coming from several heterogeneous data sources and produce an accurate diagnosis of the system's "health". In particular, dealing with signals that can be very different (e.g., in terms of sampling rate, data format, protocols, etc.) represents one of the major challenges during the selection process of the more informative signals [2].

In such a context, the use of Artificial Intelligence (AI) techniques based on sensor data fusion can be considered a viable solution to address this problem. Through machine and deep learning approaches, we realize intelligent algorithms capable of separately extracting features from a set of heterogeneous signals and then fuse this information to improve the predictive performance [3].

Cloud computing paradigm has played a key role in smart factories, acting as a central entity providing the storage and computing power necessary for the execution of complex algorithms [4]. However, when working with industrial applications aiming to predict the state of a system, time becomes a crucial component. Indeed, being able to perform a timely detection of a fault condition can avoid more severe consequences while having a strong impact in terms of time and money.

Today, Edge computing became a common solution to address this problem by shifting the computation closer to the data, thus improving the security, response times, and connection stability [5]. As a side effect, the use of this paradigm poses significant restrictions on the type of algorithms that can be performed. In this sense, an Edge device is typically equipped with a constrained hardware with limited resources, which makes it unsuitable for the execution of advanced algorithms (e.g., Deep Neural Networks) in an effective way [6].

Reservoir Computing (RC) [7] is a promising framework deriving from the Recurrent Neural Networks (RNNs) characterized by a container (called *reservoir*) that remains fixed during the training procedure and whose task is to learn complex input dynamics [8,9]. Among the models belonging to the RC, Echo State Networks (ESNs) are a family of neural networks particularly suitable for the analysis of time series, which embody the power of RNNs into models with a low number of trainable parameters [10], thus exhibiting a reduced memory footprint and complexity that can meet the strict requirements of an Edge device. In the context of fault detection and diagnosis, ESNs have demonstrated their effectiveness. Most of the approaches available in the literature involve the use of RNNs models for the analysis of time series. However, some RNNs suffer from unstable gradient problems that can affect the training process; ESNs address this issue by keeping the fixed weights contained in the reservoir and training only the output ones [11].

In this paper, we propose an ESN model which exploits sensor data fusion to extract the dynamics from a set of heterogeneous sensors and to perform a fault detection of a real-scale replica industrial plant. Through the implementation of a Cloud dashboard, we are able to collect and label the data generated by the plant and orchestrate the Edge devices attached to it to perform a real-time prediction. The experimental results demonstrate that the proposed approach benefits from the merging of separate source of information to improve its predicting capabilities while keeping the number of parameters low thanks to the use of an RC model. We also show the effectiveness of our solution when compared with other approaches.

The paper contributions are manifold. (i) We propose a fault detection model which exploits sensor data fusion running on a real industrial testbed. (ii) We adopt ESNs such that the proposed algorithm is suitable to be deployed on the constrained hardware of an Edge device. (iii) We realize a Cloud/Edge continuum architecture that allows the management of the data produced by the industrial system and the orchestration of the Edge devices executing the fault detection. (iv) We produce a comparison with other machine/deep learning methods in terms of predictive performance, model complexity and inference time.

The rest of the paper in organized as follows. Section 2 reviews the literature. Section 3 provides a description about the concepts at the base of ESNs. Section 4 presents the proposed proposed approach we implemented to assess the health condition of the industrial plant. Section 5 describes the experimental results we obtained from testing our technique. Section 6 concludes the paper and provides some insights for the future works.

2. Related Works

Sensor data fusion is becoming an important resource for the realization of smart systems. Today, we are living in an era where data of various types coming from a set of multiple sources are becoming more and more frequent, making the realization of data fusion approaches a priority. In a context characterized by a multitude of information, the swarm intelligence paradigm is another important player for extracting important features from a given dataset [12] or to solve high-dimensional optimization problems [13].

In this section, we review the works found literature, and we highlight the main differences with our solution.

The authors in [14] present a review about sensor data fusion methods for real-time analysis, putting in evidence the main concepts, the core steps and the research challenges.

In [15], the benefits that derive from the use of sensor data fusion as an effective technique to produce more accurate and reliable results through the combination of data coming from multiple sources are described. In this work the authors also present diverse data fusion schemes, highlighting the differences among them.

In [16], a Deep Enhanced Fusion Network (DEFN) is proposed for the fault diagnosis of wind turbine gearboxes using three-axial vibration signals. In this work, the authors adopt Sparse Autoencoders (SAEs) to perform the feature extraction process, and such features are then passed to an ESN to perform the actual fault detection. Although the obtained results are good, the technique is not meant to be run on the Edge due to the model's complexity. In this sense, our model has been designed to run on constrained devices thanks to its low number of trainable parameters.

The authors in [17] describe a multisensor approach for chattering detection in milling processes. Starting from sound and vibration signals, they are passed to several feature extraction processes such as Wavelet Package Decomposition (WPD), WPD parameters optimization, and the extraction of time/frequency features. Then, these features are passed as input to an ensemble method made of five algorithms, namely, Random Forest (RF), Extreme Gradient Boost (XGBoost), Support Vector Machines (SVMs), K-Nearest Neighbor (KNN) and Artificial Neural Network (ANN), to perform the chatter identification. The experimental results demonstrated the effectiveness of the authors approach; however, the large number of processing steps together with the use of an ensemble method produce a bottleneck during the inference time. When working in fault detection applications, the response time is a key factor. Our approach benefits from the use of ESNs, which exhibit a fast inference time suitable for industrial applications.

In [18], the authors propose a vibro-acoustic data fusion approach based a one-dimensional Convolutional Neural Network (1D-CNN) for bearing fault detection. The model presented in this work adopts CNNs to separately extract the features from vibration and acoustic signals. Then, these features are combined together and passed to a series of fully connected layers that perform the feature fusion. The approach obtained very good results; however, compared to ESNs, CNNs require a larger number of trainable parameters, which could not fit the memory requirements. Moreover, if not properly parametrized CNNs are more prone to overfitting. A similar approach is described in [19], where the authors adopt a 2D-CNN to fuse data coming from multiple current sensors and to perform an automatic feature extraction process. Although the results obtained are good, these method suffers from the same problem highlighted in [18].

The work presented in [20] makes use of a sensor data fusion approach to predict the Remaining Useful Life (RUL) of electromagnetic pumps. Given the pressure and vibration signals collected from the pump, the prognostic algorithm adopts a multi-objective genetic algorithm based on Long Short-Term Memories (MOGA-LSTM). The authors have successfully demonstrated the effectiveness of their technique, but exactly as in the previous cases, this algorithm is not designed to be executed of an Edge device. In our application context, this aspect is of fundamental relevance in terms of quick response times, privacy and security.

In [21], a sensor data fusion algorithm based on NARX neural networks is proposed to predict the mass flow in sugarcane harvesters. In this case, the input data are passed to a single hidden neuron for the feature extraction process; if, on the one hand, this reduces the model complexity, on the other, it does not exploit the full potential of sensor data fusion that allows it to analyze each input separately to achieve a tailored feature extraction. Of course, when the number of inputs is large, this could lead to models with a lot of layers and parameters. To mitigate this effect in our solution, we adopted ESNs, which keep fixed the majority of their weights.

The authors in [22] discuss the misalignment fault classification and describe a solution based on SVMs and vibro-acoustic sensor data fusion. Such data are used to analyze the system under different load conditions and operative settings and reached a 100% accuracy in detecting the faults. If, on the one hand, the authors obtained impressive results, on the other, the use of SVMs does not allow for catching time dynamics that only an RNN is able to catch.

In [23], a work for gear fault classification in rotating machines is presented. Starting from a set of vibration sensors, the models adopts Coherent Composite Spectrum (CCS) for a first feature extraction. Then, Principal Component Analysis (PCA) and ANN are used to perform the dimensionality reduction and a further feature extraction. These are then merged and passed to a fault classification block for the actual prediction. The authors demonstrated the effectiveness of their approach and the benefits derived from the use of PCA that allowed them to reach a 100% accuracy.

3. Echo State Networks

ESNs are a family of neural networks which belong to the RC models class particularly suitable for the analysis of time series data [24,25]. From an architectural point of view, an ESN is equivalent to an RNN, except for the presence of a sparse and randomly connected recurrent structure called a reservoir performing the feature extraction process. A peculiar aspect of this layer is the fact that the reservoir weights are fixed and do not change during the training process. Figure 1 depicts an ESN architecture, where the green and the black arrows are the trainable and the not-trainable weights, respectively, while the red lines indicate possible, but not required, connections.

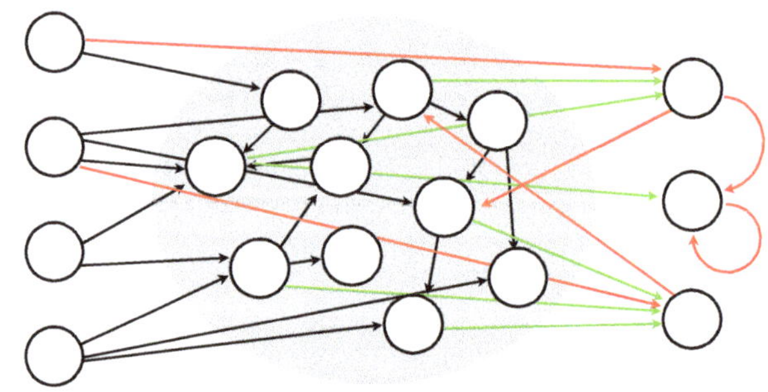

Figure 1. ESN architecture.

Given an ESN with K input units, N reservoir units and L output units, we can define the equations which govern the ESN as follows:

$$x(t+1) = f(W_{in} \cdot u(t+1) + W_r \cdot x(t) + W_b \cdot y(t)), \tag{1}$$

$$y(t+1) = g(W_{out} \cdot [u(t+1), x(t+1), y(t)]), \tag{2}$$

where $x(t+1)$ is the new computed reservoir state and is a function of the new input $u(t+1)$, the state $x(t)$ and the output $y(t)$. As far as W_{in}, W_r and W_b, they are matrices of $N \times K$, $N \times N$ and $N \times L$ elements, which store the weights between the input and reservoir layers, the reservoir weights and the backwards connections between the output and reservoir layers, respectively. Finally, $f(\cdot)$ is the activation function wrapping the equation (typically the sigmoid or the tanh).

With respect to the output equation, it is obtained as the concatenation of the new state $x(t+1)$, the input $u(t+1)$ and the previous output $y(t)$; in this sense, the output is "affected" by the past history of the model. The W_{out} is an $L \times (K+N+L)$ matrix containing the only trainable weights of the network (i.e., those connecting the reservoir layer to the output). Also in this case, an activation function $g(\cdot)$ (typically a sigmoid or the identity) wraps the equation to generate the actual output of the network.

Compared to other RNN models such as Gated Recurrent Units (GRUs) and LSTMs, ESNs can exploit the sparse recurrent structure of reservoir layer to achieve comparable results using much less parameters. In this sense, a model with a reduced number of parameters is less prone to overfitting, a well-known problem when working with complex Deep Neural Networks (DNNs), where a huge number of parameters produces large models. Evidently, this involves a reduction in the model memory footprint, a key aspect that should not be underestimated especially in this period where the majority of applications is migrating towards the constrained hardware of Edge devices. Another advantage of ESNs is represented by their faster training/inference time thanks to the fact that the majority of the weights are kept fixed. Considering our industrial scenario where the system response time is a crucial requirement, together with a small model footprint such that it can fit the hardware of an Edge device, the above-mentioned features motivated the choice of ESNs as a preferable approach over other solutions.

However, if, on the one hand, the sparse and fixed reservoir weights allow reducing the ESN model complexity and speed-up the training process due to the reduced number of parameters, on the other, the impossibility of optimizing these weights can lead to unstable results due to the random initialization [26]. Such an effect can be mitigated by using large reservoir structures to incentivize the creation of subnetworks that can catch a large number of input dynamics [8,24], but this still remains one of the main limitations of these networks.

When working with ESNs, it is also very important to satisfy the *separation* and *echo state* properties to make them properly work. With respect to the first one, it ensures the generation of separate states given different inputs. Such a condition is fundamental to avoid the extraction of the same patterns for several inputs that would inevitably cause the collapse of the ESN. In this sense, the adoption of large and sparse reservoirs is necessary to meet this property by "encouraging" the formation of mixed connections with the input layer that stimulate the production of different states and features. The echo state property (from here, the name of these networks) states that the effects of the inputs and the previous states in Equation (1) should gradually vanish over time. Indeed, this is a very important feature for an ESN since it allows balancing the present and past history of the model, so that the output is equally affected. Unlike the previous case, the satisfaction of this property comes from empirical tests, where it has been demonstrated that a W_r weight matrix with a spectral radius (i.e., the largest eigenvalue in absolute value) lower than one is a necessary (but not sufficient) requisite. The use of weight scaling techniques produce suitable matrices which meet this condition; however, it is worth to mention that for specific types of inputs, this could be neither necessary nor sufficient. For this reason, the echo state property is still the object of study.

4. Proposed Data Fusion Approach

In this subsection, we introduce the scale replica plant used as an industrial testbed, and we present the proposed approach which merges sensor data fusion and ESNs to assess its working conditions.

4.1. Industrial Testbed

The industrial testbed we used is a scale replica of an assembly plant for the transportation of car pieces adopted in automobile factories. It is powered with two gear motors, a set of six belts, and a mobile cart that can move back and forth. We instrumented the plant with vibration and current sensors to enable a mechanical and electrical monitoring, and we equipped it with a set of Edge devices for the real-time execution of our fault-detection algorithm. In particular, we used a VTV-122 sensor by IFM electronics to measure anomalous vibrations due to the brake system or cart proximity switch malfunctioning. The sensor operates with a power supply between 9 and 32 V and outputs a 4–20 mA signal.

For the electrical part, we used an inverter of the S100 series by LS Industrial Systems that we directly attached to one of the gear motors. The inverter uses a power supply between 200 and 240 V and generates an analog current output in the range 0 to 20 mA. On the Cloud side, we realized a platform, accessible via a web dashboard, through which we can collect and label the data produced by the sensors and orchestrate the Edge side deploying new deep learning models, thus creating a Cloud/Edge continuum architecture.

In such a context, the possibility to manually inject faults in our scale replica testbed (e.g., the introduction of external vibrations along the plant structure, the failure of the brake system, the increment in gears friction, the change of belts tension, etc.) has been fundamental to enable the data labeling of anomalous patterns, and gave us the opportunity to study the testbed also when subject to a faulty condition.

Figure 2 depicts the Cloud dashboard we realized to orchestrate the Edge devices connected to the industrial plant. It is divided into four sections. The first one allows the labeling of the data collected through industrial plant and their storage under different operating conditions of the system. The second one defines the starting and ending dates and time that the system will use for making a query to the internal Cloud database and to retrieve the training data. At this step, the user can also define the windows size to split the input signals into smaller time sequences to be analyzed. The third section is used to start the inference at the Edge in order to assess the working conditions of the industrial plant. Here the window size performs the same task as in the training process. Finally, the fourth section returns a report of the trained model by showing its performance and informing the user which model is going to be injected at the Edge.

Figure 2. Dashboard running on the Cloud to orchestrate the Edge devices.

4.2. Fault Detection

Catalyzed by the advent of the Industry 4.0, modern systems faced a complete transformation that revolutionized the way to interact with them. Today, we can observe industrial plants that are equipped with a large number of sensors and actuators, becoming a "smart entity" with a self-awareness that can help the human operator during the execution of several tasks. In such a context, one of the most important aspects is the diagnosis of industrial systems conditions, which became a central research topic. Preventive maintenance is no longer an option that can considered: The huge amount of components in industrial plants makes the use of this approach unsustainable, and for this reason, modern solutions are moving towards the realization of predictive frameworks capable of detecting abnormal behaviors before the occurrence of the fault.

However, when working with complex systems, the generation of an accurate diagnosis can be very challenging due to strong nonlinearity, data heterogeneity and the large number of process variables to take into account. Deep learning can address this problem through the definition of advanced predictive models that can manage a large number of inputs and learn hidden relationships among them. Fault detection represents a key element to recognize harmful patterns that usually anticipate a failure, thus reducing the maintenance costs of a system [27].

Along with fault detection, it is worth mentioning the crucial role played by Edge computing; the wide spread of this technology (which rapidly became the core element of smart factory frameworks) is a perfect example that proves the effectiveness of this paradigm. In a context where a large part of the tasks operated by an industrial system is executed in real time emerges the necessity to perform specific analyses very close to where the data are generated. Also in terms of security, Edge computing can be very useful to better preserve data privacy by performing the inference process on data stored locally. With particular reference to the industrial scenario, where the data can be sensitive, the capabilities of this technique allow the realization of more secure applications. However, moving the computation towards these devices poses significant limitations on the realization of algorithms that should meet their strict hardware resources to run effectively.

4.3. Sensor Data Fusion Model

The fault detection problem has been faced as a supervised binary classification problem where we considered two possible working states for the plant, namely, a normal condition and an anomalous one. Figure 3 shows a block diagram with the main steps performed in our framework. The data collection methodology we adopted is the following. In order to have a monitoring from both the electrical and mechanical point of view, we instrumented the plant with a vibration sensor directly attached on its structure and a inverter that we connected to one of the engines to measure the absorbed current. With respect to the sampling rates, we set the vibration signal to 30 Hz, which, from the empirical tests, was demonstrated to be a good trade-off in terms of computational complexity and signal reconstruction. Regarding the inverter, we set its internal sampling rate to 120 Hz in order to meet the Nyquist frequency; however, we noticed that the absorbed current remained stable for the majority of the time, exhibiting a change only when a fault was occurring. Considering that these signals are also analyzed in real time by an Edge device with limited computing capabilities, working with such a high rate would be unfeasible. For these reasons, we created a script to interrogate the inverter with a 5 Hz rate. The data collection phase lasted about 46 h; we started collecting the data under a normal operating setting for about 29 h, in which the plant was properly working and the cart was able to move back and forth. After this step, in the remaining 17 h, we started the injection of the anomalies (for about 8.5 h each and one at a time) inside the system, first increasing the friction of the gears and then changing the belts' tension. At the end of this process, we collected $\sim$5 M of vibration samples (i.e., $\sim$3 M normal, and $\sim$2 M anomalous) and $\sim$800 K current samples (i.e., $\sim$5 K normal, and $\sim$3 K anomalous).

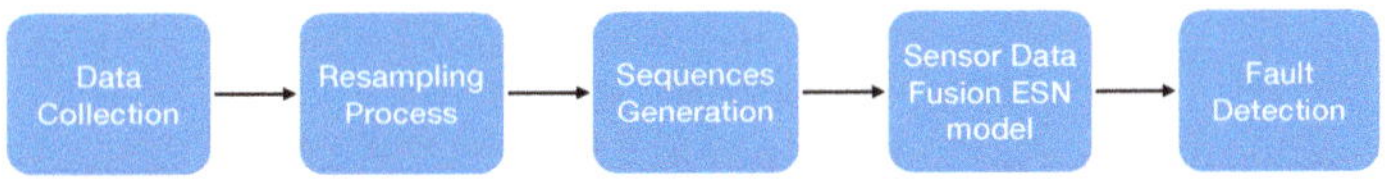

Figure 3. Block diagram showing the steps performed in our framework.

Figure 4 depicts the current and vibration signals collected from the industrial plant in correspondence of the two above-mentioned conditions. The first problem we faced working with these signals has been to make them have the same sampling rate. Specifically, we adopted the SciPy Python library, which exposes some utility APIs for signals resampling; in this sense, since the vibration exhibited a much higher number of samples than the current, we performed an upsampling process on the last one. When working with time series, having a homogeneous sampling rate is fundamental to avoid a time uncorrelation between signals that would inevitably affect the overall prediction performance.

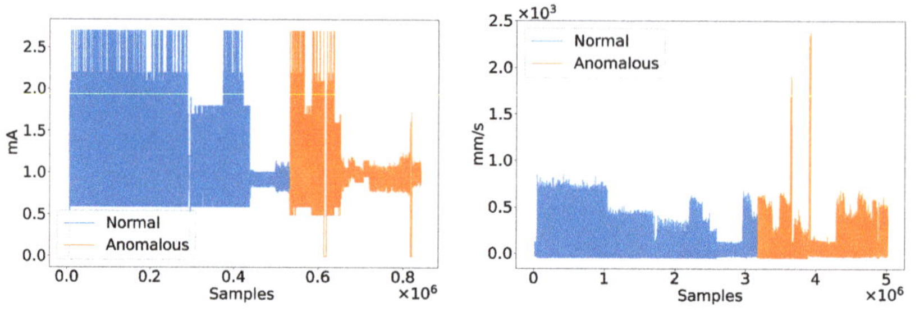

Figure 4. Current (**left**) and vibration (**right**) raw signals collected from the industrial plant in correspondence of the normal and anomalous conditions.

Then, we performed a second data preprocessing to split the raw signals into multiple time sequences by applying an overlapping sliding window. As an effect of this step, we were able to reduce the inputs complexity by focusing on smaller parts, in addition, the analysis of several sequences instead of the whole signal allows a better feature extraction. The choice of the window size strongly depends on the signals characteristics (e.g., sampling rate, variability, frequency, etc.); in our empirical experiments, we tested several windows values using as evaluation indices the resulting model prediction performance and model complexity. At the end of this step, we selected a window size of 50 samples for the input sequences that resulted to be the best trade-off. At the end of these preprocessing steps, data were ready to be passed to the sensor data fusion model.

Figure 5 shows the proposed sensor data fusion ESN model. It is structured into two parts: a feature extraction and a predictive one. Unlike a "traditional" model, where input data are typically treated as a single monolithic block, in this case, we can notice that each signal is separately managed and passed to an ESN. Such a separation has two effects: (i) It improves the model input scalability, since the addition of new signals requires a change only in the feature extraction part and not of the entire topology; (ii) it allows a tailored feature extraction according for each input. In such a context, where inputs are time series signals, the use of RNN structures represent a valid solution to find hidden time dynamics that a normal machine learning model would not be able to catch. However, when working with recurrent models such as LSTMs or GRUs, they are prone to generating a lot of parameters and require a careful parametrization. In this sense, the choice of ESNs resulted in being a good option since they work very well as time features extractors [28] and use a low number of trainable parameters which make them lightweight, fast to train and suitable to fit the hardware of an Edge device. At this step, each ESN analyzes only one signal separately from the other one; by means of their reservoir, they are able to extract from the input sequences a hidden vector containing the temporal features. The main idea behind this model is that the reservoir represents a latent space. In fact, thanks to its large dimensionality, this non-linear part of the system can extract relevant features (at a higher abstraction layer) deriving from the input series.

The predictive part of the model starts with the Fusion layer which acts as a connecting bridge between the input branches. At this level, the features extracted in the previous layers are concatenated together performing the actual sensor data fusion. Finally, the ESN layers connected in cascade have the task to perform a further feature extraction process necessary for the prediction whose results are stored in the output layer. According to the prediction task, the output layer can have various shapes and adopt different activation functions. For example, in the case of a multiclass classification problem, the output layer will have as many neurons as the number of classes to be predicted and will adopt a *softmax* activation function to produce the probabilities for a given input to belong to each of them. On the other hand, if the task to be addressed is a binary classification, then in this case is sufficient to use as output a single neuron with a sigmoid activation to evaluate the class to which the input belongs to.

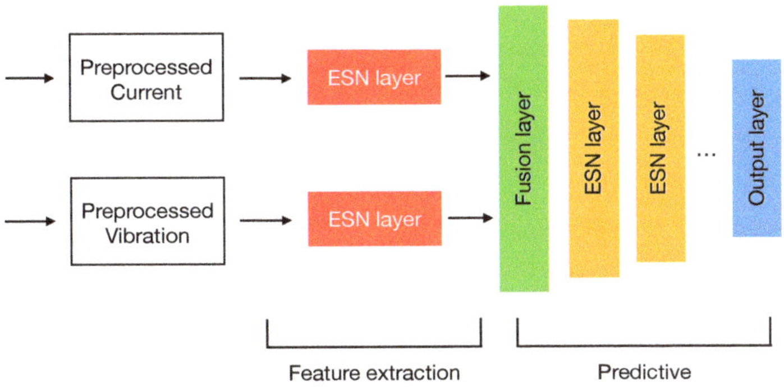

Figure 5. Proposed Sensor data fusion ESN model.

Let us define the mathematical model of our sensor data fusion ESN. If we consider a model with m separate inputs, then we can write the state and output equations of the generic ith ESN as follows:

$$\begin{cases} x(t+1)^i = f(W_{in}^i \cdot u(t+1)^i + W_r^i \cdot x(t)^i) \\ h(t+1)^i = f(W_h^i \cdot x(t+1)^i), \end{cases} \tag{3}$$

where the equations reported in Equation (3) are a variation of the ones shown in Section 3 and obtained by removing the not required connections (i.e., the red arrows of Figure 1). In such a context, the output of each ESN is a hidden vector $h(t+1)^i$ storing the temporal features extracted from the reservoir layer. Unlike Equation (2), in this case, the W_h matrix does not contain trainable weights as the output of the ESN is in turn a hidden vector:

$$\begin{cases} F(t) = [h(t+1)^1, h(t+1)^2, ..., h(t+1)^m] \\ X(t+1) = f(W_F \cdot F(t+1) + W_R \cdot X(t)) \\ y(t+1) = \phi(W_{out} \cdot X(t+1)). \end{cases} \tag{4}$$

In Equation (4), we report the equations modeling the predictive part, where $F(t)$ is the output of the fusion layer obtained as the concatenation of the features extracted in previous part. Since we adopted ESNs also in this part of the model, the equations are equivalent to the ones used in Equation (3), with the main difference being that, in this case, the input is no longer represented by the input signals (e.g., current and vibration) but by the features extracted from them. This is evident from the second equation, where the new computed state $X(t+1)$ depends on $F(t+1)$ (i.e., the features extracted and fused together at the $t+1$ timestep) and on the state $X(t)$. W_F and W_R matrices play the same role of W_{in} and W_r of Equation (1), and for this reason, their weights are kept fixed. Finally, $y(t+1)$ represents the actual output of the sensor fusion model, storing the prediction value; it is equivalent to the output of an ESN since it only depends on the new computed state $X(t+1)$ and the W_{out} matrix containing the only trainable weights of the entire architecture. As usual, the entire expression is wrapped by an activation function ϕ, which can be either the softmax or sigmoid based on the classification problem being addressed.

5. Experimental Results

In this section, we present the results obtained from testing the proposed sensor data fusion ESN model. The dataset extracted from the industrial plant is provided with binary labels such that normal and anomalous samples are labeled as 0 and 1, respectively. The splitting of the dataset has been performed using the Scikit-learn framework [29] by means of the *train_test_split* function, which takes as input the data and the test size percentage and returns the dataset split into train and test sets. In our experiments, we adopted a test size of 20%, leaving the rest (i.e., the 80%) for the training process. For the hyperparameters' tuning and validation, we extracted 15% from the training data so that the test set contains samples used exclusively for the model's evaluation. Moreover, it is worth mentioning that the following results have been averaged over a set of experiments in order to obtain a more precise evaluation.

For a better understanding, in Table 1 we report the proposed model configuration adopted in our experimentation. Such parameters derive from a grid search approach implemented using Keras Tuner (https://www.tensorflow.org/tutorials/keras/keras_tuner, accessed on 14 March 2022), during which we evaluated the network predictive performance in terms of accuracy when varying the number of ESN neurons in each layer. With respect to the feature extraction part, we used 2 ESNs (one for each input) with 20 neurons and the tanh as the activation function (representing the standard for these type of networks). Regarding the prediction part, it is composed by 2 ESNs connected in cascade with 16 and 4 neurons, respectively, with the tanh as activation. In both cases, we adopted a spectral radius of 0.9 and a connectivity of the 10% in order to meet the separation and echo state properties (see Section 3). For the trainable weights' optimization, we used the Adam optimizer with a learning rate of 0.001, and we set the limit of the training epochs to 2000. Moreover, to avoid model overfitting, we used an early stopping approach to halt the training process if the model is not able to reduce the loss for a consecutive number of epochs defined by the patience term that we set to 10. Such a technique avoids a network "overtraining" that would inevitably cause the memorization (instead of the learning) of the input–output relationship and a consequent reduction in the model's generalization capabilities. Finally, the output layer is defined by a single neuron associated with a sigmoid activation function to perform the binary classification.

Table 1. Proposed model configuration.

Training Parameters	
ESN neurons feature extraction part	20
ESN neurons predictive part	16, 4
ESN spectral radius	0.9
ESN connectivity	0.1
Output neurons	1
Activation functions	tanh, sigmoid
Learning rate	0.001
Training epochs limit	2000
Optimizer	Adam
Patience term	10

Figure 6 depicts the validation curves derived from the training procedure of the proposed sensor data fusion ESN model.

The proposed approach performed very well on the test set producing results comparable to the ones obtained during the training. With a precision of 1.0, a recall of 0.993 and a *F1-score* of 0.996, our solution is able to correctly predict the conditions of the testbed with a very low number of false positives and false negatives, a very important aspect for an algorithm running in an industrial plant. The use of ESNs for both the feature extraction and prediction parts result in an effective choice that allowed us to maintain a low amount of trainable parameters (i.e., 201) while achieving a good level of predictive performance. This, of course, has an impact not only in terms of memory footprint but also in terms of training performance that could be affordable even for an Edge device. In this sense, being able to perform an on-device training directly at the Edge could pave the way to industrial application scenarios where the plant would learn in real time the occurrence of new fault patterns.

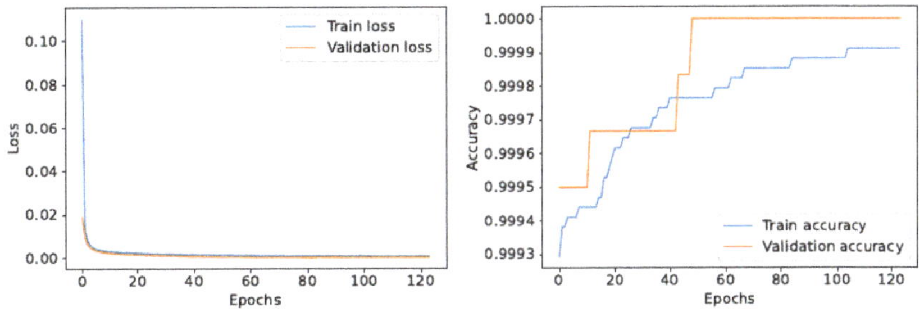

(**a**) Loss (**b**) Accuracy

Figure 6. Validation curves extracted from the proposed model after the training process. In (**a**), we can observe that both the train and validation losses follow a decreasing trend as the number of training epochs increases. Both the curves converge at a very low value of the loss around zero, thus demonstrating that the model correctly learned the relationship between the input and the output with a very good level of generalization. Such a condition is also proven by (**b**), where, in this case, the training and validation accuracy increase with the training epochs, reaching very high values around one.

To demonstrate the effectiveness of the proposed approach, we performed a series of comparisons with other models and techniques. Specifically, we considered six different approaches, namely: a model involving both ESN and fully connected layers (ESN + FC) which does not adopt sensor data fusion, a One Class Support Vector Machine (OCSVM) approach, an Isolation Forest (IF) model, a fully connected DNN (FC-DNN), a 1D-CNN approach like the one presented in [18] and an LSTM network. All the models have been implemented using TensorFlow and Scikit-learn frameworks [29,30].

For a better vision, we report in Table 2 the results for each of the above-mentioned models in terms of precision, recall, *F1-score* and the model complexity according to the number of trainable parameters.

Table 2. Models' comparison.

Model	Precision	Recall	*F1-Score*	*#Trainable Params.*
ESN+FC	1.0	0.976	0.987	925
OCSVM	0.753	0.753	0.753	-
IF	0.843	0.294	0.435	-
FC-DNN	0.992	0.973	0.982	2,873
LSTM	1.0	0.993	0.996	7,705
1D-CNN [18]	1.0	0.994	0.996	1,077
Proposed Sensor data fusion ESN	1.0	0.993	0.996	201

With respect to the ESN+FC model, we implemented it using an architecture similar to the one shown in Figure 5. In particular, we used a single ESN of 40 neurons and the tanh as activation. Since this model does not adopt sensor data fusion, both the current and vibration signals are treated as a single block and passed to the ESN to perform the feature extraction process. For the prediction part, we used 3 fully connected layers of 16 and 4 neurons with Rectified Linear Unit (ReLU) as activation and a single neuron (i.e., the output) with sigmoid as activation. The model achieved very good results in terms of precision, recall and *F1-score* (i.e., 1.0, 0.976 and 0.987, respectively), comparable to the ones reached by our model, but using 925 trainable parameters.

For the OCSVM, Scikit-learn exposes several parameters that can be set. In our experiments, we set a penalization term (i.e., an L2 penalty) C of 0.8, and we left the gamma coefficient to "auto", a special value of the library that automatically sets this coefficient to $1/(\#features)$. The most important parameter is the kernel, which specifies the type of transformation to be used by the algorithm when fitting the training data. Scikit-learn proposes different alternatives such as linear, polynomial, radial-basis (RBF) and sigmoid. From our tests, the RBF returned the best results with a precision, recall and *F1-score* of 0.753, which, however, are sensibly lower than the other approaches. In terms of trainable weights, neither the algorithm nor the library provide this information, and for this reason, we did not report it in Table 2.

The IF required a lower number of parameters to be set. To select the best number of estimators to be used in the ensemble, we performed multiple experiments when varying this parameter in the range 5–100 with a step of 5. What we noticed is that the performance remained (on average) very similar with few oscillations starting from 50 estimators, and for this reason, we adopted this value. The model reached a good level of precision equal to 0.843 but resulted in the worst recall and *F1-score* among all the approaches with 0.294 and 0.435, respectively. Like the OCSVM, the algorithms do not make use trainable weights to execute.

With regards to the FC-DNN, in this case, our goal was to test the performance of a model not having recurrent connections and therefore unable to detect time relationships. For the realization of this topology, we used 5 hidden layers of 64, 32, 16, 4 and 1 neurons connected in cascade using the ReLU as the activation function, except for the output neuron, which adopts as usual the sigmoid. The obtained results are comparable with the ones reached by the first ESN model with a precision of 0.992, a recall value of 0.973 and a *F1-score* of 0.982. On the other hand, it resulted in being the model with the second largest number of trainable parameters (i.e., 2873) due to the only presence of fully connected layers which contain a lot of parameters to be trained, thus making it unsuitable for the hardware resources of an Edge device.

We also realized a model exploiting sensor data fusion LSTM, which represent the state-of-the-art in terms of RNNs. To make a fair comparison, in the feature extraction part, we used 2 LSTMs (i.e., 1 for the current signal and 1 for the vibration) with 20 neurons and tanh activation function. With respect to the predictive part, we used 2 LSTMs connected in cascade with 16 and 4 neurons with tanh activation and terminated by a layer with 1 neuron using the sigmoid. Although the LSTM reached the same results achieved by the proposed method, it resulted in being the largest model with 7705 trainable parameters. Such a result proves the power of this type of RNN, which, however, requires the optimization of a huge number of variables.

In the 1D-CNN model, each input is passed to an one-dimensional convolutional layer with 20 filters, a kernel size of 3, a stride of 1 and ReLU as the activation function. The features extracted at this stage are then passed to 2 fully connected layers connected in cascade with 16 and 4 neurons and with ReLU as activation function. Also in this case, the output layer consists of one neuron using the sigmoid to perform the binary classification. The overall performance reached by the model was good, with a precision of 1.0, a recall of 0.994 and a *F1-score* of 0.996; on the other hand, it resulted in being the third largest model after the FC-DNN and LSTM models with 1077 trainable parameters.

As already mentioned, time is a crucial component, especially in fault detection applications, where a prompt response can save a system from a total breakdown. To this aim, we conducted a comparison of the models average inference time, focusing our attention only on the neural network models which returned the best performance when compared with IF and OCSVM approaches.

Figure 7 shows the average inference time of the five neural network models we considered in our experimentation. With respect to the ESN + FC model, it reached an inference time of 0.37 s; in such a context, we can observe that the use of the ESNs in the feature extraction part is beneficial, making this approach one of the fastest. The FC-DNN is one of the slowest models due to its large number of trainable parameters, with an average inference time of 0.57 s. Regarding the 1D-CNN, we observed a slight improvement with respect to the DNN with an inference time of 0.45 s. From an architectural point of view, this model is very similar to the ESN+FC one: They share, in fact, the same number of fully connected layers, the same neurons and the same activation functions. The only difference is represented by the feature extraction process, which is performed through an ESN and a convolutional layer. Such a result, further demonstrates the effectiveness of ESNs in reducing the inference time thanks to their sparse structure and low number of parameters. As we would expect, the LSTM model resulted in being the slowest model, with an average inference time of 1.03 s. Finally, the proposed sensor data fusion ESN, which in the previous analysis exhibited the lowest amount of parameters, in this case, resulted in being the fastest among the other models with an average inference time of 0.16 s.

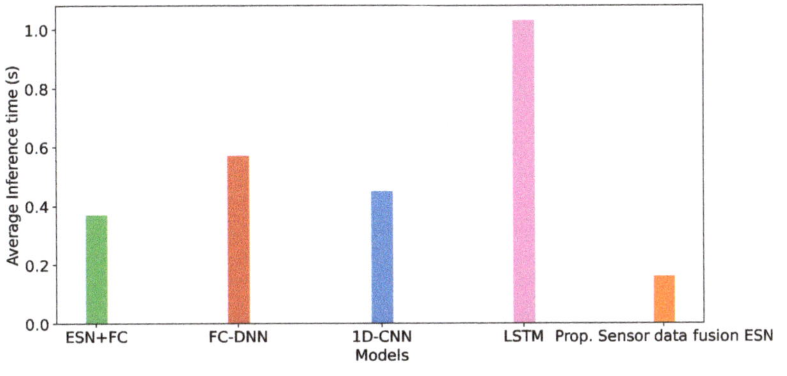

Figure 7. Models' average inference time.

The results derived from these comparisons demonstrate the effectiveness of the proposed approach, which reached a good level of performance comparable with the ones achieved by other neural network models while using a very low number of trainable parameters and exhibiting the fastest inference time. The combination of these results makes our model sufficiently small to be deployed on an Edge device with a response time suitable for the detection of faults in an industrial application context. However, we should point out that even if nowadays most of the industrial plants are provided with historical databases, it can happen that the data labeling process it not available or possible. In this sense, the supervised nature of the proposed approach poses a limitation to those contexts where it is possible to gather labeled data.

6. Conclusions

In this paper, we proposed a sensor data fusion model that exploits ESNs to perform the fault detection at the Edge of a real-scale replica industrial plant. Although the supervised nature of our method limits its applicability, the experiments demonstrated the feasibility of the proposed approach. Thanks to the use of ESNs we were able to obtain a model which achieved a very good level of performance with a precision of 1.0, a recall of 0.993 and a *F1-score* of 0.996, while keeping a limited number of trainable parameters. Such a result makes our approach suitable to be deployed on the constrained hardware of an Edge device; moreover, the fast inference time exhibited by our solution allows us to perform a real-time detection of faults in industrial systems.

Future works will be devoted to the improvement of the proposed technique through the realization of an unsupervised scheme capable of extracting useful features from the inputs and perform a fault detection on unlabeled data, to the use of a larger number of signals for a better plant monitoring and to the implementation of an on-device training procedure in order to enable a continuous learning application where the industrial plant can autonomously learn the emergence of new fault patters.

Author Contributions: Conceptualization, D.B. and F.D.V.; software, F.D.V.; validation D.B. and F.D.V.; investigation, D.B. and F.D.V.; writing—original draft preparation, F.D.V.; writing–review and editing, D.B. and F.D.V.; supervision, D.B. All authors have read and agreed to the published version of the manuscript.

Funding: This research was partially funded by the Italian Ministry of University and Research, grant Programma Operativo Nazionale (PON) "Ricerca e Innovazione" number PON 2014-2020 CCI2014IT16M2OP005.

Informed Consent Statement: Not applicable.

Data Availability Statement: Not applicable.

Conflicts of Interest: The authors declare no conflict of interest.

References

1. De Vita, F.; Bruneo, D.; Das, S.K. A Novel Data Collection Framework for Telemetry and Anomaly Detection in Industrial IoT Systems. In Proceedings of the 2020 IEEE/ACM Fifth International Conference on Internet-of-Things Design and Implementation (IoTDI), Sydney, NSW, Australia, 21–24 April 2020; pp. 245–251. [CrossRef]
2. Li, N.; Gebraeel, N.; Lei, Y.; Fang, X.; Cai, X.; Yan, T. Remaining useful life prediction based on a multi-sensor data fusion model. *Reliab. Eng. Syst. Saf.* **2021**, *208*, 107249. [CrossRef]
3. De Vita, F.; Bruneo, D.; Das, S.K. On the use of a full stack hardware/software infrastructure for sensor data fusion and fault prediction in industry 4.0. *Pattern Recognit. Lett.* **2020**, *138*, 30–37. [CrossRef]
4. Merlino, G.; Dautov, R.; Distefano, S.; Bruneo, D. Enabling Workload Engineering in Edge, Fog, and Cloud Computing through OpenStack-based Middleware. *ACM Trans. Internet Technol.* **2019**, *19*, 28. [CrossRef]
5. Shi, W.; Cao, J.; Zhang, Q.; Li, Y.; Xu, L. Edge computing: Vision and challenges. *IEEE Internet Things J.* **2016**, *3*, 637–646. [CrossRef]
6. De Vita, F.; Nardini, G.; Virdis, A.; Bruneo, D.; Puliafito, A.; Stea, G. Using Deep Reinforcement Learning for Application Relocation in Multi-Access Edge Computing. *IEEE Commun. Stand. Mag.* **2019**, *3*, 71–78. [CrossRef]
7. Lukoševičius, M.; Jaeger, H. Reservoir computing approaches to recurrent neural network training. *Comput. Sci. Rev.* **2009**, *3*, 127–149. [CrossRef]
8. Jaeger, H. Echo state network. *Scholarpedia* **2007**, *2*, 2330. [CrossRef]
9. Rodan, A.; Tino, P. Minimum complexity echo state network. *IEEE Trans. Neural Netw.* **2010**, *22*, 131–144. [CrossRef]
10. Zhang, Q.; Qian, H.; Chen, Y.; Lei, D. A short-term traffic forecasting model based on echo state network optimized by improved fruit fly optimization algorithm. *Neurocomputing* **2020**, *416*, 117–124. [CrossRef]
11. Bala, A.; Ismail, I.; Ibrahim, R.; Sait, S.M.; Oliva, D. An Improved Grasshopper Optimization Algorithm Based Echo State Network for Predicting Faults in Airplane Engines. *IEEE Access* **2020**, *8*, 159773–159789. [CrossRef]
12. Zivkovic, M.; Stoean, C.; Chhabra, A.; Budimirovic, N.; Petrovic, A.; Bacanin, N. Novel Improved Salp Swarm Algorithm: An Application for Feature Selection. *Sensors* **2022**, *22*, 1711. [CrossRef] [PubMed]
13. Long, J.; Zhang, S.; Li, C. Evolving Deep Echo State Networks for Intelligent Fault Diagnosis. *IEEE Trans. Ind. Inf.* **2020**, *16*, 4928–4937. [CrossRef]
14. Kashinath, S.A.; Mostafa, S.A.; Mustapha, A.; Mahdin, H.; Lim, D.; Mahmoud, M.A.; Mohammed, M.A.; Al-Rimy, B.A.S.; Fudzee, M.F.M.; Yang, T.J. Review of Data Fusion Methods for Real-Time and Multi-Sensor Traffic Flow Analysis. *IEEE Access* **2021**, *9*, 51258–51276. [CrossRef]
15. Liu, Z.; Xiao, G.; Liu, H.; Wei, H. Multi-Sensor Measurement and Data Fusion. *IEEE Instrum. Meas. Mag.* **2022**, *25*, 28–36. [CrossRef]
16. Pu, Z.; Li, C.; Zhang, S.; Bai, Y. Fault Diagnosis for Wind Turbine Gearboxes by Using Deep Enhanced Fusion Network. *IEEE Trans. Instrum. Meas.* **2021**, *70*, 1–11. [CrossRef]
17. Tran, M.Q.; Liu, M.K.; Elsisi, M. Effective multi-sensor data fusion for chatter detection in milling process. *ISA Trans.* **2021**. [CrossRef]
18. Wang, X.; Mao, D.; Li, X. Bearing fault diagnosis based on vibro-acoustic data fusion and 1D-CNN network. *Measurement* **2021**, *173*, 108518. [CrossRef]
19. Azamfar, M.; Singh, J.; Bravo-Imaz, I.; Lee, J. Multisensor data fusion for gearbox fault diagnosis using 2-D convolutional neural network and motor current signature analysis. *Mech. Syst. Signal Process.* **2020**, *144*, 106861. [CrossRef]
20. Akpudo, U.E.; Jang-Wook, H. An Automated Sensor Fusion Approach for the RUL Prediction of Electromagnetic Pumps. *IEEE Access* **2021**, *9*, 38920–38933. [CrossRef]
21. Lima, J.d.J.A.d.; Maldaner, L.F.; Molin, J.P. Sensor Fusion with NARX Neural Network to Predict the Mass Flow in a Sugarcane Harvester. *Sensors* **2021**, *21*, 4530. [CrossRef]
22. Patil, S.; Jalan, A.; Marathe, A. Support Vector Machine for Misalignment Fault Classification Under Different Loading Conditions Using Vibro-Acoustic Sensor Data Fusion. In *Experimental Techniques*; Springer: Berlin/Heidelberg, Germany, 2022; pp. 1–15.

23. Cao, R.; Yunusa-Kaltungo, A. An Automated Data Fusion-Based Gear Faults Classification Framework in Rotating Machines. *Sensors* **2021**, *21*, 2957. [CrossRef] [PubMed]
24. Jaeger, H. The "echo state" approach to analysing and training recurrent neural networks-with an erratum note. *Bonn Ger. Ger. Natl. Res. Cent. Inf. Technol. GMD Tech. Rep.* **2001**, *148*, 13.
25. Gallicchio, C.; Micheli, A. Echo state property of deep reservoir computing networks. *Cogn. Comput.* **2017**, *9*, 337–350. [CrossRef]
26. Patanè, L.; Xibilia, M.G. Echo-state networks for soft sensor design in an SRU process. *Inf. Sci.* **2021**, *566*, 195–214. [CrossRef]
27. Jalayer, M.; Orsenigo, C.; Vercellis, C. Fault detection and diagnosis for rotating machinery: A model based on convolutional LSTM, Fast Fourier and continuous wavelet transforms. *Comput. Ind.* **2021**, *125*, 103378. [CrossRef]
28. Li, Z.; Tanaka, G. Multi-reservoir echo state networks with sequence resampling for nonlinear time-series prediction. *Neurocomputing* **2022**, *467*, 115–129. [CrossRef]
29. Pedregosa, F.; Varoquaux, G.; Gramfort, A.; Michel, V.; Thirion, B.; Grisel, O.; Blondel, M.; Prettenhofer, P.; Weiss, R.; Dubourg, V.; et al. Scikit-learn: Machine Learning in Python. *J. Mach. Learn. Res.* **2011**, *12*, 2825–2830.
30. Abadi, M.; Agarwal, A.; Barham, P.; Brevdo, E.; Chen, Z.; Citro, C.; Corrado, G.S.; Davis, A.; Dean, J.; Devin, M.; et al. TensorFlow: Large-Scale Machine Learning on Heterogeneous Systems. 2015. Available online: https://www.tensorflow.org (accessed on 1 April 2022).

sensors

MDPI

Article

Data Fusion Approaches for the Characterization of Musts and Wines Based on Biogenic Amine and Elemental Composition

Aina Mir-Cerdà [1,2], Biel Granell [1], Anaïs Izquierdo-Llopart [1], Àngels Sahuquillo [1], José Fermín López-Sánchez [1], Javier Saurina [1,2,*] and Sonia Sentellas [1,2,3]

1 Department of Chemical Engineering and Analytical Chemistry, University of Barcelona, Martí i Franquès 1-11, E08028 Barcelona, Spain; ainamir17@gmail.com (A.M.-C.); biel.granell@gmail.com (B.G.); anais.izquierdo.llopart@gmail.com (A.I.-L.); angels.sahuquillo@ub.edu (À.S.); fermin.lopez@ub.edu (J.F.L.-S.); sonia.sentellas@ub.edu (S.S.)

2 Research Institute in Food Nutrition and Food Safety, University of Barcelona, Av. Prat de la Riba 171, Edifici Recerca (Gaudí), E08921 Santa Coloma de Gramenet, Spain

3 Serra Húnter Lecturer, Generalitat de Catalunya, E08007 Barcelona, Spain

* Correspondence: xavi.saurina@ub.edu

Abstract: Samples from various winemaking stages of the production of sparkling wines using different grape varieties were characterized based on the profile of biogenic amines (BAs) and the elemental composition. Liquid chromatography with fluorescence detection (HPLC-FLD) combined with precolumn derivatization with dansyl chloride was used to quantify BAs, while inductively coupled plasma (ICP) techniques were applied to determine a wide range of elements. Musts, base wines, and sparkling wines were analyzed accordingly, and the resulting data were subjected to further chemometric studies to try to extract information on oenological practices, product quality, and varieties. Although good descriptive models were obtained when considering each type of data separately, the performance of data fusion approaches was assessed as well. In this regard, low-level and mid-level approaches were evaluated, and from the results, it was concluded that more comprehensive models can be obtained when joining data of different natures.

Keywords: sparkling wine; biogenic amines; elemental composition; wine quality; winemaking practices; principal component analysis; data fusion approach

Citation: Mir-Cerdà, A.; Granell, B.; Izquierdo-Llopart, A.; Sahuquillo, À.; López-Sánchez, J.F.; Saurina, J.; Sentellas, S. Data Fusion Approaches for the Characterization of Musts and Wines Based on Biogenic Amine and Elemental Composition. *Sensors* **2022**, *22*, 2132. https://doi.org/10.3390/s22062132

Academic Editor: Natividad Duro Carralero

Received: 31 January 2022
Accepted: 7 March 2022
Published: 9 March 2022

Publisher's Note: MDPI stays neutral with regard to jurisdictional claims in published maps and institutional affiliations.

1. Introduction

Data fusion approaches have been considered an excellent way to enrich datasets, particularly for improving the descriptive performance of the method and the overall quality of the information [1,2]. The original datasets obtained with different analytical methods can be simply joined in a global matrix according to so-called low-level data augmentation to be further analyzed with the arsenal of chemometric methods available for characterization, classification, and quantification purposes. In addition, raw data from various sources can be pretreated specifically using chemometric methods, and the resulting individual post-processed matrices can be combined by using mid- or high-level data fusion. Hence, data augmentation approaches are especially fruitful in food analysis for dealing with characterization, classification, and authentication issues [3–5].

More specifically, in the case of wines and related alcoholic beverages, several studies have been conducted based on data fusion. Ranaweera and coworkers published an overview of wine authentication based on spectroscopic data and further chemometric analysis [6]. In a broader sense, Arslan and coworkers reviewed the characterization and authentication of alcoholic beverages by nondestructive instrumental techniques and chemometrics [7], and da Costa et al. focused on beers [8]. Among the analytical techniques used to generate data of high quality, near- and mid-infrared spectroscopies have been widely used for the authentication of grappa and other spirits [9,10], wine vinegar [11],

Mexican tequila [12], and beers [13,14]. In addition to vibrational data, other spectroscopies, such as UV–vis, excitation–emission fluorescence, and nuclear magnetic resonance, have also been suggested for different classification and characterization purposes [10–16]. For fast, selective, and sensitive analyses, high-resolution mass spectrometry combined with flow injection analysis has offered new opportunities for red wine discrimination and classification [17]. Electronic devices, including e-noses and e-tongues, have also been introduced for wine characterization, and the data gained from these techniques have often been combined to obtain more comprehensive descriptions [18–21]. For instance, colorimetric sensor arrays based on dyes exhibiting different cross-selectivities towards the analytes were used for the discrimination of alcoholic beverages, including beers and spirits [22–24]. These devices, when operated in the gas phase, resulted in optoelectronic noses in which sample recognition relied on volatile species such as aldehydes and ketones. An interesting review of optoelectronic noses can be found elsewhere [25]. Another widely used approach to generate instrumental data to assess wine features relies on separation techniques, including capillary electrophoresis, gas chromatography, and especially liquid chromatography. Several studies using chromatographic data have been reported in the scientific literature for the classification of Lambrusco wines [26], grappa spirits [10], and rums [27]. Multi-sensor data integrating information from a wide variety of analytical techniques have also been introduced for wine analysis. For instance, Izquierdo-Llopart et al. studied the classification of sparkling wines as a function of grape variety and coupage using concentrations of organic acids, phenolic compounds, antioxidant capacity, pH, total acidity, ethanol, or reducing sugars [28]. In another recent publication, Cavdaroglu and Ozed developed a strategy for the prediction of vinegar processing parameters based on UV–visible and mid-infrared spectra, pH, Brix, total acidity, total flavonoid content, total and individual phenolic contents, organic acid, sugar, and ethanol concentrations [29].

In this publication, we explore the combination of biogenic amine (BA) and elemental compositions in an attempt to find markers of winemaking practices and product quality. BAs are low-molecular-weight nitrogenous compounds arising from the decarboxylation of amino acids, which are especially abundant in wine, cheese, meat, fish, and spoiled products [30]. BAs provide valuable information on wine quality and oenological factors [31,32]. In wine and cava, concentrations of BAs can provide valuable information on product quality, as they are also a good indicator of food safety. BA contents depend on the agricultural practices involved in the production of grapes, the grape variety used, and the processes of vinification and aging, especially if the wine is exposed to the activity of microorganisms or free amino acids are present. Indeed, alcoholic and especially malolactic fermentations are principal processes involved in BA generation due to the presence of yeasts and bacteria. Some microorganisms can decarboxylate amino acids with specific enzymes to form biogenic amines, thus significantly increasing their content. Once the BAs are formed, they are relatively stable in the wine samples. Putrescine, ethanolamine, tyramine, and histamine are abundant in the final products to be commercialized. Some representative recent examples dealing with the relevance of BAs as biomarkers of wine quality are summarized in Table 1.

Table 1. Recent examples illustrating the potential role of BAs as quality markers.

Analytes	Sample Type	Method	Remarks	Ref.
Putrescine, ethanolamine, histamine, tyramine, cadaverine, phenylethylamine, agmatine, tryptamine, spermine, and spermidine	Musts, base wines, and sparkling wine; Xarel·lo and Pinot Noir varieties	HPLC-FLD with precolumn derivatization using dansyl-Cl	Putrescine, ethanolamine, tyramine, and histamine are the most important in wine quality	[33]

Table 1. *Cont.*

Analytes	Sample Type	Method	Remarks	Ref.
Isopenthylamine, ethanolamine, methylamine, ethylamine, spermidine, spermine, putrescine, tyramine, histamine, cadaverine, and tryptamine	Red and white Croatian wines from Hrvatsko zagorje and Dalmatia regions	HPLC-UV with precolumn derivatization using dansyl-Cl	BAs are a discriminating factor for a grape variety and geographical origin for red wines	[34]
Putrescine, histamine, tyramine, cadaverine, phenylethylamine, tryptamine, spermine, and spermidine	Red and white wines from Chinese markets	HPLC-FLD with precolumn derivatization using dansyl-Cl; liquid–liquid extraction of derivatives	Predominant BAs were putrescine, tyramine, and 2-phenylethylamine	[35]
Putrescine, ethanolamine, histamine, tyramine, cadaverine, phenylethylamine, tryptamine, and agmatine	Red Spanish wines; monovarietal 'Tempranillo' wines (young, oak, and aged)	UHPLC-FLD with precolumn derivatization using OPA	Storage time, temperature, and bottle closing influence BA levels. Cork stopper and refrigeration are the best conditions to prevent the increase in histamine and tyramine	[36]
Volatile compounds, amino acids, and amines; agmatine, histamine, spermidine, tyrosine, phenylethylamine, isoamylamine, putrescine, tyramine, and tryptamine	Spanish Sparkling wines from Verdejo, Viura, Malvasia, Albarin, Godello, Prieto Picudo, and Garnacha; "Champenoise" method	HPLC-UV with precolumn derivatization using diethyl ethoxymethylene-malonate	Albarin and Prieto Picudo wines showed the highest BA content	[37]
Methylamine, ethylamine, putrescine, cadaverine, histamine, spermidine, spermine, phenylethylamine, tyramine, and tryptamine	Alcoholic beverages including red and white wine	Ion-pair chromatography with butane-sulfonic acid; HPLC-potentiometric detection; multi-walled carbon nanotube sensing membrane	Tyramine and tryptamine are the most abundant in red wine; spermidine, spermine, and tryptamine are the most abundant in white wine	[38]
Histamine, putrescine, cadaverine, and tyramine	"Refosk" wine from Slovenian-Italian Karst region	HPLC-UV with precolumn derivatization using dansyl-Cl	Some strains of *Lactobacillus* have the ability to produce BAs	[39]
Cadaverine, hexylamine, histamine, phenylethylamine, putrescine, and tyramine	Chinese wines	Direct separation and detection by UHPLC-QqQ-MS/MS; QuEChERS for sample treatment	Histidine is correlated with alcoholic degree and grape variety; phenylethylamine is correlated with pH and storage time	[40]

The elemental composition has also been exploited to characterize and authenticate wines as a function of geographical, varietal, and oenological factors [41–43]. For instance, elements such as Al, B, Ca, Cu, Fe, K, Mg, and Mn are relevant markers of some wine features. Despite the apparent disparity between the two types of analytes, they are used to characterize the quality of food products, particularly wines, thus providing complementary information from inorganic and organic species. In addition to the simple combination of BA and elemental profiles in a sample-wise augmented data matrix, other data fusion approaches are explored.

2. Materials and Methods

2.1. Chemicals and Solutions

Unless otherwise specified, all reagents used were of analytical grade. General reagents for biogenic amine profiling and elemental analysis were as follows: nitric acid (65% (*w*/*w*), PanReac ApplyChem, Castellar del Vallès, Spain), sodium tetraborate (Merck KGaA, Darmstadt, Germany), dansyl chloride (98%, Thermo Fisher Scientific, Waltham, MA, USA), formic acid (>96%, Merck), acetonitrile (UHPLC PAI-ACS SuperGradient, PanReac), and chloroform (≥99.8%, Fluka, Buchs, Switzerland). Purified water was generated with an Elix3 system (Millipore, Bedford, MA, USA). Solutions for BA derivatization were dansyl-Cl reagent, prepared at a concentration of 10 mg mL^{-1} in acetone, and 0.1 mol L^{-1} sodium tetraborate as the buffer solution (pH 9.2). For ICP-OES and ICP-MS, samples were diluted with 1% (*v*/*v*) HNO_3.

Biogenic amine standards were as follows: 1,5-diaminopentane (cadaverine, 98%), 1,4-diaminobutane dihydrochloride (putrescine, 99%), spermidine trihydrochloride (99%), and spermine tetrahydrochloride (99%) from Alfa Aesar (Kandel, Germany); histamine hydrochloride (≥99%), 2-phenylethylamine hydrochloride (≥99%), tryptamine hydrochloride (≥98%), tyramine hydrochloride (≥97%), octopamine hydrochloride (≥99%), and agmatine sulfate (≥99%) from Fluka (Buchs, Switzerland); ethanolamine hydrochloride (≥98%) and hexylamine (≥98%) from TCI (Tokyo, Japan), the latter of which was used as the internal standard. Each amine was prepared as a 1000 mg L^{-1} stock solution in the laboratory by dissolution in Milli-Q water. Calibration standards were prepared by appropriately diluting stock solutions in a range from 0.1 to 50 mg L^{-1}. Stock and working solutions were stored at 4 °C until use.

Certified ICP grade single-element standards of 1000 mg L^{-1} in 1% (*v*/*v*) HNO_3 were purchased from Inorganic Ventures (Christiansburg, VA, USA). Calibration standards for ICP-MS and ICP-OES measurements were prepared by carrying out the required dilutions with 1% (*v*/*v*) HNO_3.

2.2. Samples

Musts, wines, and sparkling wines were kindly provided by Codorníu SA (Sant Sadurní d'Anoia, Spain). Table 2 describes the set of 20 monovarietal products from Xarel·lo grapes and 20 monovarietal products from Pinot Noir grapes, produced in Penedès and Costers del Segre regions (Catalonia, Spain). Each grape variety had five oenological classes, comprising musts, base wines, stabilized wines, 3-month-aged sparkling wines, and 7-month-aged sparkling wines, and four quality levels were defined for each class: A, B, C, and D, where A is the top quality and D is the lowest one. A more detailed description of the quality features is given elsewhere [33]. Sample coding is detailed in Table 2.

Table 2. List of samples under study. Sample codes are as follows: M, must; BW, base wine; SW, stabilized wine; C3, 3 months in rhyme cava wine (sparkling wine); C7, 7 months in rhyme cava wine (sparkling wine); P, Pinot Noir; X, Xarel·lo; A, quality A; B, quality B; C, quality C; D, quality D (reproduced from Ref. [33]).

Grape Variety	Quality	Must	Base Wine	Stabilized Wine	3-Month Sparkling Wine	7-Month Sparkling Wine
Pinot Noir	A	MPA	BWPA	SWPA	C3PA	C7PA
	B	MPB	BWPB	SWPB	C3PB	C7PB
	C	MPC	BWPC	SWPC	C3PC	C7PC
	D	MPD	BWPD	SWPD	C3PD	C7PD
Xarel·lo	A	MXA	BWXA	SWXA	C3XA	C7XA
	B	MXB	BWXB	SWXB	C3XB	C7XB
	C	MXC	BWXC	SWXC	C3XC	C7XC
	D	MXD	BWXD	SWXD	C3XD	C7XD

A quality control (QC) sample was prepared to assess the reproducibility of the analytical methods and the significance of the PCA models by mixing 1 mL of each must/wine/cava sample.

2.3. Instruments

The chromatographic system consisted of an Agilent 1100 Series HPLC instrument from Agilent Technologies (Waldbronn, Germany), with degasser (G1379A), binary pump (G1312A), automatic injector (G1392A), diode-array UV–vis detector (G1315B), and fluorescence detector (FLD, G1321A). Instruments for elemental analysis were Optima 3200RL ICP-OES and Nexlon 350D ICP-MS spectrometers (both from Perkin Elmer, Waltham, MA, USA) equipped with Ar plasma. Rh was used as the internal standard in ICP-MS.

2.4. Analytical Procedures

2.4.1. Biogenic Amine Determination

The method for the determination of BAs based on offline derivatization and liquid chromatography with fluorescence detection was established and validated elsewhere [33]. Briefly, BAs were derivatized offline by mixing 250 μL of sample (or biogenic amine standard), 250 μL of dansyl-Cl reagent solution, and 250 μL of buffer solution (pH 9.2). The reaction was developed at 40 °C in a thermostatic water bath (Tectron 473-100, J.P. Selecta, Barcelona, Spain) for 60 min. Derivatives were further extracted by adding 750 μL of chloroform and applying mechanical shaking for 10 min (Vortex 3 IKA, Staufen, Germany). The organic fraction was evaporated to dryness and redissolved in 600 μL of acetonitrile/water (50:50, *v*/*v*).

Derivatized samples were analyzed by HPLC-UV-FLD using a core–shell column (Kinetex C18, 150 mm × 4.6 mm I.D., 2.6 μm particle size) from Phenomenex (Torrance, CA, USA). A 0.1% (*v*/*v*) formic acid aqueous solution and acetonitrile (ACN) were used as the mobile phase components under an elution gradient program based on increasing the percentage of ACN (see Reference [33]). The flow rate was 0.7 mL min^{-1}, and the injection volume was 10 μL. UV detection was at 254 nm, and FLD was at 320 nm for excitation and 523 nm for emission.

Samples were processed in triplicate and analyzed randomly, injecting the QC sample every 10 samples.

2.4.2. Elemental Composition Determination

Samples diluted at a 1/10 ratio with 1% HNO_3 were directly analyzed by ICP-OES and ICP-MS, as explained elsewhere [43]. A blank solution (1% HNO_3) and the quality control (QC) sample were analyzed every 15 samples to check for cross-contamination and assess the repeatability of the results. All samples were analyzed in triplicate.

2.5. Data Analysis

ANOVA was performed with Microsoft Excel (Microsoft, Redmon WA, USA), and multivariate data analysis was conducted with SOLO software (Eigenvector Research, Inc. Manson, WA, USA).

Principal component analysis was applied for an exploratory characterization of musts, wines, and sparkling wines to try to identify patterns of oenological steps, product quality, and varieties using compositional data as the source of analytical information. Further details on PCA and other chemometric methods can be found elsewhere [44,45].

For each instrumental technique, data matrices (X-matrices) were generated, in which each row corresponded to a sample replicate and each column corresponded to a given analyte. X-matrix dimensions were 120 × 11 for BAs and 120 × 38 for the elemental composition. Hence, the low-level data fusion matrix was 120 × 49. The individual matrices of BA and elemental composition were pretreated by PCA to extract nine PCs, which were further combined in the mid-level approach, thus resulting in an augmented matrix of 120 × 18.

3. Results and Discussion

The performance of BA profiles and elemental composition as a source of potential descriptors of must, base wine, stabilized wine, and sparkling wine was previously assessed by Mir-Cerdà et al. and Granell et al., respectively [33,43]. In those studies, the two datasets were studied separately, so independent conclusions were drawn. In this work, however, we wanted to combine information from the two compositional profiles in order to try to improve the quality of the description and provide more comprehensive knowledge of sample features and the influence of the oenological practice, grape variety, and product quality. Hence, data matrices generated in the previous studies were fused using low- and mid-level approaches for further chemometric analysis.

First, we present the most important results extracted from the previous studies to highlight the outcomes of their independent chemometric characterizations of must, wine, and sparkling wine samples. As indicated in Section 2 (Materials and Methods), ethanolamine, putrescine, tyramine, histamine, cadaverine, spermine, spermidine, tryptamine, octopamine, lysine, and phenylethylamine were quantified by the HPLC-FLD method, and up to 36 elements were determined by ICP-OES or ICP-MS (Al, As, B, Ba, Ca, Cd, Ce, Co, Cr, Cs, Cu, Fe, Ga, K, La, Li, Mg, Mn, Mo, Na, Nd, Ni, P, Pb, Rb, S, Sb, Sc, Si, Sn, Sr, Ti, U, V, W, Y, Zn, and Zr). Table 3 shows the concentrations of some of the most significant components, which were selected because of their relevance as potential descriptors of sample type or sample quality features. As can be seen, potassium is overexpressed in musts, putrescine and histamine levels are increased in samples of C or D quality, sulfur is remarkably higher in base and stabilized wines, and ethanolamine and sodium concentration are, in general, slightly higher in sparkling wines. Similar patterns can be observed for other compounds.

Quantification errors were estimated according to Mir-Cerdà and Granell (see References [33,43]) from the analysis of the QC sample with the respective methods. The quantification errors of putrescine and ethanolamine (the most abundant amines) were 3.4% and 3.2%, respectively. For the other compounds, errors were below 10% (e.g., 5.5 for agmatine, 8.4 for tryptamine, 9.0 for phenylethylamine, 7.8 for cadaverine, 9.7 for histamine, and 6.6 for tyramine), except for spermidine and spermine (ca. 25%, caused by higher derivatization and stability issues). For the elemental composition, the errors in ICP-OES values of metals occurring at concentrations of the order of magnitude of 1 mg L^{-1} were lower than 2% (e.g., 0.9 for Mg, 1.2 for Ca, 1.5 for P, and 1.7 for Na), except for K (3.5%). For other important descriptors determined by ICP-MS, errors were ca. 5% (e.g., 4.3 for Fe, 2.7 for B, 6.3 for Cs, 7.3 for S, 6.4 for Sr, 5.6 for Ba, 5.0 for Mn, 4.5 for Cu, and 5.7 for Al). Trace elements occurring at concentrations between 1 and 100 µg L^{-1} showed errors ranging from 7 to 20% (e.g., Li, Mo, Ni, Zn, and V). The descriptive performance of these elements was more limited, thus mainly contributing to the noise.

In the case of BAs, PCA models showed well-defined clusters for each sample type, and the loading plot highlighted putrescine and ethanolamine as the best descriptors of the winemaking process. It was found that concentrations in must samples were, in general, low (except for ethanolamine). A remarkable rise in BAs was observed at the base wine stage, i.e., after the first alcoholic fermentation. This increase was even more dramatic for wines subjected to malolactic fermentation since this process has been identified as a major factor in the generation of BAs. After this stage, BA levels remained constant or slightly decreased with stabilization, second fermentation, and aging. This pattern was also observed for other amines, such as tyramine, histamine, and cadaverine. Regarding product quality, differences among high- and low-quality products were noticeable.

Regarding elemental composition, interesting patterns in the evolution of the composition of elements such as K, Cu, Ca, S, and Mg during the vinification process were found. Furthermore, some elements were recognized as potential markers of product quality. For instance, the top-quality (A) samples displayed lower contents of some elements, such as K and Ca. Other elements such as Mg, Mn, Na, Ni, Sr, and Zn also appeared in higher levels in C and D products since they were introduced from additives used in different technological processes. PCA showed two separated clusters corresponding to musts and fermented samples, thus confirming the noticeable differences due to the addition of several substances during fermentation, clarification, and stabilization processes. Subsequently, during the aging process, they tended to precipitate together with the lees, so their concentrations typically decreased in aged products.

Given the remarkable conclusions preliminarily extracted from the analysis of BA and elemental composition datasets, in the following analysis, we aimed to assess the performance of the combination of the two types of descriptors to obtain a more comprehensive characterization of the samples.

Table 3. Determination of various relevant compounds in the different samples. Concentrations are expressed in mg L^{-1}. Bold numbers denote samples with higher values.

Sample	Ethanolamine	Putrescine	Histamine	S	K	Na
MPA	2.99	2.52	0.16	2.68	**93.3**	1.22
MPB	2.70	1.42	0.14	5.86	**151.0**	2.77
MPC	3.85	4.84	0.17	8.04	**124.1**	2.05
MPD	3.49	2.14	0.13	3.83	**151.3**	2.59
MXA	2.72	1.29	0.11	3.48	**72.9**	1.87
MXB	4.01	0.43	0.10	2.91	**87.6**	2.16
MXC	5.30	3.29	0.11	3.46	**120.8**	1.78
MXD	4.09	2.81	0.12	3.31	**95.2**	1.40
BWPA	3.14	4.01	0.19	**32.5**	47.4	0.59
BWPB	5.21	3.42	0.18	**33.9**	79.0	0.50
BWPC	5.35	**24.10**	**4.00**	**32.7**	96.6	2.02
BWPD	6.13	**21.43**	**3.68**	**22.5**	77.5	3.38
BWXA	3.86	1.81	0.11	**17.7**	38.8	0.50
BWXB	5.14	3.05	0.11	**43.0**	78.9	1.37
BWXC	5.75	**10.79**	**1.76**	**57.9**	63.7	3.27
BWXD	6.51	**13.07**	**1.94**	**40.9**	75.7	2.44
SWPA	3.43	3.77	0.20	**37.5**	34.6	1.12
SWPB	5.49	2.80	0.31	**25.9**	37.6	2.23
SWPC	4.75	**10.81**	**1.20**	**24.3**	46.0	2.82
SWPD	6.57	**15.87**	**1.78**	**16.4**	30.4	5.18
SWXA	3.22	0.95	0.11	**16.9**	34.0	0.70
SWXB	6.29	2.43	0.21	**22.7**	27.2	2.00
SWXC	5.94	**14.37**	**2.32**	**21.6**	28.8	3.98
SWXD	7.13	**10.26**	**1.77**	**21.5**	35.7	4.63
C3PA	2.89	2.11	0.18	14.2	26.2	2.48
C3PB	6.05	2.69	0.30	25.9	37.1	2.28
C3PC	4.94	**12.00**	**1.40**	20.8	44.7	3.43
C3PD	6.42	**15.90**	**2.32**	16.8	25.4	4.69
C3XA	3.41	1.11	0.13	11.7	30.7	2.05
C3XB	7.14	3.14	0.26	24.4	14.2	2.40
C3XC	7.25	**17.83**	**2.64**	21.2	25.9	5.23
C3XD	6.08	**9.47**	**1.69**	19.9	39.3	5.16
C7PA	2.73	1.39	0.14	14.5	30.6	2.42
C7PB	5.50	2.28	0.26	25.3	40.7	2.20
C7PC	5.04	**11.47**	**1.38**	21.5	45.5	3.46
C7PD	6.74	**18.47**	**2.90**	19.1	21.8	5.70
C7XA	3.44	0.94	0.12	12.2	32.4	1.99
C7XB	5.55	3.11	0.26	23.4	30.4	3.42
C7XC	5.88	**10.84**	**1.87**	20.8	39.7	5.41
C7XD	6.06	**18.50**	**2.79**	21.2	41.0	5.04

3.1. Low-Level Data Fusion

In this analysis, BA and elemental composition datasets were joined by row-wise matrix augmentation, in which each row corresponded to a replicate of a given sample (it should be noted that samples were analyzed in triplicate), and each column was associated with a compositional variable (i.e., BA or element). The contents of BAs and target elements differed in both the magnitude and amplitude of concentrations, so data autoscaling was applied to equalize their influence in the models.

PCA showed a clear structure of samples according to the winemaking stage or process, regardless of other features such as quality or variety (see Figure 1). As a result, the scatter plot of PC1 vs. PC2 scores, which retained more than 43% of data variance, demonstrated clusters according to the sample type, with musts located on the left side, base wines in the upper-left area, stabilized wines predominating in the upper-right quadrant, and sparkling wines mainly in the lower-right part. Sparkling wines were also distinguished based on the aging period, with 3-month-aged wines above 7-month-aged ones. It is important to highlight that this noticeable class separation was not observed when BA and elemental datasets were analyzed separately (see References [33,43]). PC3 retained ca. 15% of the variance, thus providing some additional discrimination patterns (not shown here since the information from 3D plots was more difficult to visualize).

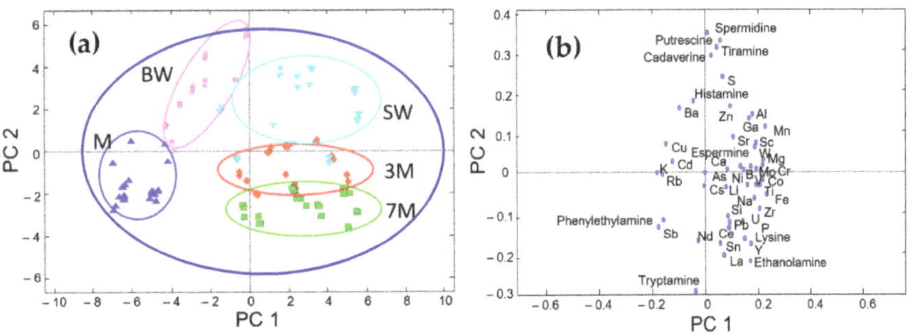

Figure 1. Plots of scores (**a**) and loadings (**b**) from the low-level data fusion. (**a**) Scatter plot of PC1 vs. PC2 scores; (**b**) scatter plot of PC1 vs. PC2 loadings. Sample assignment: M = must (blue); BW = base wine (light purple); SW = stabilized wine (light blue); 3M = sparkling wine with 3 months aging (red); 7M = sparkling wine with 7 months aging (green).

The study of leading descriptors revealed the occurrence of combined or hybrid markers. Elements such as K, Cu, Rb, and Ba were higher in musts compared with the other classes. Some of them (e.g., K and Cu) are the result of agricultural practices such as soil fertilization or mildew treatment. Various BAs—including putrescine, tyramine, spermidine, cadaverine, and histamine—were the dominant features of base and stabilized wines, as the contents of these BAs dramatically increased in the fermentations, especially when malolactic fermentation was applied to reduce the strong malic acidity of wines. A wide range of elements were characteristic of all fermented samples, as their concentrations increased because of oenological treatments with technological additives such as yeasts, tirage liquors, bentonite, and other agents. Among them, Zn, Al, Mn, Fe, Ni, and V can be cited. The BA composition of sparkling wines was different from that of stabilized wines. In this regard, ethanolamine was identified as a marker of sparkling wines since their higher alcoholic degree, achieved through the second fermentation, induced the generation of this compound. As another pattern, BA levels (e.g., cadaverine, histamine, and tyramine) slightly decreased with aging.

As a general conclusion, the BA and elemental compositions of grapes, which were assumed to be similar to the composition of musts, underwent remarkable changes after their transformation in wines. At this step, variations were statistically significant ($p < 0.05$) in all cases. In particular, levels of BAs significantly increased from musts to wines, and then they remained almost constant throughout winemaking processes and barely decayed with aging (except for ethanolamine, which slightly increased with the second fermentation). Similarly, changes in the elemental composition during vinification were dramatic as well due to the set of additives introduced to trigger fermentation and clarification processes.

Thus, an improvement in the characterization and discrimination performance was obtained after applying the data fusion approach.

3.2. Mid-Level Data Fusion

As mentioned in the experimental section, the individual matrices of BA and elemental profiles were pretreated by PCA filtering. The scores of three PCs were extracted, as they were a rich source of concentrated information, while irrelevant or ambiguous contributions were excluded from the model.

The row-wise augmented data matrix was evaluated by PCA, and the results obtained are summarized in Figure 2. The scatter plot of PC1 vs. PC2 scores shows a group of musts in the upper-left quadrant, meaning that this class is discriminated from the others. Wines and sparkling wines are scattered throughout the other sectors, without a clear separation among classes but with a certain predominance in some areas. For instance, sparkling wines tend to be across a diagonal (from the bottom left to top right), while base wine samples are located below this area. The other classes are mainly located in intermediate positions. Despite these patterns, the overall performance of this approach was lower than that obtained using the low-level model. On the other hand, the clustering of samples according to quality or variety was not detected either.

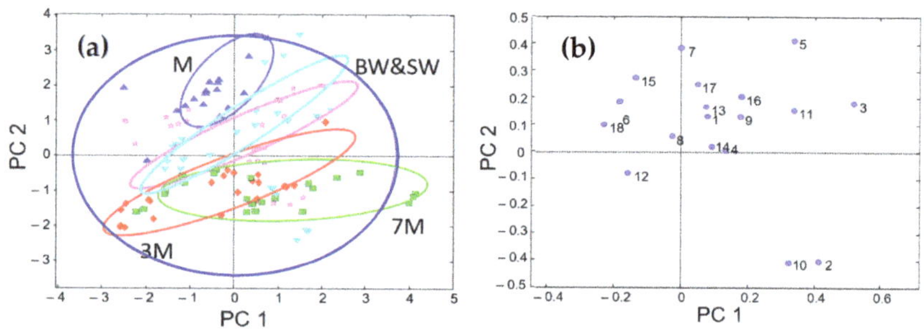

Figure 2. Plots of scores (**a**) and loadings (**b**) from mid-level data fusion. Sample assignment: M = must (blue); BW = base wine (light purple); SW = stabilized wine (light blue); 3M = sparkling wine with 3 months of aging (red); 7M = sparkling wine with 7 months of aging (green). Variable assignment: 1–9 = 1–9 PC of biogenic amines PCA; 10–18 = 1–9 PC of elemental composition PCA.

3.3. Sample Classification

Supervised studies of sample classification focused on low-level data fusion as, in this particular case, this approach has demonstrated excellent performance in describing the behavior of the samples. Given the natural dependence of sample features with respect to the vinification process, this section investigates the application of the PLS-DA to the classification of samples into the following classes: must, base wine, stabilized wine, 3-month-aged sparkling wine, and 7-month-aged sparkling wine.

The first model was established using all of the samples, in which the optimal number of latent variables (LVs) was 3. The plot of scores of LV1 versus LV2 (see Figure 3) shows a remarkable concentration of samples of each class in some specific areas. For example, musts are separated from the others and located in the bottom-right part. The base wines occupy the upper-right part of the graph, also separated from the other classes. Stabilized wines are located in the upper-left quadrant, while sparkling wines are in the lower-left part, showing two groups for 3- and 7-month-aged samples.

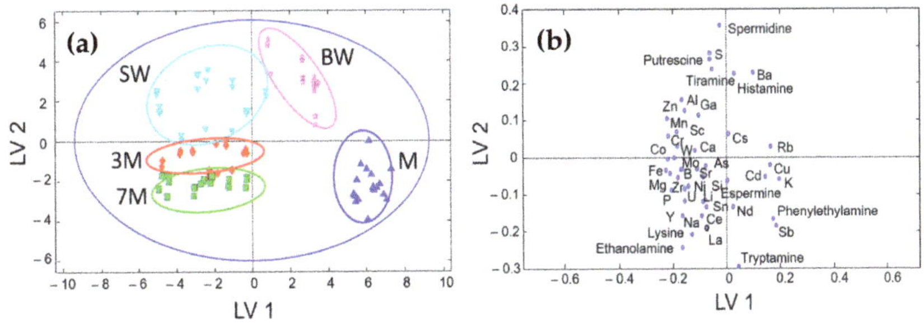

Figure 3. Plots of scores (**a**) and loadings (**b**) from low-level data fusion. (**a**) Scatter plot of LV1 vs. LV2 scores; (**b**) scatter plot of LV1 vs. LV2 loadings. Sample assignment: M = must (blue); BW = base wine (light purple); SW = stabilized wine (light blue); 3M = sparkling wine with 3 months of aging (red); 7M = sparkling wine with 7 months of aging (green).

For the external validation of the sample classification performance, a set of 120 samples (40 different combinations of class, quality, and variety analyzed in triplicate) was divided into two subsets for training and prediction, with 60 and 40% of the samples randomly selected, respectively. Table 4 summarizes the results using the PLS-DA model using three LVs for the multiclass classification of samples (i.e., classification of each given sample to one of the five classes). The calibration results showed that samples in the must and base wine classes were perfectly classified. For stabilized wine, two samples were incorrectly predicted to be base wine. For sparkling wines, a certain degree of confusion was found between 3- and 7-month-aged samples (two and three samples were misclassified, respectively). For the validation step, no unassigned samples were obtained. Additionally, musts, stabilized wines, and sparkling wines were correctly classified, while a certain amount of confusion was found for base wines (two base wines were classified as stabilized ones). Despite this confusion, which was attributed to the compositional similarities between these two classes, this method opens up promising possibilities for the study of the evolution of oenological samples throughout the vinification process. Furthermore, the performance of classification models based on the integration of BA and elemental profiles was definitively superior to that obtained from the use of each type of data separately.

Table 4. Summary of classification results with the percentages of correctly classified samples in both calibration and validation steps using PLS-DA.

	Classification Rate				
Step	**Must**	**Base Wine**	**Stabilized Wine**	**3-Month Sparkling Wine**	**7-Month Sparkling Wine**
Calibration	100%	100%	90% [1]	87% [2]	75% [3]
Validation	100%	70% [4]	100%	100%	100%

Misclassifications are as follows: [1] predicted as base wine; [2] predicted as 7-month-aged sparkling wine; [3] predicted as 3-month-aged sparkling wine; [4] predicted as stabilized wine.

4. Conclusions

In previous papers, it was found that the study and interpretation of descriptive models using biogenic amine or elemental profiles separately provided an incomplete depiction of the evolution of oenological samples throughout the production process of sparkling wines. In this study, this issue was fully solved by combining both sources of information via data fusion. The descriptive performance of a low-level approach using concentrations of BAs and metals was, in this case, superior to that of the mid-level counterpart using PCA scores as the fused data. The oenological process was found to be the principal factor affecting the composition of the studied analytes, while, in this set of samples, quality and variety issues had a lower influence on the description. In particular, excellent discrimination of musts, base wines, stabilized wines, and sparkling wines was realized, thus suggesting that this data combination can be used for successful sample characterization. Some increases in markers of the different classes were identified as well. For instance, high levels of lysine, K, and Cu were detected in musts, while BAs such as putrescine, tyramine, and histamine and elements such as Zn, Al, Mn, and Fe were predominant in base and stabilized wines, and ethanolamine was identified as a biomarker of sparkling wines. Furthermore, PLS-DA successfully classified samples as musts, base wines, stabilized wines, and sparkling wines, meaning that this method can be applied to accurately follow the evolution of oenological samples throughout the winemaking process.

Author Contributions: Conceptualization, A.I.-L. and J.S.; methodology, A.M.-C., B.G. and A.I.-L.; investigation, A.M.-C., B.G., and A.I.-L.; resources, J.S.; data curation, A.M.-C., B.G., and J.S.; writing—original draft preparation, A.M.-C., S.S., and J.S.; writing—review and editing, À.S., J.F.L.-S., J.S., and S.S.; supervision, À.S., J.F.L.-S., J.S., and S.S. All authors have read and agreed to the published version of the manuscript.

Funding: This research has been funded by the "Agencia Estatal de Investigación" of the Spanish Ministry of Science and Innovation, grant number PID2020-114401RB-C22.

Institutional Review Board Statement: Not applicable.

Informed Consent Statement: Not applicable.

Data Availability Statement: Data are contained within the article.

Acknowledgments: The authors thank Codorníu SA for providing the samples.

Conflicts of Interest: The authors declare no conflict of interest.

References

1. Borras, E.; Ferre, J.; Boque, R.; Mestres, M.; Acena, L.; Busto, O. Data fusion methodologies for food and beverage authentication and quality assessment—A review. *Anal. Chim. Acta* **2015**, *891*, 1–14. [CrossRef] [PubMed]
2. Bevilacqua, M.; Bro, R.; Marini, F.; Rinnan, A.; Rasmussen, M.A.; Skov, T. Recent chemometrics advances for foodomics. *Trends Anal. Chem.* **2017**, *96*, 42–51. [CrossRef]
3. Callao, M.P.; Ruisanchez, I. An overview of multivariate qualitative methods for food fraud detection. *Food Control* **2018**, *86*, 283–293. [CrossRef]
4. Biancolillo, A.; Marini, F.; Ruckebusch, C.; Vitale, R. Chemometric strategies for spectroscopy-based food authentication. *App. Sci.* **2020**, *10*, 6544. [CrossRef]
5. Di Rosa, A.R.; Leone, F.; Cheli, F.; Chiofalo, V. Fusion of electronic nose, electronic tongue and computer vision for animal source food authentication and quality assessment—A review. *J. Food Eng.* **2017**, *210*, 62–75. [CrossRef]
6. Ranaweera, R.K.R.; Capone, D.L.; Bastian, S.E.P.; Cozzolino, D.; Jeffery, D.W. A review of wine authentication using spectroscopic approaches in combination with chemometrics. *Molecules* **2021**, *26*, 4334. [CrossRef]
7. Arslan, M.; Tahir, H.E.; Zareed, M.; Shi, J.Y.; Rakha, A.; Bilal, M.; Huang, X.W.; Li, Z.H.; Zou, X.B. Recent trends in quality control, discrimination and authentication of alcoholic beverages using nondestructive instrumental techniques—A review. *Trends Food Sci. Technol.* **2021**, *107*, 80–113. [CrossRef]
8. Da Costa, N.L.; da Costa, M.S.; Barbosa, R. A Review on the application of chemometrics and machine learning algorithms to evaluate beer authentication. *Food Anal. Methods* **2021**, *14*, 136–155. [CrossRef]
9. Schiavone, S.; Marchionni, B.; Bucci, R.; Marini, F.; Biancolillo, A. Authentication of Grappa (Italian grape marc spirit) by Mid and Near Infrared spectroscopies coupled with chemometrics. *Vib. Spectrosc.* **2020**, *107*, 103040. [CrossRef]
10. Giannetti, V.; Mariani, M.B.; Marini, F.; Torrelli, P.; Biancolillo, A. Grappa and Italian spirits: Multi-platform investigation based on GC-MS, MIR and NIR spectroscopies for the authentication of the Geographical Indication. *Microchem. J.* **2020**, *157*, 104896. [CrossRef]
11. Rios-Reina, R.; Callejon, R.M.; Savorani, F.; Amigo, J.M.; Cocchi, M. Data fusion approaches in spectroscopic characterization and classification of PDO wine vinegars. *Talanta* **2019**, *98*, 560–572. [CrossRef] [PubMed]
12. Perez-Beltran, C.H.; Zuniga-Arroyo, V.M.; Andrade, J.M.; Cuadros-Rodriguez, L.; Perez-Caballero, G.; Jimenez-Carvelo, A.M. A sensor-based methodology to differentiate pure and mixed white tequilas based on fused infrared spectra and multivariate data treatment. *Chemosensors* **2021**, *9*, 47. [CrossRef]
13. Biancolillo, A.; Bucci, R.; Magri, A.L.; Magri, A.D.; Marini, F. Data-fusion for multiplatform characterization of an Italian craft beer aimed at its authentication. *Anal. Chim. Acta* **2014**, *820*, 23–31. [CrossRef] [PubMed]
14. Cavallini, N.; Savorani, F.; Bro, R.; Cocchi, M. Fused adjacency matrices to enhance information extraction: The beer benchmark. *Anal. Chim. Acta* **2019**, *1061*, 70–83. [CrossRef] [PubMed]

15. Rios-Reina, R.; Azcarate, S.M.; Camina, J.M.; Goicoechea, H.C. Multi-level data fusion strategies for modeling three-way electrophoresis capillary and fluorescence arrays enhancing geographical and grape variety classification of wines. *Anal Chim. Acta* **2020**, *1126*, 52–62. [CrossRef]
16. Tan, J.; Li, R.; Jiang, Z.T.; Zhang, Y.; Hou, Y.M.; Wang, Y.R.; Wu, X.; Gong, L. Geographical classification of Chinese Cabernet Sauvignon wines by data fusion of ultraviolet-visible and synchronous fluorescence spectroscopies: The combined use of multiple wavelength differences. *Aust. J. Grape Wine Res.* **2016**, *22*, 358–365. [CrossRef]
17. Campmajo, G.; Saurina, J.; Nunez, O. FIA-HRMS fingerprinting subjected to chemometrics as a valuable tool to address food classification and authentication: Application to red wine, paprika, and vegetable oil samples. *Food Chem.* **2022**, *373*, 131491. [CrossRef]
18. Ceto, X.; Capdevila, J.; Puig-Pujol, A.; del Valle, M. Cava wine authentication employing a voltammetric electronic tongue. *Electroanalysis* **2014**, *26*, 1504–1512. [CrossRef]
19. Garcia-Hernandez, C.; Salvo-Comino, C.; Martin-Pedrosa, F.; Garcia-Cabezon, C.; Rodriguez-Mendez, M.L. Analysis of red wines using an electronic tongue and infrared spectroscopy, Correlations with phenolic content and color parameters. *LWT-Food Sci. Technol.* **2020**, *118*, 108785. [CrossRef]
20. Zhang, H.H.; Shao, W.Q.; Qiu, S.S.; Wang, J.; Wei, Z.B. Collaborative analysis on the marked ages of rice wines by electronic tongue and nose based on different feature data sets. *Sensor* **2020**, *20*, 1065. [CrossRef]
21. Gutierrez, M.; Domingo, C.; Vila-Planas, J.; Ipatov, A.; Capdevila, F.; Demming, S.; Buttgenbach, S.; Llobera, A.; Jimenez-Jorquera, C. Hybrid electronic tongue for the characterization and quantification of grape variety in red wines. *Sens. Actuators B Chem.* **2011**, *156*, 695–702. [CrossRef]
22. Zhang, C.; Bailey, D.P.; Suslick, K.S. Colorimetric sensor arrays for the analysis of beers: A feasibility study. *J. Agric. Food Chem.* **2006**, *54*, 4925–4931. [CrossRef] [PubMed]
23. Li, Z.; Fang, M.; LaGasse, M.K.; Askim, J.R.; Suslick, K.S. Colorimetric recognition of aldehydes and ketones. *Angew. Chem. Int. Ed.* **2017**, *56*, 9860–9863. [CrossRef] [PubMed]
24. Li, Z.; Suslick, K.S. A hand-held optoelectronic nose for the identification of liquors. *ACS Sens.* **2018**, *3*, 121–127. [CrossRef] [PubMed]
25. Li, Z.; Suslick, K.S. The optoelectronic nose. *Acc. Chem. Res.* **2021**, *54*, 950–960. [CrossRef]
26. Silvestri, M.; Elia, A.; Bertelli, D.; Salvatore, E.; Durante, C.; Vigni, M.L.; Marchetti, A.; Cocchi, M. A mid level data fusion strategy for the Varietal Classification of Lambrusco PDO wines. *Chemom. Intell. Lab. Sys.* **2014**, *137*, 181–189. [CrossRef]
27. Belmonte-Sanchez, J.R.; Romero-Gonzalez, R.; Vidal, J.L.M.; Arrebola, F.J.; Garrido Frenich, A. H-1 NMR and multi-technique data fusion as metabolomic tool for the classification of golden rums by multivariate statistical analysis. *Food Chem.* **2020**, *317*, 126363. [CrossRef]
28. Izquierdo-Llopart, A.; Saurina, J. Multi-sensor characterization of sparkling wines based on data fusion. *Chemosensors* **2021**, *9*, 200. [CrossRef]
29. Cavdaroglu, C.; Ozen, B. Prediction of vinegar processing parameters with chemometric modelling of spectroscopic data. *Microchem. J.* **2021**, *171*, 106886. [CrossRef]
30. Simon Sarkadi, L. Amino acids and biogenic amines as food quality factors. *Pure Appl. Chem.* **2019**, *91*, 289–300. [CrossRef]
31. Saurina, J. Characterization of wines using compositional profiles and chemometrics. *TrAC Trends Anal. Chem.* **2010**, *29*, 234–245. [CrossRef]
32. Sentellas, S.; Núñez, O.; Saurina, J. Recent advances in the determination of biogenic amines in food samples by (U)HPLC. *J. Agric. Food Chem.* **2016**, *64*, 7667–7678. [CrossRef] [PubMed]
33. Mir-Cerda, A.; Izquierdo-Llopart, A.; Saurina, J.; Sentellas, S. Oenological processes and product qualities in the elaboration of sparkling wines determine the biogenic amine content. *Fermentation* **2021**, *7*, 144. [CrossRef]
34. Mitar, I.; Ljubenkov, I.; Rohtek, N.; Prkic, A.; Andelic, I.; Vuletic, N. The content of biogenic amines in Croatian wines of different geographical origins. *Molecules* **2018**, *23*, 2570. [CrossRef]
35. Liu, Y.; Han, F.L.; Liu, Y.J.; Wang, W.N. Determination of biogenic amines in wine using modified liquid-liquid extraction with high performance liquid chromatography-fluorescence detector. *Food Anal. Methods* **2020**, *13*, 911–922. [CrossRef]
36. Palomino-Vasco, M.; Rodriguez-Caceres, M.I.; Mora-Diez, N.; Pardo-Botello, R.; Acedo-Valenzuela, M.I. Biogenic amines profile in red wines regarding aging and storage conditions. *J. Food Compos. Anal.* **2019**, *83*, 103295. [CrossRef]
37. Perez-Magarino, S.; Ortega-Heras, M.; Martinez-Lapuente, L.; Guadalupe, Z.; Ayestara, B. Multivariate analysis for the differentiation of sparkling wines elaborated from autochthonous Spanish grape varieties: Volatile compounds, amino acids and biogenic amines. *Eur. Food Res. Technol.* **2013**, *236*, 827–841. [CrossRef]
38. Gil, R.L.; Amorim, C.G.; Montenegro, M.C.B.S.M.; Araujo, A.N. HPLC-potentiometric method for determination of biogenic amines in alcoholic beverages: A reliable approach for food quality control. *Food Chem.* **2022**, *372*, 131288. [CrossRef]
39. Bozic, J.T.; Butinar, L.; Marusic, M.B.; Korte, D.; Vodopivec, B.M. Determination of biogenic amines formation by autochthonous lactic acid bacteria from 'Refosk' grapes using different analytical methods. *LWT Food Sci. Technol.* **2022**, *156*, 112908. [CrossRef]
40. Han, S.Y.; Hao, L.L.; Shi, X.; Niu, J.M.; Zhang, B. Development and application of a new QuEChERS method in UHPLC-QqQ-MS/MS to detect seven biogenic amines in Chinese wines. *Foods* **2019**, *8*, 552. [CrossRef]
41. Gonzalvez, A.; Llorens, A.; Cervera, M.L.; Armenta, S.; de la Guardia, M. Elemental fingerprint of wines from the protected designation of origin Valencia. *Food Chem.* **2009**, *112*, 26–34. [CrossRef]

42. Katerinopoulou, K.; Kontogeorgos, A.; Salmas, C.E.; Patakas, A.; Ladavos, A. Geographical origin authentication of agri-food products: A review. *Foods* **2020**, *9*, 489. [CrossRef] [PubMed]
43. Granell, B.; Izquierdo-Llopart, A.; Sahuquillo, A.; López-Sánchez, J.F.; Saurina, J. Characterization of musts, wines, and sparkling wines based on their elemental composition determined by ICP-OES and ICP-MS. *Beverages* **2022**, *8*, 3. [CrossRef]
44. Kozak, M.; Scaman, C.H. Unsupervised classification methods in food sciences: Discussion and outlook. *J. Sci. Food Agric.* **2008**, *88*, 1115–1127. [CrossRef]
45. Granato, D.; Putnik, P.; Kovacevic, D.B.; Santos, J.S.; Calado, V.; Rocha, R.S.; Da Cruz, A.G.; Jarvis, B.; Rodionova, O.Y.; Pomerantsev, A. Trends in chemometrics: Food authentication, microbiology, and effects of processing. *Compr. Rev. Food Sci. Food Saf.* **2018**, *17*, 663–677. [CrossRef] [PubMed]

sensors

MDPI

Article

A Multiblock Approach to Fuse Process and Near-Infrared Sensors for On-Line Prediction of Polymer Properties

Lorenzo Strani [1], Raffaele Vitale [2], Daniele Tanzilli [1], Francesco Bonacini [3], Andrea Perolo [3], Erik Mantovani [3], Angelo Ferrando [3] and Marina Cocchi [1,*]

1 Department of Chemical and Geological Sciences, University of Modena and Reggio Emilia, Via 4 Campi 103, 41125 Modena, Italy; lostrani@unimore.it (L.S.); daniele.tanzilli@unimore.it (D.T.)
2 Centre National de la Recherche Scientifique (CNRS), Laboratoire de Spectroscopie pour les Interactions, la Réactivitè et l'Environnement (LASIRE), Cité Scientifique, University Lille, F-59000 Lille, France; raffaele.vitale@univ-lille.fr
3 Research Center, Versalis (ENI) S.p.A., Via Taliercio 14, 46100 Mantova, Italy; francesco.bonacini@versalis.eni.com (F.B.); andrea.perolo@versalis.eni.com (A.P.); erik.mantovani@versalis.eni.com (E.M.); angelo.ferrando@versalis.eni.com (A.F.)
* Correspondence: marina.cocchi@unimore.it; Tel.: +39-059-205-8554

Abstract: Petrochemical companies aim at assessing final product quality in real time, in order to rapidly deal with possible plant faults and to reduce chemical wastes and staff effort resulting from the many laboratory analyses performed every day. In order to answer these needs, the main purpose of the current work is to explore the feasibility of multiblock regression methods to build real-time monitoring models for the prediction of two quality properties of Acrylonitrile-Butadiene-Styrene (ABS) by fusing near-infrared (NIR) and process sensors data. Data come from a production plant, which operates continuously, and where four NIR probes are installed on-line, in addition to standard process sensors. Multiblock-PLS (MB-PLS) and Response-Oriented Sequential Alternation (ROSA) methods were here utilized to assess which of such sensors and plant areas were the most relevant for the quality parameters prediction. Several prediction models were constructed exploiting measurements provided by sensors active at different ABS production process stages. Both methods provided good prediction performances and permitted identification of the most relevant data blocks for the quality parameters' prediction. Moreover, models built without considering recordings from the final stage of the process yielded prediction errors comparable to those involving all available data blocks. Thus, in principle, allowing final ABS quality to be estimated in real-time before the end of the process itself.

Keywords: Acrylonitrile-Butadiene-Styrene; low-level data fusion; multiblock-partial least squares (MB-PLS); multivariate statistical process control; polymer production; quality prediction; real-time monitoring; response-oriented sequential alternation (ROSA)

Citation: Strani, L.; Vitale, R.; Tanzilli, D.; Bonacini, F.; Perolo, A.; Mantovani, E.; Ferrando, A.; Cocchi, M. A Multiblock Approach to Fuse Process and Near-Infrared Sensors for On-Line Prediction of Polymer Properties. *Sensors* **2022**, *22*, 1436. https://doi.org/10.3390/s22041436

Academic Editor: Natividad Duro Carralero

Received: 11 January 2022
Accepted: 7 February 2022
Published: 13 February 2022

1. Introduction

Nowadays, in several different domains like precision agriculture as well as pharmaceutical, food and chemical manufacturing, it is very common to utilize many analytical sensors to comprehensively characterize complex systems under study and to monitor processes while they evolve over time [1]. Analyzing the data yielded by such sensors by means of appropriate statistical tools is challenging but crucial in order to obtain meaningful physico-chemical information and design efficient production monitoring and control schemes. In particular, in industrial applications, a relevant issue is how to integrate or fuse the data resulting from sensors of different nature, potentially installed at different locations in the plant and in real time.

Multivariate Statistical Process Control (MSPC) is a well-established tool to accomplish real time monitoring and control of industrial production, in particular Latent Variables-Based MSPC (LV-MSPC) [2–7]. Most LV-MSPC relies on so-called engineering process variables [8], i.e., measured by on-line sensors controlling machinery settings (such as flow-meters, temperature and pressure probes, etc.) to build reference multivariate models for normal operating conditions (NOC), which are afterwards used to derive multivariate control charts and/or predicting quality attributes of finite product. More recently, thanks to technological developments, spectroscopic probes, especially near-infrared (NIR) ones, are extensively exploited [6,7,9–13] to monitor process evolution, or, in other words, to determine intermediate and final product quality parameters. Many studies in literature report on these aspects. Their results mainly refer to pilot scale plants [9,11,12,14] as well as to batch types of processes and seldom are engineering process variables and NIR measurements combined for constructing LV-MSPC models [6,10,14,15].

Fusing spectra with engineering variables is not a trivial task. However, process monitoring and control can greatly benefit from fusing these diverse data types, since, in this way, chemical composition-related information and physical and mechanical behavior/properties can be integrated.

This work focuses on a continuous styrenic polymer production process [16], monitored by means of NIR probes installed on-line in a production plant, as well as by standard process sensors. The main aim is to build real-time monitoring models to predict two of the main quality attributes of the final polymeric product by fusing NIR and process sensors' data. A preliminary feasibility study was recently conducted by the authors at the pilot-plant level [14].

Two aspects are particularly relevant for industry: (i) the possibility of estimating in real time the quality of a finite product, thus reducing the operational time and the amount of chemicals commonly required for laboratory off-line assessments by reference methods; and (ii) to reach the anticipated assessment of departure from desired quality before the end of production itself, in order to plan possible early modifications of the operating settings.

To this end, we investigated the application of multiblock chemometric methods [17–25] which are suitable to accomplish data fusion at low-level [26,27] and might bring interesting advantages with respect to alternative mid-level and high-level data integration strategies [26] especially in terms of model training, maintenance and interpretability. In fact, original variables are directly used without any compression steps, and it is possible to assess the salience of each block/type of sensors in the model, i.e., inspecting their degree of uniqueness or redundancy.

In particular, we compared a well-established multiblock MSPC approach, such as MultiBlock Partial Least Squares (MB-PLS) regression [21], with Response-Oriented Sequential Alternation (ROSA) [22]. The distinctive features of ROSA, which is also based on PLS regression [28,29], are: (i) to be invariant to block scaling and not to be affected by the spurious bias resulting from the combination of data blocks of different size (similarly to sequential orthogonal PLS (SO-PLS) [20]); and (ii) to be computationally efficient and capable of dealing with any number of blocks, also a very high number (differently from SO-PLS).

We tested models constructed on measurements yielded by sensors that were active at all different process stages (up to the process production end), as well as models where measurements from the last stage were excluded. This was in order to evaluate if polymer quality could be forecasted prior to the end of production. The results achieved, by both MB-PLS and ROSA, show satisfactory predictive performance for the determination of the two quality parameters investigated. At the same time, the most relevant data blocks were assessed.

2. Materials and Methods

2.1. Process Description

Data presented in the current work were collected on-line in an Acrylonitrile-Styrene-Butadiene (ABS) industrial production plant (full scale) operating in continuous process, owned by Versalis (ENI group). For the sake of simplicity, the plant can be regarded as divided into five different areas: (i) pre-poly/mixer, where the three precursor monomers (acrylonitrile, styrene and butadiene) are mixed together; (ii) reaction point A; (iii) reaction point B; (iv) reaction point C; and (v) devolatilizer/cut zone, where the finite product is cut. Throughout all these areas seventy process sensors (PS), which measure temperatures, pressures, flow rates and motor speed, and four NIR probes are installed. The NIR probes are placed in four specific and crucial areas of the production plant: one where dissolution of butadiene in styrene occurs, before the addition of acrylonitrile; one in the pipe for the recovery of condensed reagents; one between the first and the second reaction points; and one at the very end of the process, just before the cut zone. Overall, both PS and NIR probes record data/spectra with a frequency of about one minute. In this study, data registered from January 2020 to May 2021 were analyzed, even if not all the data recorded during this period were considered in model building, due to production pauses and deviations from the operative conditions relevant for the current study.

2.2. Reference Analysis

Two different parameters have been considered for the evaluation of ABS quality. Nonetheless, because of confidential agreement restrictions with the company, their actual names will not be disclosed, but they will be referred to as Property 1 and Property 2. Properties 1 and 2 are assessed off-line by collecting ABS samples, i.e., final product, two (Property 1) and three (Property 2) times per day. Property 1 is related to ABS composition, i.e., the percentage of a certain chemical compound in the final product. On the other hand, Property 2 gives information about physical features of the product and the values of the related reference analysis are expressed in grams. In the period covered by this study 597 and 904 laboratory tests (homogeneously distributed all over the time period) were carried out to determine Property 1 and Property 2, respectively. Property 1 values ranged from 20 to 21.8%; Property 2 values ranged from 3.9 to 6.1 g.

2.3. NIR Spectroscopy

A Matrix FT-NIR spectrometer (Bruker Optics, Milan, Italy) was used to acquire spectra in the four different acquisition sites. The instrument was equipped with optical fibers (length: 100 m, diameter: 600 μm), whose probes (HT immersion probe, Drawing-no. 661.2350_1, Hellma GmbH and Co. KG, Müllheim, Germany) were directly connected to the four different acquisition sites located on the process pipe. Spectra were collected in transmission mode over the 12,500–4000 cm^{-1} spectral range, with a nominal resolution of 4 cm^{-1} (64 scans per sample).

2.4. Data Analysis

2.4.1. Data Block and Multiblock Arrangement

The ensemble of collected data was arranged into nine distinct data blocks, according to the data type and the acquisition area along the process: on the one hand, PS measurements were gathered in five blocks, one per every area of the plant (see also Section 2.1); on the other hand, NIR spectra were arranged into four blocks, each corresponding to an individual optical probe. In Table 1, the names and abbreviations (which will be hereafter used) of all the blocks are shown, together with their size and the location along the plant. This is also an indication of how they are ranked in time, being a continuous process.

For both multiblock approaches, the data blocks were assembled considering the chronological progression of the ABS production process and, therefore, based on the location of the different sensors along the production line. In other words, each data point present in the datasets refers to information collected at different times, but it is correctly matched to the same processed material (i.e., data are synchronized).

Figure 1 displays a schematic representation of the low-level data fusion strategy adopted.

Table 1. Data block description.

Block Full Name	Block Abbreviated Name	Data Type	No. of Variables [1]	Order
NIR dissolution	NIR-diss	NIR Spectra	390	1
Prepoli/Mixer	Prep/mix	PS	7	2
NIR condensation	NIR-cond	NIR Spectra	390	3
Reaction Point A	RP-A	PS	15	4
NIR Reaction Point A	NIR-RP-A	NIR Spectra	390	5
Reaction Point B	RP-B	PS	10	6
Reaction Point C	RP-C	PS	8	7
Devolatilizer/cut zone	Devo/cut	PS	30	8
NIR cut zone	NIR-cut	NIR Spectra	390	9

[1] For NIR data blocks, the number of variables is equal to the spectra wave numbers, whereas for PS data blocks it is equal to the number of PS present in the respective plant area. The column "Order" highlights how the process evolves chronologically.

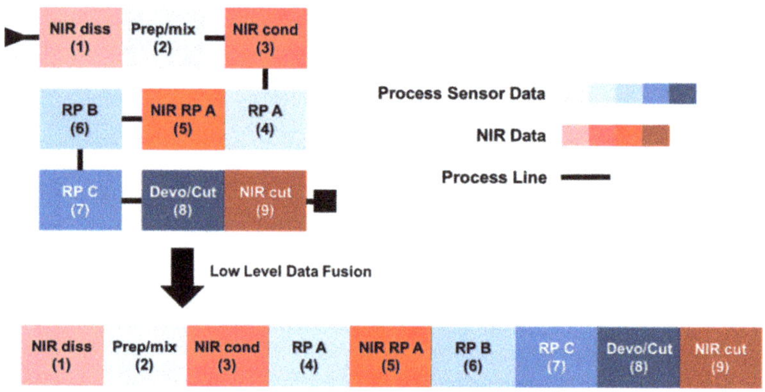

Figure 1. Schematic representation of the low-level data fusion approach resorted to in this study. Values in brackets indicate the chronological order of the data blocks.

2.4.2. Preprocessing

Individual block preprocessing

Prior to the multiblock modeling phase, each data set was preprocessed individually. In particular, variables in each PS data block were scaled to unit variance (different in nature and scales) whereas spectra, in each NIR data block, were baseline-corrected by using automatic weighted least squares [30]. Moreover, only the spectral range from 6500 to 5000 cm^{-1} (the sole one exhibiting spectral bands ascribable to either reactants or products) was taken into account for subsequent model training. Figure 2 shows the effect of the baseline correction executed on the NIR spectra of the NIR-RP-A data block.

Multiblock preprocessing

After the individual preprocessing of the single blocks, each data set was scaled to unit block variance (including column mean-centering) prior to MB-PLS [21]. In fact, MB-PLS operates directly on row-wise concatenated data blocks and a fair block contribution has to be assured.

Concerning ROSA, the individual pre-processed blocks were just mean-centered since such a method treats one block at a time, as it will be detailed in the following sections.

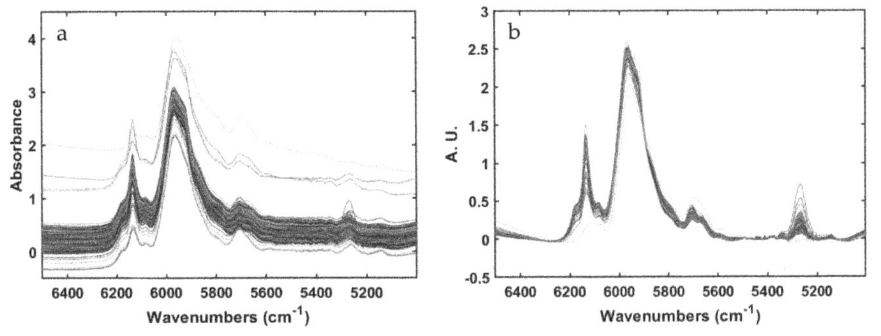

Figure 2. Spectra collected at NIR-RP-A, data block before (**a**) and after (**b**) baseline correction using automatic weighted least square method.

2.4.3. MB-PLS

We exploited here the MB-PLS implementation originally proposed by Westerhuis and Coenegracht [31] which can be looked at as standard PLS with appropriate block scaling steps as described in [21]. Thus, MB-PLS is an extension of the classical PLS regression [28] for applications involving different data blocks that share the same number of rows (observations), relating to the data matrix **X**, resulting from the row-wise concatenation of N different data blocks (Equation (1)):

$$\mathbf{X} = [\mathbf{X}_1, \mathbf{X}_2, \ldots, \mathbf{X}_N] \tag{1}$$

to the response(s) of interest.

This method provides global (also called *super-*) scores, weights, loadings and regression coefficients, as well as local (also called *block-*) scores and weights for each data block, as it is shown in Equations (2)–(5):

$$\mathbf{w}_b = \mathbf{X}^{\mathrm{T}}{}_b * \mathbf{u}/\mathbf{u}^{\mathrm{T}}\mathbf{u} \tag{2}$$

$$\mathbf{t}_b = (\mathbf{X}_b * \mathbf{w}_b)/\sqrt{\mathbf{n}_b} \tag{3}$$

$$\mathbf{w} = \mathbf{T}^{\mathrm{T}} * \mathbf{u}/\mathbf{u}^{\mathrm{T}}\mathbf{u} \tag{4}$$

$$\mathbf{t} = \mathbf{T} * \mathbf{w}/\mathbf{w}^{\mathrm{T}}\mathbf{w} \tag{5}$$

where $\mathbf{n}_b$ is the number of variables in a given block, $\mathbf{t}_b$ and $\mathbf{w}_b$ are the local scores and weights, respectively, whereas **t** and **w** are the global (*super*) scores and weights. **T** is yielded by the concatenation of all $\mathbf{t}_b$.

This way, it is possible to assess the contribution of each data block (analyzing $\mathbf{w}_b$ for the prediction of the response variable/s $\mathbf{y}/\mathbf{Y}$, improving the process understanding).

2.4.4. ROSA

Response-Oriented Sequential Alternation (ROSA) is a multiblock regression method proposed by Liland et al. [22] that is also based on PLS regression. Different from MB-PLS, in that ROSA is a sequential algorithm, similar to, e.g., SO-PLS [20], which renders the method invariant with respect to block-scaling (blocks are just mean centered), as well as to block ordering, differently from SO-PLS. These features allow dealing with a large number of blocks of different dimensions.

Moreover, ROSA exhibits a high computational efficiency, as it does not require the iterative convergence of an optimization criterion, and because only the response is deflated, not all the blocks. In fact, each PLS component is selected from a single block, picking among the various covariance-maximizing candidate components, estimated from each data block, the one returning the smallest prediction residuals. Successive components are constrained to be orthogonal to the subspace spanned by the previously winning components. Thus, scores' and loadings' orthogonality is ensured.

The ROSA algorithm for a single response variable, $\mathbf{y}$, is summarized in the following equations:

$$\mathbf{w}_b = \mathbf{X}_b{}^{\mathrm{T}} * \mathbf{y} \tag{6}$$

$$\mathbf{t}_b = \mathbf{X}_b * \mathbf{w}_b \tag{7}$$

$$\mathbf{r}_b = \mathbf{y} - \mathbf{t}_b\, \mathbf{t}_b{}^{\mathrm{T}}\mathbf{y} \tag{8}$$

where $\mathbf{X}_b$ is a single data block, while $\mathbf{w}_b$, $\mathbf{t}_b$ and $\mathbf{r}_b$ are block weights, scores and residuals, respectively. The first component is selected as the one computed from the b_{th}-block yielding the smallest residuals ($\mathbf{r}_b$), and $\mathbf{t}_1$ are taken to be equal to $\mathbf{t}_b$ of the winning block. The corresponding weights and scores are normalized (and also orthogonalized with respect to the preceding components from the second component on). The $\mathbf{y}$-loadings are finally estimated as:

$$\mathbf{q}_a = \mathbf{y}^{\mathrm{T}}\, \mathbf{t}_a \tag{9}$$

where $\mathbf{t}_a$ are the scores previously selected for the a_{th} LV.

$\mathbf{X}$-loadings ($\mathbf{P}$) and PLS regression coefficients ($\mathbf{b}$) (and possibly a constant term b_0) can be estimated according to the Equations (10)–(12), after selecting the number of optimal LVs and collecting the corresponding scores, weights, $\mathbf{y}$-loadings in matrix array $\mathbf{T}$, $\mathbf{W}$ and $\mathbf{Q}$.

$$\mathbf{P} = \mathbf{X}^{\mathrm{T}}\, \mathbf{T} \tag{10}$$

$$\mathbf{b} = \mathbf{W}(\mathbf{P}^{\mathrm{T}}\mathbf{W})^{-1}\mathbf{Q} \tag{11}$$

$$\mathbf{b}_0 = \mathbf{y}_{\mathrm{m}} - \mathbf{x}_{\mathrm{m}} * \mathbf{b} \tag{12}$$

where $\mathbf{y}_{\mathrm{m}}$ is the mean of $\mathbf{y}$ and $\mathbf{x}_{\mathrm{m}}$ is a vector with the mean for each variable of $\mathbf{X}$.

Thus, each selected LV in ROSA encodes information proceeding only from the winning b_{th}-block (the one achieving smallest residuals according to Equation (8)), and all LVs are orthogonal. It is important to notice that all blocks are always candidates at each algorithmic step. Therefore, consecutive LVs can depict information from the same block previously selected, or from a different one.

2.4.5. Multiblock Models Building

With the aim of developing predictive models for the two parameters taken into account in this study and assessing which are the most important data blocks for their estimation, both MB-PLS and ROSA were investigated.

All the available data were split into calibration and validation sets for both Property 1 and Property 2. In order to assess models' performance in a scenario mimicking a real-time application, the calibration sets comprised data collected during the year 2020 (~70% of total data), whereas the validation sets comprised data collected in 2021. Clearly, only samples, i.e., time points, for which the offline reference measurement were available were taken into account.

The two optimized best-performing models were finally utilized for assessing the values of Property 1 and 2 at time points where no reference data were acquired, in order to check whether the resulting estimations spanned a similar properties values range with respect to close time points.

In order to establish the complexity, i.e., number of PLS components, of each model, venetian blinds cross-validation with ten cancellation groups for Property 1 and four cancellation groups for Property 2 was resorted to. Model reliability was determined in terms of both root mean square error in cross-validation (RMSECV) and root mean square error in prediction (RMSEP).

Data blocks were preprocessed as described in Section 2.4.2.

For both MB-PLS and ROSA, the contribution of each block and block variables in the final predictive model was assessed by investigating the PLS regression coefficients and

Variable Importance in Prediction (VIP) [32,33]. PLS block-weights were also inspected but, for the sake of brevity, the related figures are not reported, as the provided information was similar to that obtained by regression coefficients.

2.5. Software

All the chemometric analyses were performed using routines and toolboxes implemented in the MATLAB environment (the Mathworks Inc., Natick, MA, USA).

MB-PLS has been calculated through the PLS-Toolbox version 8.9 (Eigenvector Research Inc., Wenatchee, WA, United States).

ROSA (with options for venetian blind cross-validation, VIP calculation and validation sample response prediction) was implemented by the authors based on the MATLAB code provided in ref. [22].

3. Results

3.1. Property 1 Prediction

When all the available data blocks (PS and NIR measurements for all plant areas) were simultaneously modelled ROSA resulted to be the most performant method for the prediction of Property 1, yielding a RMSEP of 0.14%. On the other hand, MB-PLS returned a RMSEP value of 0.2%. This difference, however, is not substantial. The results are shown in Table 2 and Figure 3. ROSA selected only three of the nine blocks under study, two of which, Devo/cut and NIR-cut, relate to the last stage of the process, where the polymerization is over and the product is ready to be cut. Furthermore, among the 13 latent variables selected through the cross-validation procedure (aimed at minimizing RMSECV), eight were calculated from the NIR-cut block, which highlights a crucial relevance of the final NIR sensor, in this case, for the quality prediction. Figure 3a shows how the predictions for the objects of the validation set are homogeneously distributed within the expected range of the quality parameter concerned. In Figure 3b–d the PLS regression coefficients associated to the three blocks selected by ROSA are represented (the red stars denote variables/spectral regions whose VIP scores were higher than one). In the RP-A data block (selected only one time out of 13) only three temperature sensors were found relevant for Property 1 prediction, whereas in Devo/cut and NIR-cut data blocks all the sensors and nearly all the spectral regions sampled were somewhat important. In Figure 3d it is evident that the largest (in absolute value) regression coefficients are those corresponding to bands centered at 5900 cm^{-1} and 5250 cm^{-1} that can be ascribed to the investigated ABS compound.

Table 2. Results yielded by MB-PLS and ROSA for the prediction of Property 1.

Model ID	Blocks Entering the Model	LVs	RMSEC (%)	RMSECV (%)	RMSEP (%)
MB PLS all	All	11	0.12	0.16	0.20
MB PLS no cut zone	1 to 7	11	0.13	0.17	0.23
MB PLS only PS	2–4–6–7–8	11	0.24	0.26	0.38
MB PLS only NIR	1–3–5–9	10	0.13	0.15	0.22
MB PLS only NIR no cut zone	1–3–5	8	0.14	0.15	0.22
ROSA all [1]	4(1)–8(4)–9(8)	13	0.11	0.14	0.13
ROSA no cut zone	3(6)–4(1)–5(3)–6(2)	12	0.15	0.18	0.2
ROSA only PS	2(1)–4(6)–7(3)	10	0.23	0.25	0.31
ROSA only NIR	9(8)	8	0.12	0.13	0.14
ROSA only NIR no cut zone	3(12)–5(2)	14	0.16	0.18	0.19

[1] the values in brackets indicate the number of times a certain block was selected by the ROSA algorithm.

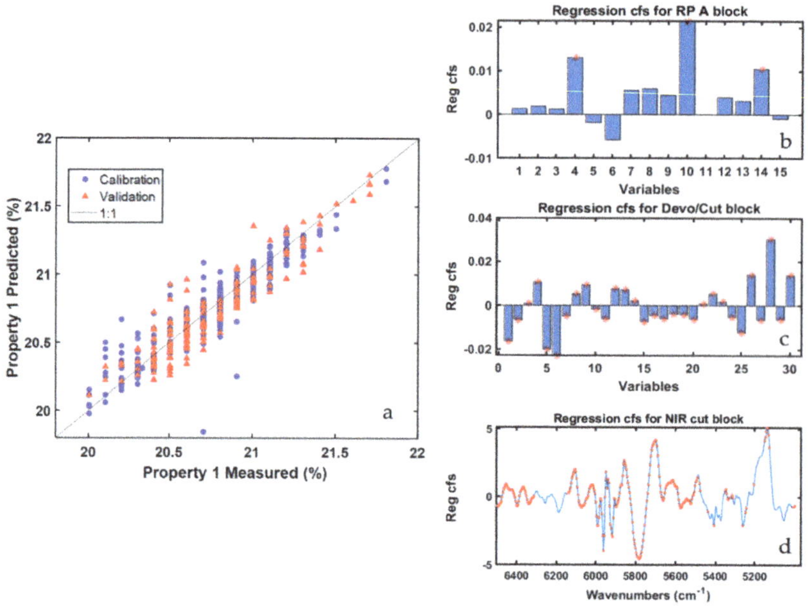

Figure 3. ROSA results for Property 1 prediction (all data blocks were modelled simultaneously). Predicted vs. measured value plot (**a**); regression coefficients for the RP-A (**b**); Devo/cut (**c**); and NIR cut (**d**) data blocks. Red stars indicate variables having VIP scores higher than one.

Although such results might already be considered relatively satisfactory from a predictive point of view, two additional aspects would be worth investigating: (i) whether reasonably good quality prediction of Property 1 values could be obtained before the product is cut (i.e., without relying on sensors installed within the cut area); and (ii) whether the exclusive use of spectral sensors or process sensors could be sufficient for a reliable estimation of this quality index. To this end, in addition to the dataset containing all the blocks, MB-PLS and ROSA models were calculated using fused datasets comprising only the blocks before the cut zone, only PS data and only NIR data (both including and excluding the spectra contained in the NIR-cut block), respectively.

Table 2 reports the results of all the computed multiblock prediction models related to Property 1. It is possible to observe that prediction errors resulting from ROSA are systematically lower than the one obtained by means of MB-PLS. It is also clear how NIR data are far more important for the prediction of Property 1 than PS data. In fact, when ROSA is run on both block types, components from NIR data sets are more often selected than those computed from PS data sets. Moreover, in MB-PLS models, variables related to NIR blocks are always relevant for Property 1 prediction. In addition, the RMSEP of models that are calculated using only NIR data is comparable to that of models using both PS and NIR data, while using only PS data blocks results in a significant increase of the prediction error in calibration, cross-validation and external validation. This is somehow expected, as Property 1 is linked to ABS chemical composition and, therefore, an analytical technique like NIR spectroscopy is definitely more suitable for its determination than more standard engineering PS probes, which only indirectly reflect how fluctuations in the process operating conditions may affect the polymer characteristics.

Since ROSA models always selected components estimated from the blocks located on the plant cut area, i.e., blocks eight and nine, we also decided to calibrate ROSA models (using both PS and NIR data and only NIR data) excluding completely such blocks from the computational procedure (see 'ROSA no cut zone' and 'ROSA only NIR no cut zone' in Table 2, respectively). In both cases, RMSEP values for models not including the cut area, were found higher, yet acceptable by process operators. This clearly makes it possible to retrieve reasonable Property 1 value estimate before the completion of the ABS production process. Moreover, similar prediction errors were obtained by using only NIR blocks or when combining NIR and PS blocks. Hence, two possible pathways can be envisioned for the real-time prediction and control of Property 1: (i) resorting to both data types and getting a clearer idea of the important process areas/sensors all along the production plant; or (ii) just exploiting NIR spectra for more efficient data management and to deal with less noisy data.

In order to evaluate the role of all types of sensors, Figure 4 displays the results yielded by the 'ROSA no cut zone' model. It is worth mentioning that half of the blocks selected by the ROSA algorithm relate to the reaction points A and B, whereas the other half to the NIR-cond data block, whose respective probe is right before these reaction points. Looking at the order (not reported for the sake of brevity) in which blocks were selected by ROSA, it can be observed how the winning blocks for the first five latent variables were RP-B (picked only one time) and NIR-RP-A (picked four times). For the remaining model dimensions, NIR-cond was selected six times in a row, while RP-A and RP-B one each. Details about the selection order are useful to assess which blocks, i.e., areas of the plant, encode the most important information for the prediction of the investigated quality parameter.

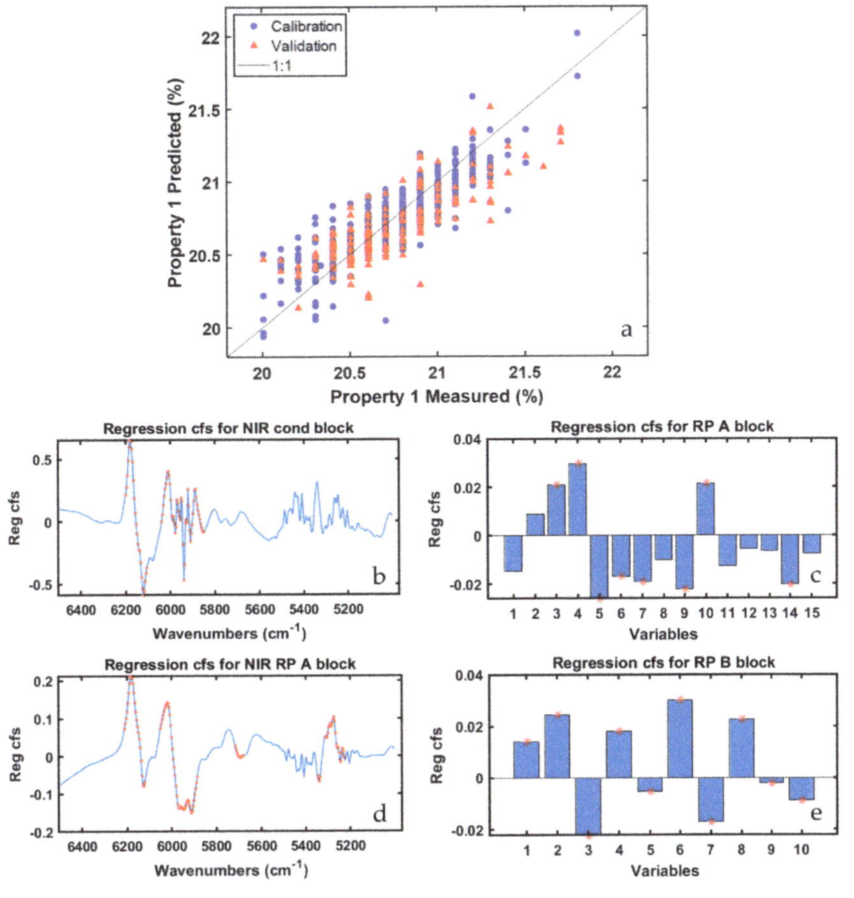

Figure 4. ROSA results for Property 1 prediction ('ROSA no cut zone' model). Predicted vs. measured value plot (**a**); regression coefficient for NIR cond (**b**); RP A (**c**); NIR RP A (**d**); and RP B (**e**) data blocks. Red stars indicate variables having VIP scores higher than one.

Figure 4b,d show the regression coefficients for the two aforementioned NIR blocks, with NIR-RP-A exhibiting a larger number of spectral variables characterized by VIP scores higher than one, especially in the region between 5400 cm^{-1} and 5250 cm^{-1}, that are ascribable to the stretching of a functional group of one of the three precursor compounds on which Property 1 directly depends. Conversely, in Figure 4c,e the regression coefficients for the PS data blocks are graphed: the most significant variables, according to their respective VIP values, are almost all related to temperature and motor speed sensors installed in different subzones of the reaction points A and B.

3.2. Property 2 Prediction

The same model building strategy described before was finally followed for the prediction of Property 2. Table 3 reports the results obtained by means of both MB-PLS and ROSA. ROSA, when all the available data blocks were simultaneously modelled, did not select any cut area block, therefore the 'ROSA no cut zone' model was not trained in this case.

Table 3. Results yielded by MB-PLS and ROSA for the prediction of Property 2.

Model ID	Blocks Entering the Model	LVs	RMSEC (g)	RMSECV (g)	RMSEP (g)
MB PLS all	All	10	0.25	0.27	0.34
MB PLS no cut zone	1 to 7	8	0.27	0.29	0.37
MB PLS only PS	2–4–6–7–8	9	0.27	0.29	0.35
MB PLS only NIR	1–3–5–9	7	0.34	0.34	0.48
MB PLS only NIR no cut zone	1–3–5	6	0.36	0.37	0.5
ROSA all [1]	2(1)–4(1)–5(1)–6(1)	4	0.32	0.33	0.46
ROSA only PS	2(1)–4(1)–6(1)	3	0.32	0.33	0.45
ROSA only NIR	5(6)–9(3)	9	0.33	0.34	0.52
ROSA only NIR no cut zone	5(8)	8	0.33	0.34	0.52

[1] The values in brackets indicate the number of times a certain block was selected by the ROSA algorithm.

MB-PLS models calibrated by using (i) all the data blocks or (ii) only PS data returned the most satisfactory results, contrary to the results obtained for Property 1. In fact, the influence NIR spectra have on the estimation of Property 2 prediction is not predominant, except for the NIR-RP-A block, which was selected many times by the ROSA algorithm and whose variables always showed VIP scores higher than one in MB-PLS. These results can be interpreted in the light of the fact that Property 2 is not linked to the chemical composition of ABS but evaluates the performance of the finite product as determined by mechanical/physical tests. Subsequently, it is undoubtedly more affected by variability occurring in the processing steps, and can change significantly even if the aforementioned chemical composition does not change. RMSEP increased up to 0.52 g when no PS block was considered. However, for models built without PS data, MB-PLS achieved a slightly better performance than ROSA (0.48–0.5 g vs. 0.52 g). These results suggested how the exclusive use of NIR sensors is not sufficient for a reliable estimation of Property 2.

Overall, MB-PLS showed a better prediction performance for Property 2. The best results were obtained by the 'MB PLS all' model (RMSEP = 0.34 g), even though 'MB PLS no cut zone' and 'MB PLS only PS' provided similar results.

In Figure 5 is where the predicted vs. measured value plot resulting from the 'MB-PLS all' model is shown. By inspecting the corresponding residuals plot (not shown for the sake of brevity) it can be observed that, on average, the 2021 production campaign (validation set), yielded lower values of Property 2 than that conducted in 2020 (calibration set). This deviation explains the relatively high difference between RMSEP and RMSEC and RMSECV. However, the presence of a reasonable amount of validation samples in the whole calibration range was guaranteed and the company deemed the prediction error acceptable for routine monitoring.

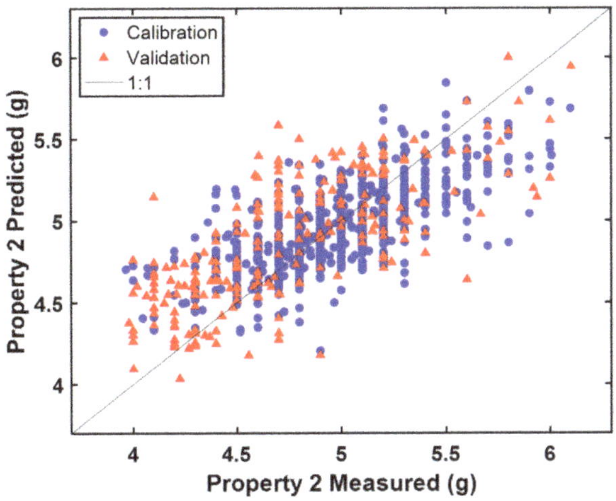

Figure 5. Predicted vs. measured value plot resulting from the 'MB-PLS all' model.

In Figure 6 the 'MB-PLS all' model regression coefficients are reported. All PS were found to be important for the prediction of Property 2 based on their VIP scores values. For what concerns the NIR blocks regression coefficients, the NIR-RP-A is confirmed to be the block with the largest number of highly predictive spectral regions, which are mainly related to the three precursors monomers of ABS. For the other NIR blocks, relevant regions of interest were found in correspondence of the absorption bands centered at 5900 cm^{-1} and 6100 cm^{-1}, respectively.

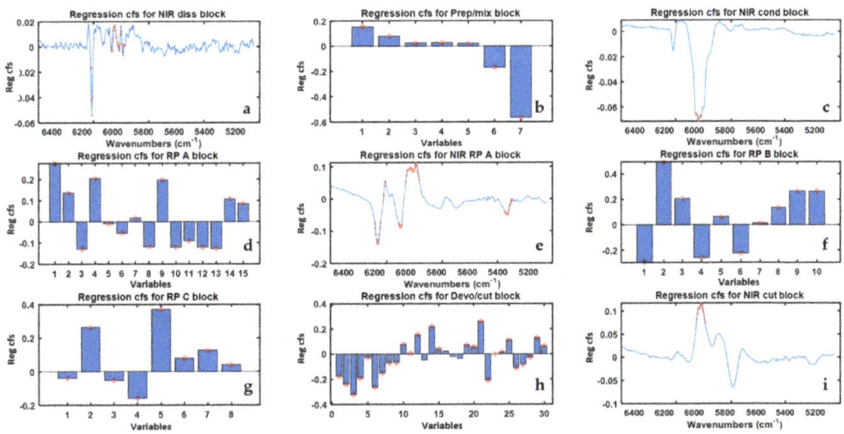

Figure 6. Regression coefficients resulting from the 'MB-PLS all' model for each data block the letters (**a–i**) refer to the different block whose name is reported on top. Red stars indicate variables exhibiting VIP scores higher than one.

3.3. Real-Time Predictions

Finally, Figure 7 illustrates the predicted values of Property 1 obtained through the ROSA model constructed on all data blocks (Table 2, row 1) for the time points for which reference response measurements were not acquired.

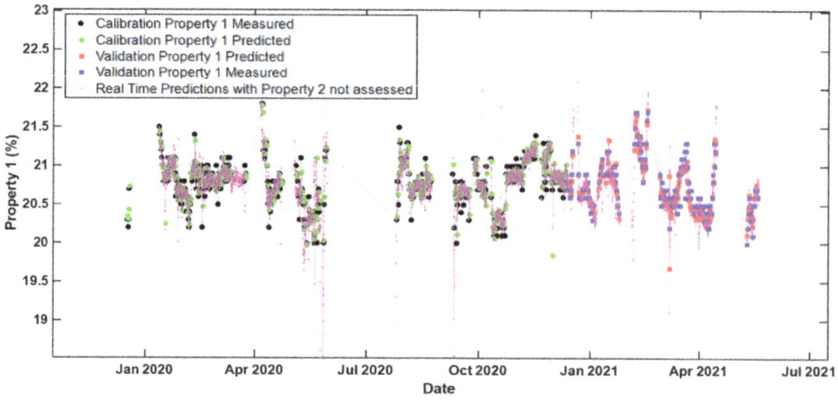

Figure 7. Real time predictions of Property 1 (i.e., time evolution of the measured and predicted values). The predictions were obtained by means of the 'ROSA all' model. Legend: black circles—calibration set measured values; green circles—calibration set predicted values; blue squares—validation set measured values; red squares—validation set predicted values; magenta dots—predicted values related to time points for which no reference response measurements were available. For ease of visualization only every 2 h predictions during the considered time period are shown.

These predicted values span a range very similar to that covered within both the calibration and the validation set. A few slight deviations were observed, interestingly right after specific shut-down time periods: such deviations may, in fact, arise from the fact that many industrial processes (including polymerization processes) take a certain time to readapt to NOC conditions after particular external interventions (e.g., cleaning, maintenance, etc.).

Similar results were obtained for real-time predictions with the model 'MB-PLS no cut zone' for Property 2, as shown in Figure 8.

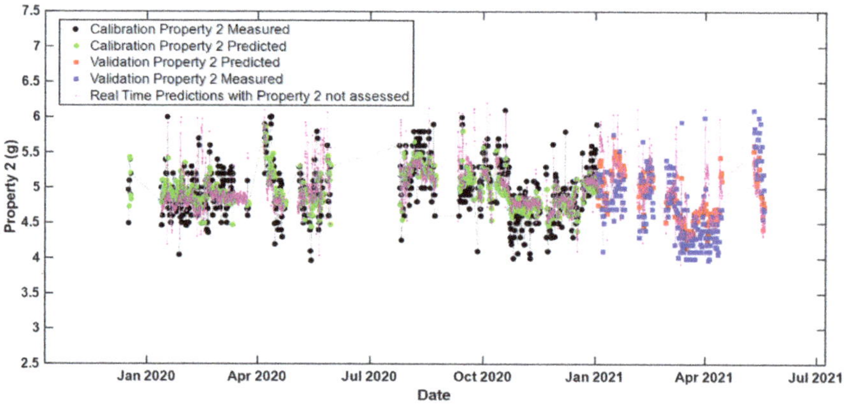

Figure 8. Real time predictions of Property 2 (i.e., time evolution of the measured and predicted values). The predictions were obtained by means of the 'MB PLS no cut zone' model. Legend: black circles—calibration set measured values; green circles—calibration set predicted values; blue squares—validation set measured values; red squares—validation set predicted values; magenta dots—predicted values related to time points for which no reference response measurements were available. For ease of visualization only every 2 h predictions during the considered time period are shown.

4. Conclusions

This work demonstrated how multiblock approaches could be used for the construction of reliable and robust real-time monitoring models for the on-line prediction of industrial quality parameters of ABS. In fact, the data partition in different blocks and the low-level data fusion strategy adopted here permitted to improve ABS production process understanding, enabling the assessment of the most crucial plant areas and the relevant sensors for the prediction of such specific parameters. Moreover, the application of these approaches is essential when two or more different analytical platforms of different nature, like the NIR spectrometer and more standard engineering process sensors, are simultaneously used to control any generic production process.

More specifically, in this article, both MB-PLS and ROSA allowed performant predictive models to be constructed for the two properties under study (i.e., Property 1 and 2). In particular, for the prediction of Property 1, ROSA resulted in a lower RMSEP compared to MB-PLS, highlighting the importance of NIR data over process sensor data when a chemical composition-related quality index is to be estimated. On the other hand, Property 2 was more efficiently predicted by a MB-PLS method, which pointed out a higher relevance of process sensors compared to NIR data when, instead, physical features need to be assessed.

Furthermore, models computed without taking into account measurements related to the final area of the plant (cut zone) provided comparable prediction errors with respect to the best models built on all the ensemble of available data. This is of great industrial interest, since, in principle, ABS quality could be determined before its production is completed, which might allow possible modifications of the plant settings and/or changes in the operating conditions to be planned in advance and with reduced costs.

In conclusion, these approaches could help in: (i) accelerating decision making and troubleshooting; (ii) reducing the amount of chemical waste generated in full-scale plants; (iii) decreasing the number of off-line laboratory tests required for quality control; and (iv) facilitating any type of operation along the production line as well as possible fault detection and diagnosis.

Author Contributions: Conceptualization, R.V., F.B., E.M., A.F. and M.C.; methodology, L.S., R.V., D.T., F.B., A.P., E.M., A.F. and M.C.; software, L.S., D.T., A.P. and M.C.; validation, R.V., F.B., A.P., E.M., A.F. and M.C.; investigation, L.S., R.V., D.T., F.B., A.P., E.M., A.F. and M.C.; resources, F.B., A.P., E.M. and A.F.; data curation, L.S., D.T., F.B., E.M. and A.P.; writing—original draft preparation, L.S. and D.T.; writing—review and editing, L.S., R.V., F.B., A.F. and M.C.; supervision, R.V., F.B. and M.C.; funding acquisition, A.F. and M.C. All authors have read and agreed to the published version of the manuscript.

Funding: LS post doctoral position was granted by Emilia Romagna region under POR FSE project "Data analytics per la REAlizzazione di sistemi predittivi e Monitoraggio real TIME di processi produttivi in industria 4.0 (DREAMTIME)" PA n° 2019-13551/RER.

Institutional Review Board Statement: Not applicable.

Informed Consent Statement: Not applicable.

Data Availability Statement: The datasets presented in this article are not readily available because of confidential agreement restrictions with the company. Requests to access the datasets should be directed to erik.mantovani@versalis.eni.com.

Acknowledgments: The authors acknowledge Angelo Ferrando of Versalis (ENI) Company for supplying data used for the current study and fruitful discussion of the results, and Federico Marini for useful suggestions and discussion concerning ROSA.

Conflicts of Interest: Authors A.F., F.B., A.P. and E.M. are employed by Versalis (ENI) SpA. The remaining authors declare that the research was conducted in the absence of any commercial or financial relationships that could be construed as a potential conflict of interest.

References

1. Bowler, A.L.; Bakalis, S.; Watson, N.J. A review of in-line and on-line measurement techniques to monitor industrial mixing processes. *Chem. Eng. Res. Des.* **2020**, *153*, 463–495. [CrossRef]
2. MacGregor, J.F.; Bruwer, M.J.; Miletic, I.; Cardin, M.; Liu, Z. Latent Variable Models and Big Data in the Process Industries. *IFAC-PapersOnLine* **2015**, *48*, 520–524. [CrossRef]
3. Kourti, T. Multivariate Statistical Process Control and Process Control, Using Latent Variables. In *Comprehensive Chemometrics*, 2nd ed.; Brown, D.S., Tauler, R., Walczak, B., Eds.; Elsevier B.V.: Amsterdam, The Netherlands, 2020; Volume 4, pp. 275–303.
4. Ferrer-Riquelme, A.J. Statistical Control of Measures and Processes. In *Comprehensive Chemometrics*, 2nd ed.; Brown, D.S., Tauler, R., Walczak, B., Eds.; Elsevier B.V.: Amsterdam, The Netherlands, 2020; Volume 4, pp. 215–236.
5. Wold, S.; Kettaneh-Wold, N.; MacGregor, J.F.; Dunn, K.G. Batch Process Modeling and MSPC. In *Comprehensive Chemometrics*, 2nd ed.; Brown, D.S., Tauler, R., Walczak, B., Eds.; Elsevier B.V.: Amsterdam, The Netherlands, 2020; Volume 4, pp. 305–332.
6. Morris, J.; Martin, E.; Stewart, D. Batch Process Monitoring through the integration of Spectral and Process Data. *IFAC-PapersOnLine* **2005**, *38*, 3–18. [CrossRef]
7. Aguado, D.; Ferrer, A.; Seco, A.; Ferrer, J. Comparison of different predictive models for nutrient estimation in a sequencing batch reactor for wastewater treatment. *Chemom. Intel. Lab. Syst.* **2006**, *84*, 75–81. [CrossRef]
8. Kadlec, P.; Gabrys, B.; Strandt, S. Data-driven Soft Sensors in the process industry. *Comput. Chem. Eng.* **2009**, *33*, 795–814. [CrossRef]
9. Gabrielsson, J.; Jonsson, H.; Trygg, J.; Airiau, C.; Schmidt, B.; Escott, R. Combining process and spectroscopic data to improve batch modeling. *AICHE J.* **2006**, *52*, 3164–3172. [CrossRef]
10. Lourenço, N.D.; Lopes, J.A.; Almeida, C.F.; Sarraguça, M.C.; Pinheiro, H.M. Bioreactor monitoring with spectroscopy and chemometrics: A review. *Anal. Bioanal. Chem.* **2012**, *404*, 1211–12137. [CrossRef] [PubMed]
11. Avila, C.; Mantzaridis, C.; Ferré, J.; de Oliveira, R.R.; Kantojärvi, U.; Rissanen, A.; Krassa, P.; de Juan, A.; Muller, F.L.; Hunter, T.N.; et al. Acid number, viscosity and end-point detection in a multiphase high temperature polymerization process using an online miniaturised MEMS Fabry-Pérot interferometer. *Talanta* **2021**, *224*, 121735. [CrossRef] [PubMed]
12. Sousa, B.V.; Silva, F.; Reis, M.A.M.; Lourenço, N.D. Monitoring pilot-scale polyhydroxyalkanoate production from fruit pulp waste using near-infrared spectroscopy. *Biochem. Eng. J.* **2021**, *176*, 108210. [CrossRef]
13. Grassi, S.; Strani, L.; Casiraghi, E.; Alamprese, C. Control and monitoring of milk renneting using FT-NIR spectroscopy as a process analytical technology tool. *Foods* **2019**, *8*, 405. [CrossRef]
14. Strani, L.; Mantovani, E.; Bonacini, F.; Marini, F.; Cocchi, M. Fusing NIR and Process Sensors Data for Polymer Production Monitoring. *Front. Chem.* **2021**, *9*, 748723. [CrossRef] [PubMed]
15. De Oliveira, R.R.; Avila, C.; Bourne, R.; Muller, F.; de Juan, A. Data fusion strategies to combine sensor and multivariate model outputs for multivariate statistical process control. *Anal. Bioanal. Chem.* **2020**, *412*, 2151–2163. [CrossRef] [PubMed]
16. Scheirs, J.; Priddy, D. *Modern Styrenic Polymers: Polystyrenes and Styrenic Copolymers*; Scheirs, J., Priddy, D., Eds.; J. Wiley and Sons, Ltd.: Chichester, UK, 2003.
17. Mishra, P.; Roger, J.M.; Jouan-Rimbaud-Bouveresse, D.; Biancolillo, A.; Marini, F.; Nordon, A.; Rutledge, D.N. Recent trends in multi-block data analysis in chemometrics for multi-source data integration. *TrAC* **2021**, *137*, 116206. [CrossRef]
18. Vitale, R.; de Noord, O.E.; Westerhuis, J.A.; Smilde, A.K.; Ferrer, A. Divide et impera: How disentangling common and distinctive variability in multiset data analysis can aid industrial process troubleshooting and understanding. *J. Chemometr.* **2020**, *35*, e3266. [CrossRef]
19. Campos, M.P.; Reis, M.S. Data preprocessing for multiblock modelling—A systematization with new methods. *Chemom. Intel. Lab. Syst.* **2020**, *199*, 103959. [CrossRef]
20. Biancolillo, A.; Næs, T. The sequential and orthogonalized PLS regression for multiblock regression: Theory, examples, and extensions. In *Data Fusion Methodology and Applications*; Cocchi, M., Ed.; Elsevier: Amsterdam, The Netherlands, 2019; Volume 31, pp. 157–177.
21. Westerhuis, J.A.; Kourti, T.; MacGregor, J.F. Analysis of multiblock and hierarchical PCA and PLS models. *J. Chemom.* **1998**, *12*, 301–321. [CrossRef]
22. Liland, K.H.; Næs, T.; Indahl, U.G. ROSA—A fast extension of partial least squares regression for multiblock data analysis. *J. Chemom.* **2016**, *30*, 651–662. [CrossRef]
23. El Ghaziri, A.; Cariou, V.; Rutledge, D.N.; Qannari, E.M. Analysis of multiblock datasets using ComDim: Overview and extension to the analysis of (K + 1) datasets. *J. Chemom.* **2016**, *30*, 420–429. [CrossRef]
24. Löfsted, T.; Trygg, J. OnPLS—A novel multiblock method for the modelling of predictive and orthogonal variation. *J. Chemometr.* **2011**, *25*, 441–455. [CrossRef]
25. Tauler, R.; Maeder, M.; de Juan, A. Multiset data analysis: Extended multivariate curve resolution. In *Comprehensive Chemometrics*, 2nd ed.; Brown, D.S., Tauler, R., Walczak, B., Eds.; Elsevier: Amsterdam, The Netherlands, 2020; Volume 2, pp. 305–336.
26. Cocchi, M. Introduction: Ways and Means to Deal with Data from Multiple Sources. In *Data Fusion Methodology and Applications*; Cocchi, M., Ed.; Elsevier: Amsterdam, The Netherlands, 2019; Volume 31, pp. 1–25.
27. Smilde, A.K.; van Mechelen, I. A Framework for Low-Level Data Fusion. In *Data Fusion Methodology and Applications*; Cocchi, M., Ed.; Elsevier: Amsterdam, The Netherlands, 2019; Volume 31, pp. 27–50.
28. Geladi, P.; Kowalski, B.R. Partial least-squares regression: A tutorial. *Anal. Chim. Acta* **1986**, *185*, 1–17. [CrossRef]

29. Ferrer, A.; Aguado, D.; Vidal-Puig, S.; Prats, J.M.; Zarzo, M. PLS: A versatile tool for industrial process improvement and optimization. *Appl. Stoch. Models Bus. Ind.* **2008**, *24*, 551–567. [CrossRef]
30. Wise, B.M.; Gallagher, N.B.; Bro, R.; Shaver, J.; Winding, W.; Scott-Coch, R. *Chemometrics Tutorial for PLS Toolbox and Solo*; Eigenvector Research, Inc.: Wenatchee, WA, USA, 2008; p. 173.
31. Westerhuis, J.A.; Coenegracht, P.M. Multivariate modelling of the pharmaceutical two-step process of wet granulation and tableting with multiblock partial least squares. *J. Chemom.* **1997**, *11*, 379–392. [CrossRef]
32. Wold, S.; Johansson, E.; Cocchi, M. PLS: Partial least squares projections to latent structures. In *3D QSAR in Drug Design. Theory, Methods and Applications*; Kubinyi, H., Ed.; ESCOM Science Publishers: Leiden, The Netherlands, 1993; pp. 523–550.
33. Favilla, S.; Durante, C.; Li Vigni, M.; Cocchi, M. Assessing feature relevance in NPLS models by VIP. *Chemom. Intell. Lab. Syst.* **2013**, *129*, 76–86. [CrossRef]

sensors

MDPI

Article

Placement of Optical Sensors in 3D Terrain Using a Bacterial Evolutionary Algorithm

Szilárd Kovács [1,*], Balázs Bolemányi [1] and János Botzheim [2]

1 Department of Mechatronics Optics and Mechanical Engineering Informatics, Faculty of Mechanical Engineering, Budapest University of Technology and Economics, 4-6 Bertalan Lajos Street, 1111 Budapest, Hungary; bolemanyibalazs@edu.bme.hu

2 Department of Artificial Intelligence, Faculty of Informatics, Eötvös Loránd University, Pázmány P. Sétány 1/A, 1117 Budapest, Hungary; botzheim@inf.elte.hu

* Correspondence: kovacs.szilard@mogi.bme.hu

Abstract: This paper proposes an optimization framework for terrain large scale optical sensor placement to improve border protection. Compared to the often used, maximal coverage of an area approach, this method minimizes the undetected passages in the monitored area. Border protection is one of the most critical areas for sensor placement. Unlike traditional border protection solutions, we do not optimize for 2D but for 3D to prevent transit. Additionally, we consider both natural and built environmental coverings. The applied environmental model creates a highly inhomogeneous sensing area for sensors instead of the previously used homogeneous one. The detection of each sensor was provided by a line-of-sight model supplemented with inhomogeneous probabilities. The optimization was performed using a bacterial evolutionary algorithm. In addition to maximizing detection, minimizing the number of the applied sensors played a crucial role in design. These two cost components are built on each other hierarchically. The developed simulation framework based on ray tracing provided an excellent opportunity to optimize large areas. The presented simulation results prove the efficiency of this method. The results were evaluated by testing on a large number of intruders. Using sensors with different quantities and layouts in the tested $1 \times 1 \times 1$ km environment, we reduced the probability of undetected intrusion to below 0.1% and increased the probability of acceptable classification to 99%.

Keywords: sensor placement; route detection; evolutionary computing

Citation: Kovács, S.; Bolemányi, B.; Botzheim, J. Placement of Optical Sensors in 3D Terrain Using a Bacterial Evolutionary Algorithm. *Sensors* **2022**, *22*, 1161. https://doi.org/10.3390/s22031161

Academic Editor: Natividad Duro Carralero

Received: 5 January 2022
Accepted: 1 February 2022
Published: 3 February 2022

1. Introduction

This study aims at an optimization framework for terrain large scale optical sensor placement. Appropriate sensors are crucial for the operation of any automated system. The right number and position of sensors are essential for a sensor system's efficient and reliable operation. The camera is one of the essential sensors in terms of surveillance. It was chosen for this study since it is reasonably priced and provides an information-rich and easy-to-understand signal. It is also crucial to consider the environment in which sensors are placed. The uniqueness of this article is not to cover the entire area but to detect any intruders. It is not necessary to cover all points of the area for detection, but it is enough to find the targets at one point in their route.

The real-life importance of this topic is protecting critical areas such as power plants, military facilities, and country borders. To be more efficient, it is high time to reconsider the previously used 2D models. Hence, these models focused on people, objects located on the ground, and land vehicles. In the case of flying targets, the 3D environment study is inevitable. Due to their relatively easy availability, flying targets such as drones or kites are more common as potential intruders. They can both be used for smuggling goods and gathering confidential information [1–3]. Equipped with a camera, they can be sent as an outpost to avoid the border patrols and find the proper timeframe for crosses without being caught. In addition to drone threat, summary studies were conducted on the applied sensors for drone monitoring and countermeasures against drones [3]. In line with the European Horizon projects [4], border protection is a priority for future years. One possible way to protect borders is through aerial reconnaissance. Long-term aerial reconnaissance can be achieved using airships and solar panels. Wind loading and easy visibility are the main disadvantages of airships, but their operation in a multimodal system can provide significant advantages. Sensor placement plays an important role in several other areas. Well-equipped sensor nodes with different sensors to monitor the environment and structure can be effectively used for social distancing and emergency management in a sensor network. Such a system has been tested in a park to manage visitors and create a favorable route for them to avoid crowding and plan an escape route in the event of an emergency [5].

This article has developed a new simulation-based approach to border management. The simulation uses ray tracing for effective application in highly inhomogeneous environments. Optimization was performed in 3D to prevent undetected ground and air crossings.

In Section 2, topic-related literature is reviewed. In Section 3, the simulation model is detailed. Section 4 describes the optimization method. Experimental results are presented in Section 5. Lastly, the conclusions and further works are summarized.

2. Related Literature

There are several sub-areas of border management, the two largest being physical border protection and psychological border management. The second's purpose is the study of people's behavior and psychological profiling [6]. The first is for the detection, localization, and prevention of illegal border crossings. In addition to high-altitude air traffic and ground crossings, the threat posed by low-altitude drones has increased over the last decade. The literature review can be divided into four main parts. The first part presents the sensors applicable to the boundary and area protection task and the related drone detection literature. The second topic is dynamic border protection methods. The third topic is static sensor placement. The last part contains sensor placement methods for more specialized tasks. Several studies have addressed the surveillance of drones [7]. The applied sensors for drone surveillance: camera (mono, RGB [8–14], multi-, hyperspectral, short-/longwave infrared [14]), radar [15], radio direction finder [16–18], acoustic (single [14,19,20], array, matrix) and laser detection and ranging. Various fusion techniques [14,21,22] have also appeared, mainly with visual, infrared, and acoustic sensors. In the case of ground transit, the use of geophones is also common [23,24]. There are static, dynamic solutions for boundary surveillance. Dynamic solutions include different patrol mechanisms [25,26]. Generally, short—a few hours—flight times are typical for drones [27], but developments aiming at flight times of several hours [28] also exist.

The border surveillance is often concentrated only on ground intruders, so 1D line arrangements are common in theoretical sensor placement methods. In most cases, sensors had uniform disk-shaped sensibility decreasing towards the edges [29–33]. Radars are well suited for area protection due to their large field of view [31]. Two dimensional (2D) coverage studies are more common in the literature than articles on border protection. Akbarzadeh et al. investigated an optimal sensor coverage in 3D elevation terrain and the built environment. The optimal 2D coordinates and horizontal and vertical angular positions of each sensor were optimized by simulated annealing, the Limited-memory Broyden–Fletcher–Goldfarb–Shanno algorithm, and the Covariance matrix adaptation evolution strategy [34]. Unlike in the natural homogenous non-flat environments in the built-in environment, the sensor placing goal was to cover inhomogeneous flat surfaces. Altahir et al. developed a weighted coverage model for installing camera surveillance systems. The placement was based on a 2D risk map in 3D space. Inversely, the sensors were placed based on 2D weighted coverage demand [35]. As a continuation of their work, they used dynamic programming as a discrete optimization for 2D generated urban layout [36]. Various aspects, such as power supply, energy efficiency, sensor lifetime, reliability, greedy coverage and the placement of the controllers in the sensor network, can be considered for the sensors' placement [32,37,38]. An obvious solution for a wireless sensor network is to use renewable energy with a rechargeable battery [32]. The energy efficiency can also be maintained by the timing of the sensors [39]. The goal is to ensure adequate coverage even in the event of outages. In the case of a fail, relocating nodes can provide an excellent solution to hold the uniformity of 2D coverage [40]. Energy supply is also an important aspect of border surveillance. Dong et al. implemented a boundary monitoring procedure with solar-powered sensors. In addition to the surface coverage, time optimization was also applied due to the limited energy of the sensor's battery [29,30]. The method was later expanded to include adaptive sensing range adjustment for energy-efficient, time-aligned alignment of sensors [41]. Another aspect of sensor placement is localization, which requires signal strength, time, or direction data from different sensors. Xu et al. investigated ideal sensor placement for single target localization based on circular time of arrival. The optimality criterion is to minimize the trace of the inverse Fisher information matrix [42]. Xu's hybrid localization procedure study is for single static target localization using the hybrid received-signal-strength, angle-of-arrival, and time-of-arrival measurements on the 2D plane [43]. Akbarzadeh et al. examined a new optimization approach for temporal coverage. The essence of temporal coverage is to cover the area around the most probable position of the target point with the available sensors. It was concluded that individual control of each sensor in series works better all at once [44]. After detection, target tracking and localization is the next important task [45–47]. Another exciting research area is the replacement of a temporarily failed sensor for localization. Pedrollo et al. trained a neural network to be a virtual sensor, replace unavailable sensors, and generate synthetic but still realistic data [48]. Another similar task is to observe 3D objects. De Rainville et al. created a framework for mobile robotic sensor placement with covariance matrix adaptation evolution strategy optimization. The mobile robots were equipped with optical sensors. The optimization goal was maximization of the pixel density on the area [49]. Herguedas et al. examined the optimal sensor placement for deformable bodies [50]. The procedure was later improved using RGB-D cameras [51]. An important area of the Optimal Sensor Placement Problem is the vibration measurements in various structures such as bridges [52]. The problem examines small dimensionally discrete sensor placement. Zhang et al. examined the coverage-based optimization of different bodies with different evolutionary algorithms [53]. Spielberg et al. performed a sensor placement task during the soft robotics simulation to monitor the inside of the soft robot [54].

3. Modeling

Simulation-based optimizations are becoming more common. Creating a suitable simulation environment has a competitive speed compared to a complex analytical solution. The studies in the simulation are very flexible and illustrative. In the field of sensor placement, simulation-based solutions are less common and are not applied for border protection. Nevertheless, the simulation-based approach has many advantages compared to the traditional analytical methods. Simulation is much more flexible, making it easy to examine even dynamic environments, and is easier to expand and apply for new tasks.

During the simulation, the signal's path between the object and the sensor was emphasized. Reflections were not considered during ray tracing because, in the studied natural environment and the use of optical sensors, this is not significant. Absorption and transparency were calculated for signal propagation. Algorithm 1 contains the process of applied signal propagation.

Algorithm 1 Ray Tracing.

```
function RAY TRACING(Sensors, Intruders, Environment)
    for Intruders' Route do
        function FIELD OF VIEW CHECK(Sensors, Intruders)
            return true/false, Ray strengths, Distance and Angular differences
        if Intruders in the sensors' field of view then
            function GROUND CROSSING AND BACKGROUND(Positions, Ground)
                return true/false, Closest ground backgrounds for sensors and intruders
            if Signal not cross the ground then
                function BUILT CROSSING AND BACKGROUND(Positions,Built elements)
                    return true/false, Closest Built element backgrounds for sensors and intruders
                if Signal not cross built elements then
                    function VEGETATION CROSSING AND BACKGROUND(Positions, Vegetation, Ray strength)
                        return Ray strengths, Closest Vegetation backgrounds for sensors and intruders
                    if Signal greater then the cut value then
                        function CLOUD CROSSING AND BACKGROUND(Positions, Clouds, Ray strength)
                            return Ray strengths, Closest Cloud backgrounds for sensors and intruders
            function SELECT CLOSES BACKGROUND(backgrounds)
                return Closest background
    return Ray strengths, Backgrounds, Distance and Angular differences
```

The disadvantages of simulation are that it approximates reality. Simulation neglections tend to produce better results than reality, so neglections must be considered in evaluating the results. The main difficulties in object detection are the visibility of the object, the background, the weather conditions, and the influence of the sun. In this article, the visibility and background attenuation of the object has been taken into account. Weather can significantly degrade detection for some sensors [55,56]. Three different effects can occur: reduction of visibility, particles appearing in the image, and, in the case of optical sensors, particles on the detector can obscure or blur regions. The first is the inevitable decrease in most optical sensors' detection, which can be considered as distance decreasing [55]. Particles can be filtered [57]. It is difficult to consider particles' blur and covering effect. One possibility for optimization is to punish the high angular position of the sensors. Another solution is simulating weather conditions [58]. The sun degrades the detection to different degrees depending on the quality of the optics. The sun's position can be calculated according to its geographic location [59]. In most cases, only the distance from the sensor is used for goodness calculation. In the prepared simulation, the ray passaged through the medium in the environment weakens the signal further. In addition to the strength of the signal, the signal-to-noise ratio from a detection perspective is more important. No model has been developed for optical sensors to test the signal-to-noise ratio for detection. This area is the best developed in the case of radar [60]. Signal-to-noise ratio based detection can be divided into four main parts: transmitter noise, receiver noise, the signal of the object to be observed, and background noise. For optical sensors in the field, the transmitter noise can be considered as the slow variation of sunlight. The receiver noise is mostly considered as the optical signal-to-noise ratio, containing dark current noise and spatial frequency transmission of optics. The signal is the visibility of the object. Background noise can be taken into account by the background environment. In this paper, the detection model has been compiled with the background in addition to object visibility, and the transmitter and receiver noise is not discussed. The receiver noise is sensor-specific and can be added to the discrete properties of the sensors. Based on the implemented ray tracing, the transmitter noise can be estimated by adding sun and other light sources. Based on the type of background, the estimated detection probability decreased. Table 1 shows the effect of background on the signal strength.

Table 1. Background impacts on the signal.

Environmental Element	Sign Decrease [%]
Clear sky	0
Clouds	20 · cloud's density
Ground	[0...50] predefined
Walls	[0...50] predefined
Vegetation	50 · vegetation's density

3.1. Environment Model

Terrain, clouds, vegetation and artificial built elements formed the modeled environment. Accurate elevation data were loaded during the design of the environment. The vegetation was loaded from a random vegetation map. The clouds and buildings were randomized. Automated loading of the entire environment is not yet implemented. The elements of the environment have different properties that are required for the placement of optical sensors. Clouds were shaped as orbs with position, size, absorption, and transmission properties. The vegetation was modeled with orbs shapes with position, size, absorption, and transparency. The constructed elements were considered rectangles with the property of position, orientation, and size. The simple shapes allow a quick parallel calculation, and more complex elements can be built. Figure 1 shows how the examined elements can construct the environment. A random vegetation map was added to the real elevation map. Clouds were generated at random locations and transparency in each iteration. Changing environmental elements increase robustness and are more realistic.

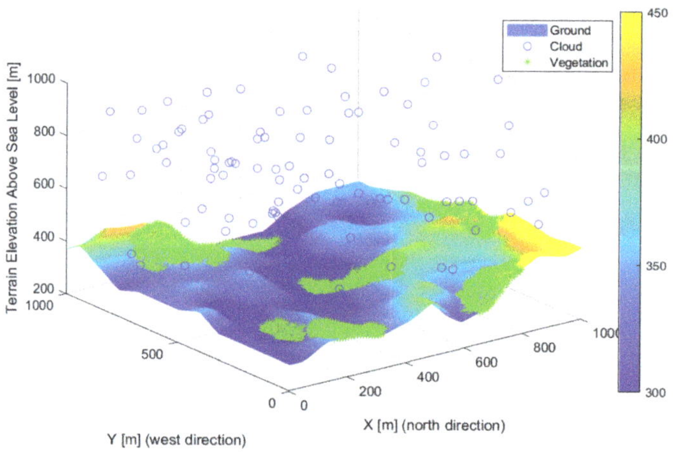

Figure 1. Model of the environment used for the studies, with elevation map, vegetation, clouds.

3.2. Sensor Model

Pinhole camera models were used with focus distance. The field of view and distance were calculated based on the focal length and sensor-specific typical detector size and resolutions. The object's size and its minimal pixels representation have an essential role in calculating visibility distance. Maximal signal strength is considered half of the visibility distance, and then it decreases linearly. The value calculated from the distance gave the initial value of the signal propagation. A crop value can be set for the detectable signal minima. The sensors have a position, orientation, and focal length properties for optimization. The detection range of a sensor plotted black is shown in Figure 2 illustrating the optimization result when only one sensor is used. A red line indicates a possible intruder route. The maximum visibility distance was determined during the optimization up to two-pixel imaging of a one-meter target. The real visibility is much smaller visibility due to the environment and background. The four-pixel projections are plotted in figures, which is better related to the visibility. A larger pixel representation is recommended for a basic classification.

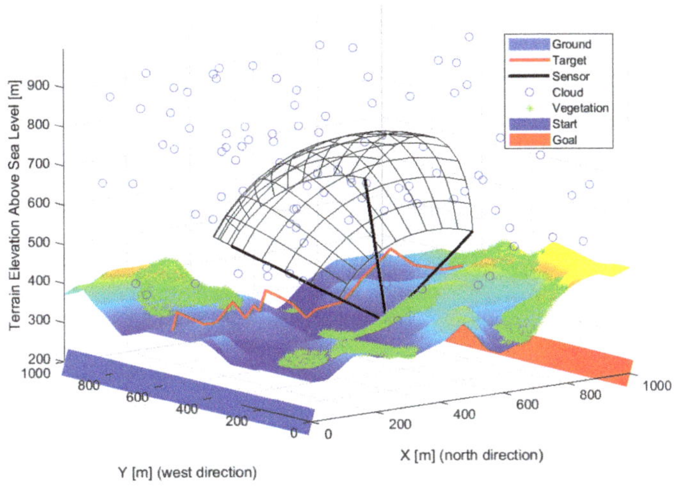

Figure 2. The detection field of a sensor is plotted black. The red line exemplifies a possible route of an object.

3.3. Target Model

Flying objects were defined in the simulation, so it was necessary to examine the 3D space instead of the usual 2D. These objects move from one edge of the simulation space (x = 0) to the other (x = xmax). Any point above the surface of the starting plane of the space (x = 0) was a possible starting point. During the route planning, the area was divided evenly along the transit direction. For the other coordinates, the maximum deflection from the previous position was calculated based on the object's top speed. The object moved randomly above the surface, between the maximum deflection and the straight direction. Objects have a size, minimal pixel representation, initial position, maximum speed, and a random path between the two edges of the study range properties. Based on the initial properties, a random path was calculated and added to the properties. Straight diagonal and mixed routes were generated in each iteration, a sample is plotted in Figures 3–5. The maximum observability is sought on each route's evaluation. The different paths (red lines) of the objects quasi evenly filled the study area. Due to the gaps, previous observations were also weighted, giving momentum to the optimization and smoothing the cost function. Some of the routes were complex despite the simple generation. Lateral, curved movements have also appeared, in addition to straight and near-straight passing attempts. Using multiple intruders in simulations with random patches results in some complex patches. The cost function considers the undetected paths with greater weight, so optimization better secures "quasi" intelligent routes. An excellent way to reduce the number of simulations is to use smarter intruders. Pinball and flood-fill algorithms can be a good solution for intelligent intruders [61].

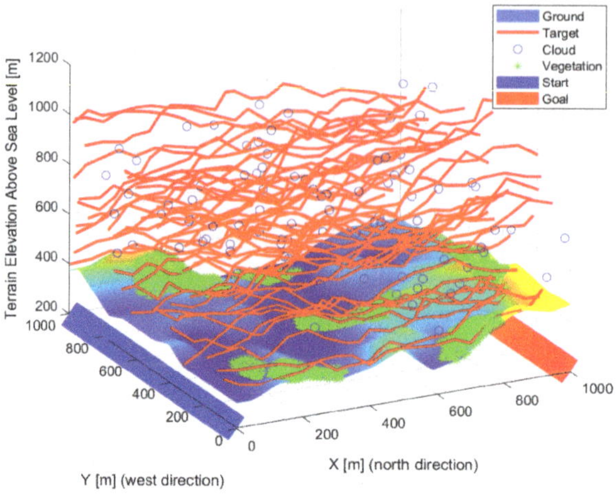

Figure 3. New routes (red lines) were generated at each iteration step.

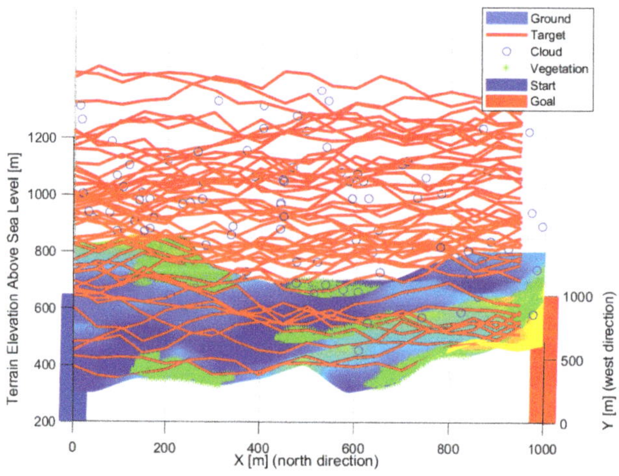

Figure 4. The different paths (red lines) of the objects quasi evenly filled the study area.

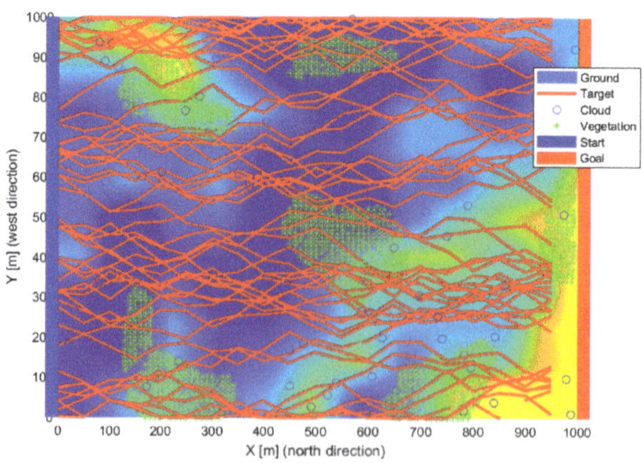

Figure 5. Some of the routes were complex despite the simple generation.

4. Optimization

4.1. Objective

The goal is to prevent unnoticed passage with minimal sensor use. It is necessary to estimate the total detection probability for the entire sensor system to recognize the hidden paths. The resulting detection is difficult to determine in sensor systems. The strongest single detection was considered the detection probability of the system. For different targets and environments, different correlation tensors can be applied. This tensor can be determined from measurements. A simple and obvious solution is to use the maximum single detection probability when combining detections of multiple sensors. This case is close to the worst case, which increases the system's reliability, but there may be negative values in the correlation tensor so that it can be better than the actual worst case. The objective can be written as the minimum observations taken from each route's maximum observation in Equation (1), where p_d is a single detection probability.

$$O_{min} = 1 - min_{|object}(max_{|route}(max_{|sensor}(p_d))). \tag{1}$$

Applying the objective as a cost function resulted in highly variable costs and frequent population exchange during optimization. The cost function has been modified to improve the quality of the optimization. The number of intruders was increased, and as a standard solution in the low batch sample proportion teachings, the principle of momentum was applied. In Equation (2), the average detection of the objects was calculated. The minimum detection (Equation (1)) and the mean detection (Equation (2)) were weighted (w_o) in Equation (3). The previously computed values have also been considered in the resulting cost (C) in Equation (4). Thus, the momentum principle was realized with the weight factor (w_c). In Equation (4) n denotes the number of simulations. Compared to fixed-structure optimization tasks, it is not practical to consider historical values with greater or uniform weighting. By changing the structure, some sensors can be replaced or combined for a more optimal result despite their excellent performance.

$$O_{mean} = 1 - mean_{|object}(max_{|route}(max_{|sensor}(p_d))) \tag{2}$$

$$O_{ext} = w_o \cdot O_{min}(p_d) + (1 - w_o) \cdot (O_{mean}(p_d)) \tag{3}$$

$$C(n) = w_c \cdot C(n-1) + (1 - w_c) \cdot O_{ext}(n). \tag{4}$$

A secondary goal of the hierarchical task is to minimize the number of sensors used. Based on the hierarchy, two levels of cost function were applied. The first level of the cost function is to secure the hidden passage, and the second level minimizes the number of sensors. The two-level design was implemented using two cost levels. Any arrangement that can permanently minimize passage will be one step lower in cost, and its goals will be expanded by reducing the number of sensors applied.

4.2. Individuals

Individuals consisted of varying amounts of sensors. Each sensor had fixed and variable parameters. Fixed parameters were detector size, resolution, and the range of the variable parameters. Variable parameters are position, vertical and horizontal orientation, and focal length. During optimization, the specified parameters could only take discrete values, and the variable parameters were continuous. The structure of the individuals is shown in Figure 6.

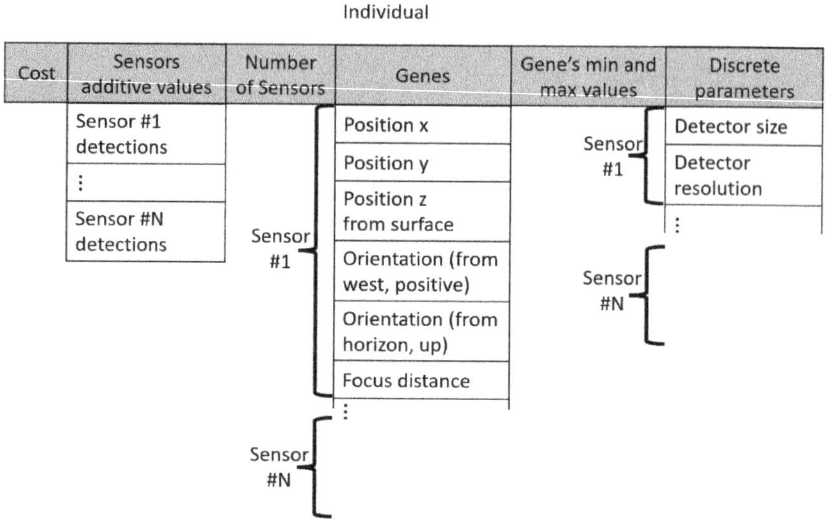

Figure 6. Structure of an individual.

4.3. Optimization Method

The Bacterial Evolutionary Algorithm (BEA) was used for optimization [62]. The BEA consists of bacterial mutation and gene transfer, shown in Algorithm 2. BEA has been applied to a wide range of problems, for instance, optimizing the fuzzy rule bases [62,63], feature selection [64], data clustering [65], and combinatorial optimization problems [66].

Algorithm 2 Bacterial Evolutionary Algorithm.

function PARAMETER INITIALIZATION(params)
  $N_{pop} \leftarrow$ Number of individuals
  $criteria \leftarrow$ Stop criteria
  $opt_{pi} \leftarrow$ Options of population initialization
  $opt_{bm} \leftarrow$ Options of bacterial mutation
  $opt_{gt} \leftarrow$ Options of gene transfer
function POPULATION INITIALIZATION(opt_{pi}, N_{pop})
  parallel(N_{pop}) create and evaluate population
while *criteria* **do**
  function BACTERIAL MUTATION ($opt_{bm}, Population$)
  function GENE TRANSFER ($opt_{gt}, Population$)
return *Best individual*

During bacterial mutation, random sensors or sensors with the lowest added value were chosen. Sensors may have left the individual, new sensors may have joined the individual, or sensors may have been replaced. During gene transfer, one or more sensor from the better-performing individual was transferred to a lower-performing individual or replaced with a sensor from the better individual. The replaced sensors were with the most negligible additive value or selected randomly. The algorithm and the operators used are shown in Figure 7. The optimization process is shown in Figure 8.

A sensor's additive value is calculated as the sum difference between the maximum detection of each intruder in the whole sensor system and the detection without that sensor. The total additive value of a sensor cannot be calculated due to the infinite possible path and the change of the sensor placement. Still, it can be estimated and summed with the previous values in each simulation.

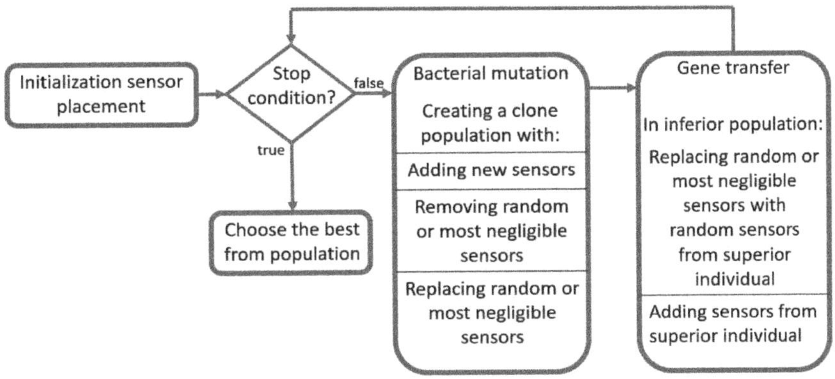

Figure 7. Bacterial Evolutionary Algorithm used for sensor placement.

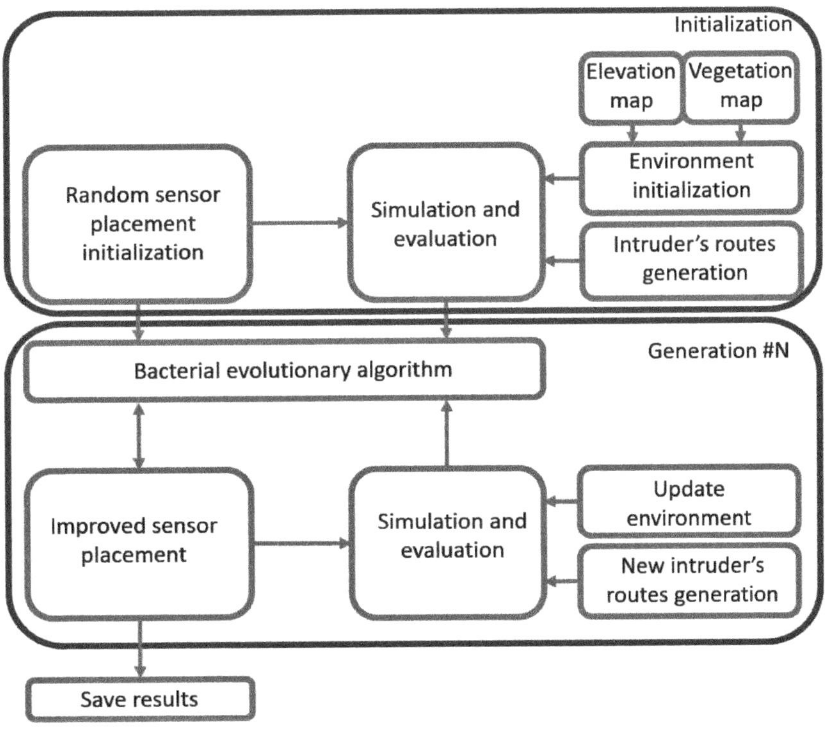

Figure 8. Optimization flow chart.

5. Experimental Results

During the experiments, a $1 \times 1 \times 1$ km area was simulated. In each iteration, the cloud map has been updated, and new paths have been initialized. 20%–80% division of random sensors or the ones with the estimated lowest added values were modified. The spontaneous mutation has weighed less due to the lack of local search. First, the sensors were selected for clone creation, and then the listed mutations occurred on them, in 30%–30%–40% of cases, a new sensor was added, removed, or replaced. The sensor's numbers were ranged and considered in selecting the mutation operation. During gene transfer, the chance of sensor transfer and sensor replacement was 50%–50%. A sensors' additive value was calculated based on a 50%–50% weighting of the previously accumulated and current additive values. Thus, the current value got more significant weight. In general, high-resolution sensors came to the fore during optimization. Several acceptable solutions have emerged for different amounts of sensors. Low-resolution sensors can be helpful in some cases. Due to the mutation operator used, sensors with minimal added value were included temporarily and for more iteration in the case of a larger number of sensors. This error can be fixed by using a local search method. The Bacterial Memetic Algorithm (BMA) [67] complements BEA with local search. This algorithm has several variants with different local search techniques such as the Levenberg–Marquardt algorithm [67], Simulated Annealing [68], Hill climbing, and discrete local search [69,70]. In depicting the results, the sensors' detection distance was plotted based on the 4-pixel projection of a one-meter target. In general, the sensors had difficulty detecting objects at high altitudes.

Figures 9–11 show the solution for three sensors. Two sensors facing crossed, the fields of views meet this to cover most of the space. The third sensor is located independently and covers a path at the edge. The layout is able to cover high altitudes with low detection probability. It is not always possible to detect intruders flying low near the ground. Each sensor looks slightly upwards, but not so much that it is greatly affected by the weather. The effects of the sun can degrade detection based on the orientation of the map and the travel direction of the intruder.

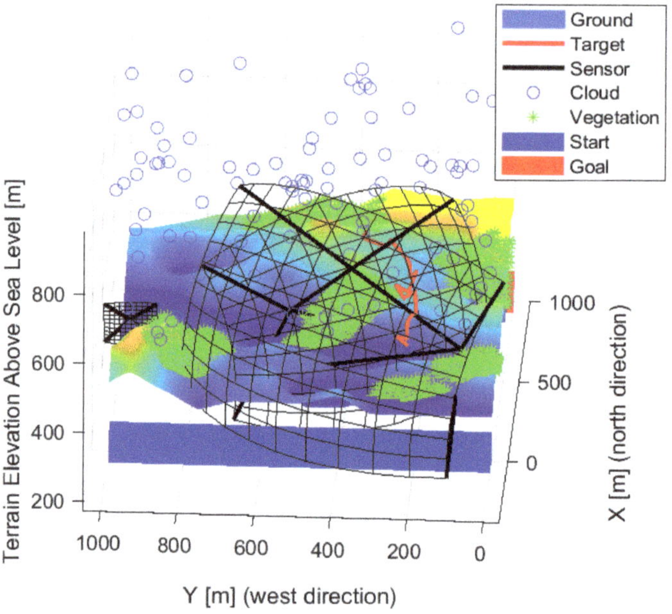

Figure 9. The solution in the case of the three sensors front-wise.

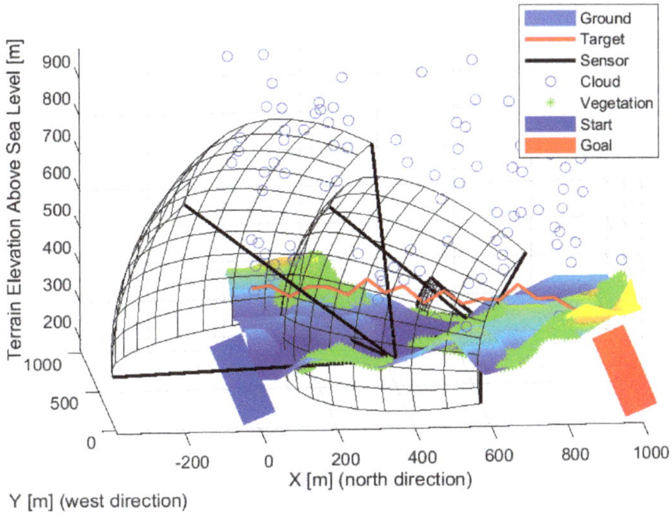

Figure 10. The solution in the case of the three sensors from the side view.

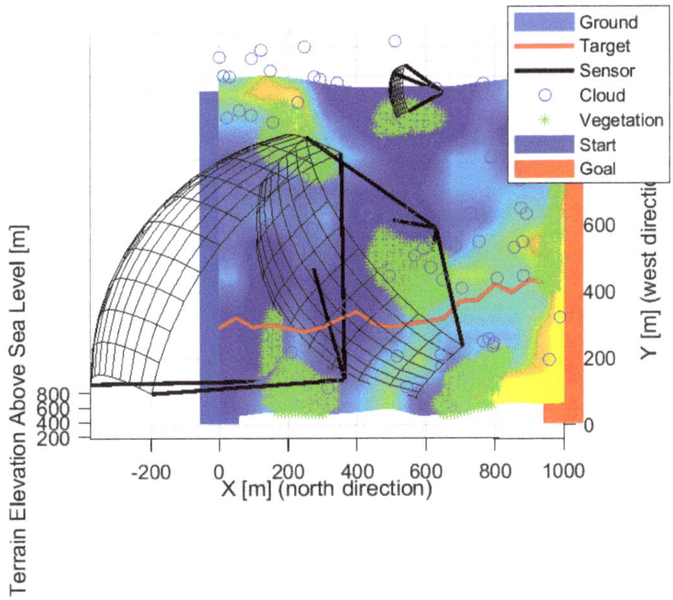

Figure 11. The solution in the case of the three sensors from above.

Figures 12–14 show the solution for four sensors. Three sensors were facing up and one forward. The sensors in the valley facing upwards are optimal as the sky background provides better detection. Upward-facing sensors can cover a large area, and the background gives them better detection. Their disadvantage is greater exposure to the weather. The forward-facing sensor is on the edge of the test area, but it covers the edge of the site and as far as possible at the edge of the forest/vegetation to see objects at low altitudes.

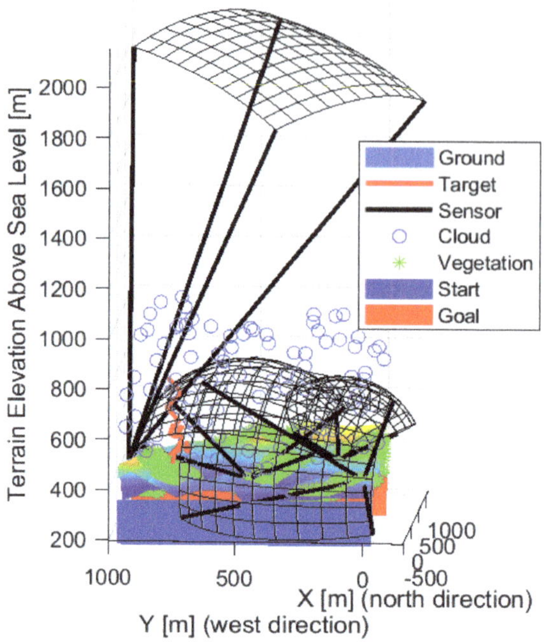

Figure 12. The solution in the case of the four sensors front-wise.

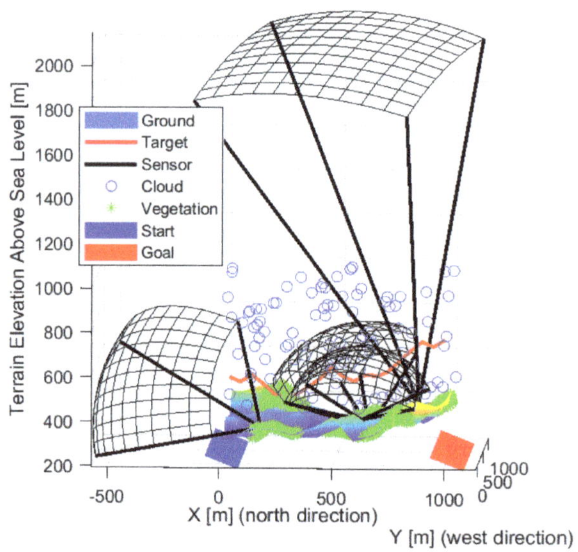

Figure 13. The solution in the case of the four sensors from the side view.

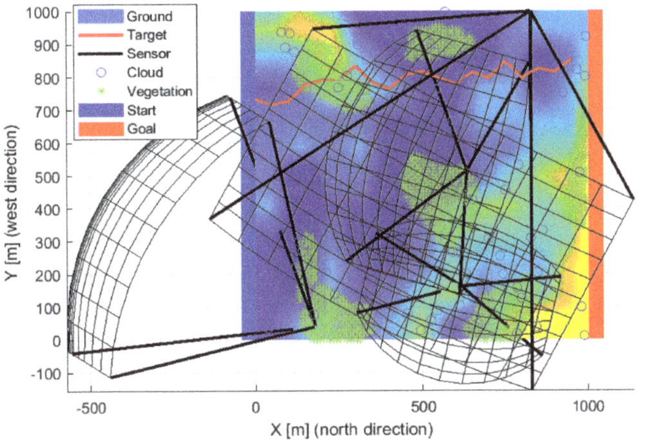

Figure 14. The solution in the case of the four sensors from above.

Figure 13 shows the sensors looking up and located in the valley are more optimal as the objects are more recognizable in the sky background. The shape of the detection space shows the task's difficulty, as, from a distance, a sensor can detect less. On the contrary, it sees a smaller area closer. Three sensors look forward and two up. A lower resolution sensor is also included in the middle. Overall, the sensors are better distributed. There are some gaps, but it covers well overall.

Figures 15–17 show solution for five sensors. They present that we can approach the optimal solution with forward and upward facing sensors.

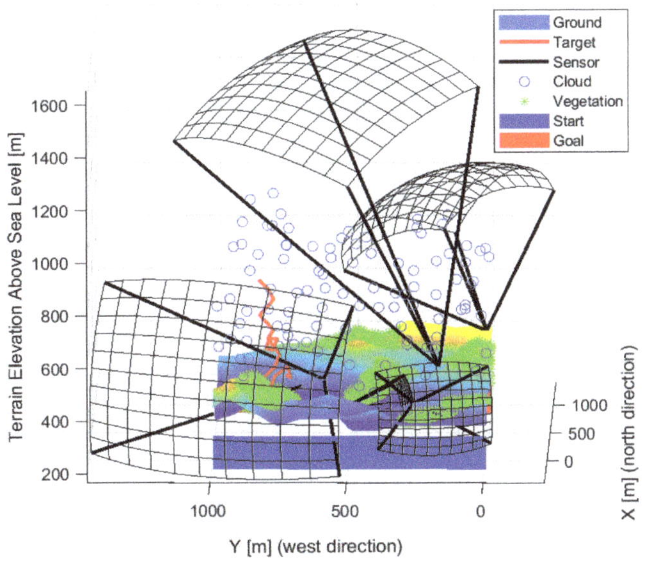

Figure 15. The solution in the case of the five sensors front-wise.

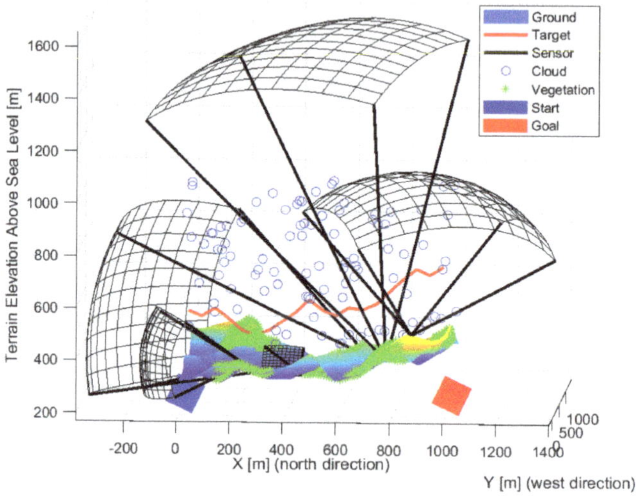

Figure 16. The solution in the case of the five sensors from the side view.

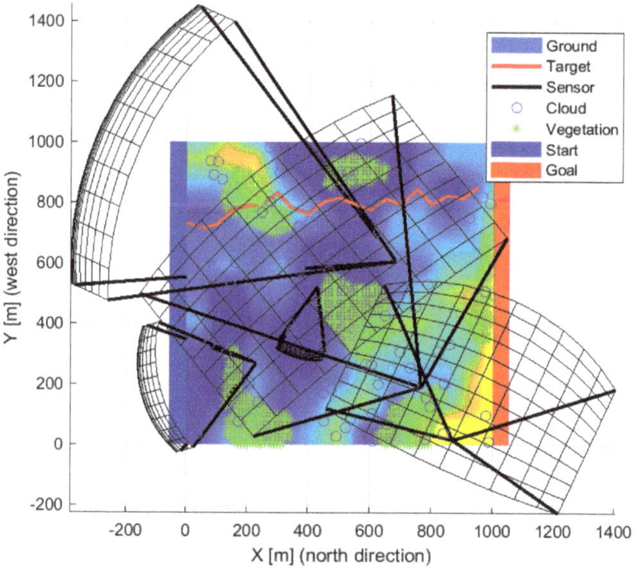

Figure 17. The solution in the case of the five sensors from above.

An arrangement using thirteen sensors is shown in the following Figures 18–20, which managed to prevent unnoticed passage. The sensors are located in almost one group, so there is no gap between them. Three upward-sensing sensors are more exposed to the effects of the weather. The visibility ranges of the upward-facing sensors touch each other and cover high-altitude routes well. Lower resolution sensors also appear mainly in valleys.

Figure 21 shows the change in cost during optimization for a different number of sensors. Each individual received an initial value of 0.5. The re-initialization of the intruder route has caused a constant fluctuation in costs. Figure 21 shows that the threshold cost is consistently reached in time with different numbers of sensors. Different threshold cost levels have been set for each case. Keeping the threshold cost in two iterations will result in a one-step lower cost per individual and an additional cost per number of sensors.

Some individuals in the population also inherit the cost of the original individual through bacterial mutation and gene transfer, and the optimization can run in parallel at the two cost levels. The inheritance of the cost gene is smaller due to the variable number of sensors, and the result of the current simulation is given more weight.

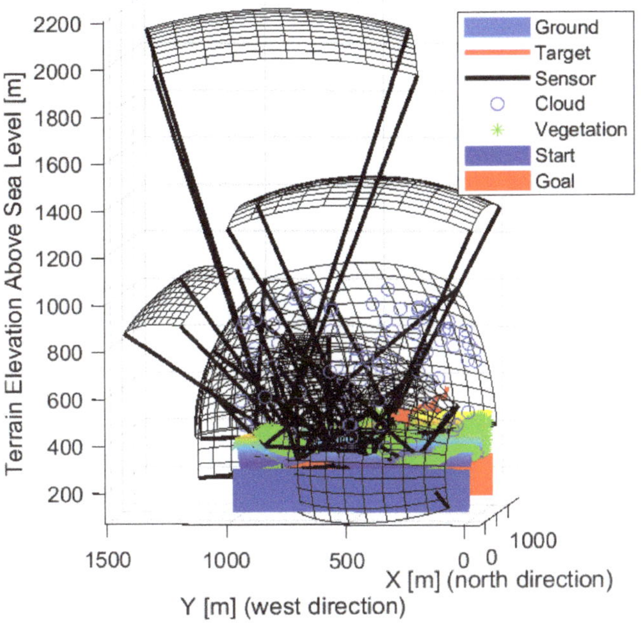

Figure 18. The solution in the case of the thirteen sensors front-wise.

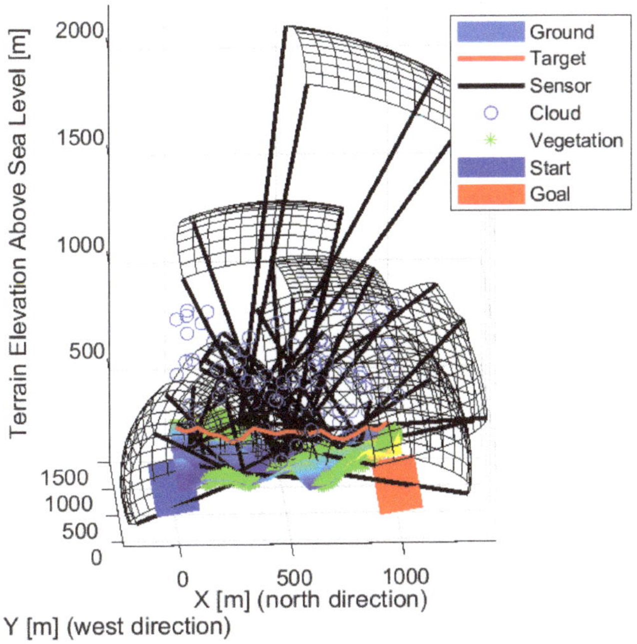

Figure 19. The solution in the case of the thirteen sensors from the side view.

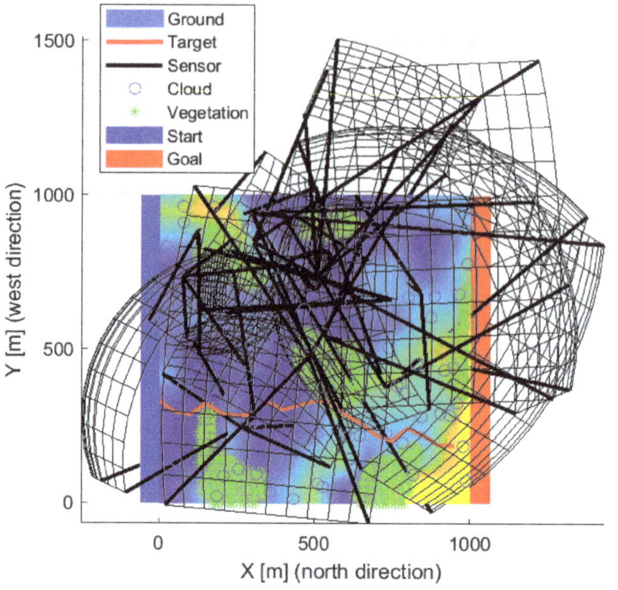

Figure 20. The solution in the case of the thirteen sensors from above.

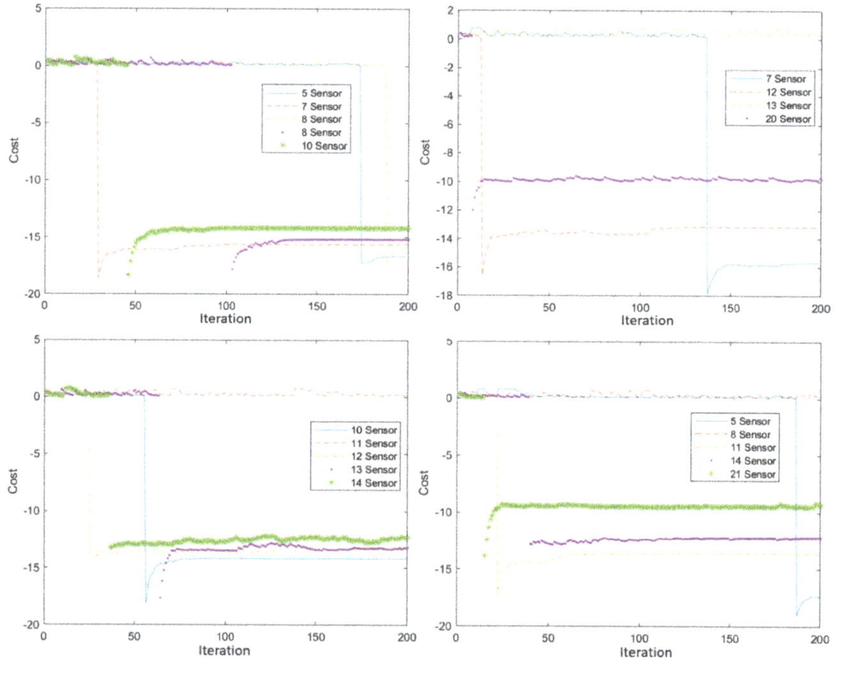

Figure 21. The two-level cost function for different numbers of sensors.

Sensor placements were tested with 10,000 random paths. Figure 22 shows an object detection probability distribution for different sensor numbers and layouts. Most detections fall into the high-reliability range of 90. . .100%. By using intelligent intruders [61], simulations can be made more efficient. There were both superior and inferior solutions for smaller and larger numbers of sensors. At the trend level, it can be seen that more sensors means higher reliability.

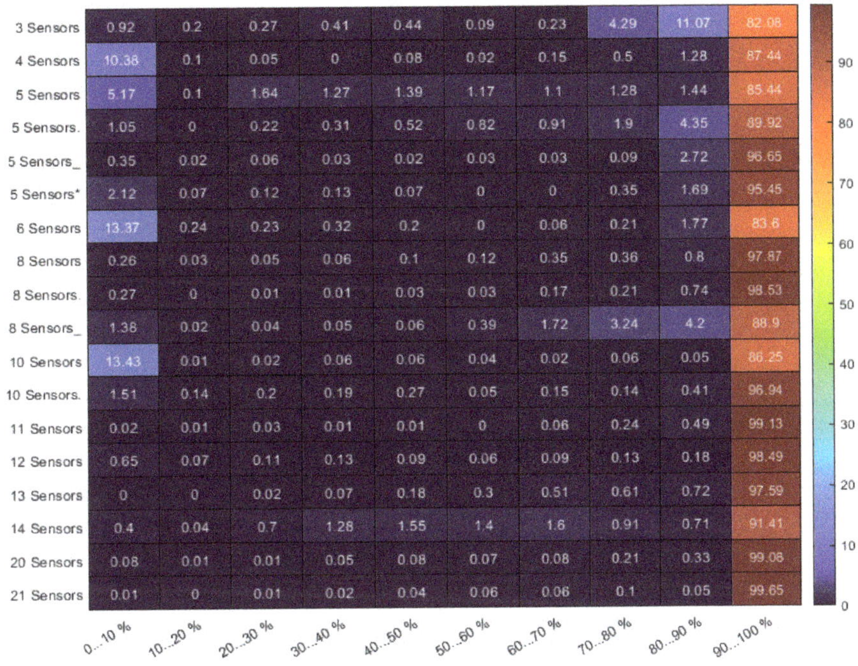

Figure 22. The detection probability distribution for different arrangements and sensor numbers.

6. Conclusions and Further Work

We have developed a method based on three-dimensional simulation compared to the currently available quasi-two-dimensional methods to prevent detection-free passage. Instead of the homogeneous or near-homogeneous detection models, we optimized a highly inhomogeneous detection model in a dynamic environment. We have developed a method based on 3D simulation compared to the currently available quasi-two-dimensional methods to prevent detection-free passage. Instead of the homogeneous/near-homogeneous detection models, we optimized a highly inhomogeneous detection model in a dynamic environment. We tested a stepped cost function for hierarchical multi-purpose optimization. The applied bacterial evolutionary algorithm was able to optimize the sensor placement. The sensors were positioned correctly in a complex environment even without local operators. For random routes, the majority of intruders (>80%) were detected with a high probability (>90% detection). Due to the simple path generation of intruders, it is more important to investigate 'quasi' intelligent intruders in the low detection categories. With the presented optimization, we succeeded in preventing undetected intrusions for 10,000 trials in the studied environment. Above the use of five sensors, undetected intrusions (<10% detection) and uncertain detections (<50% detection) can be reduced to less than 1% for intruders at low (>0.5 m) and high (<700 m) altitudes. In addition to flying intruders, the case of standing humans is also included in the parameter range. By using ten sensors, undetected intrusions and uncertain detections can be reduced to <0.1%. In addition to detection, classification is another crucial aspect. Classification is only possible with a high detection probability (>70%). The results show that a high percentage (>80%) of the intruders has a high probability of detection (>90%) could be classified with reasonable confidence. A high percentage (99%) of the intruders are expected to classify using more than twenty sensors with an acceptable probability.

Further research aims to extract environmental data based on image segmentation automatically. Other goals are to improve the environment model with the effects of weather and sun, apply intelligent intruders, and implement appropriate local search. The implemented simulation is suitable for extracting gradient approximation. Due to discrete simulations, a local search procedure using momentum would be the most efficient [71].

Author Contributions: Conceptualization, S.K., B.B. and J.B.; Methodology, S.K. and J.B.; Software, S.K.; Supervision, J.B.; Validation, S.K.; Visualization, B.B. All authors have read and agreed to the published version of the manuscript.

Funding: Project no. 2020-00163 has been implemented with the support provided from the National Research, Development and Innovation Fund of Hungary, financed under the 2020-1.1.2-PIACI-KFI funding scheme.

Institutional Review Board Statement: Not applicable.

Informed Consent Statement: Not applicable.

Conflicts of Interest: The authors declare no conflict of interest.

References

1. Jackman, A. Consumer drone evolutions: Trends, spaces, temporalities, threats. *Def. Secur. Anal.* **2019**, *35*, 362–383. [CrossRef]
2. Yaacoub, J.P.; Noura, H.; Salman, O.; Chehab, A. Security analysis of drones systems: Attacks, limitations, and recommendations. *Internet Things* **2020**, *11*, 100218. [CrossRef]
3. Chamola, V.; Kotesh, P.; Agarwal, A.; Gupta, N.; Guizani, M. A Comprehensive Review of Unmanned Aerial Vehicle Attacks and Neutralization Techniques. *Ad Hoc Netw.* **2021**, *111*, 102324. [CrossRef] [PubMed]
4. EU Research Horizon Projects. Available online: https://frontex.europa.eu/future-of-border-control/eu-research/horizon-projects/ (accessed on 1 January 2022).
5. Fedele, R.; Merenda, M. An IoT System for Social Distancing and Emergency Management in Smart Cities Using Multi-Sensor Data. *Algorithms* **2020**, *13*, 254. [CrossRef]
6. Khan, W.; Crockett, K.; O'Shea, J.; Hussain, A.; Khan, B.M. Deception in the eyes of deceiver: A computer vision and machine learning based automated deception detection. *Expert Syst. Appl.* **2021**, *169*, 114341. [CrossRef]
7. Taha, B.; Shoufan, A. Machine Learning-Based Drone Detection and Classification: State-of-the-Art in Research. *IEEE Access* **2019**, *7*, 138669–138682. [CrossRef]
8. Nalamati, M.; Kapoor, A.; Saqib, M.; Sharma, N.; Blumenstein, M. Drone Detection in Long-Range Surveillance Videos. In Proceedings of the 2019 16th IEEE International Conference on Advanced Video and Signal Based Surveillance (AVSS), Taipei, Taiwan, 18–21 September 2019; pp. 1–6. [CrossRef]
9. Lee, D.H. CNN-based single object detection and tracking in videos and its application to drone detection. *Multimed. Tools Appl.* **2021**, *80*, 34237–34248. [CrossRef]
10. Seidaliyeva, U.; Akhmetov, D.; Ilipbayeva, L.; Matson, E.T. Real-Time and Accurate Drone Detection in a Video with a Static Background. *Sensors* **2020**, *20*, 3856. [CrossRef]
11. Singha, S.; Aydin, B. Automated Drone Detection Using YOLOv4. *Drones* **2021**, *5*, 95. [CrossRef]
12. Jin, R.; Jiang, J.; Qi, Y.; Lin, D.; Song, T. Drone Detection and Pose Estimation Using Relational Graph Networks. *Sensors* **2019**, *19*, 1479. [CrossRef]
13. Behera, D.K.; Bazil Raj, A. Drone Detection and Classification using Deep Learning. In Proceedings of the 2020 4th International Conference on Intelligent Computing and Control Systems (ICICCS), Madurai, India, 13–15 May 2020; pp. 1012–1016. [CrossRef]
14. Svanström, F.; Englund, C.; Alonso-Fernandez, F. Real-Time Drone Detection and Tracking With Visible, Thermal and Acoustic Sensors. In Proceedings of the 2020 25th International Conference on Pattern Recognition (ICPR), Milan, Italy, 10–15 January 2021; pp. 7265–7272. [CrossRef]
15. Park, J.; Park, S.; Kim, D.H.; Park, S.O. Leakage Mitigation in Heterodyne FMCW Radar for Small Drone Detection With Stationary Point Concentration Technique. *IEEE Trans. Microw. Theory Tech.* **2019**, *67*, 1221–1232. [CrossRef]
16. Basak, S.; Rajendran, S.; Pollin, S.; Scheers, B. Combined RF-based drone detection and classification. *IEEE Trans. Cogn. Commun. Netw.* **2021**. [CrossRef]

17. Al-Sa'd, M.F.; Al-Ali, A.; Mohamed, A.; Khattab, T.; Erbad, A. RF-based drone detection and identification using deep learning approaches: An initiative towards a large open source drone database. *Future Gener. Comput. Syst.* **2019**, *100*, 86–97. [CrossRef]
18. Sciancalepore, S.; Ibrahim, O.A.; Oligeri, G.; Di Pietro, R. PiNcH: An effective, efficient, and robust solution to drone detection via network traffic analysis. *Comput. Netw.* **2020**, *168*, 107044. [CrossRef]
19. Anwar, M.Z.; Kaleem, Z.; Jamalipour, A. Machine Learning Inspired Sound-Based Amateur Drone Detection for Public Safety Applications. *IEEE Trans. Veh. Technol.* **2019**, *68*, 2526–2534. [CrossRef]
20. Al-Emadi, S.; Al-Ali, A.; Al-Ali, A. Audio-Based Drone Detection and Identification Using Deep Learning Techniques with Dataset Enhancement through Generative Adversarial Networks. *Sensors* **2021**, *21*, 4953. [CrossRef]
21. Aledhari, M.; Razzak, R.; Parizi, R.M.; Srivastava, G. Sensor Fusion for Drone Detection. In Proceedings of the 2021 IEEE 93rd Vehicular Technology Conference (VTC2021-Spring), Helsinki, Finland, 25–28 April 2021; pp. 1–7. [CrossRef]
22. Milani, I.; Bongioanni, C.; Colone, F.; Lombardo, P. Fusing active and passive measurements for drone localization. In Proceedings of the 2020 21st International Radar Symposium (IRS), Warsaw, Poland, 5–8 October 2020; pp. 245–249. [CrossRef]
23. Arjun, D.; Indukala, P.K.; Unnikrishna Menon, K.A. PANCHENDRIYA: A Multi-sensing framework through Wireless Sensor Networks for Advanced Border Surveillance and Human Intruder Detection. In Proceedings of the 2019 International Conference on Communication and Electronics Systems (ICCES), Coimbatore, India, 17–19 July 2019; pp. 295–298. [CrossRef]
24. Arjun, D.; Indukala, P.; Menon, K.A.U. Integrated Multi-sensor framework for Intruder Detection in Flat Border Area. In Proceedings of the 2019 2nd International Conference on Power and Embedded Drive Control (ICPEDC), Chennai, India, 21–23 August 2019; pp. 557–562. [CrossRef]
25. Qiao, Y.; Yang, J.; Zhang, Q.; Xi, J.; Kong, L. Multi-UAV Cooperative Patrol Task Planning Novel Method Based on Improved PFIH Algorithm. *IEEE Access* **2019**, *7*, 167621–167628. [CrossRef]
26. Surendonk, T.J.; Chircop, P.A. On the Computational Complexity of the Patrol Boat Scheduling Problem with Complete Coverage. *Naval Res. Logist.* **2020**, *67*, 289–299. [CrossRef]
27. Abushahma, R.I.H.; Ali, M.A.M.; Rahman, N.A.A.; Al-Sanjary, O.I. Comparative Features of Unmanned Aerial Vehicle (UAV) for Border Protection of Libya: A Review. In Proceedings of the 2019 IEEE 15th International Colloquium on Signal Processing Its Applications (CSPA), Penang, Malaysia, 8–9 March 2019; pp. 114–119. [CrossRef]
28. BorderUAS. Available online: https://frontex.europa.eu/future-of-border-control/eu-research/horizon-projects/borderuas-xFanlJ (accessed on 1 January 2022).
29. Dong, Z.; Chang, C.Y.; Chen, G.; Chang, I.H.; Xu, P. Maximizing Surveillance Quality of Boundary Curve in Solar-Powered Wireless Sensor Networks. *IEEE Access* **2019**, *7*, 77771–77785. [CrossRef]
30. Xu, P.; Wu, J.; Shang, C.; Chang, C.Y. GSMS: A Barrier Coverage Algorithm for Joint Surveillance Quality and Network Lifetime in WSNs. *IEEE Access* **2019**, *7*, 1. [CrossRef]
31. Xu, X.; Zhao, C.; Ye, T.; Gu, T. Minimum Cost Deployment of Bistatic Radar Sensor for Perimeter Barrier Coverage. *Sensors* **2019**, *19*, 225. [CrossRef] [PubMed]
32. Liu, Y.; Chin, K.W.; Yang, C.; He, T. Nodes Deployment for Coverage in Rechargeable Wireless Sensor Networks. *IEEE Trans. Veh. Technol.* **2019**, *68*, 6064–6073. [CrossRef]
33. Wang, S.; Yang, X.; Wang, X.; Qian, Z. A Virtual Force Algorithm-Lévy-Embedded Grey Wolf Optimization Algorithm for Wireless Sensor Network Coverage Optimization. *Sensors* **2019**, *19*, 2735. [CrossRef] [PubMed]
34. Akbarzadeh, V.; Gagné, C.; Parizeau, M.; Argany, M.; Mostafavi, M.A. Probabilistic Sensing Model for Sensor Placement Optimization Based on Line-of-Sight Coverage. *IEEE Trans. Instrum. Meas.* **2013**, *62*, 293–303. [CrossRef]
35. Altahir, A.A.; Asirvadam, V.S.; Hamid, N.H.B.; Sebastian, P.; Hassan, M.A.; Saad, N.B.; Ibrahim, R.; Dass, S.C. Visual Sensor Placement Based on Risk Maps. *IEEE Trans. Instrum. Meas.* **2020**, *69*, 3109–3117. [CrossRef]
36. Altahir, A.A.; Asirvadam, V.S.; Sebastian, P.; Hamid, N.H. Solving Surveillance Coverage Demand Based on Dynamic Programming. In Proceedings of the 2020 IEEE Sensors Applications Symposium (SAS), Kuala Lumpur, Malaysia, 9–11 March 2020; pp. 1–6. [CrossRef]
37. Lanza-Gutiérrez, J.M.; Caballé, N.; Gómez-Pulido, J.A.; Crawford, B.; Soto, R. Toward a Robust Multi-Objective Metaheuristic for Solving the Relay Node Placement Problem in Wireless Sensor Networks. *Sensors* **2019**, *19*, 677. [CrossRef]
38. Tahmasebi, S.; Safi, M.; Zolfi, S.; Maghsoudi, M.R.; Faragardi, H.R.; Fotouhi, H. Cuckoo-PC: An Evolutionary Synchronization-Aware Placement of SDN Controllers for Optimizing the Network Performance in WSNs. *Sensors* **2020**, *20*, 3231. [CrossRef]
39. Thomas, D.; Shankaran, R.; Sheng, Q.; Orgun, M.; Hitchens, M.; Masud, M.; Ni, W.; Mukhopadhyay, S.; Piran, M. QoS-Aware Energy Management and Node Scheduling Schemes for Sensor Network-Based Surveillance Applications. *IEEE Access* **2020**, *9*. [CrossRef]
40. Zhang, Y.; Liu, M. Regional Optimization Dynamic Algorithm for Node Placement in Wireless Sensor Networks. *Sensors* **2020**, *20*, 4216. [CrossRef]
41. Zaixiu, D.; Shang, C.; Chang, C.Y.; Sinha Roy, D. Barrier Coverage Mechanism Using Adaptive Sensing Range for Renewable WSNs. *IEEE Access* **2020**, *8*. [CrossRef]
42. Xu, S.; Ou, Y.; Wu, X. Optimal Sensor Placement for 3-D Time-of-Arrival Target Localization. *IEEE Trans. Signal Process.* **2019**, *67*, 5018–5031. [CrossRef]
43. Xu, S. Optimal Sensor Placement for Target Localization Using Hybrid RSS, AOA and TOA Measurements. *IEEE Commun. Lett.* **2020**, *24*, 1966–1970. [CrossRef]

44. Akbarzadeh, V.; Gagné, C.; Parizeau, M. Sensor control for temporal coverage optimization. In Proceedings of the 2016 IEEE Congress on Evolutionary Computation (CEC), Vancouver, BC, Canada, 24–29 July 2016; pp. 4468–4475. [CrossRef]
45. Zhang, Y.; Liang, R.; Xu, S.; Zhang, L.; Zhang, Y.; Xiao, D. A One-step Pseudolinear Kalman Filter for Invasive Target Tracking in Three-dimensional Space. In Proceedings of the 2021 IEEE International Conference on Real-time Computing and Robotics (RCAR), Xining, China, 15–19 July 2021; pp. 353–358. [CrossRef]
46. Hu, J.; Zhang, C.; Xu, S.; Chen, C. An Invasive Target Detection and Localization Strategy Using Pan-Tilt-Zoom Cameras for Security Applications. In Proceedings of the 2021 IEEE International Conference on Real-time Computing and Robotics (RCAR), Xining, China, 15–19 July 2021; pp. 1236–1241. [CrossRef]
47. Wang, S.; Guo, Q.; Xu, S.; Su, D. A Moving Target Detection and Localization Strategy Based on Optical Flow and Pin-hole Imaging Methods Using Monocular Vision. In Proceedings of the 2021 IEEE International Conference on Real-time Computing and Robotics (RCAR), Xining, China, 15–19 July 2021; pp. 147–152. [CrossRef]
48. Pedrollo, G.; Konzen, A.A.; de Morais, W.O.; Pignaton de Freitas, E. Using Smart Virtual-Sensor Nodes to Improve the Robustness of Indoor Localization Systems. *Sensors* **2021**, *21*, 3912. [CrossRef]
49. De Rainville, F.M.; Mercier, J.P.; Gagné, C.; Giguère, P.; Laurendeau, D. Multisensor placement in 3D environments via visibility estimation and derivative-free optimization. In Proceedings of the 2015 IEEE International Conference on Robotics and Automation (ICRA), Seattle, WA, USA, 26–30 May 2015; pp. 3327–3334. [CrossRef]
50. Herguedas, R.; López-Nicolás, G.; Sagüés, C. Multi-camera coverage of deformable contour shapes. In Proceedings of the 2019 IEEE 15th International Conference on Automation Science and Engineering (CASE), Vancouver, BC, Canada, 22–26 August 2019; pp. 1597–1602. [CrossRef]
51. Cuiral-Zueco, I.; López-Nicolás, G. RGB-D Tracking and Optimal Perception of Deformable Objects. *IEEE Access* **2020**, *8*, 136884–136897. [CrossRef]
52. Lee, E.T.; Eun, H.C. Optimal Sensor Placement in Reduced-Order Models Using Modal Constraint Conditions. *Sensors* **2022**, *22*, 589. [CrossRef]
53. Zhang, X.; Zhang, B.; Chen, X.; Fang, Y. Coverage optimization of visual sensor networks for observing 3-D objects: Survey and comparison. *Int. J. Intell. Robot. Appl.* **2019**, *3*, 342–361. [CrossRef]
54. Spielberg, A.; Amini, A.; Chin, L.; Matusik, W.; Rus, D. Co-Learning of Task and Sensor Placement for Soft Robotics. *IEEE Robot. Autom. Lett.* **2021**, *6*, 1208–1215. [CrossRef]
55. Zang, S.; Ding, M.; Smith, D.; Tyler, P.; Rakotoarivelo, T.; Kaafar, M.A. The Impact of Adverse Weather Conditions on Autonomous Vehicles: How Rain, Snow, Fog, and Hail Affect the Performance of a Self-Driving Car. *IEEE Veh. Technol. Mag.* **2019**, *14*, 103–111. [CrossRef]
56. Hasirlioglu, S.; Riener, A. Challenges in Object Detection Under Rainy Weather Conditions. *Intelligent Transport Systems, From Research and Development to the Market Uptake*; Ferreira, J.C., Martins, A.L., Monteiro, V., Eds.; Springer International Publishing: Cham, Switzerland, 2019; pp. 53–65.
57. Garg, K.; Nayar, S.K. Vision and Rain. *Int. J. Comput. Vis.* **2007**, *75*, 3–27. [CrossRef]
58. Hasirlioglu, S.; Riener, A. A General Approach for Simulating Rain Effects on Sensor Data in Real and Virtual Environments. *IEEE Trans. Intell. Veh.* **2020**, *5*, 426–438. [CrossRef]
59. Shapiro, F.R. The position of the sun based on a simplified model. *Renew. Energy* **2022**, *184*, 176–181. [CrossRef]
60. Stec, B.; Susek, W. Theory and Measurement of Signal-to-Noise Ratio in Continuous-Wave Noise Radar. *Sensors* **2018**, *18*, 1445. [CrossRef] [PubMed]
61. Wang, Y.; Chu, W.; Fields, S.; Heinemann, C.; Reiter, Z. Detection of Intelligent Intruders in Wireless Sensor Networks. *Future Internet* **2016**, *8*. [CrossRef]
62. Nawa, N.E.; Furuhashi, T. Fuzzy system parameters discovery by bacterial evolutionary algorithm. *IEEE Trans. Fuzzy Syst.* **1999**, *7*, 608–616. [CrossRef]
63. Botzheim, J.; Hámori, B.; Koczy, L.; Ruano, A. Bacterial algorithm applied for fuzzy rule extraction. In Proceedings of the International Conference on Information Processing and Management of Uncertainty in Knowledge-based Systems, Annecy, France, 1–5 July 2002; pp. 1021–1026.
64. Botzheim, J.; Drobics, M.; Koczy, L. Feature selection using bacterial optimization. In Proceedings of the International Conference on Information Processing and Management of Uncertainty in Knowledge-based Systems, Perugia, Italy, 15–19 June 2004; pp. 797–804.
65. Das, S.; Chowdhury, A.; Abraham, A. A Bacterial Evolutionary Algorithm for automatic data clustering. In Proceedings of the 2009 IEEE Congress on Evolutionary Computation, Trondheim, Norway, 18–21 May 2009; pp. 2403–2410. [CrossRef]
66. Luh, G.C.; Lee, S.W. A Bacterial Evolutionary Algorithm for the Job Shop Scheduling Problem. *J. Chin. Inst. Ind. Eng.* **2006**, *23*, 185–191. [CrossRef]
67. Botzheim, J.; Cabrita, C.; Koczy, L.; Ruano, A. Fuzzy Rule Extraction by Bacterial Memetic Algorithms. *Int. J. Intell. Syst.* **2009**, *24*, 312–339. [CrossRef]
68. Bódis, T.; Botzheim, J. Bacterial Memetic Algorithms for Order Picking Routing Problem with Loading Constraints. *Expert Syst. Appl.* **2018**, *105*, 196–220. [CrossRef]
69. Balázs, K.; Botzheim, J.; Koczy, L. *Comparative Investigation of Various Evolutionary and Memetic Algorithms*; Springer: Berlin/Heidelberg, Germany, 2010; Volume 313, pp. 129–140. [CrossRef]

70. Zhou, D.; Fang, Y.; Botzheim, J.; Kubota, N.; Liu, H. Bacterial memetic algorithm based feature selection for surface EMG based hand motion recognition in long-term use. In Proceedings of the 2016 IEEE Symposium Series on Computational Intelligence (SSCI), Athens, Greece, 6–9 December 2016; pp. 1–7. [CrossRef]
71. Kingma, D.P.; Ba, J. Adam: A Method for Stochastic Optimization. 2017. Available online: https://arxiv.org/pdf/1412.6980.pdf (accessed on 1 January 2022).

sensors

MDPI

Article

Deep Learning Approaches for Robust Time of Arrival Estimation in Acoustic Emission Monitoring

Federica Zonzini [1,*], Denis Bogomolov [1], Tanush Dhamija [2], Nicola Testoni [1], Luca De Marchi [2] and Alessandro Marzani [3]

1 Advanced Research Center on Electronic Systems "Ercole De Castro" (ARCES), University of Bologna, 40136 Bologna, Italy; denis.bogomolov2@unibo.it (D.B.); nicola.testoni@unibo.it (N.T.)
2 Department of Electrical, Electronic and Information Engineering (DEI), University of Bologna, 40136 Bologna, Italy; tanush.dhamija@studio.unibo.it (T.D.); l.demarchi@unibo.it (L.D.M.)
3 Department of Civil, Chemical, Environmental and Materials Engineering (DICAM), University of Bologna, 40136 Bologna, Italy; alessandro.marzani@unibo.it
* Correspondence: federica.zonzini@unibo.it

Abstract: In this work, different types of artificial neural networks are investigated for the estimation of the time of arrival (ToA) in acoustic emission (AE) signals. In particular, convolutional neural network (CNN) models and a novel capsule neural network are proposed in place of standard statistical strategies which cannot handle, with enough robustness, very noisy scenarios and, thus, cannot be sufficiently reliable when the signal statistics are perturbed by local drifts or outliers. This concept was validated with two experiments: the pure ToA identification capability was firstly assessed on synthetic signals for which a ground truth is available, showing a 10× gain in accuracy when compared to the classical Akaike information criterion (AIC). Then, the same models were tested via experimental data acquired in the framework of a localization problem to identify targets with known coordinates on a square aluminum plate, demonstrating an overreaching precision under significant noise levels.

Keywords: acoustic emissions; capsule neural network; convolutional neural network; source localization; time of arrival

Citation: Zonzini, F.; Bogomolov, D.; Dhamija, T.; Testoni, N.; De Marchi, L.; Marzani, A. Deep Learning Approaches for Robust Time of Arrival Estimation in Acoustic Emission Monitoring. *Sensors* **2022**, 22, 1091. https://doi.org/10.3390/s22031091

Academic Editor: Natividad Duro Carralero

Received: 30 December 2021
Accepted: 28 January 2022
Published: 31 January 2022

1. Introduction

Among the non-destructive testing methods for structural health monitoring (SHM), the one based on acoustic emission (AE) is particularly effective for the assessment of civil infrastructure and industrial plants, allowing the detection of active damages in structures such as buildings and bridges, pipelines, storage tanks, etc. AE testing is built on the analysis of the acoustic activity of the target structure [1], primarily due to the growth of cracking phenomena. One of the main advantages of AE relies on the possibility to localize such sources by passively capturing the acoustic response of the structure. It is, therefore, from the extraction of a batch of representative acoustic features, typically defined on a time basis, and their evolution over time, that potential dangerous defects can be detected at an early stage of degradation, and preventive alarms can be issued [2]. One of the important parameters to be extracted from a detected acoustic signal consists of the time of arrival (ToA), also known as onset time, namely the time taken by the induced wave to travel from its origin to the AE transducer. ToA, when assessed from multiple sensing positions, provides a means to localize the AE signal source. The literature about these localization methods is quite vast and comprehends, among others, approaches based on geometrical or angular relationships, such as the ones built on the angle of arrival or the difference time of arrival (DToA) [3]. Alternative strategies are based on the estimation of the signal energy [4], but this parameter is typically very sensitive to environmental and operational factors (e.g., imperfect coupling between the AE transducers and the structure), which hamper their exploitation for long-term and continuous integrity evaluation.

AE testing is typically implemented with long inspection intervals by purposely increasing the mechanical load applied to the structure [5] (https://eur-lex.europa.eu/legal-content/IT/TXT/?uri=celex:32014L0068 (accessed on Tuesday, 28 December 2021), https://ec.europa.eu/docsroom/documents/41641 (accessed on Tuesday, 28 December 2021)). However, more recently, a new approach has emerged, in which AE monitoring is performed in real time, via the permanent placement of AE equipment on the monitored structure and detecting the emissions generated during the normal load cycles. Despite the advantages in terms of responsiveness and preventive detection, the adoption of the latter approach is challenged by the difficulty in accurately identifying weak AE events in noisy environments, particularly in the early stages of defect growth [6]. Indeed, permanently installed AE systems must counteract the corruption of noise generated by operational processes, for example, vibrations of rotating machinery (e.g., a pump unit or an engine), responsible for disturbances in the surrounding environment [7] and of unpredictable noise sources, such as electromagnetic interference and ambient noise in the vicinity of the monitoring system [8,9].

The less favorable signal-to-noise ratios (SNRs), which characterize real-time monitoring acquisitions, may prevent the accurate tracking of AE features, and in particular, of the ToA. Statistical algorithms for onset time determination have proven their effectiveness at high SNRs, but the noise factors mentioned above might significantly affect the reproducibility and accuracy of ToA estimation results. Consequently, the implementation of reliable methods for signal detection, robust against operational noisy environments, is still an open research field.

The objective of the novel methods proposed in this article is (i) to build accurate and alternative models capable of handling heavily corrupted AE signals and (ii) to test them in operative SHM frameworks, such as acoustic source localization purposes. In particular, we considered the case of AE propagating in waveguides, so that the onset detection difficulty is further exacerbated by dispersion and multi-modality, as discussed in the following subsection.

1.1. Acoustic Emissions in Waveguides

When an acoustic emission event occurs as a consequence of crack, corrosion or delamination processes in a waveguide, ultrasonic guided waves (GWs) are generated and can travel long distances [10,11].

Lamb waves are the particular form of GWs which propagate in plates. Albeit showing long-range propagation and sensitivity with respect to distinct classes of damage, Lamb waves can exhibit complex behavior during propagation due to their multi-modal and dispersive nature. *Multi-modal* means that multiple Lamb waves, or guided modes, co-exist in the same frequency interval (see Figure 1). These modes are denoted as symmetric (S) (red curves) or anti-symmetric (A) (red curves) depending on the nature of the wavefield with respect to the mid plane of the plate. *Dispersive* means that Lamb waves are characterized by a frequency-dependent wave speed.

The number and type of the generated modes vary with the frequency and shape of the actuation. In general, it is possible to design the actuation in time and space to limit the effect of multi-modality and dispersion in order to reduce the complexity of the generated/received signals. In more detail, Figure 1 shows the dispersion curves [12] in terms of phase c_p (left panel) and group c_g (right panel) velocity of the Lamb waves existing in an aluminum plate of infinite extension and 1 mm thickness.

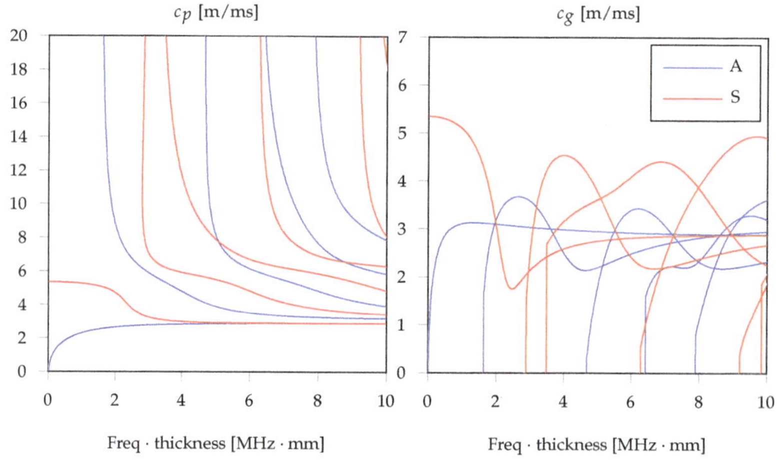

Figure 1. Dispersion curves for an aluminum square plate with 1 mm thickness and infinite extension: phase velocity on the **left** and group velocity on the **right**.

In practical cases, where it is necessary to deal with a finite plate, other complexities arise, such as the physical interaction of GWs with the mechanical boundaries, which is responsible for reflections and reverberations, a phenomenon also known as multi-path interference, the mutual interference in the case of multiple active transducers, and also the effects of environmental changes, such as temperature fluctuations, that can alter the wave propagation behavior.

1.2. ToA Estimation: From Statistical Methods to Machine Learning

By computing the similarity between two signals, cross correlation (X-Corr) can be used as a powerful tool for the estimation of the time shift of two time series [13]. In practical AE scenarios, where the monitoring network consists of a passive mesh of transducers and the true excitation source is unknown, X-Corr offers a means for DToA quantification among pairs of receivers, rather than a measure of the actual ToA. It has proved its effectiveness for the characterization of AE signals in multiple environments [14]. Combining X-Corr and frequency warping is an effective means to tackle dispersion and multi-modality in guided propagation [15]. However, X-Corr is highly susceptible to minor perturbations in the statistical properties of the input signals (e.g., residual noise sources superposed on the actual information content).

To cope with these issues, the Akaike information criterion (AIC) approaches the task of ToA identification as a pure statistical problem based on the second order statistics of the measured acoustic data. In essence, AIC leverages the concept of signal entropy to detect abrupt changes in the statistical distribution of the observed signal y [16,17]. For a discrete signal with N samples, this is achieved by computing, for each sample k, the quantity

$$\mathrm{AIC}[k] = k \log \sigma^2_{y[1:k]} + (N - k - 1) \log \sigma^2_{y[k+1:N]} \tag{1}$$

which is a logarithmic measure of the cumulative variance (σ_y^2) of the preceding ($y[1:k]$) and successive ($y[k+1:N]$) signal window with respect to the current sample index k. In other words, AIC splits the full waveform into a k dimensional and an $N-k$ dimensional smaller vector, respectively spanned by the first N and the last $N-k$ samples, and describes the level of similarity between them. The rationale is that, in correspondence of a sharp change in the signal profile, such as the one associated with the arrival of the incoming wave-front generated by the acoustic source, the divergence between the two variances increases to a large extent, generating a minimum in the overall AIC function. The time instant aligned with this minimum is the sought ToA [17].

Notably, this method can provide reliable outcomes when the processed signal presents two clearly distinct regions, e.g., a high-entropy portion where uncorrelated noise dominates, and a low-entropy segment where the acoustic signal is present [18]. Nevertheless, as discussed in [17], this might not be the case for AE monitoring scenarios, in which the mere attenuation due to signal propagation, which is responsible for low-amplitude received waveforms, is further hindered by additive operative noise, demanding for more advanced data processing solutions.

Led by the constantly increasing success of artificial intelligence (AI) in learning complex patterns hidden within signals, interesting AI solutions to tackle ToA estimation have been proposed, with particular emphasis on the seismology field. For example, a template-based artificial neural network (NN) for earthquake phase detection was proposed in [19], while [20] proposed an unsupervised fuzzy clustering logic for ToA recognition in micro-seismic waves. Another example worthy of attention was examined in the work by Zachary E. Ross [21], where the ToA of pressure waves in seismograms was considered a pattern recognition problem on top of which machine learning (ML)/deep learning (DL) models were trained. Considered among the most powerful architectures for deep and ultra-deep learning, PhaseNet and U-Net [22] were also investigated for seismic arrival time picking, reporting outstanding results. A comprehensive list of the most recent trends in this direction can be found in [23].

A close analogy exists between the seismic application domain and AEs detection. By virtue of this similarity, we drew inspiration from the solutions tested in seismology to develop the new DL solutions for AE signal processing, which are described and validated in the following sections of this manuscript.

2. Deep Learning Models for ToA Estimation

Two different neural network models were implemented for the purpose of ToA estimation, which are presented hereinafter.

2.1. Convolutional Neural Network

The convolutional neural network (CNN) [24] is a class of artificial neural networks that can extract relevant information from raw data and retain it in the form of weights and biases of the corresponding layers: the learned parameters are then used to make classifications and/or predictions as soon as new signal instances are available. A general CNN architecture consists of two main stages, i.e., *feature extraction* and *classification*, which are completely described by the following architectural blocks:

- *Convolutional layers*: it is in charge of feature extraction from input data, which are passed in a tensor form of dimensions N_{in}. A convolutional layer runs dot products between the input data and a specific set of weights (or mask), which are stored as taps of a corresponding filter, also known as *kernel*, of dimension N_{ks}. This filter is recursively applied to subsequent portions, or patches, of the input data by means of a sliding filter mask, which is shifted by a constant quantity called stride (N_{stride}).
  To increase the learning capability of the network, more than one filter is employed in a single convolutional layer: if N_{filter} is the number of total different kernels per layer, N_{filter} different maps of the the same input data are provided in the output via a proper activation function. A convolutional layer is, thus, completely determined by the tuple of values: $(N_{in}, N_{filter}, N_{ks}, N_{stride})$.
  Multiple convolutional layers are usually stacked one after the other, whose typology is dictated, in turn, by the dimensions of the manipulated data. In the case of ToA estimation, where the problem is intrinsically mono-dimensional and thought to be performed on a sensor-wise basis, 1D convolutional (Conv1D) layers are necessary.
- *Pooling layers*: this layer provides a distilled version of each feature map to shrink the computational complexity and the spatial size of the convolved features. Indeed, since

the number of points in each feature map returned at the end of a single convolutional block might be extremely large, and also since many of them only capture minor details, they can be neglected. Different pooling strategies have been proposed: max pooling (MaxPool), which only preserves the maximum value in a specific patch of the feature map; and average pooling (AvgPool), which extracts a single scalar as the average of the points falling in the same feature patch.

- *Dense layers*: once manifold representations are obtained, the sought pattern hidden within them is learned via dense fully connected (FC) layers, i.e., feed-forward layers with neurons that have full connections to all activations delivered by the previous layer. Firstly, a flattening operation is performed to unroll the feature maps provided by the last pooling layer in a uni-dimensional vector of appropriate dimension; then, these values are used as input of a standard artificial neural network, which acts either as a classifier or a regressor, depending on the desired task.

2.1.1. Large CNN Model

The first CNN architecture considered in this work is schematically represented in Figure 2 and is devised for ToA retrieval from 5000-long time series. As can be seen, five ($5000/2^{2l}$,N_{filter},10,2) Conv1D layers ($l \in \{0, \ldots, 4\}$ being the layer index), with $N_{filter} \in \{50, 100, 150, 200, 250\}$ and ReLU activation function, are stacked in cascade and followed by a MaxPool layer with a compression factor equal to 2. A global average pooling layer (Global AvgPool) is also included at the end of the convolutional block to force the regressor behavior of the network: Global AvgPool yields one single feature map out of the 250 different representations at the end of the last AvgPool layer. This single map is then passed to a first FC layer having 1024 neurons activated by ReLU; ToA can finally be retrieved from the output layer consisting of a 1×1 FC layer with one neuron and linear activation. It is worth saying that the so-far designed CNN model is characterized by 1,259,299 parameters, requiring a minimum memory space of at least 1.5 MB even in quantized form: the Adam optimizer [25] with learning rate of 0.001 and loss weight equal to 1 was used for training such parameters, and the model was trained for 15 epochs.

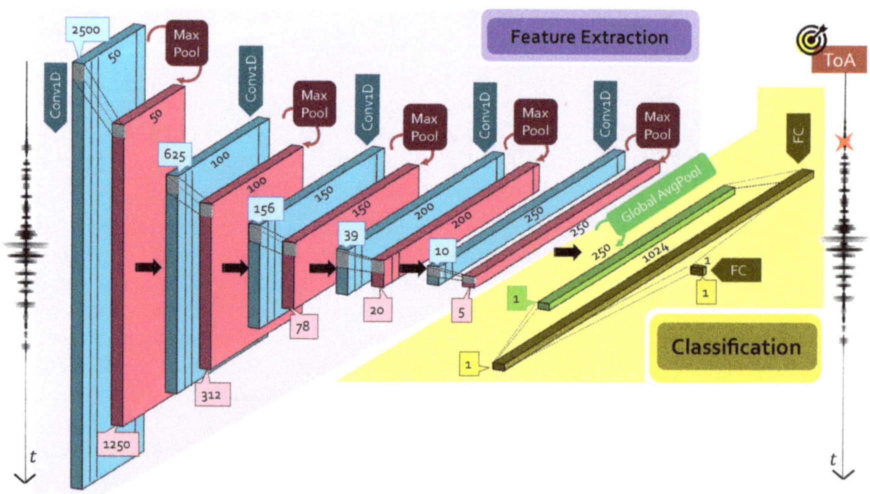

Figure 2. Proposed CNN for ToA estimation in 5000-long time series. The quantities reported in the blue and pink boxes indicate the dimensions of the output features of the corresponding layer.

2.1.2. Small CNN Model

It must be emphasized the fact that, in order to be applicable in permanent AE equipment installations with custom sensors and electronics compatible with long-term and real-time functionalities, the devised AI solutions must run on low-cost and resource-constrained devices. However, fitting the previously described CNN model to the limited capabilities of edge sensors is not practicable due to the excessive amount of memory (and, in turn, computational power) it requires. To this end, it is paramount to emphasize that the typical static and volatile memory of medium-to-high embedded devices hardly exceeds 1 MB (usually amounting to hundreds of kB) to harvest space and dynamic power consumption (https://www.st.com/en/microcontrollers-microprocessors/stm32-ultra-low-power-mcus.html (accessed on Wednesday, 29 December 2021), https://www.espressif.com/en/products/modules/esp32 (accessed on accessed on Wednesday, 29 December 2021)).

For this reason, a distilled version of the preceding model is derived as displayed in Figure 3, in which the five convolutional layers are substituted with four smaller size Cov1D+MaxPool layers with 16, 32, 64 and 64 filters while leaving unaltered all the remaining parameters. The dimensions of the GlobalAvg layer were changed accordingly. In this lighter version, only 134,481 parameters need to be learned, for a total memory occupancy of nearly 150 kB after conversion to embedded programming format, which leads not only to a complexity reduction of more than 10×, but makes the model absolutely compatible with the above-mentioned memory constraints of edge devices. Hereinafter, to differentiate the two models, this smaller one is called "small CNN".

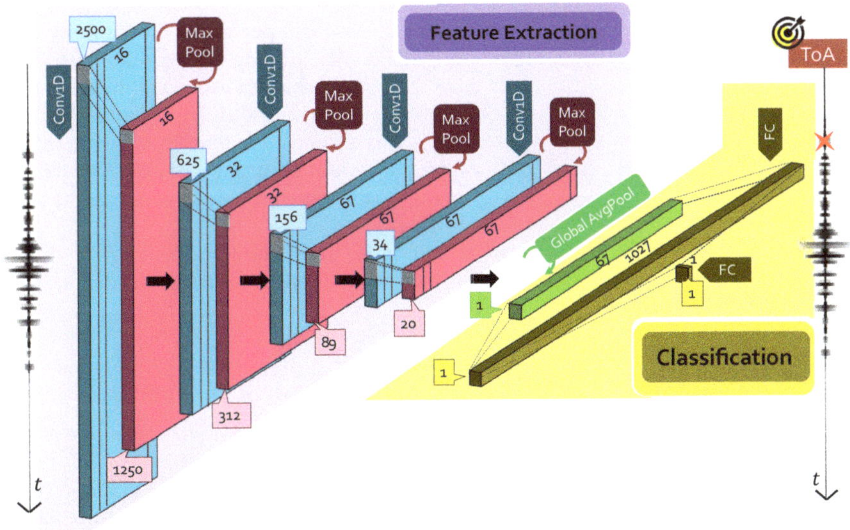

Figure 3. Proposed small CNN for ToA estimation. The quantities reported in the blue and pink boxes indicate the dimensions of the output features of the corresponding layer.

Notably, the model reduction of both the number of convolutional layers and filters per layer is preferred over other pruning strategy, given its proven advantages in terms of algorithmic complexity and memory footprint, as well as for its robustness against model over-parametrization [26] and better generalization to out-of-distribution data, as demonstrated in Section 3.3.

2.2. Capsule Neural Network

Despite their outstanding performance in multiple fields, CNNs might be ineffective under the following circumstances [27]: (i) the observed data pattern presents shifts/rotations, since CNNs are phase and shift invariant; (ii) the spatial relationship between the feature maps is an important indicator of the data distribution, since CNNs do not exploit spatial dependencies; (iii) the loss of information introduced by the pooling layers is unacceptable, especially for very deep model where pooling is mandatory.

Very recently released in the field of AI [28], CapsNet represents a powerful competitor to convolutional architectures for classification tasks. Three main reasons can be mentioned. Firstly, CapsNet transforms basic feature maps in correlated feature maps via the novel concept of capsule unit: this correlation-based approach implies the preservation of spatial dependencies between data. Secondly, albeit disregarding the pooling layer, it is capable of correct prediction, even when trained on fewer data. Thirdly, its vector-based output allows for robust classification performance by making use of simpler network architectures, favorable for implementation on edge sensors. This latter aspect is owed to the fact that CapsNet actually offers a first means for knowledge distillation, which is performed directly at an architectural level via novel machine learning operators, rather than being executed at a coding/firmware level, where most of the effort is usually spent.

Successful application of CapsNet for micro-seismic phase picking was accounted in [27], showing great performance for earthquake signal characterization. Inspired by this first attempt, an AE-oriented variant of CapsNet is proposed in this manuscript to cope with ToA prediction.

In its general form, the block diagram of a CapsNet architecture nests a capsule representation in cascade to standard convolutional layers, without pooling in between them, to learn novel representations in a lossless way. More formally, it consists of the following two elements:

- *Primary capsule*: this layer performs convolution aggregation via the so-called *capsule unit* u_i ($i \in \{1, \ldots, N_{PC}\}$ being the capsule index), corresponding to multiple combinations of the feature maps retrieved at the end of the convolution process. In their working principles, primary capsules provide an alternative form of convolutional layers: the main difference is that, in this case, a vector-based output is computed rather than working with unitary depth. As such, convolution-based processing is performed by each capsule, which is driven by an appropriate set of kernels and relative stride.
- *Digit capsule*: at this point, the agreement among different capsules has to be estimated so that it is possible to preserve the spatial dependency between those block representations with highest relevance. This concept is mathematically encoded via the weight opinion matrix W_{ij}, with $j \in \{1, \ldots, N_{class}\}$ being the number of classes, each with vector-based output of dimension N_{DC}. Hence, every capsule is judged by means of N_{class} opinions $u_{j|i}$, also called local digit capsules, to be computed as

$$u_{j|i} = W_{ij} u_i \tag{2}$$

From these local representations, a further level of feature combination is added in a spatially dependent manner, by following the routing-by-agreement protocol [29]. This procedure, also called dynamic routing, introduces the concept of *agreement*, i.e., how much the individual digit capsules agree with the combined one. The level of agreement is numerically quantified by the weight routing matrix R_{ij} via the coupling coefficient

$$c_{ij} = \frac{e^{R_{ij}}}{\sum_{c=1}^{N_{class}} e^{R_{ic}}} \tag{3}$$

As such, the final digit capsule s_j is given by $s_j = \sum_i c_{ij} u_{j|i}$. As in traditional convolutional layers, activation is required to ensure that digit capsules with low opinions shrunk to zero, since they do not convey meaningful information. However, the vector-based output of the capsules requires ad hoc functions to fulfill this task: the squashing function

$$v_j = \frac{||s_j||^2}{1 + ||s_j||^2} \frac{s_j}{||s_j||^2} \tag{4}$$

was purposely proposed in [28] to address it, where s_j and v_j are the input and output of the j-th convolutionally operated capsule. The quantity $u_{j|i} \cdot v_j$ finally yields the actual measure of agreement, i.e., the higher this product, the more preference is awarded to the corresponding primary capsule u_i. At this point, an iterative algorithm can be called to update the routing matrix, until the desired level of agreement is reached and the sought $N_{class} \times N_{DC}$ digit capsule block can be derived, which serves as the output layer for the entire neural network. Finally, it is sufficient to calculate the ℓ_2 norm of each of the N_{class} rows to obtain a corresponding value of the output probability associated to each single class.

For AE-related problems, just two classes can be considered, i.e., noise and AE signals: in this case, a high value of the output probability p_{AE} for class "AE signal" indicates that the input instance is most likely to contain a true AE event, whereas low values can be seen as indicators of noisy input.

An overview of the proposed CapsNet architecture for AE signal processing is graphically summarized in the left-hand side of Figure 4. The initial convolutional block consists of two Conv1D layers without pooling, activated by ReLU and with dimensions (500,64,9,2) and (250,128,9,2), respectively. At the output of the convolutional layer, 128 feature maps of 125 samples each are computed: these feature maps are passed to the primary capsule layer. Here, 78 primary capsules of 8 feature maps each are created, and then processed via capsule operations via kernels of size 9 and stride equal to 3. Dynamic routing is then performed, yielding to N_{class} different digit capsules with a vector size equal to 8. A last stage in which the ℓ_2 norm is applied to each row of the digit capsule block returns the two desired class probabilities (noise and AE).

In terms of model complexity, the proposed CapsNet architecture requires 301,952 parameters and allocates a memory space of 375 kB, a quantity which is 4.2× lower and 2.25× bigger than the ones required by the original and small CNN, respectively. Once again, the Adam optimizer (with a learning rate of 0.001) was used for training the model for a total of 15 epochs.

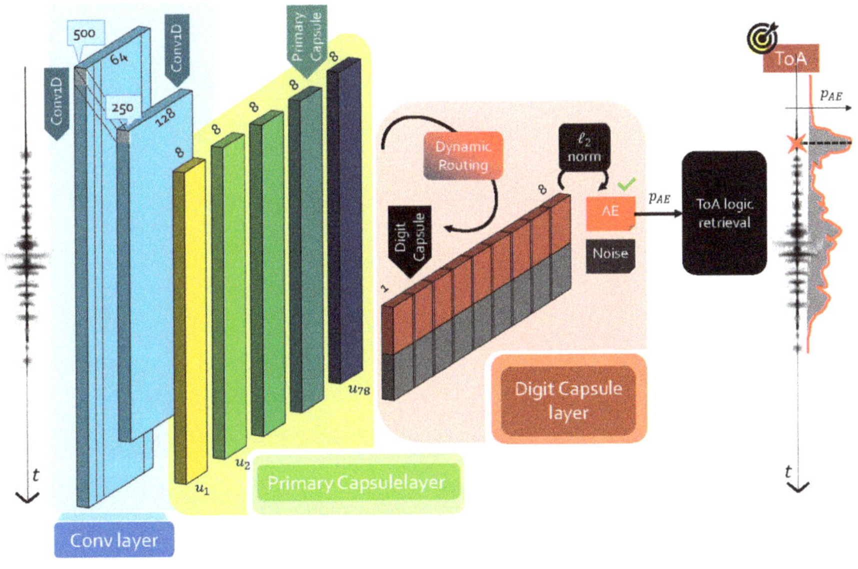

Figure 4. Proposed CapsNetToA architecture for ToA estimation.

ToA Retrieval from CapsNet: The CapsNetToA Architecture

Determining ToA with CapsNet is a two-step process. Indeed, in its definition, CapsNet acts a classifier for the input batch of data, meaning that it can only predict whether the current instance is most likely to contain low-entropy signal content (high probability) or rather noisy data (low probability). Therefore, a dedicated logic is implemented to extract one single time value out of the class probability distribution p_{AE}. Hereinafter, the entire processing flow, encompassing both CapsNet and the time retrieval logic, is named CapsNetToA.

To this end, an approach similar to the one suggested in [27] is adopted. The idea is to split the entire waveform of 5000 samples into smaller and overlapped windows, each of them identified by a unique time stamp taken as the central value of the corresponding time span. For every segment, a probability value is returned; the cumulative trend in the probability distribution can be easily obtained by concatenating, in time, the predictions related to subsequent windows.

The rationale is that the probability curve is expected to assume a low value until the signal statistics do not change. Then, when the first window containing the wave arrival is processed, the curve increases progressively, reaching its maximum (in the ideal case, unitary probability) for the exact window centered on the actual ToA.

On a first attempt, one may resort to statistical tools, such as pick-peaking or thresholding functions, to retrieve ToA as the first peak probability value. However, such a simple approach might suffer from several drawbacks, which can be listed as follows: it presents poor generalization capabilities, in the sense that the selection of a threshold or benchmark value is strictly application and environment dependent; as such, it badly conjugates with the critical variability of AE scenarios. The second reason relates to the impossibility of accurately retrieving very early onset times, i.e., the ones below or almost equal to the window length, for which the true peak probability value is unavoidably missed. In this case, in fact, all the initial windows will output a nearly unitary value and, thus, a criterion based on the leading peak selection unavoidably will estimate ToA from secondary signal arrivals.

Conversely, ML solutions can inherently handle all these sources of complexities in a very efficient and user-transparent way. For this reason, a second NN block is stacked in cascade to CapsNet to retrieve ToA from the output probability history yielded at the end of the capsule processing. It is worth observing that, in ToA terms, the problem is exactly analogous to the one faced to estimate ToA via CNN while working directly with time series data. The main difference is that, for this second scenario driven by CapsNet, probability functions are available as inputs. Coherently, in the approach presented in this work and which is novel with respect to the one in [27], it is suggested to employ the same small CNN as "ToA logical retrieval" block (see Figure 4) in a completely agnostic and general purpose manner.

The parameters of CapsNetToA were configured as follows. Assuming an operative sampling frequency of 2 MHz, the selection of a window size of 500 samples with stride equal to 10 imposed a lower bound of 5 µs to the ToA resolution. This value is compliant with the time resolution admitted for the prospective applications, where ToA usually settles around hundreds of microseconds.

3. Experimental Validation: A Numerical Framework

The effectiveness of the designed models was tested within the framework of a metallic aluminum plate, which is frequently exploited as a benchmark scenario for ToA estimation. Firstly, a preliminary phase of dataset generation was performed to train the models, whose accuracy in prediction was then assessed by comparison with ground truth labels. Noteworthy, this initial validation is of critical importance to validate the robustness of the AI solutions with respect to reference statistical methods, especially to observe how performances can scale in the presence of noise levels.

3.1. Dataset Generation

As is widely recognized, DL models require a large amount of data to be trained on to avoid erroneous predictions. Moreover, since we are dealing with prediction problems, the same data also need to be labeled. However, labeling a massive amount of experimental data is, unfortunately, practically unfeasible. Alternatively, analytical simulations could be adopted to rapidly generate the labeled dataset. In this case, as anticipated in Section 1.1, it is possible to exploit the fact that acoustic emissions travel along the mechanical medium in the form of GWs, for which the propagation pattern through the mechanical medium is well known, and a numerical simulator can be implemented.

In our scenario, a square aluminum plate with nominal thickness of 3 mm and length of 1000 mm was taken as reference, while a Gaussian modulated pulse with central frequency of 250 kHz was assumed to simulate the effect of acoustic sources: in this frequency range, only the A0 and S0 modes characterize the propagation behavior, a condition which is desired to minimize the detrimental effect of multi-modality. Multi-path effects due to reflections and reverberations were considered as well with a ray-tracing approach, purposely written in MATLAB©. Each time series consisted of 5000 samples acquired at a theoretical sampling frequency of 2 MHz: these quantities were chosen to be compatible with commercial off-the-shelf sensors for AE monitoring.

More in detail, the signal generation procedure followed the subsequent steps:

1. **Traveling distance selection**: theoretically, the number of possible propagation distances to be explored between the AE location and the receiving point is infinite. However, by exploiting the symmetry of the structure ensured by its isotropic nature, the number of useful configurations can be reduced by a large extent. A square area of 5×5 positions circumscribed to the top east corner of the plate was allocated to AE receivers, while a total amount of 10×5 AE actuation points were uniformly distributed in the left half of the plate.
2. **Noise level variation**: since the primary objective of the proposed NN alternatives is to surpass the poor estimation capabilities of reference statistical methods in the presence of noise, Gaussian noise of increasing magnitude was progressively added to the acoustic wave by sweeping the SNR from 30 dB down to 1 dB, in steps of almost 1 dB. Despite the fact that the nature of the background noise of real AE signals can indeed differ [30], additive white stationary noise (such as the one generated by electronic components) can be assumed to be the main source of SNR degradation and, consequently, was used to simulate noisy AE scenarios in this study.
3. **Pre-trigger window variation**: in real AE equipment, the starting time for data logging is triggered by the incoming wave, e.g., once it exceeds a predefined energetic threshold. However, being capable of acquiring also the moments leading up to the acoustic event is of vital importance for appropriate AE signal characterization. As such, sensors are programmed to preserve memory of the pre-trigger signal history, known as the pre-trigger window. This quantity might change widely, from hundreds to thousands of samples, depending on both the application scenario and the employed electronics.
   Although representing a deterministic parameter that does not strictly depend on the physical phenomenon at the basis of acoustic wave propagation, the pre-trigger time actually plays a crucial role during the learning stage. This observation means that, theoretically, a one-to-one correspondence should exist between one model and one pre-trigger window. This aspect not only requires time and extra computing effort, due to the fact that a new training phase must be entailed whenever a change in the network configuration occurs, but it is also not viable in practical scenarios. Therefore, a data augmentation procedure was encompassed to favor the generalization capability of the neural network models.
   To this end, acoustic signals were initially generated with a fixed pre–trigger window of 500 samples, that represents a reasonable choice for typical scenarios. Then, one time-lagged version of each signal was derived by adding randomly from 500 to 2000 samples to the initial pre-trigger window. Since the total number of samples in the time history is limited to 5000, these forward shifts required additional N_{pre} samples to be concatenated with the initial portion of the signal, while disregarding

the final $N - N_{pre}$: to avoid both discontinuities and alterations in the statistical properties, the extra portion of the signal to be added was generated in form of a white noise term drawn from a Gaussian distribution, whose variance was taken to be coincident with the one estimated for the first 400 samples in the original pre-trigger window.

Another batch of data was also generated, comprising signals with an increased pre-trigger time beyond 2500 samples, and was entirely used during the testing phase in order to probe how the neural networks could behave with respect to unforeseen delays in the signal.

4. **Label generation**: when Lamb waves are to be characterized, it is difficult to give an unambiguous definition of their time of arrival due to dispersion and multi-modality. For this reason, rather than adopting a labeling approach based on the propagation theory, a different strategy was undertaken in this work. In particular, we exploited the fact that AIC inherently provides very accurate ToA estimations when the SNR is high. As such, the label attached to each time series was taken from the output yielded by AIC when applied to noise-free signals.

A total amount of N_{inst} = 60,000 signals was generated via exhaustive combination of all the possible configurations between the propagation distance, noise level and pre-trigger window: 80% of them were used for training, 10% for validation, and the remaining 10% for testing. Each time series was then normalized and mean removed. Some exemplary signals collected at the end of the dataset generation phase are plotted in Figure 5.

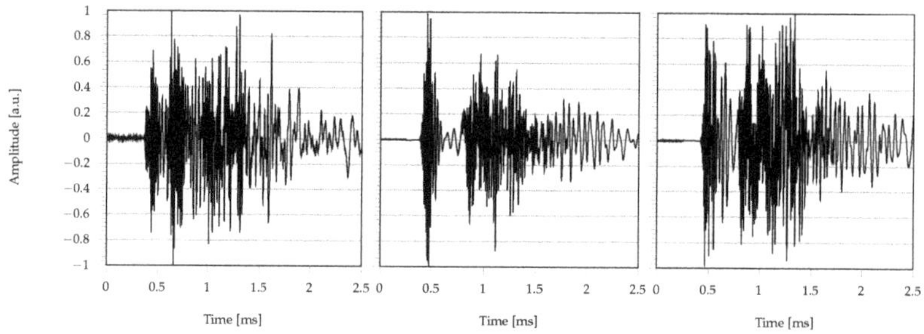

Figure 5. Example of synthetic signals generated with the ray–tracing algorithm.

3.2. Performance Metrics

Since "true" labels are available, the simplest methodology to assess the accuracy of the models is to quantify the error between predicted and actual ToAs. This strategy is efficient since it allows also to probe the accuracy of AIC in noisy scenarios: indeed, once applied to noise-corrupted variants of the same data, AIC predictions might deteriorate proportionally to the level of added perturbations.

The canonical expression of the root mean square error (RMSE),

$$\text{RMSE} = \sqrt{\frac{\sum_{i=1}^{N_{test}} (ToA_i - \hat{ToA}_i)^2}{N_{test}}} \tag{5}$$

was resorted to as an accuracy indicator: ToA_i is the true value, $\hat{ToA}_i$ is the current prediction, and N_{test} is the number of instances used during testing.

3.3. Results

The ToA estimation results for test data with a pre-trigger window below 2500 (i.e., the one used for training) are displayed in the left panel of Figure 6: they are given in terms of RMSE (log scale) for increasing SNR. From this figure, two different trends in the reported errors are evident. Firstly, the error profile of AIC is inversely proportional to the hidden noise level affecting the signal; moreover, it is characterized by a slowly decaying trend with a smoother profile. The performance results of the four approaches are in the same order of magnitude only for SNR equal to 30 dB, i.e., when, as discussed before, the signal statistics are well defined and easy to be identified, even via conventional processing tools.

Secondly, in the comparison between the AI approaches, the largest CNN shows an almost stable error around 4.5 µs, irrespective from the specific SNR apart from a negligible increment for very unfavorable noise levels. It is worth noting that the distillation operation (small CNN) is less performative for SNR below 5 dB but very effective for all the remaining SNRs. Finally, the curve error of CapsNetToA is similar to the one discussed for small CNN with slightly higher errors at high SNRs.

A graphical depiction of the output collected from the different approaches is displayed in Figure 7, which shows a zoom in the 0.1–0.4 ms window with drawn star markers indicating the predictions obtained from AIC (blue), CNN (orange), CapsNetToA (red) and small CNN (green), superimposed to the true label (yellow diamond). As can be observed, the estimates are considerably accurate in the left panel (SNR = 30 dB) for all the methods, while AIC completely fails in the case of important noise levels (same signal, but SNR = 10 dB) depicted in the right-hand side.

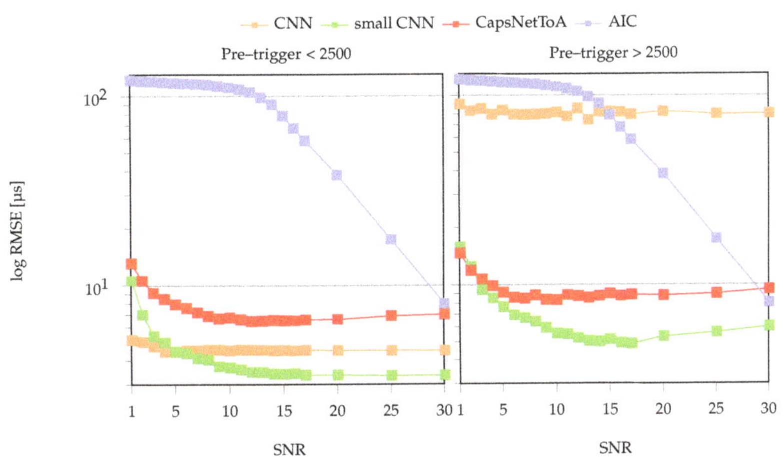

Figure 6. RMSE error for ToA estimation on synthetic test dataset for varying SNRs: Pre–trigger window used for training (i.e., window lower than 2500 samples) on the left and increased pre–trigger window on the right.

The performance achieved by varying the pre-trigger windows is reported in the right-hand side of Figure 6. As can be seen, no change is observable for AIC, owing to its time invariance. Conversely, an abrupt loss of accuracy affects the biggest CNN model, whose RMSE error maintains the same profile at the expense of a 20× increase in the magnitude. Such behavior can be attributed to over-parametrization problems as well as to poor generalization capabilities of the CNN due to the huge amount of parameters to be learned with respect to the actual amount of instances used in the training phase and the complexity of the problem at hand. In further detail, this is due to the fact that the number of learnable parameters largely exceeds the total amount of training instances. A second reason for such difference in CNN performances consists in potential *overfitting*, i.e., over-adaptation to the training dataset leading to a lack of capability to generalize during testing. Consequently, the biggest CNN model may not adequately capture the hidden ToA information in AE signals.

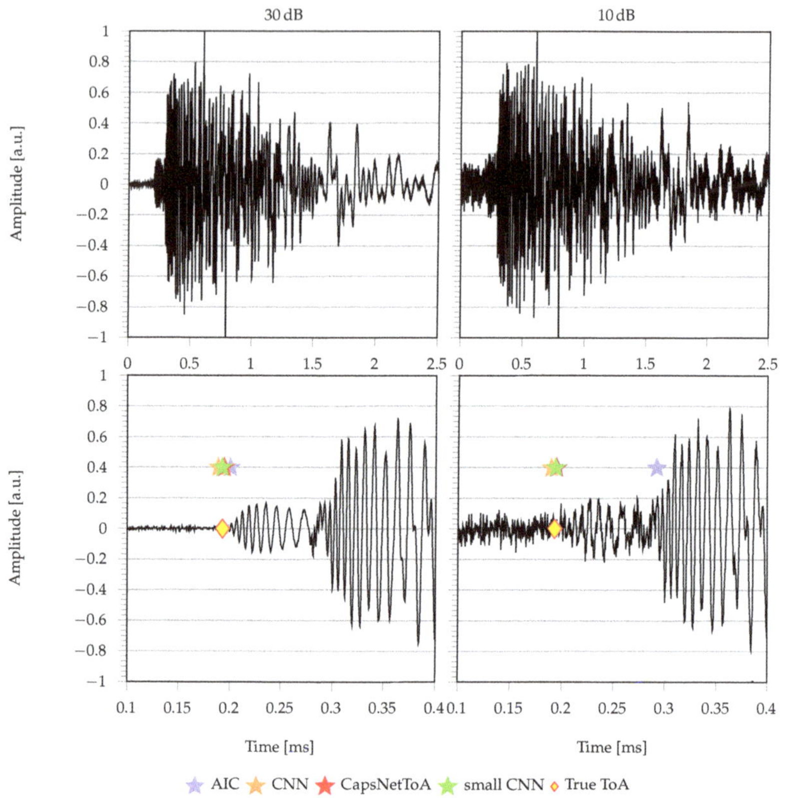

Figure 7. ToA predictions for AIC (blue), CNN (orange), CapsNetToA (red) and small CNN (green) with synthetic dataset: (**left column**) SNR = 30 dB and (**right column**) SNR = 10 dB, with second row depicting a magnified region of the time interval where ToA is located.

Conversely, as shown in Figure 6, the smallest CNN model is not prone to overfitting or over-parametrization, thanks to its more appropriate balancing between the number of trainable parameters and available data. Therefore, the distilled model is capable of generalizing from the trend hidden within training data, which is actually the expected goal of the neural network, rather than memorizing the training data themselves as happens with the largest CNN. Proof is in the fact that the small CNN model is still capable of providing coherent results, once again showing the best results among all the considered methods. For these reasons, only the small CNN model and CapsNetToA are taken into consideration in the following analyses, due to their better performance in terms of memory footprint, accuracy and generalization with respect to the length of the pre-trigger window.

Finally, the computational complexity of the designed NN models was also evaluated in terms on execution time in order to assess their actual portability on edge devices. To this end, when tested on a 1.8 GHz dual core Intel® core i-5 processor, the average inference time for a single signal was measured to be equal to 8.63 ms, 4.79 ms and 39.61 ms for CNN, small CNN and CapsNetToA (comprehensive, in this case, of both CapsNet and the time retrieval logic), respectively. As can be observed, the computational time nearly halves while moving from the biggest to the smallest CNN, while CapsNetToA shows the longest execution time. This is due to the fact that this architecture requires the ToA output probability curve to be reconstructed from several overlapped windows of the input signal, an operation which imposes CapsNet to be executed multiple times. For the specific CapsNet design configuration considered in this work, 1200 sliding windows need to be processed, each of them asking for nearly 30 µs. Note that such an amount of time can be reduced either by reducing the number of windows or by changing the length of the window itself. In line with these results, it is worth saying that, even if the performance might scale when the same algorithms are deployed on embedded devices depending on the frequency clock of the featured processor, the above reported execution times are compatible with the near-sensor implementation of the investigated models.

4. Experimental Validation: ToA for Acoustic Source Localization

In this section, experimental data for a laboratory aluminum plate, companion to the simulated one used for the sake of preliminary validation, are exploited to assess the suitability of the trained models to cope with acoustic source localization problems driven by in-field data.

4.1. Materials: The AE Equipment

The acoustic waves are actuated in form of guided elastic waves via a tone burst of two cycles, central frequency of 250 kHz and nominal amplitude of 0.6 Vpp, by means of an arbitrary waveform generator Agilent 33220A: its output was passed through a gated power amplifier RITEC GA-2500A (100× signal magnification) and then connected to an actuator, a Murata piezoelectric ceramic disk. Signals were acquired by a miniaturized acquisition system, the so-called piezoelectric 'sensor node (SN)' able to acquire, pre-process and characterize AE signals for real-time continuous monitoring. The core of the SN is the STMicroelecronics STM32F3, a low voltage 32-bit mixed signal microcontroller (MCU), which integrates both digital signal processing and floating-point unit instructions. The MCU features 40 kB SRAM and 256 kB FLASH memory, enough for the storage of data acquired by three AE sensors. The MCU embeds high precision analog components, such as four analog-to-digital converters (ADCs) with a resolution up to 12 bit via successive approximation and embedded voltage reference, four rail-to-rail input/output, low offset voltage programmable gain amplifiers (PGAs) and two 12 bit digital-to-analog converter (DAC) channels, which allow to obtain the maximum output swing. The frequency response of the device provides stable behavior from 10 kHz and 600 kHz, with a maximum attenuation of 3 dB. A complete description of the electronics is detailed in [31,32]. Each device of the network features up to three piezoelectric acquisition channels S1, S2, and S3, which passively acquire signals. Three piezoelectric sensors G150 whose frequency range lies between 60 kHz and 400 kHz were deployed at three corners of the plate, as schematically drawn in Figure 8, and a picture of the setup is presented in Figure 9.

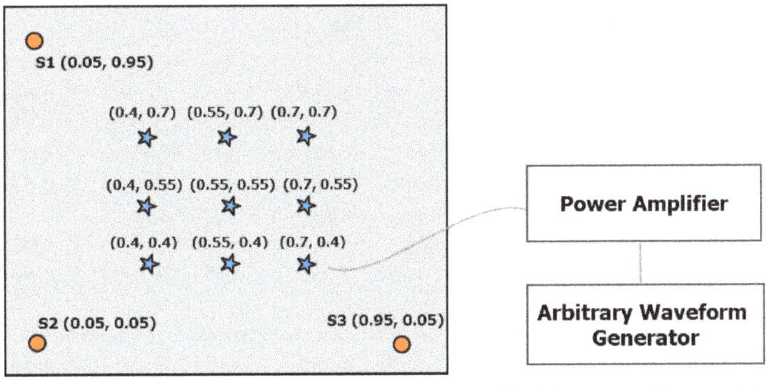

Figure 8. Sensor deployment diagram on the metallic laboratory plate used for the sake of source localization.

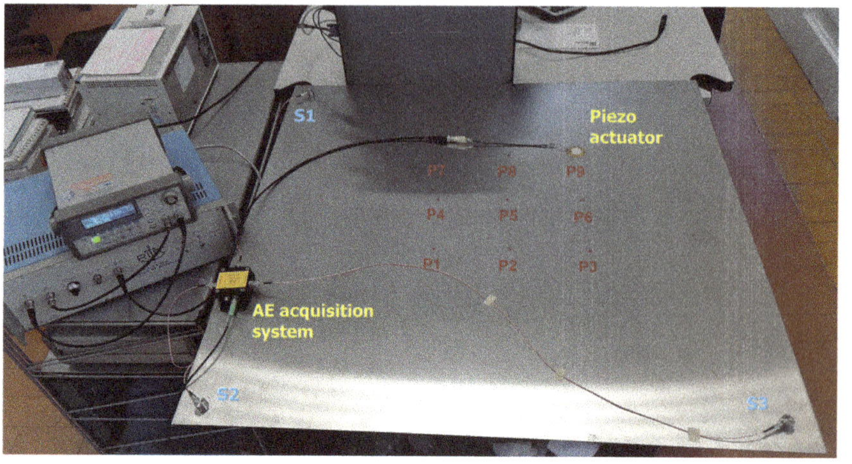

Figure 9. A photo of the experimental setup.

4.2. Methods

4.2.1. Objectives and Testing Procedures

Two main objectives were pursued with these experiments: (i) assess the time picking performance against operative disturbances responsible for important noise levels, and (ii) evaluate the suitability of the devised neural network architectures to deal with source localization. Nine different points (star markers in Figure 8) were selected for actuation, which are uniformly spaced in a square area with length equal to 0.3 m. For each point, experiments were repeated three times with a constant sample window of 5000 samples, pre-trigger window of 1500 samples and sample rate equal to 2 MHz. To evaluate the effect of noise, and following the testing procedure described in Section 3.1, white Gaussian noise was added to the gathered time series by sweeping the SNR in the interval of 2–20 dB at integer steps of 4 dB.

4.2.2. Localization Algorithm

To achieve successful object localization in 2D environments, a minimum network density of three sensing elements with known position has to be deployed on the monitored structure. Among the various strategies, the triangulation method proposed in [33] was leveraged in this work for its geometrical simplicity and well-proven functioning for isotropic/homogeneous structures, such as the one offered by the considered test-bed.

A graphical rendering of the triangulation problem is pinpointed in Figure 10, where a generic plate is instrumented with a network of three sensors, namely S1, S2 and S3. If S2 serves as a reference sensor for the network, θ_1 and θ_3 represent the relative orientation of sensors S1 and S3 with respect to a horizontal axis crossing S2 and aligned along the longitudinal dimension of the plate; similarly, D_{12} and D_{32} indicate the spatial distance between the two pairs of sensors, S1–S2 and S3–S2.

Now, if an acoustic event (red thunderbolt with label AE) occurs at a generic point of the structure, the triangulation algorithm aims at retrieving the set of polar coordinates (d, θ) uniquely identifying the acoustic source in the space. $d = d_2$ corresponds to the sought S2–AE distance, while $\theta_2 = \theta$ indicates where, among all the possible directions, the acoustic signal comes from. Once generated, the acoustic wave propagates over the structure and strikes the three sensors in three different instants of time due to the different AE-to-sensor distance: by application of the ToA estimation strategies investigated before, a guess of these onset times can be formulated. Let us denote them as ToA_1, ToA_2 and ToA_3 for S1, S2 and S3, respectively. Unfortunately, these quantities do not coincide with the physical time $t_i = d_i/c_g$ taken by the wave to travel the corresponding distances d_i at a wave speed $c_g = 5205\,\text{m/s}$ (related to the fastest propagating mode, the S0 one, at a frequency of 250 kHz); they rather represent a cumulative sum between t_i and the event actuation time T_0, which is common to all the sensors but not known to the system.

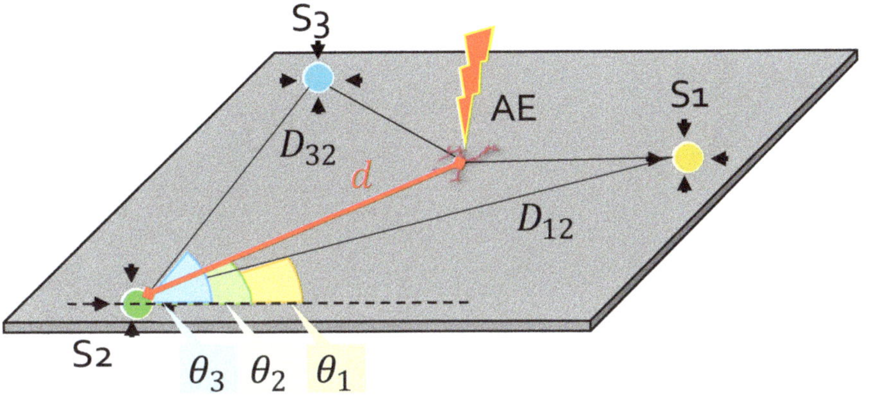

Figure 10. Geometrical representation of the adopted localization model via a three sensor array: the objective is to estimate the AE-to-sensor distance d and its relative direction θ.

Thus, even if geometrically well-posed, the estimation of d from the mere knowledge of ToA_2 is not practicable. However, by computing the DToA δ_{i2} out of the absolute ToA between the reference sensor S2 and the two remaining nodes, the T_0–independent quantities

$$\delta t_{32} = ToA_3 - ToA_2 = (t_3 + T_0) - (t_2 + T_0) = t_3 - t_2$$
$$\delta t_{12} = ToA_1 - ToA_2 = (t_1 + T_0) - (t_2 + T_0) = t_1 - t_2$$

can be obtained for sensor pairs S2–S3 and S2–S1, respectively, which coincide with the physical differences between the true wave propagation times. Analogously, the spatial difference between the traveled distances can be formulated as

$$\delta d_{32} = d_3 - d = c_g \delta t_{32}$$
$$\delta d_{12} = d_1 - d = c_g \delta t_{12}$$

The mathematical problem can thus be solved by means of the linear system:

$$\begin{cases} d_3 = d + c_g \delta t_{32} \\ d_1 = d + c_g \delta t_{12} \end{cases} \tag{6}$$

which, in this form, is not solvable since three unknowns but only two equations are available. To overcome this issue, it is sufficient to apply simple trigonometric relationships to the geometrical scheme in Figure 10: in particular, the Carnot's theorem states that

$$d_3^2 = d^2 + D_{32}^2 - 2dd_{32}\cos(\theta_3 - \theta) \tag{7a}$$
$$d_1^2 = d^2 + D_{12}^2 - 2dd_{12}\cos(\theta - \theta_1) \tag{7b}$$

Hence, by taking the square power of both sides in Equation (6) and plugging Equation (7a,b) in its first and second rows, respectively, the system can be rewritten only in terms of d and θ, which are the sought outputs of the localization process:

$$\begin{cases} d = \frac{1}{2}\frac{D_{32}^2 - c^2\delta t_{32}^2}{c_g\delta t_{32} + D_{32}\cos(\theta_3 - \theta)} \\ d = \frac{1}{2}\frac{D_{12}^2 - c^2\delta t_{12}^2}{c_g\delta t_{12} + D_{12}\cos(\theta - \theta_1)} \end{cases} \tag{8}$$

Algebraic manipulation of the system in Equation (8) yields the auxiliary solving equation for θ to coincide with:

$$r\sin(\theta+\theta_a)=K \tag{9}$$

with

$$\begin{aligned} G &= \frac{D_{32}^2 - c_g^2\delta t_{32}^2}{D_{12}^2 - c_g^2\delta t_{12}^2} \\ g_1 &= GD_{12}\cos\theta_1 - D_{32}\cos\theta_3 \\ g_2 &= GD_{12}\sin\theta_1 - D_{32}\sin\theta_3 \\ K &= -c_g(G\delta t_{12} - \delta t_{32}) \\ r &= \sqrt{g_1^2+g_2^2} \\ \theta_a &= \tan^{-1}\left(\frac{g_1}{g_2}\right) \end{aligned}$$

Finally, d can be computed back from the system in Equation (8). By knowing θ and d, the estimated AE source position $\hat{P}(\hat{x}_P;\hat{y}_P)$ can be easily derived as

$$\begin{cases} \hat{x}_P = x_{S_2} + d\cdot\cos\theta \\ \hat{y}_P = y_{S_2} + d\cdot\sin\theta \end{cases} \tag{10}$$

where x_{S_2} and y_{S_2} represents the 2D coordinates of the reference sensor S2.

4.2.3. Performance Evaluation Procedure

All the quantities appearing in Equation (9), apart from δt_{12} and δt_{32}, are deterministic once the sensor network configuration is defined. As such, achieving good localization capability is, inversely, a synonym of accurate onset time estimation and offers, in these terms, a powerful means to assess the quality of the time picking activity of the devised NN models. Note that, in this operative setting where no synchronization is present between the actuation and the reception components of the monitoring network, no possibility exists to define a true ground truth due to the uncertainties implied by the experimental setting.

During testing, three ToA estimators were considered for the purpose: besides the small CNN model and CapsNetToA trained on the synthetic data, AIC was also applied from the pool of conventional statistical approaches. The Euclidean distance $d_p = |P-\hat{P}|$ between the true $P(x_P;y_p)$ and estimated $\hat{P}(\hat{x}_P;\hat{y}_p)$ AE position returned by Equation (10) were computed to assess the accuracy for source localization task. Moreover, the collected signals were corrupted with white Gaussian noise of increasing entity, i.e., by gradually reducing the SNR from 20 dB to 2 dB. In this way, it was possible to investigate how the same model could perform in remarkably harsher propagating environments, as they could appear in real AE scenarios subjected to varying but not predictable noisy conditions.

4.3. Results

Outcomes from localization experiments are drawn in Figure 11: for each excitation point, three markers are included, corresponding to as many tests in this configuration. Additionally, it is worth specifying that points indicating a constant error of $d_P = 1$ m are included to identify all those cases in which the estimated coordinates are not compatible with the physical solutions for the structure at hand (i.e., negative or larger than 1.41 m distances, which is the maximum propagation length for this plate). Figure 12 provides an example of time domain signals for excitation point 5 in two different noisy conditions: 20 dB on the left column and 8 dB on the right, with magnified ToA values in the second row.

First of all, one remark is worthy of attention, which is related to the difference in the experimental signals with respect to the synthetic ones employed for training (see comparison between Figures 5 and 11). In fact, in the real setting, the effects of the sensor transfer function as well as the detrimental ones due to attenuation, multiple reflections and propagation in the physical medium might lead the envelope of the acquired signals to vary in a significant way. As such, being capable of obtaining an accurate prediction on real signals starting from a simulated dataset can be seen as a more severe test to be passed.

Going deeper into the results, the NN methods provide more consistent estimates, irrespective of the single excitation point and just showing a minor increment starting from SNR = 4 dB, where the errors increase up to 30 cm for 2 dB, which represents a challenging working condition for AE equipment. Conversely, d_P reduces to a large extent in all the remaining noisy configurations, the CapsNetToA model being the most accurate estimator with an average error of 5 cm, followed by small CNN, whose average error is less than 8 cm.

Moreover, it is possible to observe that the quite similar localization pattern shown by small CNN, which tends to worsen the lower the SNR, is a consequence of the noise generation procedure exposed in Section 3.1, according to which the same signal, but with different additive noise, is processed. For the same reason, even if less pronounced, an analogous trend characterizes also AIC and CapsNetToA.

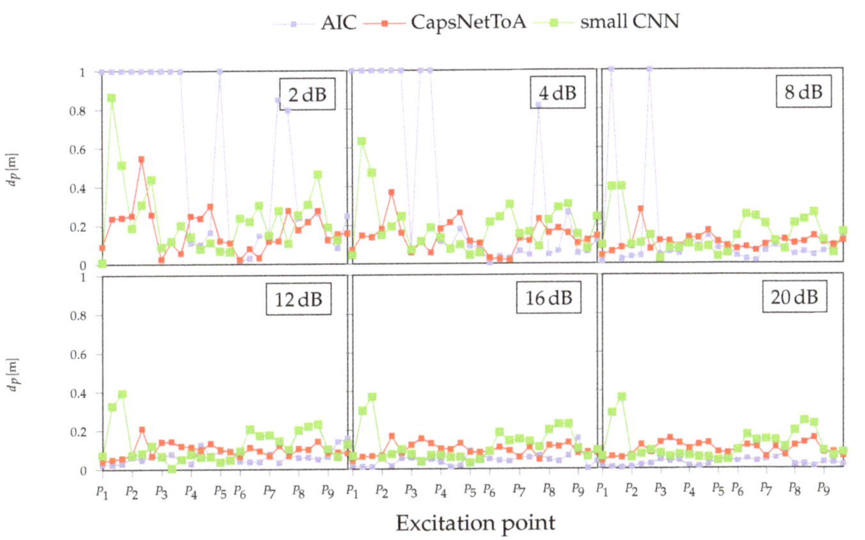

Figure 11. Absolute errors for acoustic signal localization on a laboratory metallic plate under various noise levels.

The slightly better performance of CapsNetToA with respect to the small CNN can be attributed to the superior generalization capability of the first solution, which better handles the discrepancy between the actuated pulse and the synthetic one used in the training. This generalization capability is a very desirable property in practical scenarios, where the actuated pulse is unknown.

Moreover, the same plots show that, despite AIC being highly performative in low noise conditions, as proven by a maximum deviation of 7 cm for the 20 dB configuration, the introduction of high noise levels leads AIC to completely fail ($d_P = 1$ m) in multiple positions (e.g., P_1, P_2, and P_3).

The above considerations confirm the trends already reported in Figure 6, showing that AIC remains a robust and competitive strategy for AE signal characterization for relatively low noise values affecting data, whose drawbacks in dealing with poor SNRs can be overcome via AI approaches, and which can still achieve a satisfactory level of precision in the identification of the acoustic source.

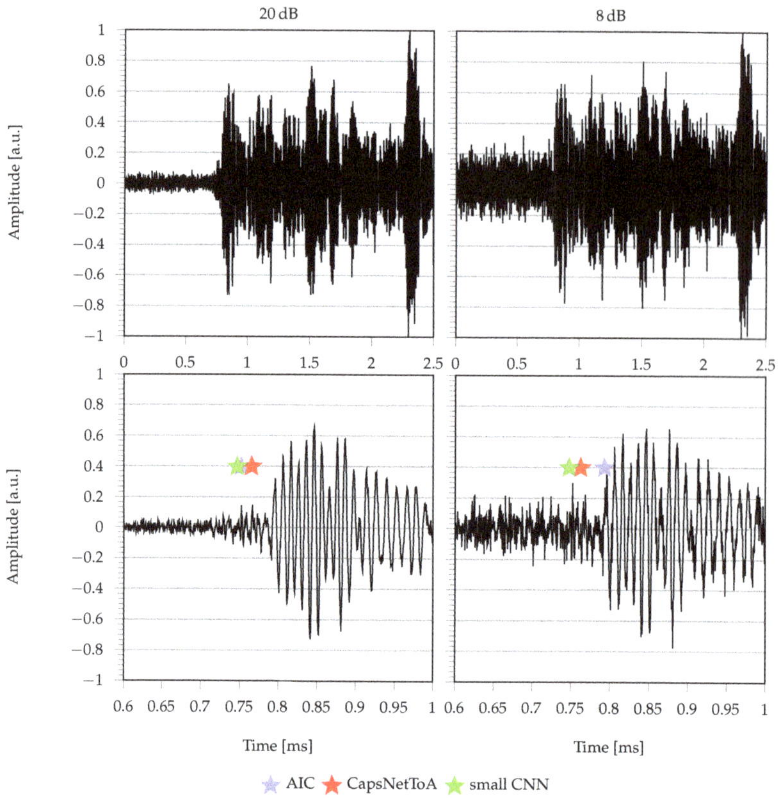

Figure 12. ToA predictions for AIC (blue), CapsNetToA (red) and small CNN (green) for excitation point 5: (**left column**) SNR = 20 dB and (**right column**) SNR = 8 dB, with second row depicting a magnified region of the time interval where ToA is located.

5. Conclusions

In this work, innovative methods taken from the AI processing field are presented for ToA estimation in AE-generated signals. The proposed solutions, based on a multi-layer CNN and on a capsule architecture with dedicated time retrieval logic, have the peculiar advantage of providing consistent results, even in the presence of significantly low SNRs. When tested on both a synthetic dataset generated for the characterization of a square aluminum plate and for source localization in laboratory conditions, they attained 10× more accurate results than the AIC algorithm, which can be considered a standard in the field. In particular, among the proposed NN methods, the novel CapsNetToA scored the highest results thanks to its correlation-based approach, achieved via the unique concept of the capsule unit, which allows spatial dependencies between data to be preserved. Secondly, by replacing pooling layers, which are the primary source for loss of information in conventional CNN architectures, with peculiar vector-based output layers, it can implicitly generalize better to out-of-distribution data and gain higher immunity to noisy outliers. Future work will be devoted to embedding of the same algorithms on edge devices, for the sensor-near retrieval of the ToA as required in practical scenarios. Furthermore, additional models and related experimental results will be investigated, in which the frequency content of the excited signals will be changed according to the different spectral bands used in AE testing.

Author Contributions: Conceptualization, F.Z., D.B. and L.D.M.; methodology, F.Z., D.B.; software, F.Z. and T.D.; validation, F.Z., D.B. and T.D.; formal analysis, F.Z., D.B. and T.D.; investigation, F.Z., D.B. and T.D.; resources, A.M. and N.T.; data curation, F.Z., D.B. and T.D.; writing—original draft preparation, F.Z., D.B. and T.D.; writing—review and editing, F.Z., D.B., N.T., L.D.M. and A.M.; visualization, F.Z. and D.B.; supervision, L.D.M. and A.M.; project administration, F.Z. and L.D.M.; funding acquisition, L.D.M. and A.M. All authors have read and agreed to the published version of the manuscript.

Funding: This work is funded by INAIL (Istituto Nazionale Italiano per l'Assicurazione contro gli Infortuni sul Lavoro) within the BRIC/2018, ID = 11 framework, project MAC4PRO ("Smart maintenance of industrial plants and civil structures via innovative monitoring technologies and prognostic approaches").

Institutional Review Board Statement: Not applicable.

Informed Consent Statement: Not applicable.

Data Availability Statement: Not applicable.

Conflicts of Interest: The authors declare no conflict of interest.

References

1. Gholizadeh, S.; Leman, Z.; Baharudin, B.H.T. A review of the application of acoustic emission technique in engineering. *Struct. Eng. Mech.* **2015**, *54*, 1075–1095. [CrossRef]
2. Berkovits, A.; Fang, D. Study of fatigue crack characteristics by acoustic emission. *Eng. Fract. Mech.* **1995**, *51*, 401–416. [CrossRef]
3. Chesné, S.; Deraemaeker, A. Damage localization using transmissibility functions: A critical review. *Mech. Syst. Signal Process.* **2013**, *38*, 569–584. [CrossRef]
4. Ainsleigh, P.L. Acoustic echo detection and arrival-time estimation using spectral tail energy. *J. Acoust. Soc. Am.* **2001**, *110*, 967–972. [CrossRef]
5. Vallen, H. AE testing fundamentals, equipment, applications. *J. Nondestruct. Test.* **2002**, *7*, 1–30.
6. Wirtz, S.F.; Söffker, D. Improved signal processing of acoustic emission for structural health monitoring using a data-driven approach. In Proceedings of the 9th European Workshop on Structural Health Monitoring, Manchester, UK, 10–13 July 2018.
7. Sikorska, J.; Mba, D. Challenges and obstacles in the application of acoustic emission to process machinery. *Proc. Inst. Mech. Eng. Part E J. Process Mech. Eng.* **2008**, *222*, 1–19. [CrossRef]
8. Ramesh, S. *The Applied Welding Engineering. Processes, Codes, and Standards*; Elsevier: Amsterdam, The Netherlands, 2012; pp. 307–319.
9. Karbhari, V.M. *Non-Destructive Evaluation (NDE) of Polymer Matrix Composites*; Elsevier: Amsterdam, The Netherlands, 2013.
10. Rose, J.L. Guided Wave Studies for Enhanced Acoustic Emission Inspection. *Res. Nondestruct. Eval.* **2021**, *32*, 192–209. [CrossRef]
11. Faisal, H.M.; Giurgiutiu, V. Theoretical and numerical analysis of acoustic emission guided waves released during crack propagation. *J. Intell. Mater. Syst. Struct.* **2019**, *30*, 1318–1338. [CrossRef]
12. Lowe, M.J.S. *Wave Propagation | Guided Waves in Structures*; Encyclopedia of Vibration; Elsevier: Amsterdam, The Netherlands, 2001; pp. 1551–1559.
13. Carotenuto, R.; Merenda, M.; Iero, D.; Della Corte, F.G. An indoor ultrasonic system for autonomous 3-D positioning. *IEEE Trans. Instrum. Meas.* **2018**, *68*, 2507–2518. [CrossRef]
14. Adrián-Martínez, S.; Bou-Cabo, M.; Felis, I.; Llorens, C.D.; Martínez-Mora, J.A.; Saldaña, M.; Ardid, M. Acoustic signal detection through the cross-correlation method in experiments with different signal to noise ratio and reverberation conditions. In *International Conference on Ad-Hoc Networks and Wireless*; Springer: Berlin/Heidelberg, Germany, 2014; pp. 66–79.
15. Perelli, A.; De Marchi, L.; Marzani, A.; Speciale, N. Frequency warped cross-wavelet multiresolution analysis of guided waves for impact localization. *Signal Process.* **2014**, *96*, 51–62. [CrossRef]
16. St-Onge, A. Akaike information criterion applied to detecting first arrival times on microseismic data. In *SEG Technical Program Expanded Abstracts 2011*; Society of Exploration Geophysicists: Tulsa, OK, USA, 2011; pp. 1658–1662.
17. Pearson, M.R.; Eaton, M.; Featherston, C.; Pullin, R.; Holford, K. Improved acoustic emission source location during fatigue and impact events in metallic and composite structures. *Struct. Health Monit.* **2017**, *16*, 382–399. [CrossRef]
18. Li, X.; Shang, X.; Morales-Esteban, A.; Wang, Z. Identifying P phase arrival of weak events: The Akaike Information Criterion picking application based on the Empirical Mode Decomposition. *Comput. Geosci.* **2017**, *100*, 57–66. [CrossRef]
19. Ross, Z.E.; Rollins, C.; Cochran, E.S.; Hauksson, E.; Avouac, J.P.; Ben-Zion, Y. Aftershocks driven by afterslip and fluid pressure sweeping through a fault-fracture mesh. *Geophys. Res. Lett.* **2017**, *44*, 8260–8267. [CrossRef]
20. Chen, Y. Automatic microseismic event picking via unsupervised machine learning. *Geophys. J. Int.* **2020**, *222*, 1750–1764. [CrossRef]
21. Ross, Z.E.; Meier, M.A.; Hauksson, E. P wave arrival picking and first-motion polarity determination with deep learning. *J. Geophys. Res. Solid Earth* **2018**, *123*, 5120–5129. [CrossRef]

22. Zhu, W.; Beroza, G.C. PhaseNet: A deep-neural-network-based seismic arrival-time picking method. *Geophys. J. Int.* **2019**, *216*, 261–273. [CrossRef]
23. Rojas, O.; Otero, B.; Alvarado, L.; Mus, S.; Tous, R. Artificial neural networks as emerging tools for earthquake detection. *Comput. Sist.* **2019**, *23*, 335–350. [CrossRef]
24. Tabian, I.; Fu, H.; Sharif Khodaei, Z. A convolutional neural network for impact detection and characterization of complex composite structures. *Sensors* **2019**, *19*, 4933. [CrossRef]
25. Zhang, Z. Improved adam optimizer for deep neural networks. In Proceedings of the 2018 IEEE/ACM 26th International Symposium on Quality of Service (IWQoS), Banff, AB, Canada, 4–6 June 2018; pp. 1–2.
26. Lin, J.; Rao, Y.; Lu, J.; Zhou, J. Runtime neural pruning. In Proceedings of the 31st International Conference on Neural Information Processing Systems, Long Beach, CA, USA, 4–9 December 2017; pp. 2178–2188.
27. Saad, O.M.; Chen, Y. CapsPhase: Capsule Neural Network for Seismic Phase Classification and Picking. *IEEE Trans. Geosci. Remote Sens.* **2021**. [CrossRef]
28. Sabour, S.; Frosst, N.; Hinton, G.E. Dynamic routing between capsules. *arXiv* **2017**, arXiv:1710.09829.
29. Mandal, B.; Dubey, S.; Ghosh, S.; Sarkhel, R.; Das, N. Handwritten indic character recognition using capsule networks. In Proceedings of the 2018 IEEE Applied Signal Processing Conference (ASPCON), Kolkata, India, 7–9 December 2018; pp. 304–308.
30. Barat, V.; Borodin, Y.; Kuzmin, A. Intelligent AE signal filtering methods. *J. Acoust. Emiss.* **2010**, *28*, 109–119.
31. Testoni, N.; De Marchi, L.; Marzani, A. A stamp size, 40 mA, 5 grams sensor node for impact detection and location. In Proceedings of the European Workshop on SHM, Bilbao, Spain, 5–8 July 2016.
32. Bogomolov, D.; Testoni, N.; Zonzini, F.; Malatesta, M.; de Marchi, L.; Marzani, A. Acoustic emission structural monitoring through low-cost sensor nodes. In Proceedings of the 10th International Conference on Structural Health Monitoring of Intelligent Infrastructure, Porto, Portugal, 30 June–2 July 2021.
33. Jiang, Y.; Xu, F. Research on source location from acoustic emission tomography. In Proceedings of the 30th European Conference on Acoustic Emission Testing & 7th International Conference on Acoustic Emission, Granada, Spain, 12–15 September 2012.

sensors

MDPI

Article

Clustering and Smoothing Pipeline for Management Zone Delineation Using Proximal and Remote Sensing

S. Hamed Javadi [1,2], Angela Guerrero [2] and Abdul M. Mouazen [2,*]

1 Interuniversity Micro-Electronics Center (IMEC), Kapeldreef 75, 3001 Leuven, Belgium; hamed.javadi@imec.be
2 Precision Soil and Crop Engineering Group (Precision SCoRing), Department of Environment, Faculty of Bioscience Engineering, Ghent University, Coupure Links 653, 9000 Gent, Belgium; angela.guerrero@ugent.be
* Correspondence: abdul.mouazen@ugent.be

Abstract: In precision agriculture (PA) practices, the accurate delineation of management zones (MZs), with each zone having similar characteristics, is essential for map-based variable rate application of farming inputs. However, there is no consensus on an optimal clustering algorithm and the input data format. In this paper, we evaluated the performances of five clustering algorithms including *k*-means, fuzzy C-means (FCM), hierarchical, mean shift, and density-based spatial clustering of applications with noise (DBSCAN) in different scenarios and assessed the impacts of input data format and feature selection on MZ delineation quality. We used key soil fertility attributes (moisture content (MC), organic carbon (OC), calcium (Ca), cation exchange capacity (CEC), exchangeable potassium (K), magnesium (Mg), sodium (Na), exchangeable phosphorous (P), and pH) collected with an online visible and near-infrared (vis-NIR) spectrometer along with Sentinel2 and yield data of five commercial fields in Belgium. We demonstrated that *k*-means is the optimal clustering method for MZ delineation, and the input data should be normalized (range normalization). Feature selection was also shown to be positively effective. Furthermore, we proposed an algorithm based on DBSCAN for smoothing the MZs maps to allow smooth actuating during variable rate application by agricultural machinery. Finally, the whole process of MZ delineation was integrated in a clustering and smoothing pipeline (CaSP), which automatically performs the following steps sequentially: (1) range normalization, (2) feature selection based on cross-correlation analysis, (3) *k*-means clustering, and (4) smoothing. It is recommended to adopt the developed platform for automatic MZ delineation for variable rate applications of farming inputs.

Keywords: clustering; feature selection; management zone delineation; precision agriculture

Citation: Javadi, S.H.; Guerrero, A.; Mouazen, A.M. Clustering and Smoothing Pipeline for Management Zone Delineation Using Proximal and Remote Sensing. *Sensors* **2022**, *22*, 645. https://doi.org/10.3390/s22020645

Academic Editor: Natividad Duro Carralero

Received: 28 November 2021
Accepted: 10 January 2022
Published: 14 January 2022

1. Introduction

Traditional agricultural practices consider fields as homogeneous management units, under which farm operations assume no within-field variability in the soil or crop. However, agricultural soils are often extremely variable in space and time, and understanding its variability is essential to successfully manage farming inputs site-specifically and dynamically at a field scale [1]. To address within-field variability, variable management solutions are adopted using precision agriculture (PA) technologies, which aim at the site-specific application of farm inputs (e.g., seeds, fertilizers, manure, pesticides, and water) according to the soil and crop requirements [2]. Variable rate applications—also referred to as site-specific applications—are implemented in practice as map-based, sensor-based, or a combination of both approaches [3]. For both map-based and map-sensor-based solutions, the within-field variability should be classified into a few zones with similar characteristics. In fact, the most widely used approach to manage the variability of fields concerns the use of management zones (MZs) [4]. MZs are sub-areas of a field that have a relatively homogeneous combination of yield-limiting factors with respect to soil–landscape attributes [5], for which a single rate of a specific crop input is appropriate to maximize outputs such as yield and yield quality [6]. The accurate delineation of MZ maps is the key requirement for the successful implementation of map-based and map-sensor-based variable rate applications.

Several approaches are introduced in [2] to delineate MZs by the data fusion of several layers of information including farmer's knowledge, terrain attributes, weather conditions, soil type, yield data from several seasons, crop growth characteristics, and soil properties. Indeed, MZ delineation considers the variables that are correlated to the yield, since the goal is to maximize the yield [7]. Guerrero et al. in [8] have shown that involving more layers of information in MZ delineation provides more robust results in terms of improving yield. De Benedetto et al. [9] delineated homogeneous areas by data fusion of electromagnetic induction sensor, a ground penetration radar, and remote sensing satellite hyperspectral images. Fleming et al. [10] combined the data of soil organic matter, clay, nitrate, potassium, zinc, electrical conductivity, and corn yield for variable-rate fertilization purposes. Pantazi et al. [4] proposed to delineate MZ maps by means of self-organizing clustering using soil data collected by an on-line soil sensing platform [11], crop normalized differential vegetation index (NDVI) of satellite imagery, and historical yield data.

Data fusion *potentially* reduces the prediction variance and hence improves the prediction precision [12]. However, when fusing different kinds of data, double counting the same information given by those data may degrade the overall performance [13–15]. Furthermore, as pointed out by Schenatto et al. [16], different kinds of data with different values' ranges can impact the MZ delineation and result in favor of just a few of the involving data. Accordingly, we analyzed the impacts of feature selection and data normalization as solutions to correlation and inconsistent data ranges, respectively, in this paper.

To characterize the within-field variability, different sampling methods can be used. The most common method of soil sampling in a field is grid sampling mostly adopted to determine the chemical and physical properties [17]. Another sampling method is the use of proximal soil-sensing technologies to collect high-density data of 1000–2000 reading per ha, which introduces complexity and sources of errors, particularly when the resolution in one direction is much different than the resolution in the other direction. Examples include the online data collected by the visible and nearinfrared (vis-NIR) spectroscopy [18,19], where a resolution of 1 m by 10 m is very common, introducing complexity during interpolation and clustering.

MZ delineation can be considered as a data-mining problem, as it contemplates either classifying or clustering the field into a number of contiguous areas [20]. For MZ delineation, numerous clustering algorithms—such as *k*-mean, mean shift, fuzzy C-means (FCM), hierarchical clustering, density-based spatial clustering of applications with noise (DBSCAN), and particle swarm algorithm (PSO) [21–23]—have been already adopted. Recently, a deep-learning-based algorithm has also been examined by Javadi et al. [24]. However, all the examined algorithms have their own peculiarities in terms of features and efficiency (Karkral et al., 2020), since clustering is a complex task owing to the large number of interrelated parameters, resulting in a nonlinear problem. One source of nonlinearity stems from the inconsistent sampling resolution, which is common with online proximal soil sensing.

Clustering techniques are mostly unsupervised and attempt to explore the inherent structure of the data, often in terms of Euclidean distance. Different normalization methods were evaluated [16] for MZ delineation, without any clear discussion on why the data should be normalized. On the other hand, an apple orchard was delineated into MZs in [25] using a geostatistics method, in which the spatial correlation of data is taken into account. The spatial correlation has been also suggested to be considered in [26], where multicollocated cokriging was used for variable-rate fertigation. Indeed, soil and crop properties in agricultural fields generally present spatial dependence; hence, it is important to use geostatistical methods (kriging interpolation after semivariogram analysis) where soil or crop properties are considered as random regionalized variables, and the gradual geographical variation is described by a spatial covariance function [25,26]. Most of the papers concerning MZ delineation have not explicitly discussed the inclusion of the geographical coordinates of the data. However, the coordinates data were explicitly used in [9,25,27–30]. However, so far, there is no consensus not only on the clustering method but also the format of input data and inclusion or exclusion of spatial correlation and locations. This is particularly true when at least one layer of data involved in the clustering is collected at inconsistent sampling resolution over space, such as the example of the online vis-NIR sensor.

In this paper, we evaluated five clustering methods, namely, *k*-means, FCM, shift mean, hierarchical, and DBSCAN, in MZ delineation of five fields with different sizes in different regions of Belgium. The goal was to determine the optimal clustering method and data inputs for the delineation of MZs, using online collected soil data with inconsistent spatial resolution in addition to yield and crop data obtained from processing the data of satellite Sentinel2. Furthermore, we proposed a clustering and smoothing pipeline (CaSP) for MZ delineation, which gives a smoothed scheme of MZs and is applicable in practice by the variable rate agricultural machinery. We examined the performance of the proposed CaSP in the delineation of MZs maps in all the five studied fields using spatial statistical indicators.

2. Materials and Methods

The flowchart of the steps performed in this study for the different MZ delineation schemes is depicted in Figure 1. Each step is elaborated in what follows.

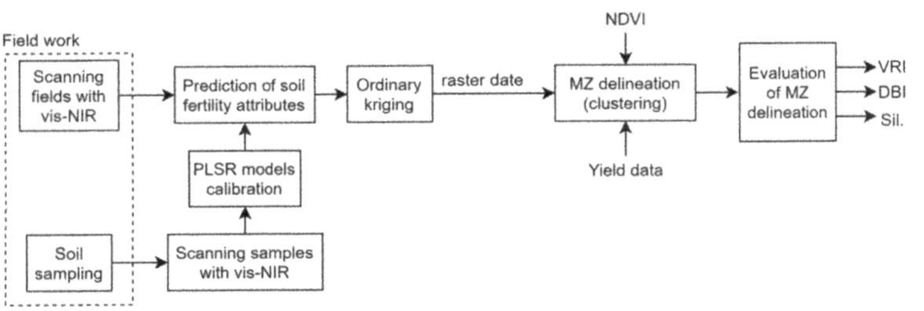

Figure 1. The flowchart of the evaluation steps different management zone (MZ) delineation schemes based on on-line collected soil fertility attributes, normalized difference vegetation index (NDVI), and yield. Different MZ delineation schemes were evaluated in terms of variance reduction index (VRI), Davies–Boulding index (DBI), and Silhouette index (Sil.). The soil fertility attributes were predicted based on visible-near-infrared (vis-NIR) readings.

2.1. Experimental Sites

The soil fertility attributes, normalized differential vegetation index (NDVI), and yield data of five commercial fields in the Flanders region in Belgium were used in this study (Figure 2). The soil fertility attributes included moisture content (MC), organic carbon (OC), calcium (Ca), cation-exchangeable content (CEC), exchangeable potassium (K), magnesium (Mg), sodium (Na), pH, and exchangeable phosphorus (P). The study fields consisted of a 21 ha field in Landen called Grootland (N 50°47′22.5″, E 5°6′48.8″), an 12 ha field in Huldenberg named Kouter (N 50°48′38.9″, E 4°34′50.0″), and three fields in Veurne: a 12 ha field called Beers (N 51°1′1.4″, E 2°34′42.8″), a 8 ha field named Fabrieke (N 51°1′53.9″, E 2°34′16.9″), and a 12 ha field designated as Krokey (N 50°59′58.3″, E 2°32′52.1″). The results of a soil texture analysis determined by means of the Robinson–Kohn pipette method (ISO 11277) indicated that soil in Grootland, Krokey, and Kouter was a silty loam, in Beers was a sandy loam, and in Fabrieke was a loam according to the United State Department of Agriculture (USDA) classification (Table 1). This region registers an annual average temperature of 10.6 °C and a monthly average precipitation of 39.68 mm. The fields have an annual crop rotation of wheat, barley, oilseed rape, sugar beet, and potatoes with a short duration autumn cover crop.

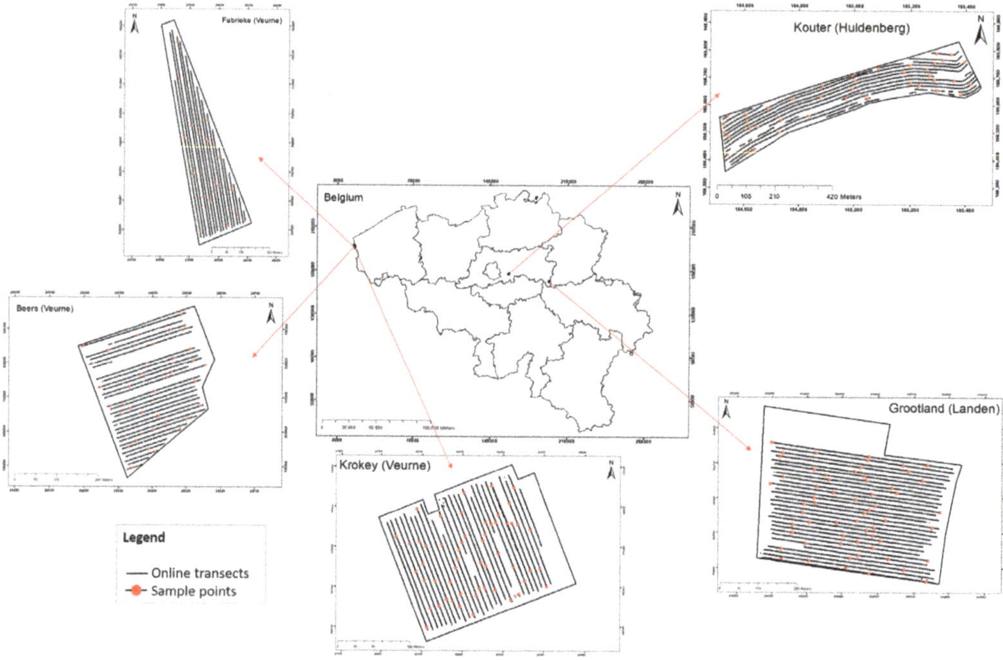

Figure 2. Locations of the five experimental sites in Flanders, Belgium, along with the online scanning lines and the locations of the random soil sampling points in Fabrieke, Beers, Krokey, Kouter, and Grootland.

Table 1. Information of the spectral library used for the development of visible and near-infrared (vis-NIR) calibration models for three farms for the prediction of key soil properties using the online spectra collected with the online multi-sensor platform. Reprinted with permission from Ref. [11]. Copyright 2021 Elsevier.

Model	Field Name	% Clay	% Sand	% Silt	Soil Texture (USDA)	No. Samples	Total Samples
Huldenberg	Kouter (Target field)	12.6	11.6	75.8	Silt Loam	40	155
	Duidelbergen	10.2	10.3	79.4	Silt Loam	24	
	Voor de Heeves	12.0	9.5	78.5	Silt Loam	43	
	Lange Weid	10.3	10.3	79.4	Silt Loam	48	
Landen	Grootland (Target field)	13.3	6.3	80.4	Silt Loam	60	179
	Gimgelomse	13.2	32.7	54.2	Silt Loam	38	
	Kattestraat	–	–	–	–	20	
	Dal	–	–	–	–	23	
	Bottelare [1]	–	–	–	–	25	
	Thierry [1]	–	–	–	–	13	
Veurne	Beers (Target field)	16.5	54.0	29.5	Sandy Loam	39	122
	Fabrieke (Target field)	16.2	37.8	46.0	Loam	25	
	Krokey (Target field) [2]					54	
	Watermachine	14.5	51.6	33.9	Loam	20	
	Bottelare [1]	–	–	–	–	25	
	Thierry [1]	–	–	–	–	13	

[1] These fields are located in Bottelare and Mouscron, respectively, but their data were used to improve the accuracy of models developed for the Landen and Veurne farm. [2] Krokey field was not included in de development of the Veurne model.

2.2. Data Acquisition

Soil data were obtained by scanning the fields with an online multi-sensor platform, as shown in Figure 3, which was designed and developed by Mouazen [11] as discussed in [8,31]. The platform included a vis-NIR spectroscopy sensor (Tec5 Ag, Germany) with a measurement range of 305–1700 nm. The platform is attached to a tractor by means of the three-point hitch and pulled along parallel lines at a distance of 12 m between neighboring lines and at an average speed of 3 km/h. By creating a trench with a subsoiler, the platform is capable of collecting the vis-NIR soil spectra at 15–25 cm depth every second. It includes a differential global positioning system (DGPS) with RTK correction and a position accuracy of ± 0.2 m (version CFX-750, Trimble, Sunnyvale, CA, USA) and a datalogger (Compact Rio 9082, National Instruments, USA) to acquire and store the collected soil spectra and DGPS readings at 1 Hz, using a custom-built Labview software (National Instruments, USA). Kouter, Beers, and Grootland were scanned in 2018, and Fabrieke and Krokey were scanned in 2019 after harvest of the previous crops (Figure 2).

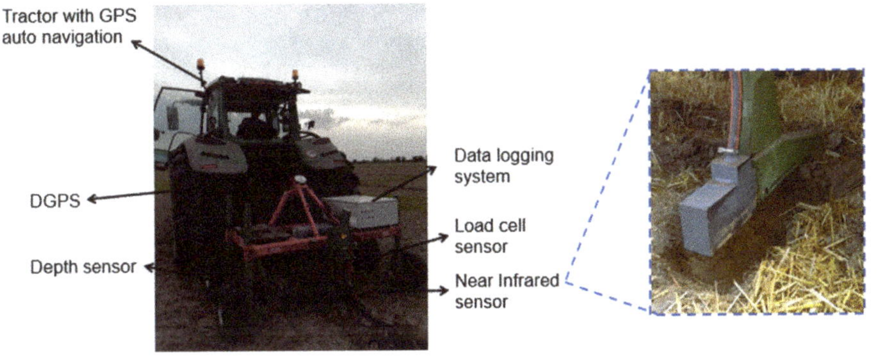

Figure 3. The multiple-sensor platform used for collecting soil data. DGPS: differential global positioning system.

In addition to the vis-NIR data, the NDVI data were obtained from the processed data of the Sentinel2 satellite imagery for Beers, Fabrieke, Krokey, and Grootland. Some soil attributes can be estimated using satellite data [32,33]; however, this was not the case in this study, since the accuracy of satellite data can degrade the high accuracy of soil attribute estimation models derived from vis-NIR spectra [13,34,35]. For the Kouter field, high-resolution NDVI data were collected using six Green Seeker sensors installed on a liquid fertilizer sprayer. To use NDVI data obtained from the satellite imagery and the Green Seeker sensor, extra data processing was performed. First, a kriging interpolation using the NDVI values along each field was performed; then, a common grid of 5×5 m was created, and finally, NDVI values were extracted for each pair of coordinates in the common grid. The yields data of cereal crops in the previous season in each field were collected using combine harvesters equipped with yield sensors providing high-resolution yield data (in Kouter field: John Deere W550, in Grootland field: Claas Lexion 740 with yield monitoring with Quantimeter, and in Fabrieke and Beers fields: Class Lexion 760 with yield monitoring with Quantimeter).

2.3. Modeling of Visible and Near-Infrared Spectra

In addition to the online measurements, random soil samples were collected manually from each field (Table 1) with the aim to build prediction models for soil attributes (pH, Ca, Mg, MC, OC, P, CEC, K, and Na), similar to what was explained in [3] (pp. 1–38) and [13,34]. A total of 155, 179, and 121 soil samples were collected from different fields in three farms, and these were used to develop three groups of models for Huldenberg, Landen, and Veurne farms, respectively. Cross-validation by using the leave-one-out technique was possible for the Kouter field (in Huldenberg farm), since limited data were available from this field to support independent validation (40 samples). In the other fields, the entire dataset was divided into calibration (70%) and validation (30%) sets. Afterwards, pre-treatment algorithms were applied to enhance the accuracy of the prediction models. These algorithms included removal of the spectral shift at 1045 nm [36], cutting noisy parts at the edges of the spectra, moving average to reduce spectral noise, standard normal variate transformation [37] or normalization, a Savitzky–Golay first derivative and a Savitzky–Golay smoothing. Finally, after performing a principal component analysis (PCA) to investigate the similarities or dissimilarities in the spectra, we developed partial least squares regression (PLSR) models for the prediction of the soil attributes in RStudio version 1.1.463 (RStudio Inc., Boston, MA, USA) with open-source libraries [38].

2.4. *Mapping of Online Measured Soil Properties*

The developed PLSR calibration models were used for estimating the soil attributes using the online collected spectra in the five fields. Then, since the attributes of soil in a field are spatial correlated [39], high-resolution maps of the soil attributes were obtained using ordinary kriging [40]. In ordinary kriging, an estimation of any attribute in any point is given by a linear combination of the available measurements while the weights of the linear combination are obtained from semivariograms [41]. After the kriging interpolation, all attributes were resampled to a common grid of 5 × 5 m and a pair of geographical coordinates was calculated for each of the grid points.

It is worth mentioning that ordinary kriging was adopted in this study since, despite simple kriging, it does not assume prior knowledge of the mean and covariance of the attributes in a land, and hence, it is the most common kriging approach in the literature of management zone delineation [2,16]. Other types of kriging, such as block kriging, universal kriging, and indicator-based kriging, also exist, in which attempts to improve the interpolation performance result in the cost of more computational power demand. However, since the focus of this study was not on kriging, we resorted to the most common approach.

2.5. *Overview of Clustering Algorithms*

In this study, the performance of five clustering algorithms was evaluated in different scenarios. The clustering algorithms used were unsupervised since the data, i.e., the geo-referenced soil attributes, were not labeled. The algorithms have briefly been discussed in the following subsections. In discussing the clustering algorithms, x_i denotes the ith d-dimensional input data, $i \in \{1, \ldots, n\}$, c_i denotes the cluster to which x_i belongs, and μ_j is the centroid of cluster $j \in \{1, \ldots, k\}$ with k being the number of the clusters.

2.5.1. k-Means

k-means divides the n-dimensional data into k categories with the objective to minimize the sum of the within-cluster variances. While simple, it is considered as an efficient clustering algorithms in many data analysis applications. It needs k to be defined and works as follows:

1. Randomly initialize k cluster centroids $\boldsymbol{\mu}_1, \ldots, \boldsymbol{\mu}_k \in \mathbb{R}^d$.
2. For $i \in \{1, \ldots, n\}$, update:

$$c_i = \arg\min_j \left\| \boldsymbol{x}_i - \boldsymbol{\mu}_j \right\|^2 .$$

3. For $j \in \{1, \ldots, k\}$, update $\mu_j =$ centroid of the data of cluster j.
4. Repeat steps 2 and 3 for a specified number of iterations (or until convergence).

2.5.2. Fuzzy C-Means (FCM)

FCM works similarly to k-means. The only difference is that it aims at minimizing the weighted sum of the within-cluster variances. The weights define the clustering fuzziness.

Indeed, FCM does not strictly assign each point to a specific cluster. Instead, the cluster membership is fuzzy. The algorithm works as discussed below.

Parameters: k, m (fuzziness coefficient—a real number greater than 1)

FCM algorithm:

1. Randomly initialize k cluster centroids $\boldsymbol{\mu}_1, \ldots, \boldsymbol{\mu}_k \in \mathbb{R}^d$.
2. For $i \in \{1, \ldots, n\}$ and $j \in \{1, \ldots, k\}$, update:

$$u_{ij} = \frac{1}{\sum_{l=1}^{k} \left(\frac{\|\boldsymbol{x}_i - \boldsymbol{\mu}_j\|}{\|\boldsymbol{x}_i - \boldsymbol{\mu}_l\|} \right)^{\frac{2}{m-1}}}.$$

3. For $j \in \{1, \ldots, k\}$, update:

$$\boldsymbol{\mu}_j = \frac{\sum_{i=1}^{n} u_{ij}^m \boldsymbol{x}_i}{\sum_{i=1}^{n} u_{ij}^m}.$$

4. Repeat steps 2 and 3 for a specified number of iteration (or until convergence).
5. After the algorithm stops, each point i joins the cluster with the highest u_{ij} value.

2.5.3. Mean Shift

Mean shift is a density-based mode-seeking algorithm. It tries to first estimate the density of the data by using a kernel and then looks for the modes of the distribution. In order to find the modes, it iteratively moves each point to its denser neighborhood. Mean shift is a non-parametric clustering algorithm, meaning that it does not need the number of clusters to be specified in advance. Instead, it tries to find the number of clustering according to the density of the data. The algorithm steps are as follows.

Parameters: h (the kernel bandwidth—note that a kernel should be chosen in advance. The mostly used kernel is the Gaussian kernel).

Mean shift algorithm:

1. Initialize seeds set $\mathcal{S}$ for calculating the density

$$f(\boldsymbol{x}) = \sum_{\boldsymbol{x}_i \in \mathcal{S}} K(\boldsymbol{x} - \boldsymbol{x}_i),$$

where $K(.)$ is a kernel function.

2. For each seed $\boldsymbol{s} \in \mathcal{S}$, calculate the mean shift:

$$m(\boldsymbol{s}) = \frac{\sum_{\boldsymbol{x}_i \in \int} K(\boldsymbol{x}_i - \boldsymbol{s})\boldsymbol{x}_i}{\sum_{\boldsymbol{x}_i \in \mathcal{N}(\int) K(\boldsymbol{x}_i - \boldsymbol{s})}},$$

where $N(s)$ denotes the neighborhood of $\boldsymbol{s}$.

3. For each seed $\boldsymbol{s} \in \mathcal{S}$, update $\boldsymbol{s} = m(\boldsymbol{s})$.
4. Repeat steps 2 and 3 for a specified number of iterations (or until convergence).
5. After the algorithm stops, the modes are considered as the centroids of the clusters, and each point joins to the closest mode.

2.5.4. Hierarchical Clustering

The hierarchical clustering algorithm seeks to build a hierarchy of clusters based on either of two approaches: agglomerative or divisive. Agglomerative is a bottom–up approach based on which pairs of clusters are merged together in order to build up the hierarchy. On the other hand, the divisive approach is a top–down method that starts from one cluster including all data and then splits the cluster recursively. In this paper, the agglomerative hierarchical clustering was adopted. The algorithm receives k as the input parameter and works as follows:

1. Assign all points an individual cluster number.
2. Merge points with the smallest distance. In other words, points with smallest distance join the same cluster.
3. Repeat step 2 until k clusters are obtained.

2.5.5. Density-Based Spatial Clustering of Applications with Noise (DBSCAN)

DBSCAN, similar to mean shift, is a non-parametric clustering algorithm. In other words, the number of clusters, k, does not need to be specified for it. Instead, it reaches a number of clusters based on the density of data and two parameters discussed below.

Parameters: ϵ (the neighborhood distance), m_p (minimum number of data points to define a cluster)

DBSCAN algorithm:

1. Select a random data point.
2. If the number of the neighbors is less than m_p, the point is marked as an outlier with label -1.
3. If the number of the neighbors is more than or equal to m_p, the point and its neighbors establish a cluster.
4. Repeat step 3 for all points within the established cluster. In other words, for all joined points, check their neighbor points and join their neighbors to the established cluster.
5. From the remaining points that have not yet been met, select a random data point. Repeat steps 2 to 5 until all data points are met.

2.6. Feature Selection (Data Decrease)

Among the soil attributes used for MZ delineation, OC and MC have direct signatures in the spectral characteristics in the vis-NIR range, while other attributes are estimated indirectly based on their correlation with MC and OC [19]. Applying correlated features to clustering models imposes computational burden and might degrade the clustering quality. Accordingly, we studied the effect of feature selection by cross-correlation analysis. To this end, the cross-correlation of the data was computed using Pearson correlation in order to remove highly correlated layers. The removal of highly correlated data before clustering in MZ delineation applications has also been recommended by [16,42].

2.7. Clustering Scenarios

We evaluated the above-mentioned clustering algorithms in different scenarios listed in Table 2. In all scenarios, except in kmeans-nn-nc, data are normalized. For normalization, the data ranges were scaled into an interval between 0 and 1. In other words, range normalization was applied to the data, since it has been shown to be the most effective normalization method in MZ delineation applications [16].

2.8. Evaluation of Clustering Algorithms

Since the clustering algorithms used in this study were unsupervised, it was not possible to evaluate their performances by comparing the clustering results against true labels. Instead, there exist heuristic metrics to assess the quality of unsupervised clustering; however, these metrics do not measure the validity of the model's predictions. In order to choose the most appropriate clustering scheme for MZ delineation, we adopted three metrics, namely, Davies–Bouldin index (DBI) [43], Silhouette index [44], and variation reduction index (VRI) [16,45].

Considering k clusters, DBI is computed by [43]:

$$DBI = \frac{1}{k}\sum_{i=1}^{k} \max \frac{s_i + s_j}{d_{ij}}, \quad (1)$$

where s_i is the average distance of the data of cluster i from its center and d_{ij} denotes the distance between the centers of cluster i and cluster j. In fact, the intuition of DBI is that the clustering schemes with denser clusters that are further from each other are better. DBI is always positive, and its fewer values indicate better clustering quality and vice versa.

Table 2. The clustering scenarios evaluated in this study.

Clustering Scenario	Clustering Method and the Conditions of Its Input Data
kmeans-nn-nc [1]	*k*-means, no data normalization, xy coordinate data not considered
kmeans-wn-nc	*k*-means, with data normalization, xy coordinate data not considered
kmeans-wn-wc	*k*-means, with data normalization, with xy coordinate data
kmeans-nc-dec	*k*-means, xy coordinate data not considered, data decreased
kmeans-wc-dec	*k*-means, with xy coordinate data, data decreased
FCM-wn-nc	FCM, with data normalization, xy coordinate data not considered
FCM-wn-wc	FCM, with data normalization, with xy coordinate data
FCM-nc-dec	FCM, xy coordinate data not considered, data decreased
FCM-wc-dec	FCM, with xy coordinate data, data decreased
MS-wn-nc	Mean shift, with data normalization, xy coordinate data not considered
MS-wn-wc	Mean shift, with data normalization, with xy coordinate data
MS-wc-dec	Mean shift, with xy coordinate data, data decreased
MS-nc-dec	Mean shift, xy coordinate data not considered, data decreased
hier-wn-nc	Hierarchical, with data normalization, xy coordinate data not considered
hierarchical-wn-wc	Hierarchical, with data normalization, with xy coordinate data
hierarchical-nc-dec	Hierarchical, xy coordinate data not considered, data decreased
hierarchical-wc-dec	Hierarchical, with xy coordinate data, data decreased
DBSCAN-wn-nc	DBSCAN, with data normalization, xy coordinate data not considered
DBSCAN-wn-wc	DBSCAN, with data normalization, with xy coordinate data
DBSCAN-wc-dec	DBSCAN, with xy coordinate data, data decreased
DBSCAN-nc-dec	DBSCAN, xy coordinate data not considered, data decreased

[1] nn: no normalization; wn: with normalization; nc: no coordinate; wc: with coordinate; dec: decreased data; xy coordinate: cartesian coordinate.

The Silhouette index quantifies the clustering quality by defining how well each data point has been assigned to its own cluster. Considering k clusters, the Silhouette index for data i is given by [44]:

$$s(i) = \frac{b(i) - a(i)}{\max\{a(i), b(i)\}}, \tag{2}$$

where $a(i) = \frac{1}{|C_i|-1}\sum_{j\in C_i, j\neq i} d(i,j)$ and $b(i) = \min_{l\neq i} \frac{1}{|C_l|}\sum_{j\in C_l} d(i,j)$, respectively, indicate the similarity of data i to its own cluster and its dissimilarity to other clusters with $|C_l|$ and $d(i,j)$ being the size of cluster l and the distance between data i and j, respectively. The overall Silhouette index is computed by averaging the indices of all data. Silhouette values range between -1 and 1 with more values indicating better clustering quality and vice versa.

DBI and Silhouette indices have emerged from the machine learning (ML) context and are applicable on any application, including those in precision agriculture. We also adopted VRI introduced by [45] specifically for evaluating the quality of MZ delineation. The VRI rational is that the variance of the soil attributes within the MZs should be less than their overall variance. The more the within-cluster variances decrease, the better the clustering quality. The VRI for soil attribute θ is given by [45]:

$$VRI_\theta = \left(1 - \frac{\sum_{i=1}^{k} A_i v_{C_i}^{\theta}}{v^\theta}\right) \times 100\%,$$

in which A_i is the proportion of the area covered by cluster i, v^θ is the overall variance of the soil attribute θ, and $v_{C_i}^{\theta}$ denotes its variance within cluster i. More values of VRI implicate better MZ delineation quality and vice versa. In this paper, the overall VRI is obtained by averaging the VRI of all studied soil attributes.

2.9. Clustering and Smoothing Pipeline (CaSP) for Management Zone Delineation

After evaluation of different clustering methods in different scenarios, the most appropriate method was specified. Nevertheless, the clustering results usually include small parts of a cluster located apart within another cluster. Let us refer to these small parts as islands, since this is what they really look like. On the other hand, the clustering results are used in form of a recommendation map for variable-rate application of farming inputs such as fertilizers [8], manure [46], and seeds [47], for which the actuators of the agricultural machines cannot respond to the small modification needed in the applied rate due to these small islands. Therefore, it was necessary to introduce an additional smoothing step to the MZ maps in order to make them appropriate for variable-rate implementation in practice, as shown in Figure 4. To this end, we used the DBSCAN algorithm and applied it to just the geographic coordinates of the data points in each cluster, since DBSCAN is in nature capable of discovering the islands and single apart data points (as outliers) [48], as discussed in Section 2.5. The pseudocode of the smoothing algorithm has been shown in Algorithm 1. This algorithm takes as input the geographic locations of the data in a Cartesian system (i.e., xy coordinate) together with their corresponding labels and the island size. The xy coordinates of the data of each cluster are clustered by DBSCAN. Then, the outliers will join to the cluster of their closest data point. If more than one cluster is obtained, it means that the cluster includes two or more separate parts. Then, the algorithm joins the parts smaller than island size to another cluster according to the majority rule. More specifically, the algorithm first finds the edge points of the island area. Then, the label of the island data is changed to the majority of the labels of the neighbors of the edge points. The overall clustering and smoothing pipeline (CaSP) for MZ delineation is shown in Figure 4.

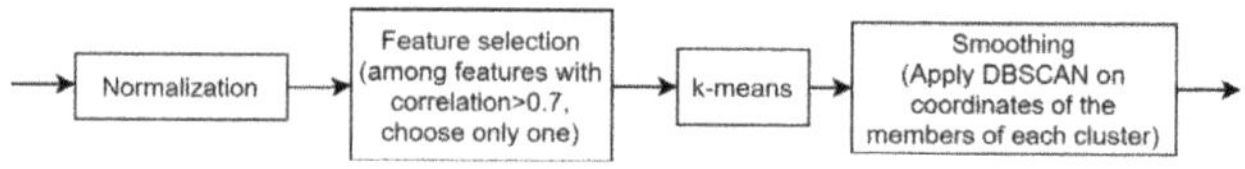

Figure 4. The clustering and smoothing pipeline (CaSP) for the management zone delineation approach of the current work.

Algorithm 1: Smoothing algorithm

```
Data: xy, label, island_size
Result: Smoothed clustering scheme
for cluster C_i, i ∈ {1,...,k} do
    intra_label = DBSCAN(xy ∈ C_i);
    for xy with intra_label = −1 (i.e., for outliers) do
        Join to the cluster to which the closest data point belongs
    end
    if max(intral_label) > 0 (i.e., if more than one cluster has obtained) then
        for each obtained cluster do
            if the cluster size < island_size then
                Find the edge points of the island;
                ℒ ← the labels of the neighbors of the edge points;
                label_majority ←majority of ℒ;
                Change the label of island points to label_majority;
            end
        end
    end
end
```

3. Results and Discussion

3.1. Evaluation of Clustering Algorithms

Mean shift and DBSCAN are non-parametric methods, meaning that the number of MZs is not needed to be specified. However, we evaluated the performance of k-means, FCM, and hierarchical methods for dividing the fields into four MZs (i.e., $k = 4$). For FCM, in addition to k, the fuzziness coefficient m needed to be specified. We used $m = 2$, since this value has been suggested in most applications [49]. Mean shift needed just the kernel bandwidth h to be set, for which we used the related function in the Sklearn package of Python (version 1.0.2) [50]. This function gave a value around $h = 0.41$. For DBSCAN, we considered $m_p = 3$ and set ϵ as the mean of the spatial scanning resolution.

The evaluation results of the clustering algorithms in the studied scenarios have been listed in Table 3. Recall that higher values of the Silhouette index and VRI and smaller DBI values indicate better clustering performance. According to their definitions in Equations (1) and (2), DBI and the Silhouette index are not defined for cases with just one cluster, for which VRI is zero. In the studied scenarios, mean shift and DBSCAN gave just one cluster in some cases, whose indices have been specified by n.v. (no value) in Table 3.

Table 3. Evaluation of clustering methods including k-means, fuzzy C-means (FCM), mean shift (MS), hierarchical, and density-based spatial clustering of applications with noise (DBSCAN) in terms of Davies–Doublin index (DBI), Silhouette index (Sil.), and variance reduction index (VRI).

Field Name	Krokey			Kouter			Grooteland			Beers			Fabrieke		
Score	**DBI**	**Sil.**	**VRI**	**DBI**	**Sil.**	**VRI**	**DBI**	**Sil.**	**VRI**	**DBI**	**Sil.**	**VRI**	**DBI**	**Sil.**	**VRI**
km. [1]-nn-nc	0.51	0.56	35.65	0.55	0.53	12.25	0.67	0.45	33.81	0.56	0.52	28.27	0.75	0.31	3.56
km.-wn-nc	1.65	0.19	44.37	1.55	0.25	33.42	1.40	0.21	47.82	1.33	0.25	45.35	1.43	0.20	23.56
km.-wn-wc	1.53	0.23	44.59	1.53	0.21	27.47	1.40	0.24	43.24	1.32	0.26	46.92	1.57	0.24	30.32
km.-nc-dec	1.51	0.21	40.33	1.33	0.27	33.31	1.45	0.20	45.86	1.36	0.24	45.19	1.21	0.31	48.50
km.-wc-dec	1.52	0.23	40.98	1.61	0.21	26.55	1.38	0.24	41.57	1.29	0.27	47.37	1.59	0.22	28.40
FCM-wn-nc	1.72	0.15	40.68	2.27	0.16	24.79	2.01	0.15	42.93	1.47	0.22	44.13	1.55	0.19	55.89
FCM-wn-wc	1.75	0.17	42.78	1.54	0.20	26.87	1.46	0.24	42.73	1.37	0.24	45.58	1.40	0.23	57.98
FCM-nc-dec	1.94	0.13	36.40	2.28	0.17	21.46	1.77	0.16	42.32	1.40	0.23	45.00	1.99	0.19	19.97
FCM-wc-dec	1.78	0.19	38.30	1.63	0.19	25.29	1.42	0.23	41.18	1.31	0.26	46.80	1.67	0.17	10.56
MS-wn-nc	1.44	0.28	21.40	1.63	0.20	25.29	1.42	0.23	41.18	1.31	0.25	46.80	1.40	0.23	58.28
MS-wn-wc	1.62	0.25	19.79	3.45	0.07	24.89	1.43	0.26	30.28	1.41	0.23	39.73	1.00	0.32	38.30
MS-wc-dec	1.37	0.29	21.72	1.47	0.02	20.75	1.69	0.22	21.45	1.32	0.28	37.78	1.43	0.10	24.67
MS-nc-dec	n.v.	n.v.	0	1.05	0.15	9.56	1.43	0.23	27.16	1.67	0.23	21.92	0.93	0.16	34.23
hier. [2]-wn-nc	1.86	0.15	39.18	2.42	0.24	16.88	2.06	0.16	36.90	1.70	0.22	37.86	1.25	0.25	55.18
hier.-wn-wc	1.65	0.22	41.84	2.17	0.18	18.18	1.65	0.23	36.65	1.51	0.22	40.16	1.30	0.27	55.58
hier.-nc-dec	1.78	0.18	35.17	1.37	0.25	31.02	1.67	0.17	41.65	1.44	0.21	41.58	1.43	0.23	29.04
hier.-wc-dec	1.53	0.22	38.20	1.52	0.23	29.67	1.38	0.22	38.91	1.38	0.25	45.43	1.49	0.21	28.58
DBS. [3]-wn-nc	1.34	-0.23	0.93	2.34	0.13	10.30	n.v.	n.v.	0	n.v.	n.v.	0	2.76	0.16	5.77
DBS.-wn-wc	1.45	-0.27	0.92	2.76	0.01	7.54	4.27	0.08	1.84	3.05	0.07	7.90	2.70	0.12	5.59
DBS.-wc-dec	2.64	0.17	0.07	1.99	0.20	10.88	1.28	0.20	0.47	0.85	0.06	1.10	2.21	0.18	14.29
DBS.-nc-dec	2.13	0.25	0.08	1.82	0.26	11.84	1.72	0.00	1.94	2.42	-0.1	4.43	2.03	0.23	14.11

[1] kmeans. [2] hierarchical. [3] DBSCAN.

The data were normalized in all scenarios except in kmeans-nn-nc. Although DBI and the Silhouette index suggest this scenario as the best, its VRI values are poor. As will be elaborated later, normalization is essential in clustering, since ignoring it will weigh more on the data with large values and neglect data with small values (e.g., weighs more on Ca and neglects OC and pH). Accordingly, the first scenario (e.g., no normalization) is not recommended. Among other scenarios, VRI has always suggested to use k-means. The Silhouette index agrees with VRI in Kouter, Beers, and Fabrieke, while it suggests mean shift for Krokey and Grooteland. On the other hand, DBI has the best values for either mean shifting or DBSCAN.

A close examination is needed into the MZs obtained by different scenarios in order to conclude on the format of the input data, i.e., normalizing or not, inclusion or exclusion of the coordinates, and whether or not to apply feature selection. For the sake of brevity, we describe the MZs results and the yield map for just field Krokey—as shown in Figure 5—since this field has more variability compared to other fields. As seen, DBSCAN gave just one cluster with several points as outliers. Indeed, DBSCAN performs clustering based on density and considers a cluster as unique as long as it is dense. Density-based clustering algorithms are helpful in object detection in machine vision applications. Since the focus in PA applications is on within-field variability, DBSCAN does not perform desirably and has to be excluded. This argument applies to mean shift as well and also has to be excluded for variable rate applications (Figure 5). Compared to DBSCAN, mean shift has two more disadvantages: (1) it demands a high computational power and (2) it is very sensitive to its parameter h.

It is worth noting that as in case of Krokey, DBSCAN and mean shift are shown to be not suitable for the other three fields of this study due to the same conclusion drawn above. FCM works similar to k-means but needs an additional hyperparameter as the fuzziness coefficient to be specified while there is no clue on how to set this parameter for variable rate applications. This makes k-means as the best method for clustering, whose results need to be analyzed further. Among the k-means results, k-means with no coordinate data (i.e., just using the soil attributes) after feature selection (i.e., kmeans-nc-dec) visually shows more correlation with the yield map, and this observation was also confirmed by the farmer. Revisiting Table 3 indicates that the VRI and Silhouette indices of kmeans-nc-dec are, if not maximum, among the highest values throughout all the five study fields. Its DBI has also the small values compared to other scenarios. In the Fabrieke case, after excluding kmeans-nn-nc, all the three indices suggest using kmeans-nc-dec. In Kouter, both DBI and the Silhouette index suggest kmeans-nc-dec, for which VRI is also very close to its maximum value among k-means scenarios with normalized inputs. The values of the indices for the k-means scenarios are very close to each other in the other two fields, which allow drawing the same conclusion as that of Krokey that Kmeans-nc-dec is the best performing scenario. This is particularly true as this scenario has performed much better than all the other k-means solutions in Fabrieke. However, the kmeans-wn-nc shows comparable performance indicators to that of the kmeans-nc-dec, although its performance deteriorated significantly for Fabrieke. This is the reason why it is concluded that MZ delineation using Kmeans-nc-dec scenario would provide the most stable solution in general.

From the computational point of view, kmeans-nc-dec is also optimal, since excluding coordinates decreases the data dimension, while the dimension is further reduced after feature selection. It is worth noting that since some of the soil attributes are spatially correlated (e.g., Ca and CEC), including spatial data in clustering is generally recommended, as they were explicitly considered in similar applications [9,25,27–30]. Since we used the raster data after spatial interpolation using kriging, the spatial correlation has been taken into consideration once, so that reconsidering it one more time means overweighing the spatial information.

Figure 6 demonstrates the scatter plot of yield vs. the soil attributes and crop NDVI in selected clustering scenarios shown for field Krokey, as an example. Figure 6a highlights the importance of normalizing the data. Since the clustering algorithms use (Euclidean) distance for establishing clusters, the data should be normalized in order for all soil attributes to impact the distance equally; otherwise, only the data with large values will produce the clustering outcome. As seen in Figure 6a, the clustering outcome was affected mainly by Ca variability, because the Ca values were much larger than the other attributes (Figure 6a). Its great overlap between the four clusters is depicted for the other soil attributes. The scatter plots in the other scenarios depict that clustering has taken all the input variables equally since they were normalized; however, each scenario gave different clustering outcomes, which is attributed to applying different algorithms on the data. After normalization, the overlap between the four different clusters was greatly reduced, and a clear separation of classes can be observed not only for Ca but also for the other soil attributes.

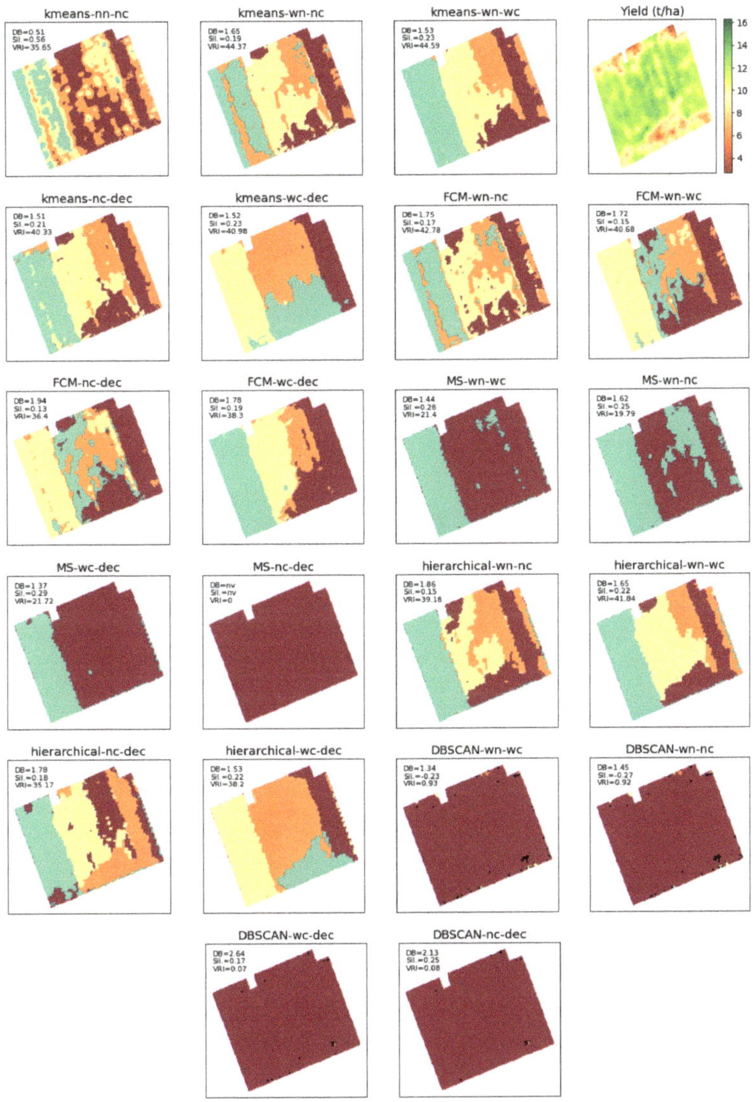

Figure 5. Comparison of different clustering schemes in delineation of management zones, shown for field Krokey as an example. When not explicitly mentioned in the scheme title, normalization is included. The clustering performances are evaluated in terms of Davies–Bouldin score (DB), Silhouette score (Sil.), and variance reduction index (VRI).

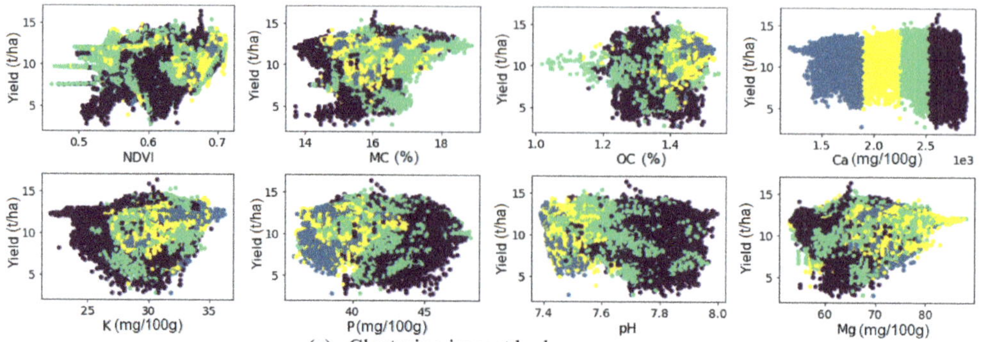

(a) Clustering impact by kmeans-nn-nc

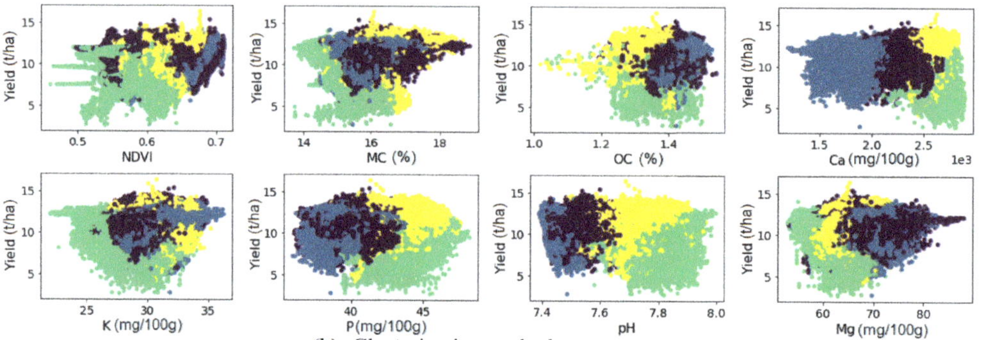

(b) Clustering impact by kmeans-wn-nc

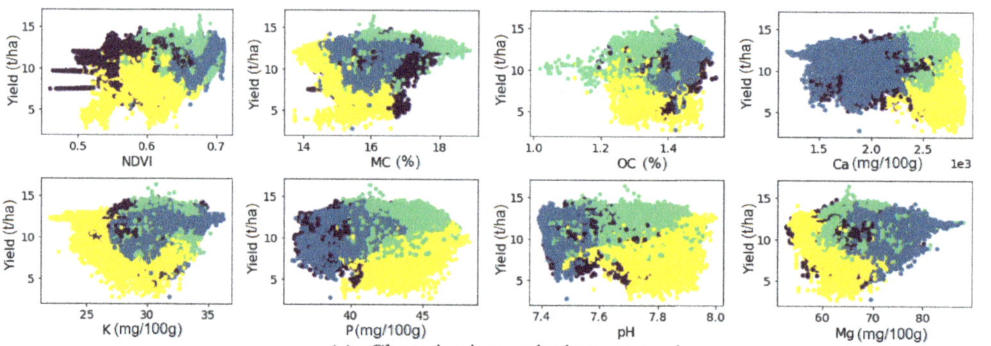

(c) Clustering impact by kmeans-nc-dec

Figure 6. *Cont.*

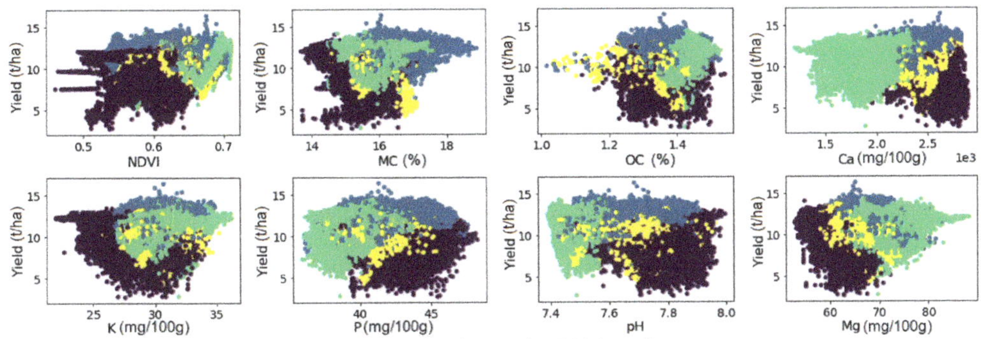

(d) Clustering impact by FCM-nc-dec

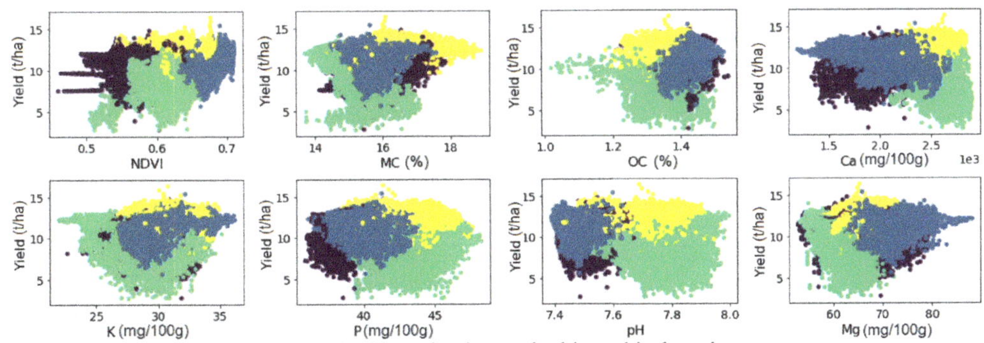

(e) Clustering impact by hierarchical-nc-dec

Figure 6. The impact of the clustering schemes on the scatter plot of yield vs. NDVI, and other soil attributes.

3.2. Evaluation of MZ Delineation by CaSP

Feature selection by covariance analysis has been recommended by [16,42]. Figure 7 shows the cross-correlation matrix of the soil attributes for field Krokey as an example. It can be observed that there were high correlations (>0.7) among the soil Ca, CEC, Na, and pH, except between pH and Na, where a correlation of 0.48 was observed. Therefore, Ca, Na, and pH were removed from the analysis, since CEC was given a higher priority from a soil fertility perspective [51,52].

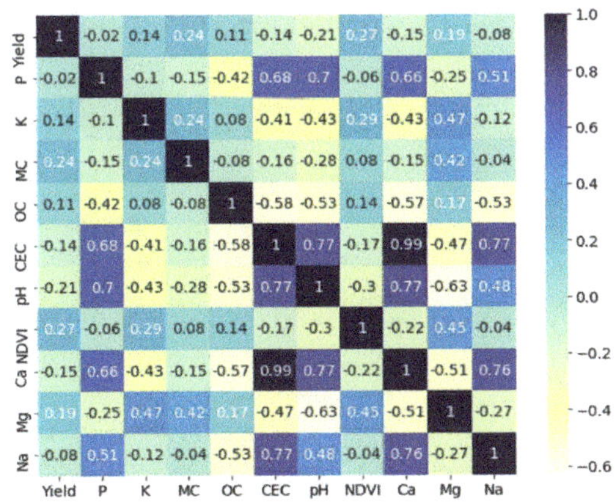

Figure 7. The cross-correlation (Pearson correlation) matrix of the soil attributes in field Krokey.

The clustering results obtained by the proposed CaSP MZ delineation scheme has been illustrated in Figure 8, where MZ maps delineated with and without smoothing were compared. As can be seen, smoothing has appropriately filtered small apart areas (islands), which cannot be accounted for in practice during variable rate applications. While varying the rate of farming inputs using agricultural machinery equipped with PA control-enabling technology, it is necessary that the size of the machinery active control unit is smaller or equal to the smallest islands, i.e., they have to support high-resolution control, which is costly. In case the islands are smaller than the agricultural machinery size, then the variable rate will not be implementable in practice. The filter designed in this study is flexible and allows the filtering of islands with different sizes (Figure 4). However, removing very large-size islands can lead to ignoring important fertility zones in the field, for which the agricultural machine can respond to correctly during field operations. This feature allows for the implementation of different agricultural machinery during variable rate applications, each of which would require smoothing islands of different sizes to match the size of the machine to be used.

As seen in Figure 8, the MZ map given by *k*-means for Krokey shows an appropriate but partial visual correlation with the yield map. Interestingly, the MZ map was able to capture the fringe lane by the road (the lane in the right side of the MZ map). While smoothing the map has kept the similarity with the yield map, it has removed the very small island parts. The MZ maps of field Kouter with smaller measurement resolution also show a very good similarity with the yield map, which is a similarity that was better than that of Krokey field. In case of Grooteland, the MZ maps indicate high variability within the field, while the yield seems to be almost uniformly distributed over the field area. This is due to enforcing the algorithm to divide the field into four MZs, so that soil fertility attributes have the major contribution on the MZ map. However, exploring the MZ maps in more detail shows it still has some indicative correlations with the yield. Specifically, the management zone indicated by the beige color has captured the low-fertility zones of the field, and these are well correlated with areas with low values in the yield map. Here, smoothing was effective in making the MZ map more suitable for practical application. In the Beers and Fabrieke cases, there exists a good visual correlation between the MZ maps and yield. Specially, there was a low-fertility zone in the central region of the Beers field, which has been captured very well by CaSP. According to the above and when high-sampling resolution data on soil and crop are considered, we recommend the CaSP based on *k*-means clustering for automatic delineation of MZ maps for the deployment of variable rate applications of farming inputs. The ideal solution should have the following successive steps of data processing: (1) range normalization, (2) feature selection based on cross-correlation analysis, (3) *k*-means clustering, and (4) smoothing by DBSCAN.

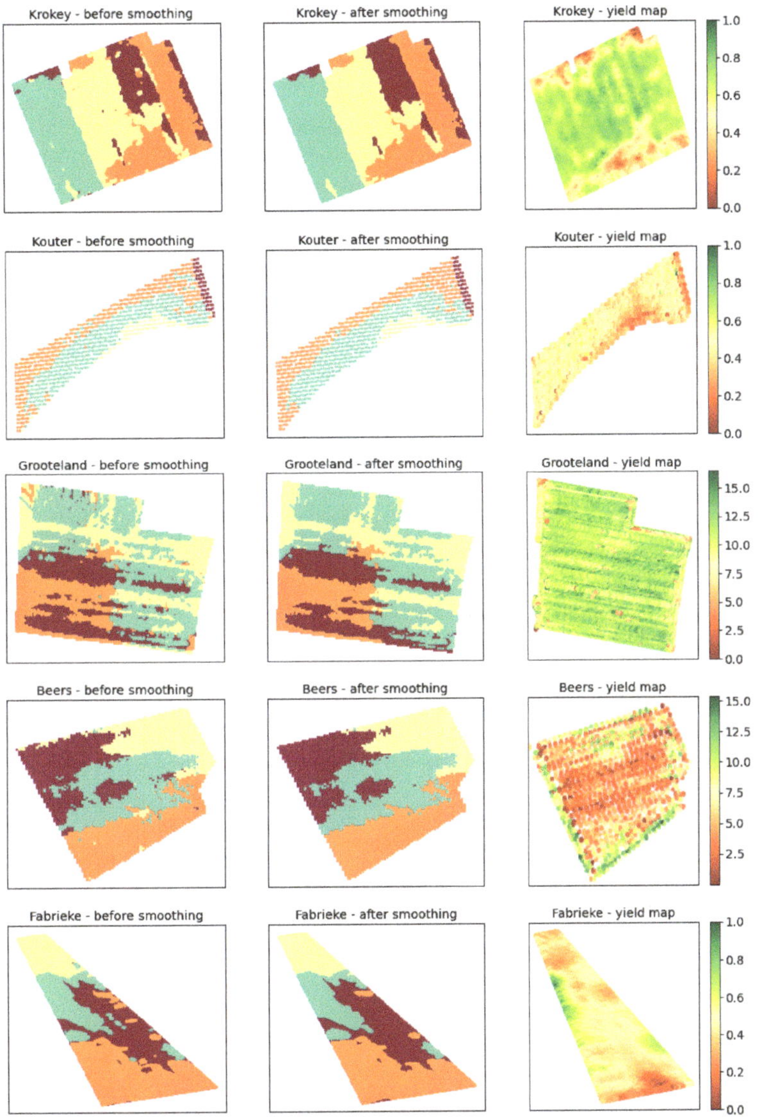

Figure 8. The clustering results before and after smoothing and comparing them with the yield maps in the study fields. The clustering schemes in the middle column are the outcome of management zone delineation by clustering and smoothing pipeline (CaSP).

4. Conclusions

In this paper, five clustering algorithms were evaluated in different scenarios for MZ delineation in five arable farming fields, with the intention to evaluate their suitability for variable rate applications. The clustering algorithms included *k*-means, FCM, mean shift, hierarchical, and DBSCAN. These algorithms were evaluated in scenarios with and without range normalization, geographical coordinates, and feature selection. On-line measured soil fertility attributes (pH, Ca, Mg, Na, P, CEC, MC, K, and OC) at high sampling resolution, along with crop NDVI and the yield data, were used as input to the clustering algorithms. Spatial interpolation using ordinary kriging was carried out in order to get high-resolution data.

The results suggested *k*-means as the optimal clustering algorithm after normalizing and exclusion of the GPS coordinates. Nevertheless, it was noted that the coordinates should be ignored, since the spatial correlations of the data had been previously considered when the data were interpolated using ordinary kriging. In general, if the data are not interpolated using any spatial interpolation algorithms, it is recommended to include the coordinates in order to account for the spatial correlation among soil attributes. Furthermore, it was concluded that feature selection optimized after cross-correlation analysis improves the MZ delineation quality while reducing computational burden.

Moreover, a smoothing algorithm was proposed based on DBSCAN for filtering out small areas of a cluster within other clusters. Overall, an MZ delineation pipeline was proposed including the following steps: (1) range normalization, (2) feature selection by cross-correlation analysis, (3) *k*-means clustering, and (4) smoothing. The effectiveness of this pipeline—to which we referred to as CaSP, standing for clustering and smoothing pipeline—was demonstrated by the practical application of MZs concerning the machinery size during variable rate applications. Future study directions may include analysis of the effect of the accuracy of soil attributes predictions on clustering quality and also improving the efficiency of the feature selection operator.

Author Contributions: Conceptualization, S.H.J.; methodology, S.H.J.; software, S.H.J.; investigation, S.H.J.; resources, S.H.J. and A.G.; writing—original draft preparation, S.H.J.; writing—review and editing, A.G. and A.M.M.; supervision, A.M.M. All authors have read and agreed to the published version of the manuscript.

Funding: This research was jointly funded by Research Foundation Flanders (FWO) for the Odysseus I SiTeMan Project (Nr. G0F9216N) and the European Commission (Horizon 2020) for SIEUSOIL project (No. 818346).

Institutional Review Board Statement: Not applicable.

Informed Consent Statement: Not applicable.

Conflicts of Interest: The authors declare no conflict of interest.

Abbreviations

The following abbreviations are used in this manuscript:

Ca	Calcium
CaSP	Clustering and smoothing platform
CEC	Cation exchange capacity
DBI	Davies–Bouldin index
DBSCAN	Density-based spatial clustering of applications with noise
DGPS	Differential global positioning system
FCM	Fuzzy C-means
K	Exchangeable potassium
MC	Moisture content
Mg	Magnesium
MS	Mean-shift
MZ	Management zone
Na	Sodium
NDVI	Normalized difference vegetation index
OC	Organic carbon
P	Exchangeable phosphorous
PA	Precision agriculture
Sil.	Silhouette
vis-NIR	Visible-near-infrared
VRI	Variance reduction index

References

1. Shanahan, J.F.; Kitchen, N.R.; Raun, W.R.; Schepers, J.S. Responsive in-season nitrogen management for cereals. *Comput. Electron. Agric.* **2008**, *61*, 51–62. [CrossRef]
2. Nawar, S.; Corstanje, R.; Halcro, G.; Mulla, D.; Mouazen, A.M. *Chapter Four-Delineation of Soil Management Zones for Variable-Rate Fertilization: A Review*; Advances in Agronomy; Academic Press: Cambridge, MA, USA, 2017; Volume 143, pp. 175–245. [CrossRef]
3. Guerrero, A.; De Neve, S.; Mouazen, A.M. *Current Sensor Technologies for In Situ and On-Line Measurement of Soil Nitrogen for Variable Rate Fertilization: A Review*; Academic Press: Cambridge, MA, USA, 2021. [CrossRef]
4. Pantazi, X.; Moshou, D.; Mouazen, A.; Alexandridis, T.; Kuang, B. Data fusion of proximal soil sensing and remote crop sensing for the delineation of management zones in arable crop precision farming. In Proceedings of the CEUR Workshop, Kavala, Greece, 17–20 September 2015; pp. 765–776.
5. Haghverdi, A.; Leib, B.G.; Washington-Allen, R.A.; Ayers, P.D.; Buschermohle, M.J. Perspectives on delineating management zones for variable rate irrigation. *Comput. Electron. Agric.* **2015**, *117*, 154–167. [CrossRef]
6. Vrindts, E.; Mouazen, A.M.; Reyniers, M.; Maertens, K.; Maleki, M.R.; Ramon, H.; De Baerdemaeker, J. Management Zones based on Correlation between Soil Compaction, Yield and Crop Data. *Biosyst. Eng.* **2005**, *92*, 419–428. [CrossRef]
7. Doerge, T. *Management Zone Concepts*; Potash and Phosphate Institute: Norcross, GA, USA, 2000.
8. Guerrero, A.; De Neve, S.; Mouazen, A.M. Data fusion approach for map-based variable-rate nitrogen fertilization in barley and wheat. *Soil Tillage Res.* **2021**, *205*, 104789. [CrossRef]
9. De Benedetto, D.; Castrignanò, A.; Rinaldi, M.; Ruggieri, S.; Santoro, F.; Figorito, B.; Gualano, S.; Diacono, M.; Tamborrino, R. An approach for delineating homogeneous zones by using multi-sensor data. *Geoderma* **2013**, *199*, 117–127. [CrossRef]
10. Fleming, K.; Westfall, D.; Wiens, D.; Brondahl, M.C. Evaluating Farmer Defined Management Zone Maps for Variable Rate Fertilizer Application. *Precis. Agric.* **2000**, *2*, 201–215. [CrossRef]
11. Mouazen, A.M. *Soil Survey Device*; International Publication Published under the Patent Cooperation Treaty (PCT); World Intellectual Property Organization, International Bureau: Brussels, Belgium, 2006; International Publication Number: WO2006/015463; PCT/BE2005/000129; IPC: G01N21/00; G01N21/00.
12. Nawar, S.; Cipullo, S.; Douglas, R.K.; Coulon, F.; Mouazen, A.M. The applicability of spectroscopy methods for estimating potentially toxic elements in soils: State-of-the-art and future trends. *Appl. Spectrosc. Rev.* **2019**, *55*, 1–33. [CrossRef]
13. Javadi, S.H.; Mouazen, A.M. Data Fusion of XRF and Vis-NIR Using Outer Product Analysis, Granger–Ramanathan, and Least Squares for Prediction of Key Soil Attributes. *Remote Sens.* **2021**, *13*, 2023. [CrossRef]
14. Javadi, S.; Mohammadi, A. Plackett fusion of correlated decisions. *AEU-Int. J. Electron. Commun.* **2019**, *99*, 341–346. [CrossRef]
15. Javadi, S.; Mohammadi, A. Fire detection by fusing correlated measurements. *J. Ambient. Intell. Humaniz. Comput.* **2017**. [CrossRef]
16. Schenatto, K.; de Souza, E.G.; Bazzi, C.L.; Gavioli, A.; Betzek, N.M.; Beneduzzi, H.M. Normalization of data for delineating management zones. *Comput. Electron. Agric.* **2017**, *143*, 238–248. [CrossRef]
17. Song, X.; Wang, J.; Huang, W.; Liu, L.; Yan, G.; Pu, R. The delineation of agricultural management zones with high resolution remotely sensed data. *Precis. Agric.* **2009**, *10*, 471–487. [CrossRef]
18. Kodaira, M.; Shibusawa, S. Mobile Proximal Sensing with Visible and Near Infrared Spectroscopy for Digital Soil Mapping. *Soil Syst.* **2020**, *4*, 40. [CrossRef]
19. Mouazen, A.M.; Baerdemaeker, J.D.; Ramon, H. Effect of Wavelength Range on the Measurement Accuracy of Some Selected Soil Constituents Using Visual-Near Infrared Spectroscopy. *J. Near Infrared Spectrosc.* **2006**, *14*, 189–199. [CrossRef]
20. Janrao, P.; Palivela, H. Management zone delineation in Precision agriculture using data mining: A review. In Proceedings of the 2015 International Conference on Innovations in Information, Embedded and Communication Systems (ICIIECS), Coimbatore, India, 19–20 March 2015; pp. 1–7. [CrossRef]
21. Fu, Q.; Wang, Z.; Jiang, Q. Delineating soil nutrient management zones based on fuzzy clustering optimized by PSO. *Math. Comput. Model.* **2010**, *51*, 1299–1305. [CrossRef]
22. Janrao, P.; Mishra, D.; Bharadi, V. Clustering Approaches for Management Zone Delineation in Precision Agriculture for Small Farms. In Proceedings of the International Conference on Sustainable Computing in Science, Technology and Management (SUSCOM), Jaipur, India, 26–28 February 2019.
23. Karkra, R.; Kaur, S.; Kaur, M.; Sharma, R.; Upadhyay, R.R. Management zone delineation in precision agriculture using machine learning algorithms. *J. Nat. Remedies* **2020**, *21*, 22–29.
24. Javadi, S.H.; Guerrero, A.; Mouazen, A.M. Source localization in resource-constrained sensor networks based on deep learning. *Neural Comput. Appl.* **2020**, *33*, 4217–4228. [CrossRef]
25. Aggelopooulou, K.; Castrignanò, A.; Gemtos, T.; Benedetto, D.D. Delineation of management zones in an apple orchard in Greece using a multivariate approach. *Comput. Electron. Agric.* **2013**, *90*, 119–130. [CrossRef]
26. Landrum, C.; Castrignanò, A.; Mueller, T.; Zourarakis, D.; Zhu, J.; De Benedetto, D. An approach for delineating homogeneous within-field zones using proximal sensing and multivariate geostatistics. *Agric. Water Manag.* **2015**, *147*, 144–153. [CrossRef]
27. Cordero, E.; Longchamps, L.; Khosla, R.; Sacco, D. Joint measurements of NDVI and crop production data-set related to combination of management zones delineation and nitrogen fertilisation levels. *Data Brief* **2020**, *28*, 104968. [CrossRef]
28. Gavioli, A.; de Souza, E.G.; Bazzi, C.L.; Schenatto, K.; Betzek, N.M. Identification of management zones in precision agriculture: An evaluation of alternative cluster analysis methods. *Biosyst. Eng.* **2019**, *181*, 86–102. [CrossRef]

29. Paccioretti, P.; Córdoba, M.; Balzarini, M. FastMapping: Software to create field maps and identify management zones in precision agriculture. *Comput. Electron. Agric.* **2020**, *175*, 105556. [CrossRef]
30. Li, X.; Pan, Y.-C.; Zhong, Q.; Zhao, C.-J. Delineation and Scale Effect of Precision Agriculture Management Zones Using Yield Monitor Data over Four Years. *Agric. Sci. China* **2007**, *6*, 180–188. [CrossRef]
31. Nawar, S.; Abdul Munnaf, M.; Mouazen, A.M. Machine Learning Based On-Line Prediction of Soil Organic Carbon after Removal of Soil Moisture Effect. *Remote Sens.* **2020**, *12*, 1308. [CrossRef]
32. Kim, J.; Grunwald, S.; Rivero, R.G. Soil Phosphorus and Nitrogen Predictions Across Spatial Escalating Scales in an Aquatic Ecosystem Using Remote Sensing Images. *IEEE Trans. Geosci. Remote Sens.* **2014**, *52*, 6724–6737. [CrossRef]
33. Ji, F.; Meng, J.; Cheng, Z.; Fang, H.; Wang, Y. Crop Yield Estimation at Field Scales by Assimilating Time Series of Sentinel-2 Data Into a Modified CASA-WOFOST Coupled Model. *IEEE Trans. Geosci. Remote Sens.* **2021**, *60*, 4400914. [CrossRef]
34. Javadi, S.H.; Munnaf, M.A.; Mouazen, A.M. Fusion of Vis-NIR and XRF spectra for estimation of key soil attributes. *Geoderma* **2021**, *385*, 114851. [CrossRef]
35. Tavares, T.R.; Molin, J.P.; Javadi, S.H.; Carvalho, H.W.; Mouazen, A.M. Combined Use of Vis-NIR and XRF Sensors for Tropical Soil Fertility Analysis: Assessing Different Data Fusion Approaches. *Sensors* **2021**, *21*, 148. [CrossRef]
36. Mouazen, A.M.; Maleki, M.R.; Cockx, L.; Van Meirvenne, M.; Van Holm, L.H.J.; Merckx, R.; De Baerdemaeker, J.; Ramon, H. Optimum three-point linkage set up for improving the quality of soil spectra and the accuracy of soil phosphorus measured using an on-line visible and near infrared sensor. *Soil Tillage Res.* **2009**, *103*, 144–152. [CrossRef]
37. Barnes, R.J.; Dhanoa, M.S.; Lister, S.J. Standard Normal Variate Transformation and De-Trending of Near-Infrared Diffuse Reflectance Spectra. *Appl. Spectrosc.* **1989**, *43*, 772–777. [CrossRef]
38. Stevens, A.; Ramirez-Lopez, L. *An Introduction to the Prospectr Package;* R Package Vignette; 2020. Available online: https://cran.r-project.org/web/packages/prospectr/vignettes/prospectr.html (accessed on 25 November 2021).
39. Usowicz, B.; Lipiec, J. Spatial variability of soil properties and cereal yield in a cultivated field on sandy soil. *Soil Tillage Res.* **2017**, *174*, 241–250. [CrossRef]
40. Webster, R.; Oliver, M. *Spatial Variability and Affecting Factors of Soil Nutrients in Croplands of Northeast China;* John Wiley & Sons Ltd.: The Atrium, UK, 2007.
41. Bhattacharjee, S.; Mitra, P.; Ghosh, S.K. Spatial Interpolation to Predict Missing Attributes in GIS Using Semantic Kriging. *IEEE Trans. Geosci. Remote Sens.* **2014**, *52*, 4771–4780. [CrossRef]
42. Bazzi, C.L.; Souza, E.G.; Uribe-Opazo, M.A.; Nóbrega, L.H.P.; Rocha, D.M. Management Zones Definition Using Soil Chemical and Physical Attributes in a Soybean Area. *Eng. Agríc.* **2013**, *33*, 952–964. [CrossRef]
43. Davies, D.L.; Bouldin, D.W. A Cluster Separation Measure. *IEEE Trans. Pattern Anal. Mach. Intell.* **1979**, *PAMI-1*, 224–227. [CrossRef]
44. Rousseeuw, P.J. Silhouettes: A graphical aid to the interpretation and validation of cluster analysis. *J. Comput. Appl. Math.* **1987**, *20*, 53–65. [CrossRef]
45. Dobermann, A.; Ping, J.L.; Adamchuk, V.I.; Simbahan, G.C.; Ferguson, R.B. Classification of Crop Yield Variability in Irrigated Production Fields. *Agron. J.* **2003**, *95*, 1105–1120. [CrossRef]
46. Zhang, J.; Guerrero, A.; Mouazen, A.M. Map-based variable-rate manure application in wheat using a data fusion approach. *Soil Tillage Res.* **2021**, *207*, 104846. [CrossRef]
47. Munnaf, M.A.; Haesaert, G.; Van Meirvenne, M.; Mouazen, A.M. Map-based site-specific seeding of consumption potato production using high-resolution soil and crop data fusion. *Comput. Electron. Agric.* **2020**, *178*, 105752. [CrossRef]
48. Lang, H.; Xi, Y.; Zhang, X. Ship Detection in High-Resolution SAR Images by Clustering Spatially Enhanced Pixel Descriptor. *IEEE Trans. Geosci. Remote Sens.* **2019**, *57*, 5407–5423. [CrossRef]
49. Pal, N.R.; Bezdek, J.C. On cluster validity for the fuzzy c-means model. *IEEE Trans. Fuzzy Syst.* **1995**, *3*, 370–379. [CrossRef]
50. Pedregosa, F.; Varoquaux, G.; Gramfort, A.; Michel, V.; Thirion, B.; Grisel, O.; Blondel, M.; Prettenhofer, P.; Weiss, R.; Dubourg, V.; et al. Scikit-learn: Machine learning in Python. *J. Mach. Learn. Res.* **2011**, *12*, 2825–2830.
51. Hazelton, P.; Murphy, B. *Interpreting Soil Test Results: What Do All the Numbers Mean;* CSIRO Publishing: Melbourne, Australia, 2007.
52. Sharma, A.; Weindorf, D.C.; Wang, D.; Chakraborty, S. Characterizing soils via portable X-ray fluorescence spectrometer: 4. Cation exchange capacity (CEC). *Geoderma* **2015**, *239–240*, 130–134. [CrossRef]

sensors

MDPI

Article

Delicar: A Smart Deep Learning Based Self Driving Product Delivery Car in Perspective of Bangladesh

Md. Kalim Amzad Chy [1], Abdul Kadar Muhammad Masum [1], Kazi Abdullah Mohammad Sayeed [1] and Zia Uddin [2,*]

[1] Department of Computer Science and Engineering, International Islamic University Chittagong, Chittagong 4210, Bangladesh; kalim.amzad.chy@gmail.com (M.K.A.C.); akmmasum@iiuc.ac.bd (A.K.M.M.); s.titas244@gmail.com (K.A.M.S.)

[2] Software and Service Innovation Department, SINTEF Digital, 0316 Oslo, Norway

* Correspondence: zia.uddin@sintef.no

Abstract: The rapid expansion of a country's economy is highly dependent on timely product distribution, which is hampered by terrible traffic congestion. Additional staff are also required to follow the delivery vehicle while it transports documents or records to another destination. This study proposes Delicar, a self-driving product delivery vehicle that can drive the vehicle on the road and report the current geographical location to the authority in real-time through a map. The equipped camera module captures the road image and transfers it to the computer via socket server programming. The raspberry pi sends the camera image and waits for the steering angle value. The image is fed to the pre-trained deep learning model that predicts the steering angle regarding that situation. Then the steering angle value is passed to the raspberry pi that directs the L298 motor driver which direction the wheel should follow. Based upon this direction, L298 decides either forward or left or right or backwards movement. The 3-cell 12V LiPo battery handles the power supply to the raspberry pi and L298 motor driver. A buck converter regulates a 5V 3A power supply to the raspberry pi to be working. Nvidia CNN architecture has been followed, containing nine layers including five convolution layers and three dense layers to develop the steering angle predictive model. Geoip2 (a python library) retrieves the longitude and latitude from the equipped system's IP address to report the live geographical position to the authorities. After that, Folium is used to depict the geographical location. Moreover, the system's infrastructure is far too low-cost and easy to install.

Keywords: computer vision; self-driving car; smart product delivery; Internet of Things; convolution neural network; Raspberry Pi 3

Citation: Chy, M.K.A.; Masum, A.K.M.; Sayeed, K.A.M.; Uddin, M.Z. Delicar: A Smart Deep Learning Based Self Driving Product Delivery Car in Perspective of Bangladesh. *Sensors* **2022**, *22*, 126. https://doi.org/10.3390/s22010126

Academic Editors: Natividad Duro Carralero and Hossam A. Gabbar

Received: 30 September 2021
Accepted: 16 December 2021
Published: 25 December 2021

1. Introduction

Failure to deliver the product in time is a typical scenario of Bangladesh that affects the economy significantly. Among different reasons, the root cause of this scenario is to stay stuck in traffic congestion. According to a recent statistic, because of the congestion in Dhaka, the capital of Bangladesh, the amount of loss is around BDT 200 billion annually [1]. Investigators have reported a loss of 3.2 million working hours a day of traffic jams [2]. The Center for Economics and Business Research is projected that, by 2030, it will increase to almost BDT 300 billion [2]. Furthermore, in our country, road accidents are deeply linked with drivers' behavior. Most of them are tempted to race on the lane, neglecting the risk of an accident. Disobeying traffic regulations and signals also leads to critical accidents and disasters. This ill-mindedness has caused so many disasters, taken too many souls and caused mass destruction in the last decades across the world. At least 4138 people were killed and 4411 wounded in 4147 crashes in 2019, while 2635 were killed and 1920 wounded in 2609 accidents in 2018, according to police [1]. In cases where it is impossible for a person to avoid a car accident, self-driving cars will save millions of lives and subside the on-time product delivery failure case without road accidents.

Artificial Intelligence (AI) plays a significant role in almost every aspect of human life, in every type of industry. For example, researchers [3,4] used a support vector regression algorithm to predict the water parameters. Considering physical and operational factors, another group of researchers [5] engaged AI to assess pipe break rate and [6] decoding clinical biomarker space of COVID-19. Nowadays, AI is also broadly used in building the smart city [7,8], smart meter [9,10], agriculture [11–13], education [14,15], healthcare [16–18] and so on. Machine learning is a branch of artificial intelligence that allows machines to learn without being explicitly taught from prior data or experiences. Nowadays, the neural network is a popular type of machine learning algorithm that mimics the human brain. CNN (Convolutional Neural Networks) and other groundbreaking systems have provided tremendous results in computer vision. In the majority of cases, they improved the preceding manual extraction features and created new cutting-edge solutions for such tasks as image classification [19], captioning [20], object detection [21] or semantic segmentation [22]. A machine's reaction times and alerts are far better. In addition, these vehicles were fitted with extraordinary capabilities by long-range cameras and ultrasonic sensors. Since the last decade, extensive work has been carried out on autonomous robotics and driving systems. Many research studies focus on the classification, identification and development of decisions based on the vision to improve, evolving techniques and algorithms. There are also some off-road studies. In our comprehensive study, we have felt the need for some missing features or works in those studied works.

Our self-driving product delivery vehicle can move on a road autonomously through the deployed deep learning pre-trained model. The car's key input is real-time camera footage mounted on the roof. The system outputs the respective steering angle and drives the car accordingly. Because the camera is the only control system input, the purpose of the project is to teach the vehicle how to handle the steer. The network is trained on a different machine and then shifted to an onboard computer to regulate the vehicle. Then the autonomous product delivery vehicle is entirely independent of other machines. Furthermore, the position of the car is reported to the authority through a map to monitor. Obstacle avoidance is a different problem that can also be overcome, but it goes outside the scope of the study to combine it with the system. The current system configuration is not that capable of dealing with both steering angle prediction and obstacle avoidance. This self-driving vehicle work will significantly change traffic systems and public safety in a developing country like ours. It can also support national defense forces to perform ground monitoring or conduct rescue tasks. More particularly, the risk of an accident can be reduced dramatically. Moreover, the development cost of this system requires about BDT 30K–40K for hardware and 20K–30K for software and other experimental purposes. As a result, product delivery car owners in developing nations like Bangladesh would find the technology beneficial and economical.

The objectives of this research are to develop a self-driving car for overcoming the product delivery failure without any road accidents, to design a low-cost infrastructure with effective outcomes, to build an end-to-end deep learning model equipped in the self-driving car prototype, and to broadcast the geographical location of the vehicle through a map in real-time.

With the introduction, this paper is composed of five parts. Section 2 covers the literature review, and Section 3 contains working procedure, functional units, dataset collection, normalization, augmentation, pre-processing, deep learning model and driving instruction forwarding strategies. Section 4 shows the experimental outcomes. Finally, Section 5 addresses the analysis and future scope.

2. Related Works

Lots of significant works and research have been performed on the autonomous vehicle aspect. The NHTSA (National Highway and Traffic Safety Administration) describes five levels of autonomous vehicles [23] shown in Figure 1. In no automation (level 0), the human driver does all the driving. Lane-keeping, cruise control or assisted breaking are a few examples of level 1(driver assistance). Tesla Autopilot [24] claims at their level 2 position. The Waymo (Google) self-driving car [25] is an example of conditional automation (level 3). Waymo announced in 2017 that they are testing level 4 driving [24]. Full automation (level 5)—The driving system takes complete control over the entire driving task under all circumstances. The human driver does not need to be inside the car. Recent attacks targeting VANET (Vehicular ad hoc network) with autonomous Levels 1 to 4, which are not entirely autonomous, have been documented. Denial of service attack [26], sybil attack [27], timing attack [28], illusion attack [29], message tampering [30], and node impersonation [29] are examples of these types of attacks.

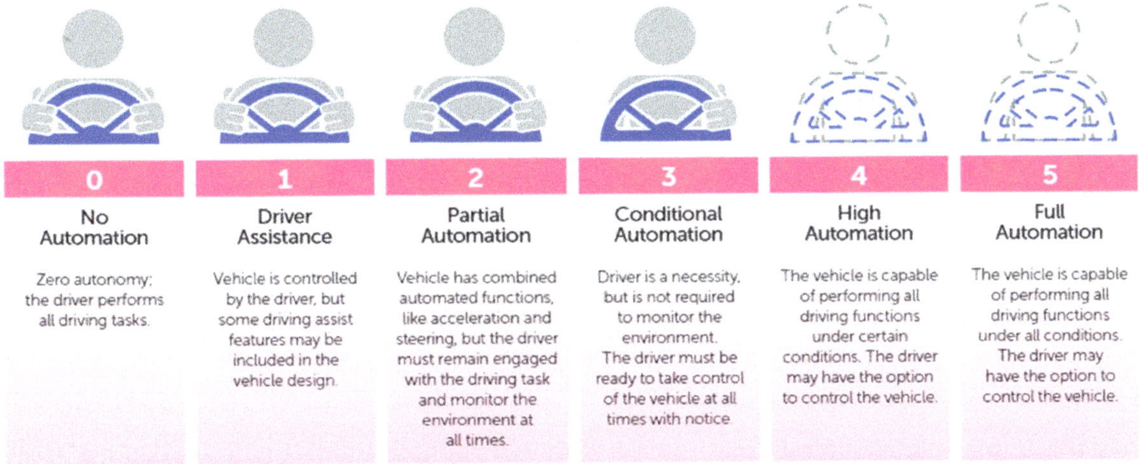

Figure 1. Levels of autonomous driving by NHTSA.

The non-AI solution practices control theory to determine a steering angle to hold the vehicle on the desired trajectory, typically identified by algorithms for computer vision. PID (Proportional Integral Derivative) controller is one of the most popular methods in control theory [31]. The controller functions in a loop that continually computes an error value $e(t)$ as a variance between the input from the vehicle and the next command signal. A correction will be measured and applied afterwards. The correction value $u(t)$ consists of three parts (proportional, integral, derivative) and, as shown in Figure 2, can be determined from the error $e(t)$.

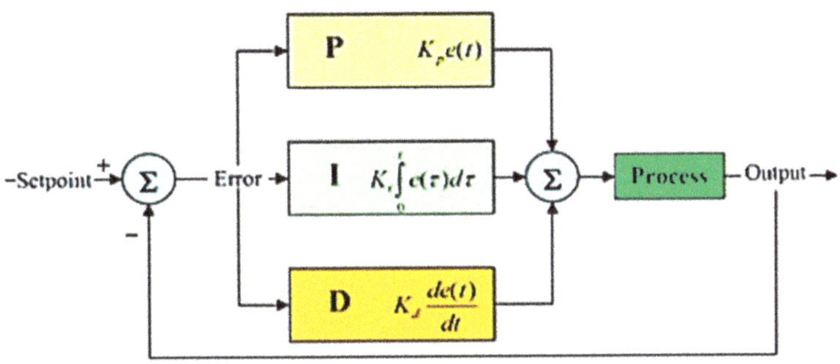

Figure 2. Calculation of correction value in a PID loop.

The whole mathematical formula is the following:

$$u(t) = k_p e(t) + k_i \int_0^t e(t)dt + k_d \frac{de(t)}{dt} \tag{1}$$

The standard approach to solving the problem of self-contained driving has divided the problem into several sub-problems, including lane marking, path planning and low-level control, which make up a processing pipeline [32]. Researchers have recently explored a new approach that simplifies the standard control pipeline dramatically through deep neural networks to produce direct control outputs from sensor inputs [33]. The gaps between the two methods are shown in Figure 3. Figure 3a visualizes the standard approach in which the system predicts the motor torques based on the observation of the image data. This approach split the problem into several sub-problems such as state estimation, modeling and prediction, motion planning, low-level controller. In contrast, to solve the same problem, Figure 3b demonstrates a deep neural network approach to predict the motor torques directly from the image observation.

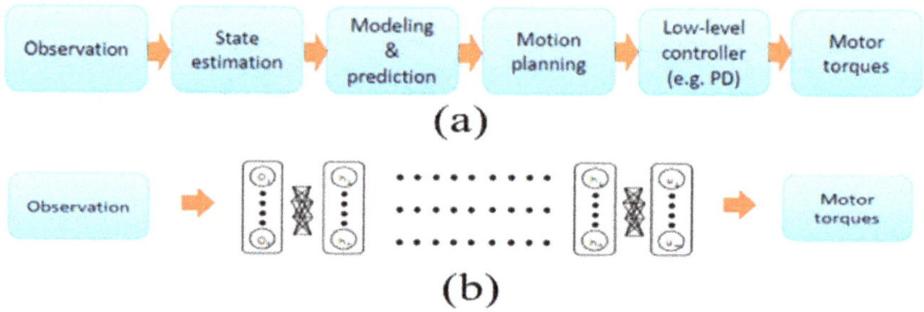

Figure 3. Standard robotics control vs DNN based end-to-end control [10]. (**a**) standard approach (**b**) deep neural network approach.

In the late 1980s [34], a modern, completely linked neural network employed neural networks to monitor automatic cars. In the late 2000's it was later demonstrated [35] using a six-stage, fully interconnected neural network (CNN) in the DARPA Autonomous Vehicle (DAVE) project and most recently in the NVIDIA DAVE-2 project [32], with a nine-layered CNN network. The training process of the NVIDIA project has been displayed in Figure 4, where the steering angle is recorded for the center camera image and the left and right camera image steering angle is shifted. Then fit into CNN architecture, calculate the error and adjust the weight via backpropagation. The architecture of the CNN model used by NVIDIA is nine-layer depth, including 5 convolution layers and three dense layers. The first three convolution layers contain 24, 36, 48 kernels and the rest two convolution layers consist of 64 kernels. This architecture includes 27 million connections and 250 thousand parameters.

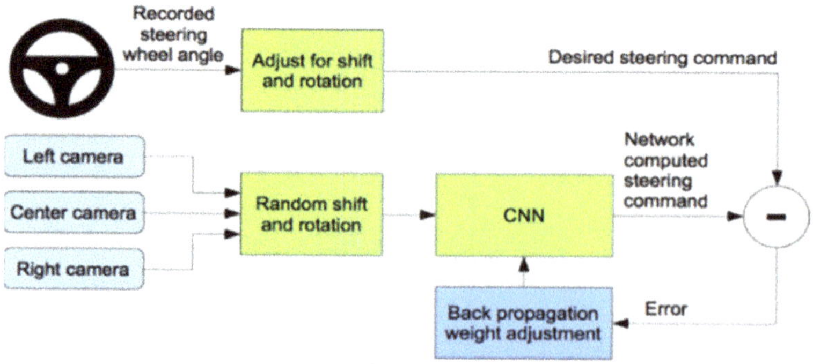

Figure 4. Training the neural network [13].

The testing procedure (Figure 5) is a sample where the weight and the CNN architecture are saved, and the camera image goes through that saved model, predicting the steering angle and the car drive by the wired interface.

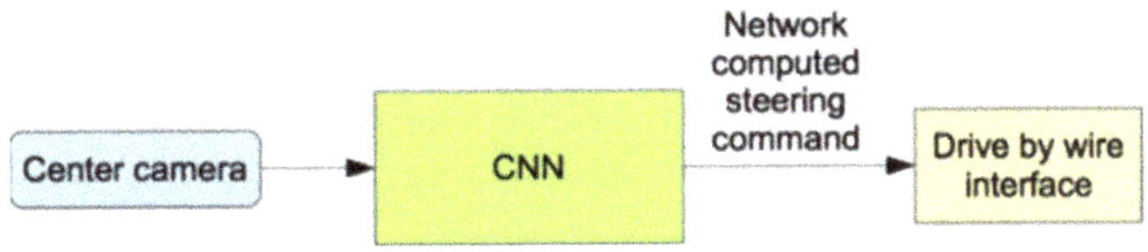

Figure 5. Testing the neural network.

There are two different phases to the use of deep neural networks [36]. The first step is training, in which the backpropagation techniques change the weights of the network. The next phase is when unseen data are fed into the network to produce the predicted output (e.g., the predicted image classification, for example) once it has been trained–i.e., network weights minimize errors in training example. The training phase is generally more computational and requires high throughput, usually not available on embedded platforms. On the other hand, the inferencing process is comparatively less computer-intensive and latent, if not more so, is as critical as software output because many case stores have strict real-time requirements. For example, with neural network and computer vision-based learning methods, Masum et al. [37] attempted to introduce an autonomous automotive program. The system predicts the steering angle learning from live images according to which the vehicle moves autonomously.

David Stavens et al. [38] have attempted to describe the ruggedness of autonomous off-road vehicles for the terrain project. They proposed a supervised machine learning approach to estimate the roughness of the terrain from laser range data. They used data from the 2005 DARPA Grand Challenge to compare nearby surface points acquired with a laser. Bajracharya et al. [6] did the same kind of work in their research. They used self-supervised training from sensors to know the near-field terrain traversability. The near-field classification was then used to direct the far-field training of terrain traversability. As part of the DARPA Learning Applied to Ground Robots (LAGR) project, the methodology developed was incorporated into a fully autonomous off-road navigation system. Problems in mobile off-road vehicles and mobile robotics caused by poor stereo vision are increasing and remain vulnerable for as long as possible. Junsoo Kim et al. [39] introduced a model focused on long-distance stereo vision to solve this problem. Training data generation on every image frame in a self-supervised way gives robust, consistent stereo module label input, ensuring success. From an input image, meaningful features are acquired, and information is learned. These features train real-time classifiers that can identify complex terrain to distance from the horizon. They claim that it exceeds the max stereo range of 12 m and can see paths and obstacles at a distance of 5 to more than 100 m [39].

The extensive usage of self-driving technology is exemplified by trains [40]. Some of such self-driving trains include the Docklands Light Railway (DLR) in London, UK [41], Yurikamome in Tokyo, Japan [40], London Heathrow airport's ultra-pods [41] and SkyTrain in Vancouver, Canada [42]. The successor of Robot Operating System (ROS) ROS2 based self-driving vehicle architecture can activate safe and reliable real-time behavior [43]. Bakioglu et al. [44] proposed VIKOR and TOPSIS algorithms to prioritize risks in self-driving cars, while another group of researchers [45] proposed a self-driving delivery robot in last-mile logistics. Navigation routes, one-way streets, speech recognition, and no-entry status are all things that self-driving vehicles require [46]. Based on an adaptive large neighborhood algorithm (ALNS), Guo et al. [47] proposed a multimodal transport distribution model for self-driving vehicles. Dommès et al. [48] investigated aged and young pedestrians' behavior in front of the conventional and self-driving car wherein mixed traffic conditions. They undertook the simulated two-way street-crossing task. When delivering commands to a self-driving vehicle Deruyttere et al. [49] developed a model that can determine uncertainty, detect the causing objects of uncertainty and generate a question for the passenger that describes the objects.

Tinghui Zhou et al. [50] used monocular video sequence networks of a single view depth and multi-view pose. They approached it as unsupervised through similar approaches were made by others as supervised. Using the images, they were tempted to train the network with a targeted view (single view) and computed losses from some multi-views (closer and distant views from the target view). Yanlei Gu et al. [51] proposed a prototype mimicking the human driving system from the actual traffic environments dataset. Again, different algorithms for different functions of the autonomous vehicle have been suggested. Such as Voronoi Diagram (complete but limited to the static environment), Occupancy Grid (low computational power but has problems vehicle dynamics), Driving Corridors (continuous collision-free space findings but costs of computation with motion constraints), etc., algorithms are used for planning for searching the best space available in the path. Driving Corridors and Non-Linear Constrained Optimization method for intersection and Multiple Criteria Decision Making for non-intersection segments planning, Mixed-Observability MDP for pedestrian crossing, etc., these implemented obstacle detection and decision making. Trajectory Planning is being worked out by Tiji Algorithm, 4th Order Polynomials, Cubic Bezier curves, etc., and many other algorithms are used [52]. Here the authors provided elaborated criticism and evaluation of such algorithms based on different factors.

In manufacturing plants, a line following robot is often used for the pick-and-place features. The robot receives the products from a position and deposits them on an intended location via a pre-specified path. This route is often specified on a black surface as a white line or on a white surface as a black line. Mostafa et al. [53] propose an amphibian line following robot, which can move in both lands and at certain water levels. A line that follows a robot reaches its target by following the predefined path as a white line over a black surface. The line IR sensor is often used to determine which emission led will emit an infrared ray, and the detector led will receive the infrared ray. By a fixed threshold, the robot will sense the rows. L293D motor driver regulates the wheel position and torque of the vehicle. For vehicle rotation, the DC motor is placed onto the wheel. The system can sense the water road and, like a speed boat, activate a propeller mechanism with an integrated water sensor. Since it is an autonomous device, the planned robot is free of any direct human intervention. The idea of the line after the robot is used for different sectors such as rescue, recreation, libraries, searches, and the army. Colak et al. [54] have developed a clever robot line to keep children entertained in shopping malls. This system uses a black line of 4.8 cm to load up to 400 kg. The control functions are remote and manual. Islam et al. [55] have also proposed a low-cost system that can travel around 500 gm without falling off the ground. To strengthen the health care system, Punetha et al. [56] used a robot concept row. If the patient requires drugs, the medicine will automatically be transported along the road, reducing human effort.

A group of researchers [53] proposes an amphibian line following robot for product delivery in Bangladesh perspective, which can move in both lands and at certain water levels. A line that follows a robot reaches its target by following the predefined path as a white line over a black surface. However, ensuring a predefined path as a white line over a black surface for a long distance is a great challenge for this system. In contrast, our proposed approach can decide the driving direction based on the existing road lane, capturing real-time road images in adverse weather such as rainy, cloudy, etc. In Bangladesh, such a kind of system will be a great addition in ensuring on-time product delivery. Many research studies focus on classification, identification and development of decisions based on the vision to improve, evolving techniques and algorithms. There are also some off-road studies. In our comprehensive analysis, we have felt the need for some missing features or works in those studied works. In traffic situations, weather plays an important role. It also influences vision-based independence.

An Extended Kalman Filter (EKF) localization technique considers adverse weather conditions while estimating the car's posture by registering 3D point clouds against gaussian mixture multiresolution maps [57]. In another study, Ahmad et al. [58] consider weather and lighting conditions in the context of road marking. They consider various messages as distinct categories, while most systems [59,60] use OCR-based algorithms to detect letters first and then write. Unlike stormy, rainy days, dark conditions are created and lighting on the bright sunny day. Such changes affect the camera or other sensors for visual input. These environmental effects are considered in various contexts such as estimating the car's posture, road markings, etc. However, it was not investigated so thoroughly in a dark, rainy environment while the sensor captured image is not clear as it is supposed to be. Land and off-road research primarily illustrated the roughness of the route, visibility and road roughness styles. Very few have examined fragile or damaged road sections (such as deep holes, damaged/broken road pieces, etc.). From a country viewpoint, damaged and broken highways are causing severe traffic and transportation havoc. Studies showing the identification of these broken sections of the road were not possible as well as other cases. Furthermore, the cost of lots of studies is not optimized. Some used Bluetooth modules to communicate and transfer data between vehicle and computer which is expensive and not required. Moreover, using a Bluetooth device reduces the power of a self-driving car's ability for a long drive as the coverage of a Bluetooth module is very limited. Furthermore, there is no feature to monitor in real-time and observe the geographical location of the vehicle.

3. Methodology and Implementation

A good design of a system has a significant impact on the successful implementation of a project. The overall architecture of the system is demonstrated in Figure 6. By supplying the power into the Raspberry pi, the heart force of the system, the system starts to initiate. A buck converter converts the 12V lipo battery power supply into 5V and 3A and continuously feeds into raspberry pi enough for raspberry pi to be operating. To program and utilize the raspberry pi despite an extra monitor, we have used a VNC viewer from a local pc. VNC viewer provides instant remote access to the target computer. As the RAM of raspberry pi is too slow to run a pre-trained deep learning predictive model, we need to choose a technique where the predictive model runs into another high-configured computer and the data transfers to the raspberry pi. The high configured local computer acts as a host, the raspberry pi as a client, and the server uses a Transmission Control Protocol (TCP). After establishing the communication between the local pc and raspberry pi, the camera module becomes active and transfers the image to the pc.

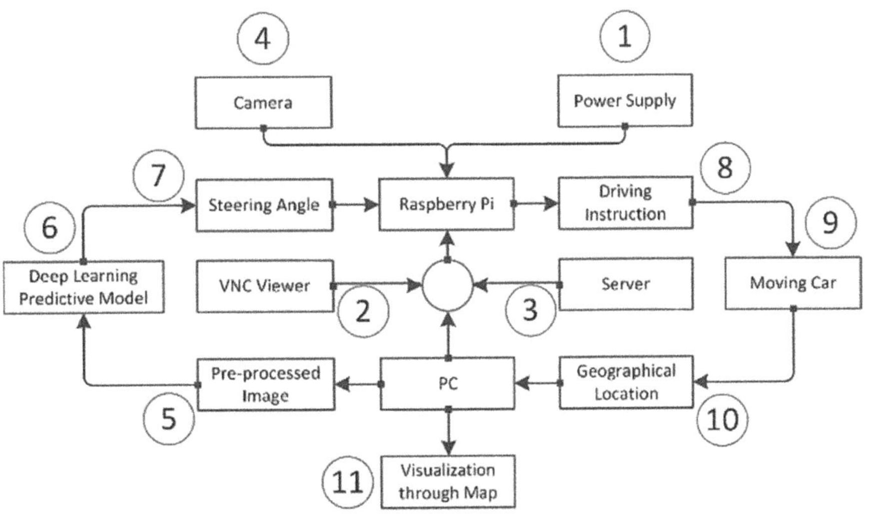

Figure 6. The overall design of the system.

Because of the low processing power of the raspberry pi, per second, only ten images have been sent to the local pc. After receiving the image, the image goes through the pre-processing steps that include removing the upper part of the image, blurring the image, transforming the image from RGB to YUV and resizing the image. Then the pre-processed image is sent to the pretrained deep learning model based upon the extended version of the Nvidia CNN model for the self-driving car. The pretrained model can predict what the steering angle for that image in that situation is. The steering angle data is transferred to the raspberry pi through the previously established communication. Based on this steering angle, the raspberry pi decides which direction it should advance, either forward, left, right, or reverse. This instruction is transferred to the self-driving product delivery car. Based on the instruction, the vehicle follows the direction. From the IP address, one can find out the geographical position of the vehicle and track it. Furthermore, the system visualizes the geographical position, i.e., longitude and latitude, through a well-organized map. Each step is discussed in upcoming sections.

This section may be divided into subheadings. It should provide a concise and precise description of the experimental results, their interpretation, and the experimental conclusions that can be drawn.

3.1. Functional Hardware Units of the System

To develop the system, we require hardware tools as well as software tools. In our project, we have used different components for controlling speed, direction, transmitting and receiving data, and showing the vehicle's speed on display. The hardware components used in our project are enlisted below:

1. Raspberry Pi 3 Model B+
2. NoIR Camera with Night Vision
3. Motor Driver IC (L298)
4. Plastic Gear motor
5. 3 cell Lipo Battery (12V)
6. Buck Converter
7. Acrylic Chassis Board
8. Connecting wires
9. Switch

3.2. Functional Software Tools of the System

To develop the system, we require software tools along with hardware tools. To drive the hardware, the software performs a leading role. Following software, programming language, library, package, etc., are used in our work:

1. Python programming language: Python is a high-level, general-purpose programming language.
2. Google Colab: Colaboratory (also known as Colab) is a free Jupyter notebook environment running in the cloud and storing on Google Drive notebooks.
3. Numpy: NumPy is a library that supports multi-dimensional arrays and matrices.
4. Pandas: Pandas is used for data manipulation and analysis.
5. Matplotlib: Matplotlib is the Python programming language plotting library.
6. Keras: Keras is an open-source neural-network library written in Python. It can run top of TensorFlow, R, Theano or PlaidML, to allow quick experimentation with deep neural networks [61].
7. Tensorflow: TensorFlow is an open and free software library for data flow used for machine learning applications like neural networks.
8. Imgaug: A library for image augmentation in machine learning experiments, particularly CNN (Convolutional Neural Networks).
9. OpenCV: OpenCV-Python is OpenCV's Python API. It integrates OpenCV C++ API's best qualities with Python language.
10. Scikit-learn: It is a free machine learning library for the Python programming language.
11. VNC viewer: VNC Viewer transforms a mobile into a virtual desktop, giving one immediate access from anywhere in the world to one's Mac, Windows and Linux computers.
12. Sublime Text 3: Sublime Text is an advanced script, markup and prose text editor.

13. Geoip2
14. Folium

3.3. Power Supply Strategy

The power supply strategy is displayed in Figure 7. A 3cell 1500 mah 12V Lipo battery is used as the primary power source that supplies the power to the raspberry pi and L298 motor driver. The raspberry pi requires 5V and 3A to come into the working state.

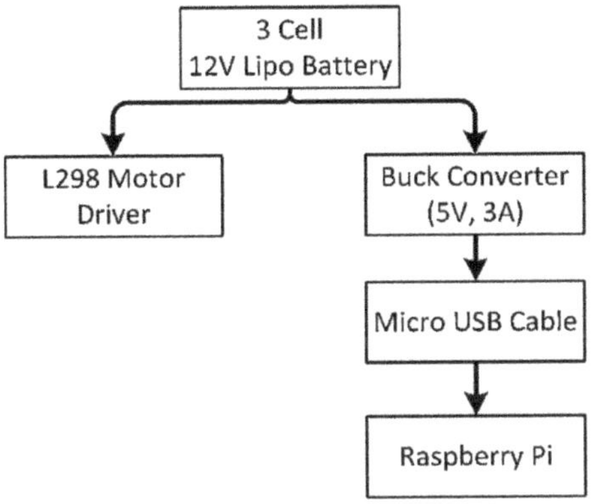

Figure 7. Power flow strategy.

A direct connection with the battery may cause the death of raspberry pi because of the overpowering supply. So, to regulate the power supply, we have placed a buck converter in between lipo battery and raspberry pi that continuously provides 5V and 3A. The raspberry pi connects with the buck converter through a micro USB cable.

3.4. Deep Learning Predictive Model

Figure 8 is a step-by-step developing process of the predictive model to forecast the steering angle based upon the given road image.

3.5. Dataset Collection

We need a dataset containing a massive collection of road images and steering angles against that image for a deep learning predictive model. Different nations' legislators (e.g., the USA, China, Australia, Singapore, and South Korea) [62–68] have established or are adopting different regulating measures to enhance the security and privacy of data utilized and sent by autonomous cars. The gathering of data on public roadways is essential for self-driving car autonomy [69]. There exist several datasets developed by individuals or organizations such as Sullychen [70], Nvidia [32], Udacity, commaai, Apollo [71], etc. However, the dataset is too large and beyond our processing capability because of our limited computational resources. For example, the opensource dataset by commaai is 45 GB in compressed and 80 GB in uncompressed [72]. In Ref. [73], the authors provide 27 publicly available vehicle datasets, assess them based on various parameters, and recommend selecting the most suited dataset for specific goals. Furthermore, Udacity published a huge open-source dataset in a sunny and overcast environment ranging from 23 GB to 183 GB in size [74]. So, for experimental purposes and considering the limited hardware resources we have developed our own dataset using an open-source Udacity simulator [75]. This simulator was designed for a Nanodegree program of Udacity in a unity environment with two moods. One is training mood and another one autonomous mode. One can drive a car in two tracks, and at the time of driving the steering angle, throttle, speed, etc., is recorded against each image. At the training mood of the Udacity simulator, one needs to set the path directory where the image will be saved and the steering angle is saved as a log file against each image.

We have collected the data on track two and saved the data into a folder shown in Table 1. There have three cameras in the Udacity simulator that track center, left, right images accordingly. Besides steering angle, it also saves the throttle, reverses the speed at that time. The images are saved in jpg format into a different folder, while a log file into CSV format tracks the image path. In this way, we have collected more than 8.4K images based on developing our predictive model.

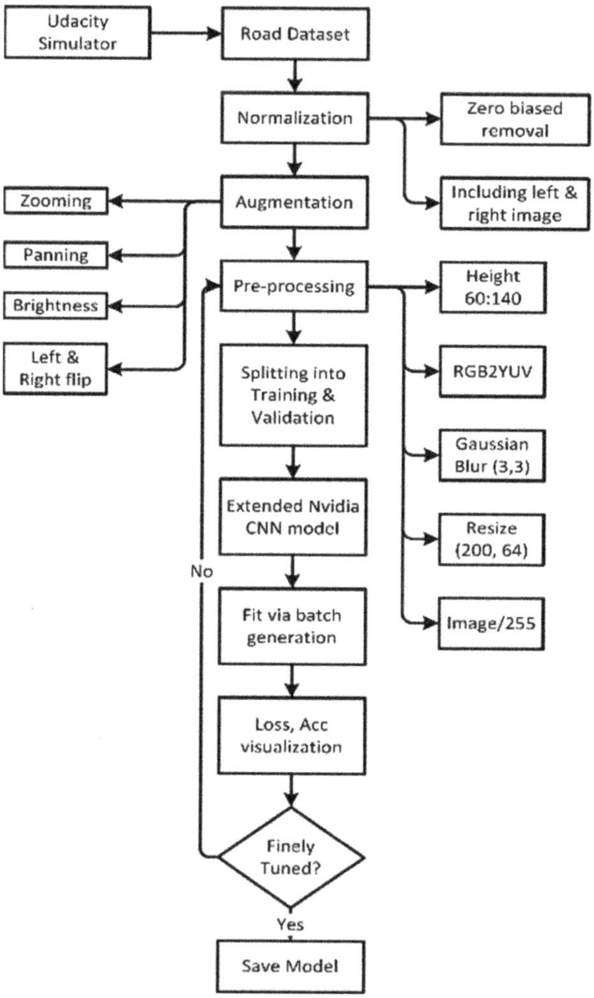

Figure 8. Flowchart of the steering angle prediction model.

Table 1. Recorded data at training mode.

Center	Left	Right	Steering	Throttle	Reverse	Speed
E:\android\curly_final\ IMG\center_2019_04_ 12_01_52_50_770.jpg	E:\android\curly_final\ IMG\left_2019_04 _12_01_52_50_770.jpg	E:\android\curly_final\ IMG\right_2019_04 _12_01_52_50_770.jpg	0	0	0	0.00014
E:\android\curly_final\ IMG\center_2019_04 _ 12_01_52_50_846.jpg	E:\android\curly_final\ IMG\left_2019_04_ 12_01_52_50_846.jpg	E:\android\curly_final\ IMG\right_2019_04_ 12_01_52_50_846.jpg	0	0	0	0.000199
E:\android\curly_final\ IMG\center_2019_04_ 12_01_52_50_917.jpg	E:\android\curly_final\ IMG\left_2019_04_ 12_01_52_50_917.jpg	E:\android\curly_final\ IMG\right_2019_04_ 12_01_52_50_917.jpg	0	0	0	0.00026

3.6. Normalization

To understand the data distribution against the steering angle, we need to visualize the dataset. Through histogram, in Figure 9, we have visualized the data across 25 bins where the zero steering angle is too high, about more than 4K. So, we need to remove zero biased data so that the model generalizes the steering angle.

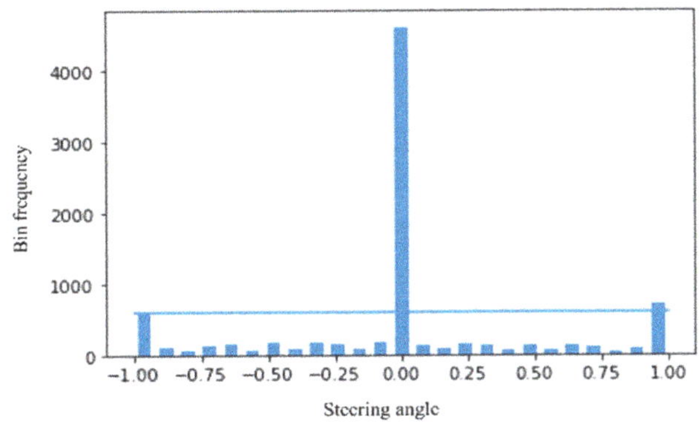

Figure 9. Visualization of the dataset.

We have considered a maximum of 600 hundred images per bin (Figure 10). So, the more than 600 images bin keeps a maximum of 600 images and removes the rest of the images. After this type of normalization, our dataset is down from 8.4K to 4K, which is too low.

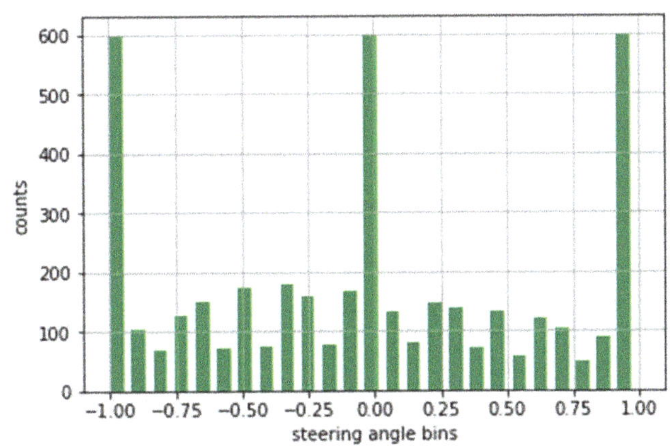

Figure 10. Normalized form of the dataset.

To increase the dataset, we have also considered the left and right images. The steering angle in the dataset is actually based upon the center image. So, the steering angle will be slightly sifted from the center for the left and right images. We have considered 0.15 positive sifted for the left image and 0.15 negative shifted for the right image. Furthermore, the left and right images help us more to generalize the dataset like this type of road image may come into a real scenario. After this technique, the size of the dataset became more than 12.8K.

3.7. Augmentation

Our dataset does not resemble real-world road data, such as gloomy environments, zoomed views, and so on, as we employed a simulator. However, even now, the size of the photograph is insufficient. Augmentation is a procedure that artificially increases a training dataset's size by modifying the images in the dataset. ImageDataGenerator, a Keras deep learning module, is mostly used in image data augmentation techniques. Among various augmentation techniques, we experimented with four approaches: zooming, panning, brightness, and random flipping that best fit our data.

In the zooming technique, the image is zoomed randomly by interpolating pixel values or adding new pixel values around the image. If a float is specified, [1−value, 1+value] will be the zoom range. So we do not zoom across the x-axis, but across the y-axis, we zoom at 1.3 scales. Figure 11 is a sample of the zoomed image.

Figure 11. Zoomed image.

To pan an image, we have selected the following parameters for x and y and a sample image is displayed in Figure 12.

$$\text{translate\_percent} = \{\text{"x"}: (-0.1, 0.1), \text{"y"}: (-0.1, 0.1)\}$$

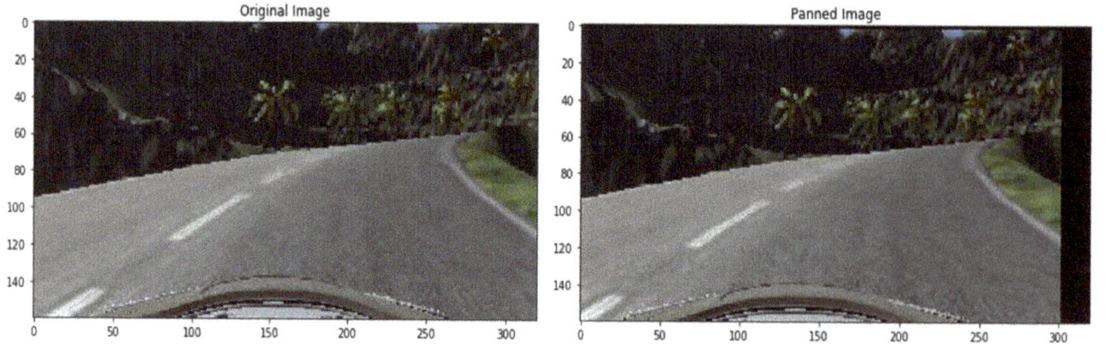

Figure 12. Panned image.

The brightness of the image can be changed either by randomly darkening images, brightening images or both. Values underneath 1.0 obfuscate the image, e.g., [0.5, 1.0], whereas values greater than 1.0 illuminate the object, e.g., [1.0, 1.5] where 1.0 does not affect illumination. We used a scale from 0.2 to 1.2, a sample shown in Figure 13.

Figure 13. Brightness altered image.

Flipping into left or right is another technique used in image augmentation. For example, the right-oriented images turned left and left-oriented into right. In previous approaches, we do not need to change the steering angle across the changing of the images. However, in the case of flipping, the road image is the opposite. So, we need to flip the steering angle across the image. A sample flipped image has been displayed in Figure 14 with the flipped steering angle.

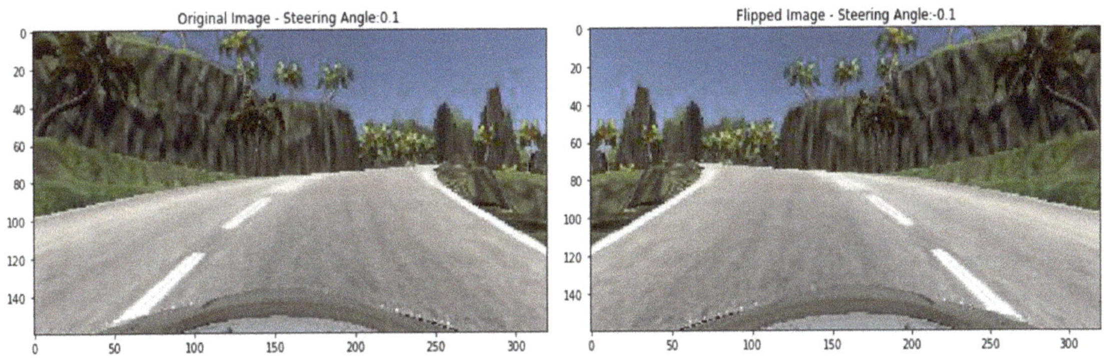

Figure 14. Flipped image.

3.8. Pre-Processing of the Dataset

Pre-processing is another crucial technique to smooth the image before feeding it into training steps. We have considered five pre-processing methods. First, an original image and after the pre-processing step, the pre-processed image is shown in Figure 15. From the original image, we have seen that the top part contains natural scenery that do not have any value in steering angle prediction. Besides, removing this part also minimize the size of the image.

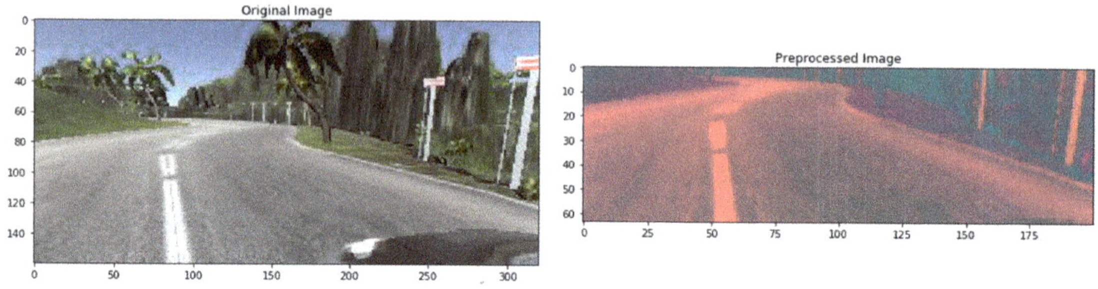

Figure 15. Pre-processed image.

The YUV color model is closer to human color perception than the standard RGB model. So, we convert the RGB image into YUV format. Then, blurring the image remove the noise and clean the image. We have used gaussian blur with 3×3 kernel size. Then we resize the image into 200×64. Finally, to normalize the pixel value, we divide each value via 255 as the value range is 0 to 255 in the original image.

3.9. Splitting of the Dataset

After pre-processing, the images need to be split into training and validation sets. The model learns the steering angle through the training set, and via the validation set, it will examine how accurately it learns. We preserve 20% of data for validation purposes so that after training, we can test the performance of the trained model how much it learns. Figure 16 clearly states that the distribution of the training and validation set is quite similar and so fit for the pass into training step.

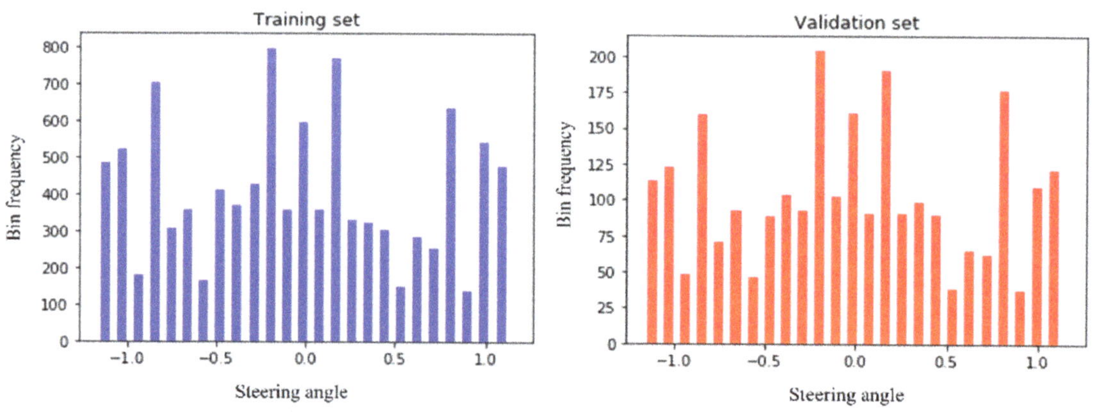

Figure 16. Splitting the images into training and validation set.

3.10. Convolution Neural Network Architecture

A deep learning Convolution Neural Network (CNN or ConvNet) is a subset of deep neural networks, most commonly used in visual image processing. To train and test deep learning CNN models, each image will go through a sequence of convolution layers with kernels, pooling layer, fully connected layers and apply activation function (softmax, tanh, ReLu etc.) to classify an object with probabilistic values. Figure 17 is a full CNN flow to analyze an image as input and identify the objects according to values.

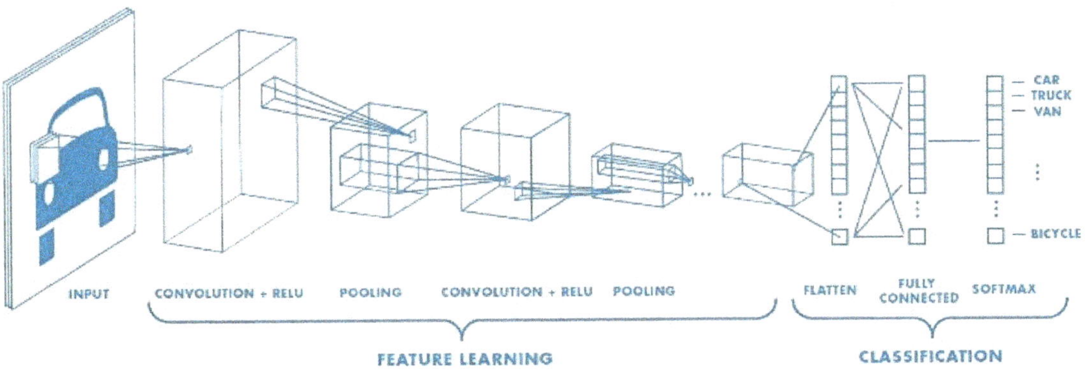

Figure 17. A CNN network with many convolutional layers.

The first layer to extract features from an input image is convolution. Convolution is a dot product of an input image with a kernel to understand the feature. To understand the feature for an image, there have to be various types of convolution kernels. Despite defining the kernel, in CNN, we represent the number of kernels with the dimension. We have followed the Nvidia CNN model to design our deep learning CNN architecture (Figure 18). Nvidia experiments with this CNN architecture in their self-driving project with more than 72-h of video data. They tuned various parameters and found better outcomes for this network. We also found promising results from this architecture. This architecture has nine layers combining five convolution layers, three densely connected layers and one output layer. The first three convolution layers are glued with 2×2 subsamples where the first layer, the second layer, third layer have 24, 36, 48 kernels, respectively, with 5×5 kernel size. The following two convolution layers include 64 kernels with 3×3 size features. Then we flatten the matrix and connect the flatten layer with a dense layer having 256 neurons. The successive two dense layers have 100, 10 neurons accordingly. The final one is the output layer with one neuron. In each layer, we have used 'elu' activation function but the output layer. To optimize the model, we have chosen the Adam optimizer. For each deep learning model, the Adam optimization algorithm, an extension to stochastic gradient descent, was used as the optimization algorithm. Recently it has seen broader adoption of computer vision and natural language processing for deep learning applications.

Because adjusting the parameter learning rates looking at the average initial moments (the mean) as in RMSProp, Adam uses the sum of the gradient's second moments (the uncentric variance). The algorithm explicitly determines an exponential growth rate of the gradient and the square gradient. The beta1 and beta2 parameters regulate the decay rates of such moving averages. The learning rate used in our model is 0.0001, and the mean squad error is a loss function. Despite fitting all the images into RAM, we have fit the dataset through a batch generator. We run the model with 20 epochs. Then after training, we visualize the loss rate and accuracy rate. If the model loss and accuracy rate are not good, we go back to the pre-processing steps and follow the same flow and again train and visualization. After several times experiment, we have found an optimized model. Then we have saved the model into hdf5 file format, where the network architecture and the weight are stored.

After preparing the trained model, we have tested the model into a Udacity simulator in an autonomous model where the model is fitted with the simulator. The input image fits the model after pre-processing steps and predicts the steering angle following which the car moves forward. We have developed a prototype to experiment with how the self-driving car model works in real life from the perspective of Bangladesh. We assemble the hardware parts, including 4-wheel chassis board, 4 motors, L298 motor driver, Lipo battery, buck converter, raspberry pi, camera, power switch, etc.

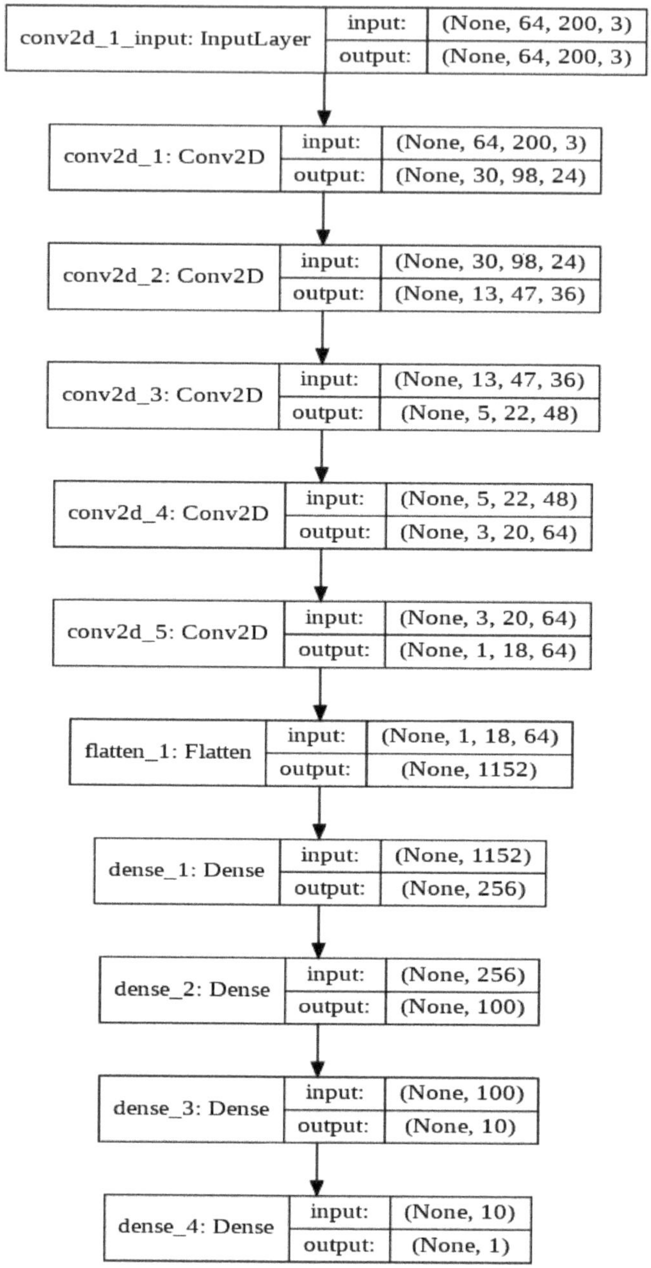

Figure 18. Experimented with Nvidia CNN architecture.

3.11. Driving Instruction through L298 Motor Driver

Four motors are connected with four wheels, and the L298 motor driver controls the direction and rotation of the motor. Battery power is distributed to the L298 motor driver and raspberry pi with a buck converter and a USB cable. A Noir Camera is placed in front of the camera and directly attached to the raspberry pi. The camera module captures the road video and passes the image to the raspberry pi at a 10 fps rate. The raspberry pi passes the image to the pc through server communication on TCP protocol. The pretrained model takes the image as input, predicts the steering angle, and transfers it back to the raspberry pi. Based on the steering angle, the raspberry pi commands the L298 motor driver to move accordingly.

The working procedure of the L298 motor driver is shown in Figure 19. The out pin 1, 2 is connected with the right motor and pin 3, 4 with the left motor. Enable pin 1 for the right motor that is connected with GPIO pin 4 of the raspberry pi. Similarly, enable pin 2 for the left motor with GPIO pin 27. Input pin 1, 2 of the L298 driver is connected with GPIO 17, 22 for the right motor and GPIO 23, 24 with input pin 3, 4 for the left motor.

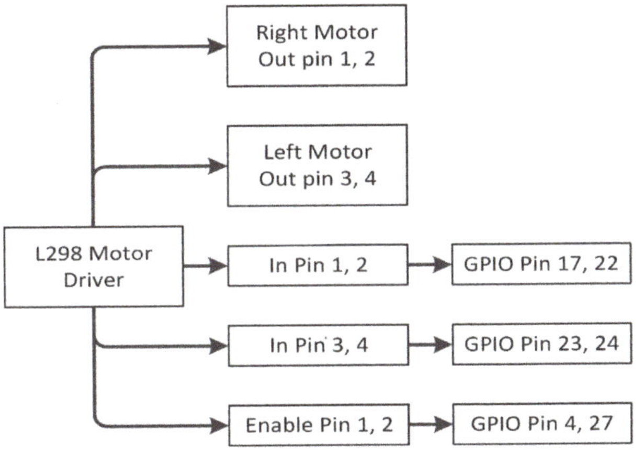

Figure 19. L298 Motor driver working process.

4. Experimental Result

We have developed a prototype to experiment with how the self-driving car model works in real life from the perspective of Bangladesh (Figure 20). First, we assemble the hardware parts, including 4-wheel chassis board, 4 motors, L298 motor driver, Lipo battery, buck converter, raspberry pi, camera, power switch etc. Then, four motors are connected with four wheels, and the L298 motor driver controls the direction and rotation of the motor. Finally, battery power is distributed to the L298 motor driver and raspberry pi with a buck converter and a USB cable.

Figure 20. Hardware assembly of self-driving car.

The performance of the car is tested on the actual road track. The vehicle is tested in both lightening and cloudy atmospheres to understand its behavior on the change of the environment. Figure 21a represents the lightening environment, and Figure 21b a little bit of a cloudy environment. In both environments, the car performs very well to maintain its track on the actual road.

The Noir camera module is placed in front of the camera and directly attached to the raspberry pi. The camera module captures the road video and passes the image to the raspberry pi at a 10 fps rate. The raspberry pi passes the image to the pc through server communication on TCP protocol. The pretrained model takes the image as input, predicts the steering angle, and transfers it to the raspberry pi. The car moves towards its direction based upon the steering angle. The loss rate of the deep learning CNN model is shown in Figure 22. From this scenery, we have seen that the loss rate is decreasing for both training and validation datasets regarding increasing the number of epochs. Validation loss and training loss difference is very well. Therefore, the model is neither overfitted nor underfit.

The accuracy of the model is measured in various environments or turning. Table 2 lists all the accuracy where on lightening conditions the model outperforms then cloudy climate. Similarly, right turning accuracy is 89.3% higher compared to straight and left turning.

In terms of accuracy, we compared our model to the previous literature in Table 3 as well. The temporal fusion process employed in the TCNN setup is temporal convolution. A fixed-length window of three (TCNN3) and nine (TCNN9) seconds was used. The performance of TCNN models continues to increase, and the larger the time horizon, the better. That's why TCNN9 accuracy is 84.6% better than TCNN3 83.3%. However, it needs a fixed size history window and is more memory intensive than the LSTM-based method. It performs similarly 84.5% to TCNN9 when using the CNN-LSTM method. While the Nvidia CNN architecture studied in our research shows an overall 89.2% that is notable than other configurations.

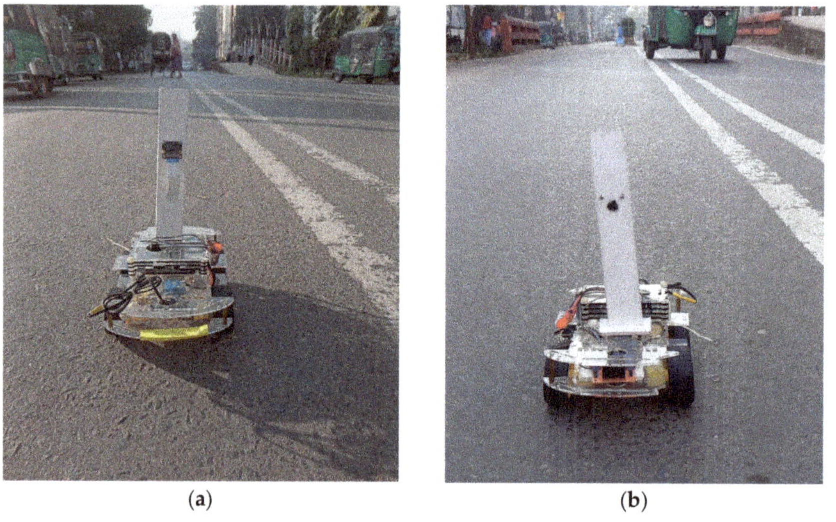

Figure 21. The performance of the car in the various environment (**a**) in lightening atmosphere (**b**) in a cloudy atmosphere.

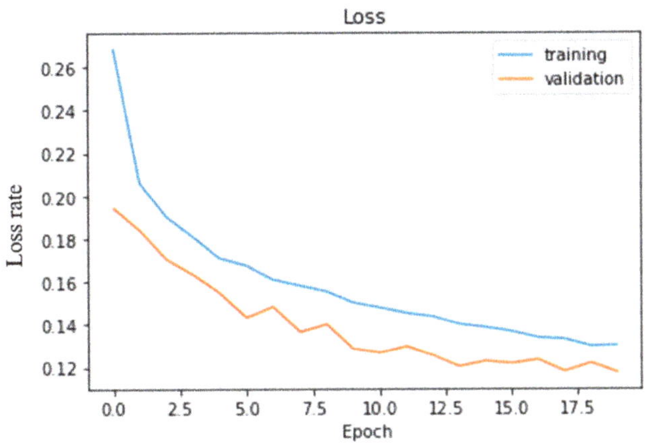

Figure 22. The loss rate of experimented CNN architecture.

Table 2. Performance of the model on various environment/turn.

Environment/Turning	Accuracy
Cloudy	88.9%
Lightening	89.6%
Left	87.1%
Right	89.3%
Straight	89.0%

Table 3. Accuracy comparison with previous literature proposed architecture.

Configuration	Accuracy
TCNN3	83.3%
TCNN9	84.6%
CNN-LSTM	84.5%
Nvidia CNN	89.2%

The performance of the whole autonomous product delivery car network is recorded on a per-frame basis. The camera sensor on the car passes 10 frames per second to the remotely connected high configuration pc via raspberry pi that requires 0.07 sec per frame. The image processing and to be predicted the steering angle requires 0.02 sec per frame. The steering angle info then sent back to the raspberry pi to drive the car accordingly requires another 0.03 sec. The network requires 0.12 sec per frame from image capture to prediction. The performance of the trained model is experimented with using the Udacity simulator in autonomous mode. A few snapshots from the various angle in autonomous mode have been demonstrated in Figure 23. Through socket programming, the Udacity simulator passes the road image to the model and predicts the steering angle. This steering angle back to the car, and the vehicle moves according to that angle. The predicted steering angle is shown in the top-left position of the Udacity simulator. Figure 23a is a sample for a left turn where we have found that the model predicts angle as $-12.16°$ and Figure 23b another rightly turned position and model predict $8.50°$ steering angle. More curved situations are also displayed in Figure 23c, complex right turn, and 23d, hill tracked right turn, where the model predicts $17.25°$ and $17.09°$, respectively. Furthermore, the predicted steering angle is shown in the command prompt at the left position of the images.

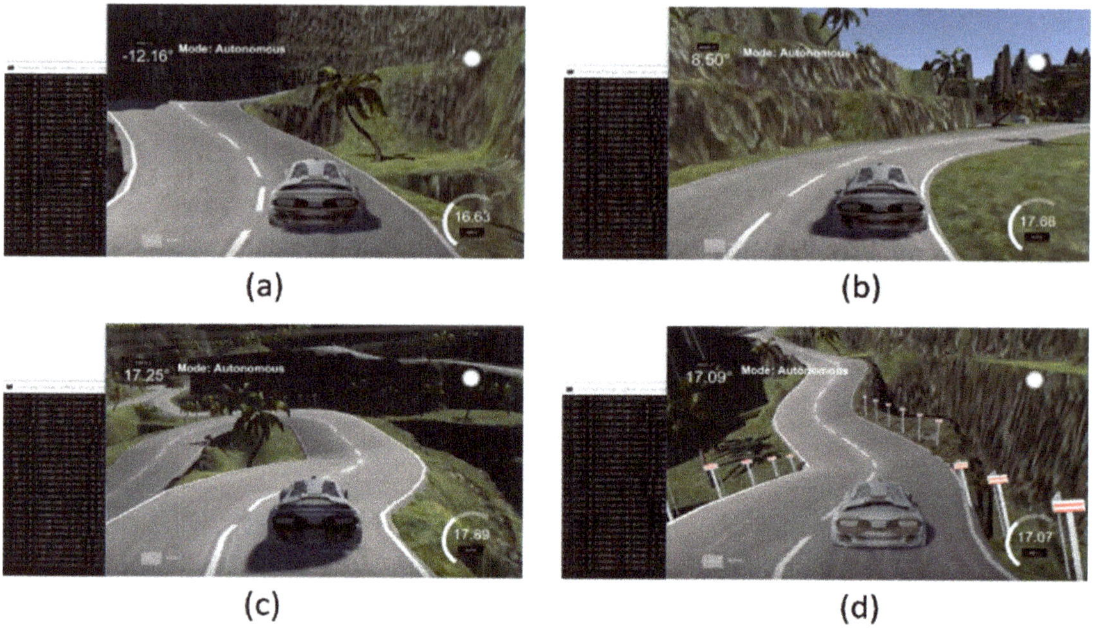

Figure 23. Performance of the model into autonomous mode (**a**) left turn prediction angle (**b**) right turn prediction angle (**c**) complex right turn prediction angle (**d**) hill tracked right turn prediction angle.

The visualization of the geographical position of the vehicle is one of the great features to track the car immediately. At the time of product delivery, the vehicle owner can track his vehicle at any time. Geoip2 library is used to track the car from its IP address.

After being given the IP address, geoip2 returns the geographical data of that vehicle. Those geographical data, i.e., longitude and latitude, are visualized through the Folium library of Python programming language. A demonstration of the current position of the self-driving product delivery vehicle is shown in Figure 24.

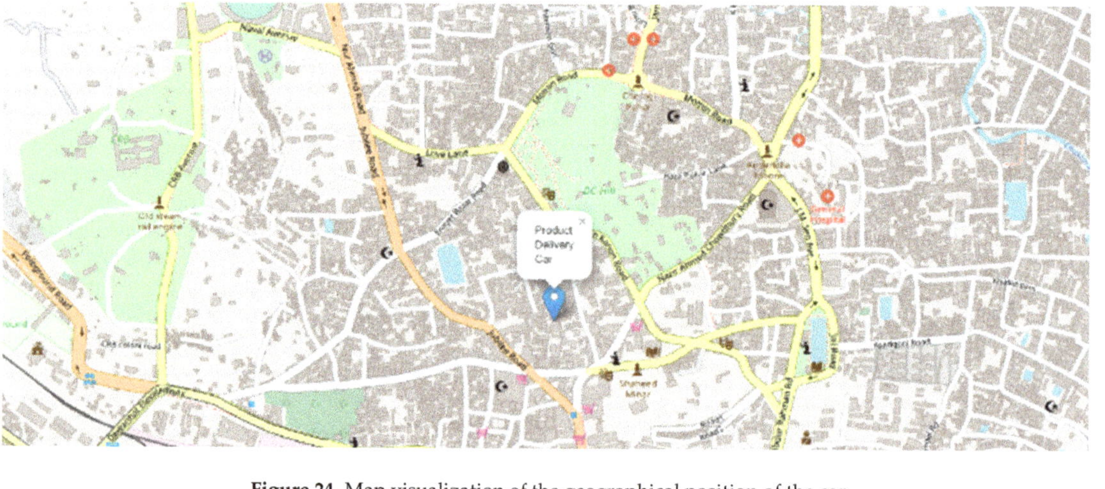

Figure 24. Map visualization of the geographical position of the car.

From the perspective of Bangladesh Road, the model is tested in several environments such as darkness, brightness distortion, gloomy atmosphere, etc., and performs at a satisfactory level. A source to destination position is shown in Figure 25. Because of the diverse environment augmentation to the original dataset, the vehicle is fit for the actual road of Bangladesh. We have experimented with the car at the Chittagong—Cox's Bazar highway at the Rahattarpul area. During the self-driving vehicle movement, we have stored several pre-processed images that are sent from the Raspberry pi attached camera module shown in Figure 26.

Figure 25. Map visualization of the source and destination place of the car.

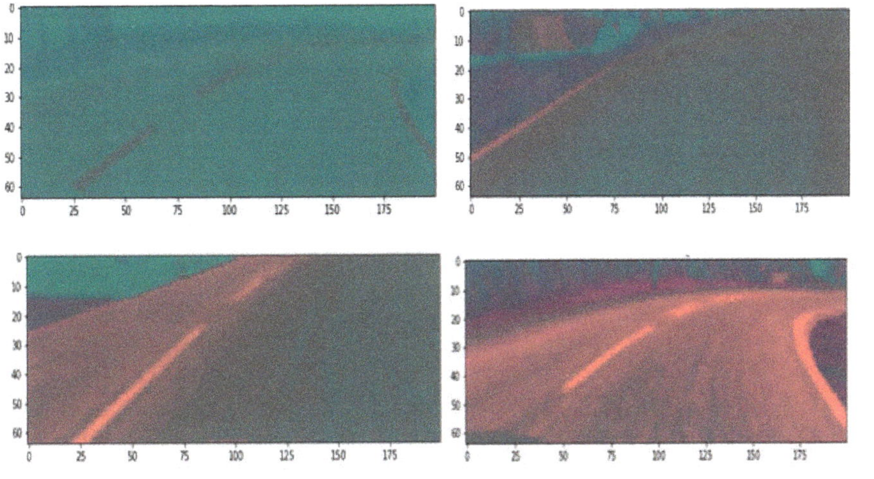

Figure 26. Sent Pre-processed image at testing time of the vehicle.

5. Discussion and Conclusions

There is a lot of trouble with on-time product delivery from Bangladesh's perspective, and human decision-making errors cause severe road accidents. Many drivers obey their feelings even though they are not correct in the moment. Thus, driving system automation will solve those problems. Therefore, autonomous vehicles can ensure on-time product delivery and reduce accident rates because of human error. We have developed Delicar, such a low-cost self-driving product delivery vehicle where the camera placed on the roof of the vehicle capture the image and raspberry pi sends the image to the pre-trained model for steering angle with respect to that image.

Moreover, it is low in cost and easy to implement. However, there does not exist any authentication system to receive the product. Anyone from the destination can receive the product that is a shortcoming of the study. Extensive chances of development in this work are kept open. Lots of essential features can be added to it in the future. To detect damages and holes in preceding the vehicles in the road using cameras and sensors and produce warning system is the future scope of our research along with double step authentication to receive the product such as password, fingerprint, etc. The future direction of the study also includes the most effective path programming and obstacle avoidance to reach the destination safely and quickly. The interplay of smart people, smart technology, and smart processes, which may be shown as the Smart Golden Triangle, eventually determines the success of smart cities. Such an intelligent product delivery car will drive the smart city to the next level. However, in Bangladesh, the traffic congestion costs 3.2 million working hours daily, BDT 200 billion annually. To ensure the traffic rules are followed and strictly avoid overtaking, the self-driving car is a great alternative. As an impact of such a solution, self-driving product delivery cars will contribute to the economy via utilizing very few human resources. This autonomous product delivery car will advance the e-commerce industry to the next level by ensuring on-time delivery. In supply chain management, self-driving product delivery cars may not only have a significant influence on logistics by lowering costs and delays, but they could also have a significant impact on distribution and manufacturing centers. Therefore, the Government should instantly install this proposed system to deliver the product in time to save Bangladesh's economic deterioration.

Author Contributions: Conceptualization and experiment design, M.K.A.C., A.K.M.M. and M.Z.U.; methodology, M.K.A.C.; data set generation, M.K.A.C. and K.A.M.S.; data validation, M.K.A.C. and K.A.M.S.; formal analysis, K.A.M.S.; investigation, M.K.A.C.; resources, M.K.A.C.; data curation, M.K.A.C.; writing—original draft preparation, M.K.A.C. and A.K.M.M.; prepared figures and/or tables, M.K.A.C., A.K.M.M.; writing—review and editing, M.K.A.C., A.K.M.M. and M.Z.U.; visualization, M.K.A.C.; supervision, A.K.M.M. and M.Z.U.; project administration, A.K.M.M. and M.Z.U. All authors have read and agreed to the published version of the manuscript.

Funding: This research received no external funding.

Data Availability Statement: Data is describing within the article. The data that support the findings of this study are available from the corresponding author upon reasonable request.

Conflicts of Interest: The authors declare no conflict of interest. The funders had no role in the design of the study; in the collection, analyses, or interpretation of data; in the writing of the manuscript, or in the decision to publish the results.

References

1. Maasum, A.K.M.; Chy, M.K.A.; Rahman, I.; Uddin, M.N.; Azam, K.I. An Internet of Things (IoT) based smart traffic management system: A context of Bangladesh. In Proceedings of the 2018 International Conference on Innovations in Science, Engineering and Technology (ICISET), Chittagong, Bangladesh, 27–28 October 2018; pp. 418–422.
2. Taj, F.W.; Masum, A.K.M.; Reza, S.T.; Chy, M.K.A.; Mahbub, I. Automatic accident detection and human rescue system: Assistance through communication technologies. In Proceedings of the 2018 International Conference on Innovations in Science, Engineering and Technology (ICISET), Chittagong, Bangladesh, 27–28 October 2018; pp. 496–500.
3. Najafzadeh, M.; Niazmardi, S. A Novel Multiple-Kernel Support Vector Regression Algorithm for Estimation of Water Quality Parameters. *Nat. Resour. Res.* **2021**, *30*, 3761–3775. [CrossRef]
4. Najafzadeh, M.; Homaei, F.; Farhadi, H. Reliability assessment of water quality index based on guidelines of national sanitation foundation in natural streams: Integration of remote sensing and data-driven models. *Artif. Intell. Rev.* **2021**, *54*, 4619–4651. [CrossRef]
5. Amiri-Ardakani, Y.; Najafzadeh, M. Pipe Break Rate Assessment While Considering Physical and Operational Factors: A Methodology Based on Global Positioning System and Data Driven Techniques. *Water Resour. Manag.* **2021**, *35*, 3703–3720. [CrossRef]
6. Saberi-Movahed, F.; Mohammadifard, M.; Mehrpooya, A.; Rezaei-Ravari, M.; Berahmand, K.; Rostami, M.; Karami, S.; Najafzadeh, M.; Hajinezhad, D.; Jamshidi, M. Decoding Clinical Biomarker Space of COVID-19: Exploring Matrix Factorization-based Feature Selection Methods. *medRxiv* **2021**. [CrossRef]
7. Kong, L. A study on the AI-based online triage model for hospitals in sustainable smart city. *Future Gener. Comput. Syst.* **2021**, *125*, 59–70. [CrossRef]
8. Lv, Z.; Qiao, L.; Kumar Singh, A.; Wang, Q. AI-empowered IoT security for smart cities. *ACM Trans. Internet Technol.* **2021**, *21*, 1–21. [CrossRef]

9. Masum, A.K.M.; Chy, M.K.A.; Hasan, M.T.; Sayeed, M.H.; Reza, S.T. Smart Meter with Load Prediction Feature for Residential Customers in Bangladesh. In Proceedings of the 2019 International Conference on Energy and Power Engineering (ICEPE), Dhaka, Bangladesh, 14–16 March 2019; pp. 1–6.
10. Masum, A.K.M.; Saveed, M.H.; Chy, M.K.A.; Hasan, M.T.; Reza, S.T. Design and Implementation of Smart Meter with Load Forecasting Feature for Residential Customers. In Proceedings of the 2019 International Conference on Electrical, Computer and Communication Engineering (ECCE), Cox'sBazar, Bangladesh, 7–9 February 2019; pp. 1–6.
11. Bhat, S.A.; Huang, N.-F. Big Data and AI Revolution in Precision Agriculture: Survey and Challenges. *IEEE Access* **2021**, *9*, 110209–110222. [CrossRef]
12. Jung, J.; Maeda, M.; Chang, A.; Bhandari, M.; Ashapure, A.; Landivar-Bowles, J. The potential of remote sensing and artificial intelligence as tools to improve the resilience of agriculture production systems. *Curr. Opin. Biotechnol.* **2021**, *70*, 15–22. [CrossRef] [PubMed]
13. Chy, M.K.A.; Masum, A.K.M.; Hossain, M.E.; Alam, M.G.R.; Khan, S.I.; Alam, M.S. A Low-Cost Ideal Fish Farm Using IoT: In the Context of Bangladesh Aquaculture System. In *Inventive Communication and Computational Technologies*; Springer: Cham, Switzerland, 2020; pp. 1273–1283.
14. Zhang, K.; Aslan, A.B. AI technologies for education: Recent research & future directions. *Comput. Educ. Artif. Intell.* **2021**, 2, 100025.
15. Elshafey, A.E.; Anany, M.R.; Mohamed, A.S.; Sakr, N.; Aly, S.G. Dr. Proctor: A Multi-modal AI-Based Platform for Remote Proctoring in Education. In *Artificial Intelligence in Education, Proceedings of the International Conference on Artificial Intelligence in Education, Utrecht, The Netherlands, 14–18 June 2021*; Springer: Cham, Switzerland, 2021; pp. 145–150.
16. Lee, D.; Yoon, S.N. Application of artificial intelligence-based technologies in the healthcare industry: Opportunities and challenges. *Int. J. Environ. Res. Public Health* **2021**, *18*, 271. [CrossRef] [PubMed]
17. Davahli, M.R.; Karwowski, W.; Fiok, K.; Wan, T.; Parsaei, H.R. Controlling Safety of Artificial Intelligence-Based Systems in Healthcare. *Symmetry* **2021**, *13*, 102. [CrossRef]
18. Apell, P.; Eriksson, H. Artificial intelligence (AI) healthcare technology innovations: The current state and challenges from a life science industry perspective. *Technol. Anal. Strateg. Manag.* **2021**, 1–15. [CrossRef]
19. Krizhevsky, A.; Sutskever, I.; Hinton, G.E. Imagenet classification with deep convolutional neural networks. *Adv. Neural Inf. Process. Syst.* **2012**, *25*, 1097–1105. [CrossRef]
20. Fang, H.; Gupta, S.; Iandola, F.; Srivastava, R.K.; Deng, L.; Dollár, P.; Gao, J.; He, X.; Mitchell, M.; Platt, J.C. From captions to visual concepts and back. In Proceedings of the IEEE Conference on Computer Vision and Pattern Recognition, Boston, MA, USA, 7–12 June 2015; pp. 1473–1482.
21. Redmon, J.; Divvala, S.; Girshick, R.; Farhadi, A. You only look once: Unified, real-time object detection. In Proceedings of the IEEE Conference on Computer Vision and Pattern Recognition, Las Vegas, NV, USA, 27–30 June 2016; pp. 779–788.
22. Long, J.; Shelhamer, E.; Darrell, T. Fully convolutional networks for semantic segmentation. In Proceedings of the IEEE Conference on Computer Vision and Pattern Recognition, Boston, MA, USA, 7–12 June 2015; pp. 3431–3440.
23. U.S. Department of Transportation. *Automated Driving Systems—A Vision for Safety*. Available online: https://www.nhtsa.gov/sites/nhtsa.dot.gov/files/documents/13069a-ads2.0_090617_v9a_tag.pdf (accessed on 3 April 2019).
24. Badue, C.; Guidolini, R.; Carneiro, R.V.; Azevedo, P.; Cardoso, V.B.; Forechi, A.; Jesus, L.; Berriel, R.; Paixao, T.M.; Mutz, F. Self-driving cars: A survey. *Expert Syst. Appl.* **2021**, *165*, 113816. [CrossRef]
25. Daily, M.; Medasani, S.; Behringer, R.; Trivedi, M. Self-driving cars. *Computer* **2017**, *50*, 18–23. [CrossRef]
26. Alam, S.; Sulistyo, S.; Mustika, I.W.; Adrian, R. Review of potential methods for handover decision in v2v vanet. In Proceedings of the 2019 International Conference on Computer Science, Information Technology, and Electrical Engineering (ICOMITEE), Jember, Indonesia, 16–17 October 2019; pp. 237–243.
27. Baza, M.; Nabil, M.; Mahmoud, M.M.E.A.; Bewermeier, N.; Fidan, K.; Alasmary, W.; Abdallah, M. Detecting sybil attacks using proofs of work and location in vanets. *IEEE Trans. Dependable Secur. Comput.* **2020**. [CrossRef]
28. Schmittner, C.; Chlup, S.; Fellner, A.; Macher, G.; Brenner, E. ThreatGet: Threat modeling based approach for automated and connected vehicle systems. In Proceedings of the AmE 2020-Automotive meets Electronics; 11th GMM-Symposium, Dortmund, Germany, 10–11 March 2020; pp. 1–3.
29. Cui, J.; Liew, L.S.; Sabaliauskaite, G.; Zhou, F. A review on safety failures, security attacks, and available countermeasures for autonomous vehicles. *Ad Hoc Networks* **2019**, *90*, 101823. [CrossRef]
30. Dibaei, M.; Zheng, X.; Jiang, K.; Maric, S.; Abbas, R.; Liu, S.; Zhang, Y.; Deng, Y.; Wen, S.; Zhang, J. An overview of attacks and defences on intelligent connected vehicles. *arXiv* **2019**, arXiv:1907.07455. preprint.
31. Levine, W.S. *The Control Handbook*; CRC Press: Boca Raton, FL, USA, 2018.
32. Bojarski, M.; Del Testa, D.; Dworakowski, D.; Firner, B.; Flepp, B.; Goyal, P.; Jackel, L.D.; Monfort, M.; Muller, U.; Zhang, J. End to end learning for self-driving cars. *arXiv* **2016**, arXiv:1604.07316. preprint.
33. Levine, S.; Finn, C.; Darrell, T.; Abbeel, P. End-to-end training of deep visuomotor policies. *J. Mach. Learn. Res.* **2016**, *17*, 1334–1373.
34. Pomerleau, D.A. *Alvinn: An Autonomous Land Vehicle in a Neural Network*; Artificial Intelligence And Psychology Project; Carnegie-Mellon University: Pittsburgh, PA, USA, 1989.
35. Lecun, Y.; Cosatto, E.; Ben, J.; Muller, U.; Flepp, B. *Dave: Autonomous Off-Road Vehicle Control Using End-to-End Learning*. DARPA-IPTO Final Report. 2004. Available online: https://cs.nyu.edu/~{}yann/research/dave/ (accessed on 15 February 2019).

36. N.T. Report. *GPU-Based Deep Learning Inference: A Performance and Power Analysis*. Available online: http://developer.download.nvidia.com/embedded/jetson/TX1/docs/jetson_tx1_whitepaper.pdf?autho=1447264273_0fafa14fcc7a1f685769494ec9b0fcad&file=jetson_tx1_whitepaper.pdf (accessed on 23 May 2019).
37. Masum, A.K.M.; Rahman, M.A.; Abdullah, M.S.; Chowdhury, S.B.S.; Khan, T.B.F.; Raihan, M.K. A Supervised Learning Approach to An Unmanned Autonomous Vehicle. In Proceedings of the 2019 International Conference on Intelligent Computing and Control Systems (ICCS), Madurai, India, 15–17 May 2019; pp. 1549–1554.
38. Stavens, D.; Thrun, S. A self-supervised terrain roughness estimator for off-road autonomous driving. *arXiv* **2012**, arXiv:1206.6872. preprint.
39. Hadsell, R.; Sermanet, P.; Ben, J.; Erkan, A.; Scoffier, M.; Kavukcuoglu, K.; Muller, U.; LeCun, Y. Learning long-range vision for autonomous off-road driving. *J. Field Robot.* **2009**, *26*, 120–144. [CrossRef]
40. Deb, S.; Strawderman, L.; Carruth, D.W.; DuBien, J.; Smith, B.; Garrison, T.M. Development and validation of a questionnaire to assess pedestrian receptivity toward fully autonomous vehicles. *Transp. Res. Part C Emerg. Technol.* **2017**, *84*, 178–195. [CrossRef]
41. Nordhoff, S.; De Winter, J.; Kyriakidis, M.; Van Arem, B.; Happee, R. Acceptance of driverless vehicles: Results from a large cross-national questionnaire study. *J. Adv. Transp.* **2018**, *2018*, 5382192. [CrossRef]
42. Robertson, R.D.; Meister, S.R.; Vanlaar, W.G.; Hing, M.M. Automated vehicles and behavioural adaptation in Canada. *Transp. Res. Part A Policy Pract.* **2017**, *104*, 50–57. [CrossRef]
43. Reke, M.; Peter, D.; Schulte-Tigges, J.; Schiffer, S.; Ferrein, A.; Walter, T.; Matheis, D. A self-driving car architecture in ROS2. In Proceedings of the 2020 International SAUPEC/RobMech/PRASA Conference, Cape Town, South Africa, 29–31 January 2020; pp. 1–6.
44. Bakioglu, G.; Atahan, A.O. AHP integrated TOPSIS and VIKOR methods with Pythagorean fuzzy sets to prioritize risks in self-driving vehicles. *Appl. Soft Comput.* **2021**, *99*, 106948. [CrossRef]
45. Chen, C.; Demir, E.; Huang, Y.; Qiu, R. The adoption of self-driving delivery robots in last mile logistics. *Transp. Res. Part E Logist. Transp. Rev.* **2021**, *146*, 102214. [CrossRef]
46. Li, L.; Lin, Y.-L.; Zheng, N.-N.; Wang, F.-Y.; Liu, Y.; Cao, D.; Wang, K.; Huang, W.-L. Artificial intelligence test: A case study of intelligent vehicles. *Artif. Intell. Rev.* **2018**, *50*, 441–465. [CrossRef]
47. Guo, Y.; Chen, X.; Yang, Y. Multimodal transport distribution model for autonomous driving vehicles based on improved ALNS. *Alex. Eng. J.* **2021**, *61*, 2939–2958. [CrossRef]
48. Dommès, A.; Merlhiot, G.; Lobjois, R.; Dang, N.-T.; Vienne, F.; Boulo, J.; Oliver, A.-H.; Cretual, A.; Cavallo, V. Young and older adult pedestrians' behavior when crossing a street in front of conventional and self-driving cars. *Accid. Anal. Prev.* **2021**, *159*, 106256. [CrossRef]
49. Deruyttere, T.; Milewski, V.; Moens, M.-F. Giving commands to a self-driving car: How to deal with uncertain situations? *Eng. Appl. Artif. Intell.* **2021**, *103*, 104257. [CrossRef]
50. Zhou, T.; Brown, M.; Snavely, N.; Lowe, D.G. Unsupervised learning of depth and ego-motion from video. In Proceedings of the IEEE Conference on Computer Vision and Pattern Recognition, Honolulu, HI, USA, 21–26 July 2017; pp. 1851–1858.
51. Gu, Y.; Hashimoto, Y.; Hsu, L.-T.; Iryo-Asano, M.; Kamijo, S. Human-like motion planning model for driving in signalized intersections. *IATSS Res.* **2017**, *41*, 129–139. [CrossRef]
52. Katrakazas, C.; Quddus, M.; Chen, W.-H.; Deka, L. Real-time motion planning methods for autonomous on-road driving: State-of-the-art and future research directions. *Transp. Res. Part C Emerg. Technol.* **2015**, *60*, 416–442. [CrossRef]
53. Mostafa, M.S.B.; Masum, A.K.M.; Uddin, M.S.; Chy, M.K.A.; Reza, S.T. Amphibious Line following Robot for Product Delivery in Context of Bangladesh. In Proceedings of the 2019 International Conference on Electrical, Computer and Communication Engineering (ECCE), Cox'sBazar, Bangladesh, 7–9 February 2019; pp. 1–6.
54. Colak, I.; Yildirim, D. Evolving a Line Following Robot to use in shopping centers for entertainment. In Proceedings of the 2009 35th Annual Conference of IEEE Industrial Electronics, Porto, Portugal, 3–5 November 2009; pp. 3803–3807.
55. Islam, M.; Rahman, M. Design and fabrication of line follower robot. *Asian J. Appl. Sci. Eng.* **2013**, *2*, 27–32.
56. Punetha, D.; Kumar, N.; Mehta, V. Development and applications of line following robot based health care management system. *Int. J. Adv. Res. Comput. Eng. Technol. (IJARCET)* **2013**, *2*, 2446–2450.
57. Wolcott, R.W.; Eustice, R.M. Robust LIDAR localization using multiresolution Gaussian mixture maps for autonomous driving. *Int. J. Robot. Res.* **2017**, *36*, 292–319. [CrossRef]
58. Ahmad, T.; Ilstrup, D.; Emami, E.; Bebis, G. Symbolic road marking recognition using convolutional neural networks. In Proceedings of the 2017 IEEE intelligent vehicles symposium (IV), Los Angeles, CA, USA, 11–14 June 2017; pp. 1428–1433.
59. Greenhalgh, J.; Mirmehdi, M. Detection and Recognition of Painted Road Surface Markings. In Proceedings of the ICPRAM (1), Lisbon, Portugal, 10–12 January 2015; pp. 130–138.
60. Hyeon, D.; Lee, S.; Jung, S.; Kim, S.-W.; Seo, S.-W. Robust road marking detection using convex grouping method in around-view monitoring system. In Proceedings of the 2016 IEEE Intelligent Vehicles Symposium (IV), Gothenburg, Sweden, 19–22 June 2016; pp. 1004–1009.
61. Chollet, F. *Keras*; Github: San Francisco, CA, USA, 2015; Available online: https://github.com/fchollet/keras (accessed on 17 April 2019).
62. Nikitas, A.; Njoya, E.T.; Dani, S. Examining the myths of connected and autonomous vehicles: Analysing the pathway to a driverless mobility paradigm. *Int. J. Automot. Technol. Manag.* **2019**, *19*, 10–30. [CrossRef]

63. Evans, J. Governing cities for sustainability: A research agenda and invitation. *Front. Sustain. Cities* **2019**, *1*, 2. [CrossRef]
64. Choi, Y.; Rhee, S.-W. Current status and perspectives on recycling of end-of-life battery of electric vehicle in Korea (Republic of). *Waste Manag.* **2020**, *106*, 261–270. [CrossRef] [PubMed]
65. Bollinger, B.L. The Security and Privacy In Your Car Act: Will It Actually Protect You? *North Carol. J. Law Technol.* **2017**, *18*, 214.
66. Lim, H.S.M.; Taeihagh, A. Autonomous vehicles for smart and sustainable cities: An in-depth exploration of privacy and cybersecurity implications. *Energies* **2018**, *11*, 1062. [CrossRef]
67. Thompson, N.; Mullins, A.; Chongsutakawewong, T. Does high e-government adoption assure stronger security? Results from a cross-country analysis of Australia and Thailand. *Gov. Inf. Q.* **2020**, *37*, 101408. [CrossRef]
68. Feng, S.; Feng, Y.; Yan, X.; Shen, S.; Xu, S.; Liu, H.X. Safety assessment of highly automated driving systems in test tracks: A new framework. *Accid. Anal. Prev.* **2020**, *144*, 105664. [CrossRef]
69. Van Brummelen, J.; O'Brien, M.; Gruyer, D.; Najjaran, H. Autonomous vehicle perception: The technology of today and tomorrow. *Transp. Res. Part C Emerg. Technol.* **2018**, *89*, 384–406. [CrossRef]
70. SullyChen. Autopilot-TensorFlow. Available online: https://github.com/SullyChen/Autopilot-TensorFlow (accessed on 3 April 2019).
71. Apollo. Apollo Data Open Platform. Available online: http://data.apollo.auto/?locale=en-us&lang=en (accessed on 3 April 2019).
72. Santana, E.; Hotz, G. Learning a driving simulator. *arXiv* **2016**, arXiv:1608.01230. preprint.
73. Yin, H.; Berger, C. When to use what data set for your self-driving car algorithm: An overview of publicly available driving datasets. In Proceedings of the 2017 IEEE 20th International Conference on Intelligent Transportation Systems (ITSC), Yokohama, Japan, 16–19 October 2017; pp. 1–8.
74. Udacity. Self-Driving-Car. Available online: https://github.com/udacity/self-driving-car/tree/master/datasets (accessed on 5 May 2019).
75. Udacity. Self-Driving-Car-Sim. Available online: https://github.com/udacity/self-driving-car-sim (accessed on 5 May 2019).

sensors

MDPI

Communication

Data Fusion Based on an Iterative Learning Algorithm for Fault Detection in Wind Turbine Pitch Control Systems

Leonardo Acho *,† and Gisela Pujol-Vázquez †

Department of Mathematics, Universitat Politècnica de Catalunya-BarcelonaTech (ESEIAAT), 08222 Terrassa, Spain; gisela.pujol@upc.edu
* Correspondence: leonardo.acho@upc.edu
† These authors contributed equally to this work.

Abstract: In this article, we propose a recent iterative learning algorithm for sensor data fusion to detect pitch actuator failures in wind turbines. The development of this proposed approach is based on iterative learning control and Lyapunov's theories. Numerical experiments were carried out to support our main contribution. These experiments consist of using a well-known wind turbine hydraulic pitch actuator model with some common faults, such as high oil content in the air, hydraulic leaks, and pump wear.

Keywords: data fusion; iterative learning; fault detection; pitch system; wind turbines

Citation: Acho, L.; Pujol-Vázquez, G. Data Fusion Based on an Iterative Learning Algorithm for Fault Detection in Wind Turbine Pitch Control Systems. *Sensors* **2021**, *21*, 8437. https://doi.org/10.3390/s21248437

Academic Editor: Natividad Duro Carralero

Received: 7 November 2021
Accepted: 16 December 2021
Published: 17 December 2021

1. Introduction

Data fusion is a mathematical discipline that deals with the acquisition, processing, and combination of synergies of information gathered from sensors [1]. Data fusion can be defined as the combination of data and information from different sources, to obtain improved information [2]. This data fusion is usually done to analyze and understand a phenomenon [3–5], for instance a system malfunction. Data fusion techniques are present in a wide range of applications, such as smart city applications [6], allowing to manage multiple data sources; food analysis context [7]; guidance and control of autonomous vehicles [8]; medical studies [9], and so on. In addition, there are different analysis methods that combine data from different sources, where the most common options are algorithms based on optimization [10], multiblock (or multitable) methods [11], and statistical data fusion [12]. In our article, we used an original statistical parametric identification to perform data fusion, where covariance of sensory information is not required, which is generally not available.

Moreover, the data sensor has been useful to detect possible failures in the pitch actuator systems of wind turbines [13–16]. Furthermore, it turns out that a parameterized plant modeling can be a key factor for efficient fault diagnosis based on the fusion of sensor data. If in addition, only a single sensor is used for the data fusion process, an option to generate data to be merged is through an iterative process. On the other hand, it is well known that the iterative learning process can improve your performance on repetitive tasks in a finite period of time [17–19]. Therefore, the main objective of this article is to develop a fusion of a sensor data based on adaptive iterative learning. This process will provide data in each periodic cycle task that will be further analyzed for fault diagnosis in the wind turbine pitch actuators.

The pitch system of a wind turbine adjusts the pitch angle of the blade by turning it. In the case of a three-bladed wind turbine, there are generally three identical pitch actuators [15,20]. This part of a wind turbine is responsible for capturing the wind to convert it into mechanical energy and then into electrical one. Therefore, if the pitch actuator system has a fault, the energy efficiency conversion will be affected, among others mechanical and structural wear. Some common faults are high oil content in the air, hydraulic leaks, and pump wear. Our data fusion approach is capable of detecting these types of failures.

The iterative learning theory used for adaptive learning of a process is a key factor in many iterative learning control frameworks [17–19]. Therefore, an appropriate mathematical model of the pitch system will be important, and as simple as possible to perform a simple adaptive iterative learning method to our main objective. Furthermore, an iterative learning control has been used to improve control performance of proportional controllers and derivative ones [21]. Simulation results are given in [21] to support this affirmation. In [22], an iterative learning control theory is employed in a first-order hyperbolic system that helps improve controller robustness on desired time-varying trajectories. This is also supported by performing numerical examples of given hyperbolic systems. In addition, [23] shows the realization of the synchronization of non-identical neural network systems that have a variable delay in time coupled by means of an iterative learning control. According to the simulation results shown in [23], the synchronization objective is satisfied. Finally, in [24], iterative learning control is applied to a novel computational fluid dynamics model to show the performance of the controller in improving the aerodynamic load of wind turbines. In the same way, we will use numerical results to support our main contribution.

The rest of the structure of this article is as follows. Section 2 presents our data fusion approach by using a simple mathematical model of the pitch system, and the use of an adaptive iterative learning framework based on Lyapunov's theory. Section 3 shows the results of numerical simulations and followed by Section 4, where the advantages of the proposed method are discussed. Finally, a summary is presented in Section 5.

2. Wind Turbine Mathematical Modeling

We use the following mathematical model of a tone actuator system [25]:

$$\dot{\beta}(t) = -\frac{1}{\tau}\beta(t) + \frac{1}{\tau}u_p(t), \tag{1}$$

where $u_p(t)$, $\beta(t)$, and τ are the pitch angle command, the pitch angle, and the system time constant, respectively. This mathematical model is a simple one of the following more exact model (see [14] and references there in):

$$\ddot{\beta}(t) = -2\zeta\omega_n\dot{\beta}(t) - \omega_n^2(\beta(t) - u_p(t)) \tag{2}$$

where, once again, $\beta(t)$ is the pitch angle and $u_p(t)$ is the pitch angle command; ω_n and ζ are the natural frequency and damping, respectively. Table 1 shows the healthy and faulty scenarios for a wind turbine. Therefore, our data fusion approach, for design, will use the simple model (1) and, in testing, the second model (2) under the different scenarios presented in Table 1.

Table 1. Common faulty scenarios [26].

Scenario	Abbreviation
No fault	H
High air oil content	F_1
Hydraulic leakage	F_2
Pump wear	F_3

3. Results

In this section, the statements of the iterative learning algorithm, applied to sensor data fusion, to detect pitch actuator failures, are stated.

3.1. Adaptive Iterative Learning Approach

The adaptive iterative learning control scheme is based on performing repetitive tasks to obtain a parameter estimation. In our case, we repeat a trajectory tracking of a wind turbine, where the unknown parameter comes from the system time constant. To do so, we rewrite system (1) as follows:

$$\dot{\beta}(t) = -\theta(\beta(t) - u_p(t)), \tag{3}$$

where $\theta = 1/\tau$ is considered an unknown parameter. Then, adaptive iterative learning deals with finding a periodic learning dynamic to observe the parameter θ that governs the pitch dynamics.

First, we have to consider the following assumptions about wind turbine pitch actuator systems:

(A1) The angle pitch dynamic is limited. That is, there exists a positive constant $\beta_M \in \mathbb{R}$, such that $0 < \beta(t) < \beta_M$ for all $t \geq 0$.
(A2) The systems (1) and (2) are bounded–input–bounded–output (BIBO)-stable dynamics. Hence, θ in (3) should be a positive constant parameter and assumed unknown.
(A3) The pitch angle command is bounded. That is, there exists a positive constant $u_M \in \mathbb{R}$, such that $| u_p(t) | \leq u_M$ for all $t \geq 0$.

We define now the next adaptive iterative learning algorithm defined over the time interval $t \in [0, T]$:

$$\begin{aligned} (1-\gamma)\hat{\theta}_k(t) = \gamma\hat{\theta}_{k-1}(t) + \gamma|\beta(t) - u_p(t)|\,, \\ \hat{\theta}_k(0) = \hat{\theta}_{k-1}(T)\,, \\ \hat{\theta}_0(t) = \theta_{ini}\,, \end{aligned} \tag{4}$$

where k denotes the k-th learning cycle, or iteration number. The rest of values are: θ_{ini} is a constant parameter; $0 < \gamma < 1$ is the parameter set by the user; and T is the time-interval of the iterative process. The above adaptive iterative dynamic is a special case to the one proposed, for instance, in [18]. Hence, this dynamic is a kind of parameter observer to θ in (3).

3.2. Data Fusion Design

Now, we present how to perform the data fusion of the experimental data, raw (4), obtaining for each iteration a significant information able to characterize the data. To do this, the boundedness of $\hat{\theta}_k(t)$ (4) must be established first, and then the data fusion can be accomplished.

To begin with the main result, let use define the following $L_{\infty e}$ norm [18]:

$$\| x(t) \|_{\infty e} = \sup_{0 \leq t \leq T} \| x(t) \|, \tag{5}$$

where $\| \cdot \|$ denotes any vector norm. If the above norm exists, then $x(t) \in L_{\infty e}[0, T]$.

Next, we have the following result that ensures the boundedness of the iteration method (4); that is, $\hat{\theta}_k(t)$ remains in a bounded region for $t \in [0, T]$ and for each iteration k.

Theorem 1. *The iterative learning process proposed in (4) linked to (3), and under the assumptions (A1)–(A3), produces $\hat{\theta}(t) \in L_{\infty e}[0, T]$ for each iteration process $k = 1, 2, \ldots$.*

The proof of Theorem 1 consists on consider an energy function related to each iteration, and show that its sequence is bounded. Then, we can ensure that the sequence of parameter $\hat{\theta}_k(t)$ is also bounded in $[0, T]$; that is, $\hat{\theta}_k(t) \in L_{\infty e}[0, T]$ for $k = 1, 2, \ldots$.

Proof. To prove the theorem, it is sufficient to see that the dynamic of (4) remains bounded. Let us define the following positive definite functional $V_k(T)$ as a Lyapunov-like function [18]:

$$V_k(T) = \alpha_1 \int_0^T \hat{\theta}_k^2(t)dt \tag{6}$$

The difference of the $V_k(t)$ is given by

$$\begin{aligned} \Delta V_k(T) &= V_k(T) - V_{k-1}(T) \\ &= \int_0^T (\hat{\theta}_k^2(t) - \hat{\theta}_{k-1}^2(t))dt \end{aligned} \tag{7}$$

Let us first simplify the integral term $\hat{\theta}_k^2(t)$ in (7). Hence, using (4), we get:

$$\begin{aligned} \hat{\theta}_k^2(t) &= \left(\tfrac{\gamma}{1-\gamma}\right)^2 (\hat{\theta}_{k-1}(t) + \mid \beta(t) - u_p(t) \mid)^2 \\ &= \left(\tfrac{\gamma}{1-\gamma}\right)^2 \left(\hat{\theta}_{k-1}^2(t) + \mid \beta(t) - u_p(t) \mid^2 + 2 \mid \beta(t) - u_p(t) \mid \hat{\theta}_{k-1}(t)\right) \end{aligned} \tag{8}$$

Then, using (8), the difference ΔV_k (7) becomes:

$$\begin{aligned} \Delta V_k(T) &= \int_0^T \hat{\theta}_k^2(t) - \hat{\theta}_{k-1}^2(t)\,dt \\ &= \int_0^T \left(\tfrac{\gamma^2}{(1-\gamma)^2} - 1\right)\hat{\theta}_{k-1}^2(t) + \tfrac{\gamma^2}{(1-\gamma)^2} \mid \beta(t) - u_p(t) \mid^2 + \\ &\quad 2\tfrac{\gamma^2}{(1-\gamma)^2} \mid -\beta(t) + u_p(t) \mid \hat{\theta}_{k-1}(t)\,dt \end{aligned} \tag{9}$$

Now, we define γ such that $\frac{\gamma^2}{(1-\gamma)^2} - 1 \leq 0$. Then, we get:

$$\begin{aligned} \Delta V_k(T) &\leq \int_0^T \tfrac{\gamma^2}{(1-\gamma)^2} \mid \beta(t) - u_p(t) \mid^2 + \\ &\quad 2\tfrac{\gamma^2}{(1-\gamma)^2} \mid \beta(t) - u_p(t) \mid \hat{\theta}_{k-1}(t)\,dt \end{aligned} \tag{10}$$

The boundedness of $V_k(T)$ (6) is concluded because $\beta(t)$ and $u_p(t)$ are bounded signals. From assumptions (A1) and (A3), (10) satisfies:

$$\Delta V_k(T) \leq \frac{\gamma^2}{(1-\gamma)^2}\left((\beta_M + u_M)^2 T + 2(\beta_M + u_M)\int_0^T \hat{\theta}_{k-1}(t)dt\right). \tag{11}$$

Finally, taking into account that $\hat{\theta}_k(t)$ is a continuous function in $[0, T]$, for each k-th iteration, we conclude that the integral term in (11) is achievable. Hence, $\Delta V_k(t) < \infty$, and $\tilde{\theta}_k(t) \in L_{\infty e}$ for all k, and then $\hat{\theta}_k(t) \in L_{\infty e}$. So, $\hat{\theta}(t) \in C[0, T]$ for each iteration. □

Once we ensure the boundedness of the parameter estimation, we can present the data fusion scheme. The data fusion process employs experimental raw data to extract its arithmetic mean that describes it. Then, a new data raw $\psi(t)$ is obtained with improved information. The data fusion block performs the following mathematical operation:

$$\psi(t) = \frac{1}{nT}\sum_{k=0}^{\lfloor nT \rceil} \hat{\theta}_k(t), t \in [0, T], \tag{12}$$

where $\lfloor nT \rceil$ is the nearest integer of nT, and it corresponds to the n iterative cycles, where each cycle is ran for $t \in [0, T]$. Therefore nT is the entire simulation time. Notice that all the sensor data obtained from the iterative process $\hat{\theta}_k(t)$ (4) are merged into a single data raw $\psi(t)$. Then, this new data $\psi(t)$ is obtained under each scenario presented in Tables 1 and 2, and must be compared to the healthy model to establish a detection fault algorithm.

Table 2. Parameters for hydraulic pitch system under common faulty scenarios [26].

Scenario	Parameter ω_n (rad/s)	Parameter ζ
H	11.11	0.6
F_1	5.73	0.45
F_2	3.42	0.9
F_3	7.27	0.74

3.3. Fault Detection Algorithm

We now propose a diagnosis of pitch actuator failures based on the fusion data theory. Based on Theorem 1, Figure 1 shows the health monitoring system proposed for the diagnosis of failures in actuator devices in wind turbines. The data employed here are presented in Table 2 [26]. The diagnosis is based on the following steps. Under each scenario in Table 1, a data fusion raw $\psi(t)$ (12) is obtained. Then, a decision parameter m is defined in each case. First, a healthy value of m for the nominal plant H (healthy scenario in Table 1) is derived, referred to as m_H in Figure 1. Secondly, under each faulty case, parameter m is evaluated and compared to m_H to decide if a failure occurs.

First, let us define parameter m: it corresponds to the regression of data fusion raw. That is, a linear relation is used to fit our data $(t, \psi(t))$ in (12) to a polynomial function of degree one, and by using minimum squares method. Therefore, the linear regression stage does this regression on the merged data and only m, the slope of the linear regression, is implemented.

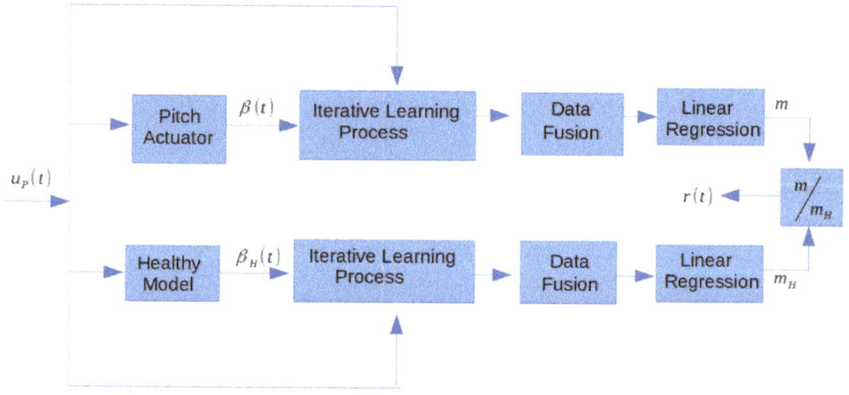

Figure 1. Residual signal $r(t)$ for fault diagnosis based on an interactive learning process. The iterative learning process is as stated in Theorem 1. The linear regression block is a post-processing unit that is enabled after the end of the elapsed time for the test. Subscript H refers to the nominal scenario H in Table 1.

Then, to detect a pitch actuator failure, the factor $r = \frac{m}{m_H}$ is evaluated. If $r >> 1$, a malfunction of the system has occurred, as showed in the next section.

4. Numerical Simulations

Table 2 shows the stages analyzed. Therefore, the healthy model in Figure 1 refers to the H scenario in Table 1. The experimental parameter considered in pitch actuator exact model (2), simpler model (3) and the iterative learning algorithm (4) are defined in Tables 1–3.

Table 3. Experimental parameters.

Name	Value
γ	0.5
T	$\frac{1}{30}$ sec
θ_{ini}	0
n	1000

For reference, the following color labeling is used: (H) blue, (F_1) red, (F_2) orange, and (F_3) green. By using the pitch command signal given in Figures 2–4 show the results of the simulation of the proposed scheme . Then, in all simulations, additive noise is attached to the pitch command signal for the robustness analysis of the proposed method. Table 4 shows the obtained regression parameters, where in the three faulty scenarios parameter r is greater than 1 and the detection algorithm works. Moreover, Figure 4 pictures data fusion variable $\psi(t)$ (4), and again the fault detection is illustrated.

Table 4. First numerical experiment results of regression parameter m.

Case	m	$r = m/m_H$
H	810.14 (m_H)	1
F_1	1732.83	2.13
F_2	2921.09	3.60
F_3	1238.93	1.52

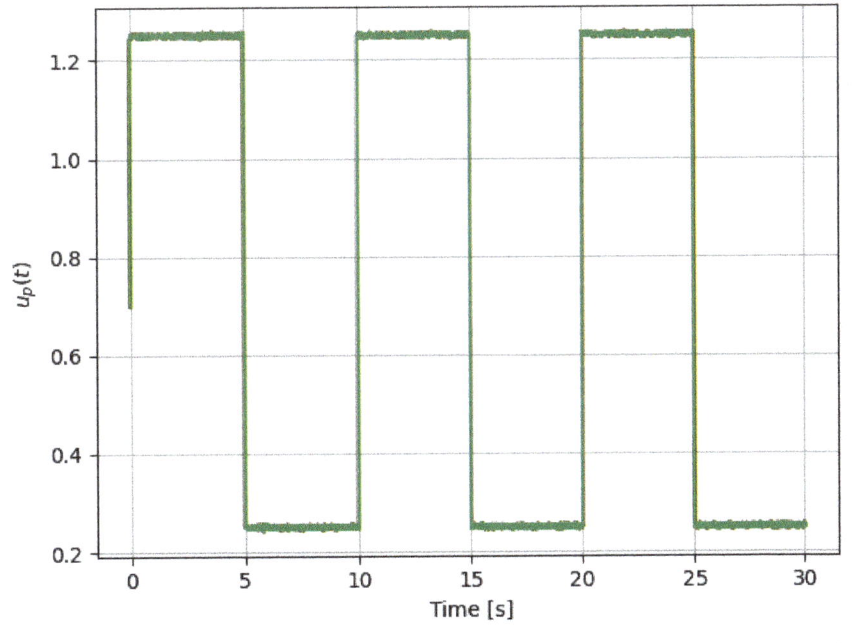

Figure 2. First simulation: pitch command signal for testing, where additive noise signal can be observed.

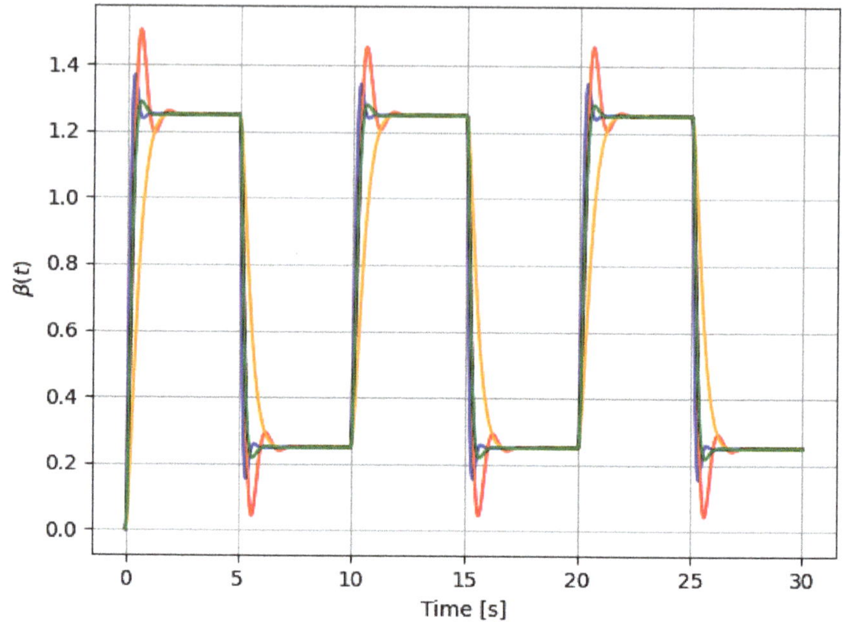

Figure 3. First simulation: pitch actuator responses for each case in Table 1: H (blue), F_1 (red), F_2 (orange), and F_3 (green). The command signal is the one in Figure 2.

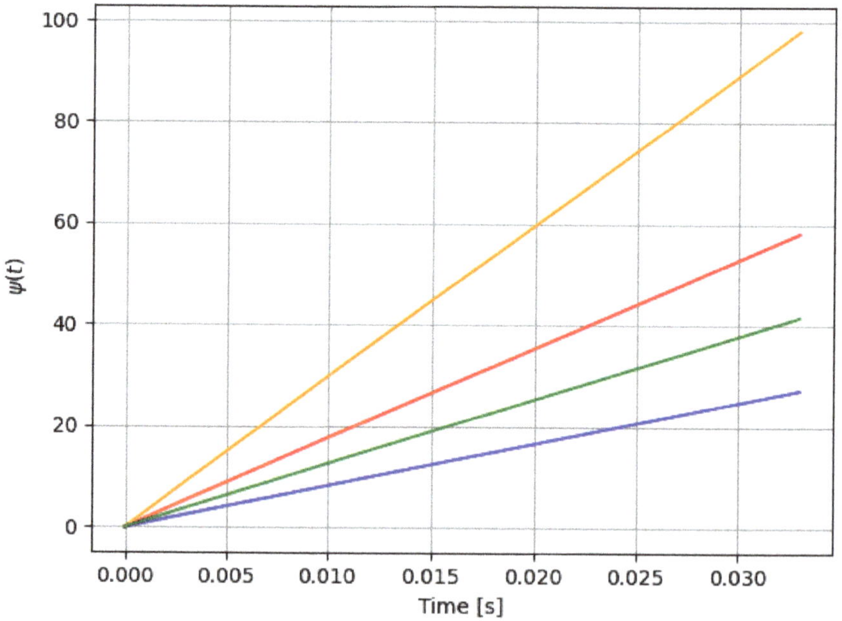

Figure 4. First simulation: data fusion raw $\psi(t)$ in (4) for each case in Table 1: H (blue), F_1 (red), F_2 (orange), and F_3 (green). Notice that the fault detection is reached.

Second experiment outcomes are shown in Figures 5–7. Once again, Table 5 gives the reading regression parameters. From Tables 1–5, a threshold to the residual signal $r(t)$ can be easily set to locate each failure. That is, despite the noise added to the data, our method is able to discern among the three different failure scenarios. As Table 5 shows, the parameter r for each case is located in a range of different values.

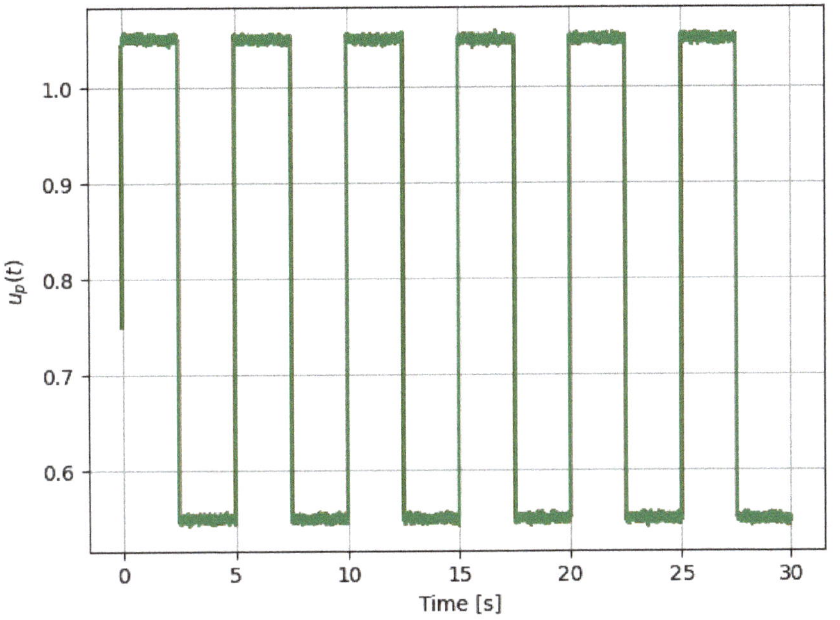

Figure 5. Second experiment: pitch command signal, where the additive noise is included to show the robustness of the proposed method.

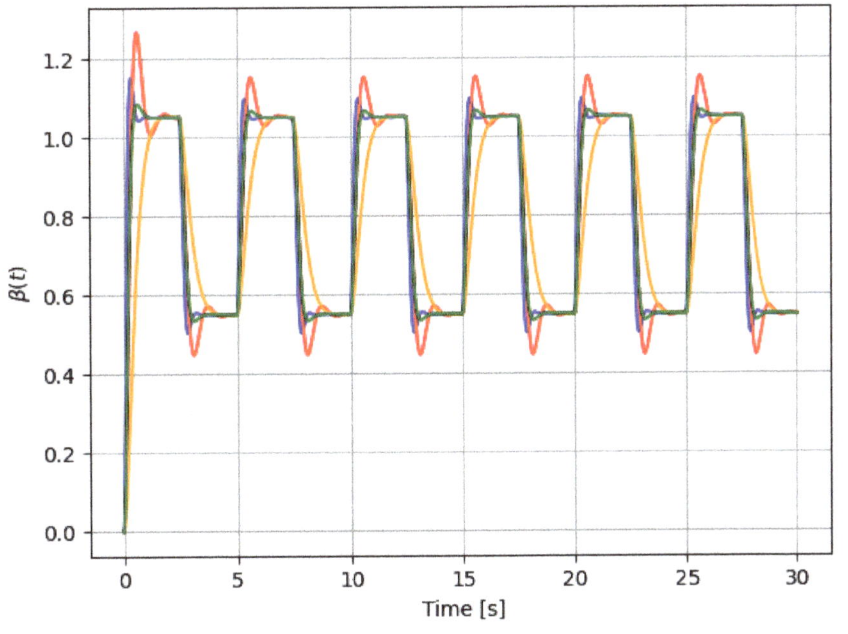

Figure 6. Pitch actuator responses to command in Figure 5: H (blue), F_1 (red), F_2 (orange), and F_3 (green).

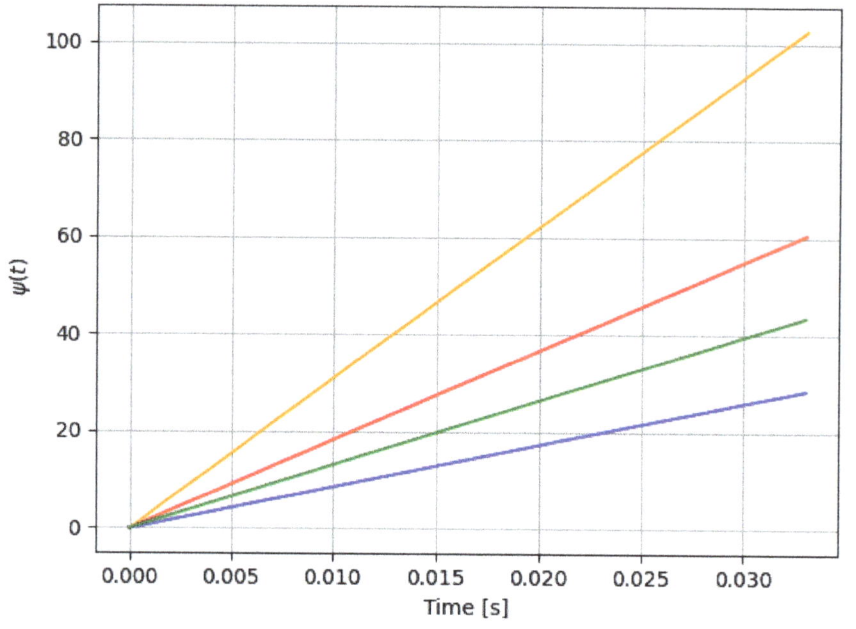

Figure 7. Data fusion for each case in Table 1: H (blue), F_1 (red), F_2 (orange), and F_3 (green), under the second experiment. This figure is related to Table 4: both indicates a pitch actuator fault detection.

Table 5. Second numerical experiment results.

Case	Regression Parameter	$r = m/m_H$
H (m_H)	848.30	1
F_1 (m)	1800.57	2.12
F_2 (m)	3053.26	3.59
F_3 (m)	1295.67	1.52

5. Discussion

Based on the simplest model used for the pitch actuator system (3), and because the iterative process identifies a parameter related to the system time-constant, the best option for the iterative process is to use a stepped pitch reference command, as shown in the previous simulations. However, to see the performance of our approach, we use a sine pitch command signal as shown in Figure 8. Numerical experiment results are shown in Figures 9 and 10. Furthermore, Table 6 gives the related results of the iterative process results. Even in this case, the system reacts differently to different failure cases. Although the sinusoidal signal is not commonly used as a reference in estimating the time constant of a system, our approach still allows us to detect variability of this parameter. Compared to Table 5, the classification is not as robust, as expected when dealing with a sinusoidal input.Therefore, future work will be to design a residual signal, as for the sinusoidal pitch command, which will do the same job.

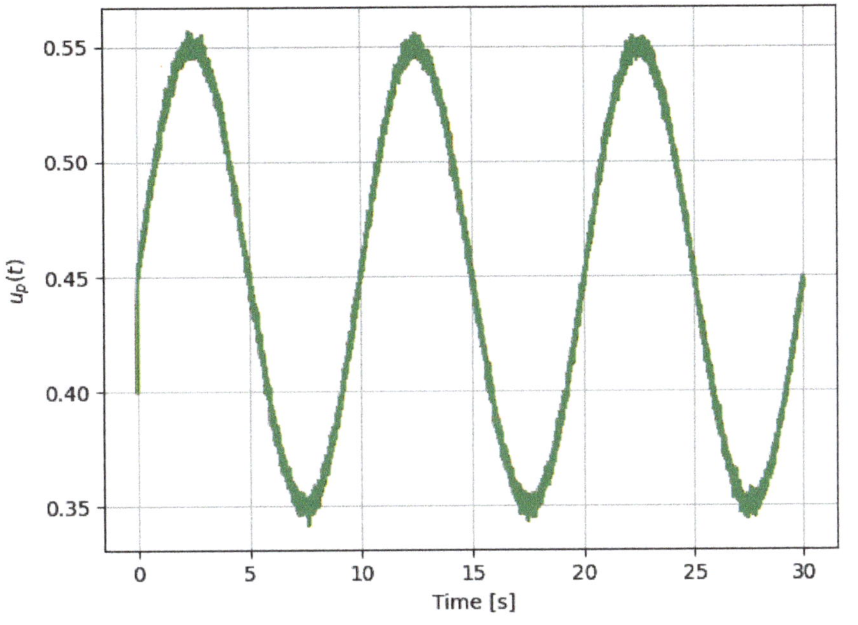

Figure 8. Sinusoidal pitch command signal.

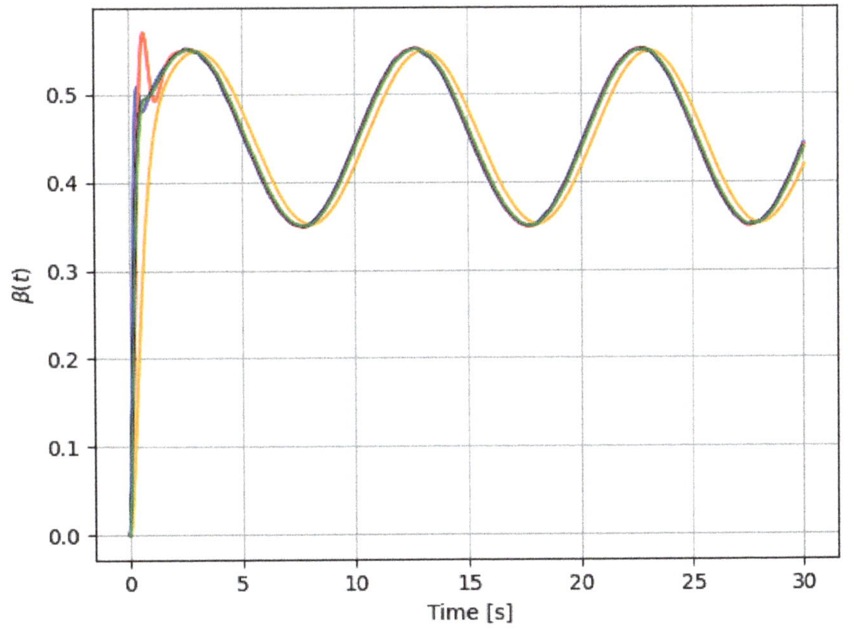

Figure 9. Actuator responses: H (blue), F_1 (red), F_2 (orange), and F_3 (green).

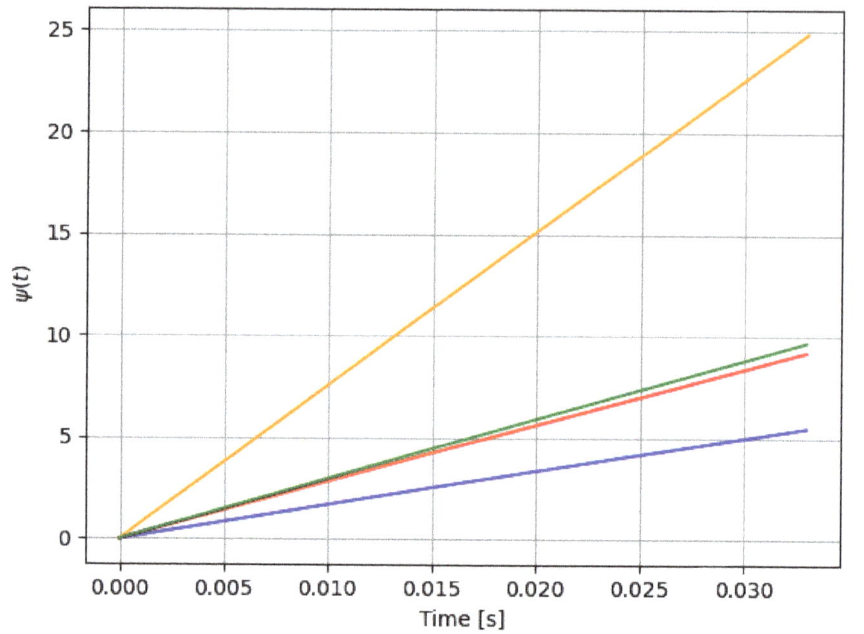

Figure 10. Data fusion profile: H (blue), F_1 (red), F_2 (orange), and F_3 (green). In this case, the fault detection is clear for F_2 faulty scenario.

Table 6. Third numerical experiment results.

Case	Regression Parameter	$r = m/m_H$
H (m_H)	162.42	1
F_1 (m)	273.60	1.68
F_2 (m)	739.94	4.55
F_3 (m)	287.80	1.77

6. Conclusions

In this article, we developed an iterative learning algorithm capable of isolating pitch actuator faults based on a square pitch command signal. The option to employ the iterative learning approach is the ability to learn from the past to arrive at a present conclusion. This is an important process in system learning based on data results. Hence, our approach can be a recent contribution of this theory, to pitch actuator analyses in wind turbines.

Author Contributions: Conceptualization, L.A.; methodology, L.A. and G.P.-V.; software, L.A.; validation, G.P.-V.; formal analysis, L.A. and G.P.-V.; investigation, L.A. and G.P.-V.; resources, L.A.; data curation, G.P.-V.; writing—original draft preparation, L.A.; writing—review and editing, G.P.-V.; visualization, G.P.-V.; supervision, L.A.; project administration, L.A. All authors have read and agreed to the published version of the manuscript.

Funding: This research received no external funding.

Institutional Review Board Statement: Not applicable.

Informed Consent Statement: Not applicable.

Data Availability Statement: Data supporting the reported results can be provided by the authors upon reasonable request.

Conflicts of Interest: The authors declare no conflict of interest.

References

1. Hall, D.L.; Llinas, J. An introduction to multisensor data fusion. *Proc. IEEE* **1997**, *85*, 6–23. [CrossRef]
2. White, F.E. *Data Fusion Lexicon*; Joint Directors of Laboratories (JDL), Technical Panel for C3; NCCOSC Research and Development Center: San Diego, CA, USA, 1991.
3. Varshney, P.K. Multisensor data fusion. *Electron. Commun. Eng. J.* **1997**, *9*, 245–253. [CrossRef]
4. Dong, X.L.; Gabrilovich, E.; Heitz, G.; Horn, W.; Murphy, K.; Sun, S.; Zhang, W. From data fusion to knowledge fusion. *arXiv* **2015**, arXiv:1503.00302.
5. Castanedo, F. A review of data fusion techniques. *Sci. World J.* **2013**, *2013*, 704504. [CrossRef] [PubMed]
6. Lau, B.P.L.; Marakkalage, S.H.; Zhou, Y.; Hassan, N.U.; Yuen, C.; Zhang, M.; Tan, U.X. A survey of data fusion in smart city applications. *Inf. Fusion* **2019**, *52*, 357–374. [CrossRef]
7. Biancolillo, A.; Boqué, R.; Cocchi, M.; Marini, F. Chapter 10—Data Fusion Strategies in Food Analysis. In *Data Fusion Methodology and Applications*; Cocchi, M., Ed.; Data Handling in Science and Technology Series; Elsevier: Amsterdam, The Netherlands, 2019; Volume 31, pp. 271–310. [CrossRef]
8. Panicker, M.; Mitha, T.; Oak, K.; Deshpande, A.M.; Ganguly, C. Multisensor data fusion for an autonomous ground vehicle. In Proceedings of the 2016 Conference on Advances in Signal Processing (CASP), Pune, India, 9–11 June 2016; pp. 507–512. [CrossRef]
9. Vavrinsky, E.; Subjak, J.; Donoval, M.; Wagner, A.; Zavodnik, T.; Svobodova, H. Application of Modern Multi-Sensor Holter in Diagnosis and Treatment. *Sensors* **2020**, *20*, 2663. [CrossRef] [PubMed]
10. Vervliet, N.; De Lathauwer, L. Chapter 4—Numerical Optimization-Based Algorithms for Data Fusion. In *Data Fusion Methodology and Applications*; Cocchi, M., Ed.; Data Handling in Science and Technology Series; Elsevier: Amsterdam, The Netherlands, 2019; Volume 31, pp. 81–128. [CrossRef]
11. Cariou, V.; Jouan-Rimbaud Bouveresse, D.; Qannari, E.; Rutledge, D. Chapter 7—ComDim Methods for the Analysis of Multiblock Data in a Data Fusion Perspective. In *Data Fusion Methodology and Applications*; Cocchi, M., Ed.; Data Handling in Science and Technology Series; Elsevier: Amsterdam, The Netherlands, 2019; Volume 31, pp. 179–204. [CrossRef]
12. Xia, Y.; Leung, H. Performance analysis of statistical optimal data fusion algorithms. *Inf. Sci.* **2014**, *277*, 808–824. [CrossRef]
13. Ruiz, M.; Mujica, L.E.; Alferez, S.; Acho, L.; Tutiven, C.; Vidal, Y.; Rodellar, J.; Pozo, F. Wind turbine fault detection and classification by means of image texture analysis. *Mech. Syst. Signal Process.* **2018**, *107*, 149–167. [CrossRef]
14. Acho Zuppa, L. A quick fault detection system applied to pitch actuators of wind turbines. *Renew. Energy Power Qual. J.* **2020**, *18*, 13–17. [CrossRef]
15. Vidal, Y.; Tutivén, C.; Rodellar, J.; Acho, L. Fault diagnosis and fault-tolerant control of wind turbines via a discrete time controller with a disturbance compensator. *Energies* **2015**, *8*, 4300–4316. [CrossRef]
16. Pujol-Vazquez, G.; Acho, L.; Gibergans-Báguena, J. Fault Detection Algorithm for Wind Turbines' Pitch Actuator Systems. *Energies* **2020**, *13*, 2861. [CrossRef]
17. Bristow, D.A.; Tharayil, M.; Alleyne, A.G. A survey of iterative learning control. *IEEE Control Syst. Mag.* **2006**, *26*, 96–114.
18. Tayebi, A. Adaptive iterative learning control for robot manipulators. *Automatica* **2004**, *40*, 1195–1203. [CrossRef]
19. Acho, L. Iterative learning control for homing guidance design of missiles. *Def. Technol.* **2017**, *13*, 360–366. [CrossRef]
20. Chaaban, R.; Ginsberg, D.; Fritzen, C.P. Structural load analysis of floating wind turbines under blade pitch system faults. In *Wind Turbine Control and Monitoring*; Springer: Cham, Switzerland, 2014; pp. 301–334.
21. Nguyen, P.; Nguyen, N. An intelligent parameter determination approach in iterative learning control. *Eur. J. Control* **2021**, *61*, 91–100. [CrossRef]
22. Gu, P.; Tian, S. Iterative learning control with high-order internal model for first-order hyperbolic systems. *ISA Trans.* **2021**, in press. doi: 10.1016/j.isatra.2021.03.006. [CrossRef] [PubMed]
23. Qiang, H.; Lin, Z.; Zou, X.; Sun, C.; Lu, W. Synchronizing non-identical time-varying delayed neural network systems via iterative learning control. *Neurocomputing* **2020**, *411*, 406–415. [CrossRef]
24. Blackwell, M.; Tutty, O.; Rogers, E.; Sandberg, R. Iterative learning control applied to a non-linear vortex panel model for improved aerodynamic load performance of wind turbines with smart rotors. *Int. J. Control* **2016**, *89*, 55–68. [CrossRef]
25. Inthamoussou, F.A.; Bianchi, F.D.; De Battista, H.; Mantz, R.J. Gain Scheduled H∞ Control of Wind Turbines for the Entire Operating Range. In *Wind Turbine Control and Monitoring*; Springer: Cham, Switzerland, 2014; pp. 71–95.
26. Nazir, M.; Khan, A.Q.; Mustafa, G.; Abid, M. Robust fault detection for wind turbines using reference model-based approach. *J. King Saud-Univ.-Eng. Sci.* **2017**, *29*, 244–252. [CrossRef]

sensors

MDPI

Review

A Review of Multisensor Data Fusion Solutions in Smart Manufacturing: Systems and Trends

Athina Tsanousa [1,*], Evangelos Bektsis [1], Constantine Kyriakopoulos [1], Ana Gómez González [2], Urko Leturiondo [2], Ilias Gialampoukidis [1], Anastasios Karakostas [1], Stefanos Vrochidis [1] and Ioannis Kompatsiaris [1]

1 Information Technologies Institute, Centre for Research and Technology Hellas, 6th km Charilaou-Thermi Road, 57001 Thessaloniki, Greece; evanbekt@iti.gr (E.B.); kyriak@iti.gr (C.K.); heliasgj@iti.gr (I.G.); akarakos@iti.gr (A.K.); stefanos@iti.gr (S.V.); ikom@iti.gr (I.K.)

2 Ikerlan Technology Research Centre, Basque Research and Technology Alliance (BRTA), P°. J. Mª. Arizmendiarrieta 2, 20500 Arrasate-Mondragón, Spain; ana.gomez@ikerlan.es (A.G.G.); uleturiondo@ikerlan.es (U.L.)

* Correspondence: atsan@iti.gr

Abstract: Manufacturing companies increasingly become "smarter" as a result of the Industry 4.0 revolution. Multiple sensors are used for industrial monitoring of machines and workers in order to detect events and consequently improve the manufacturing processes, lower the respective costs, and increase safety. Multisensor systems produce big amounts of heterogeneous data. Data fusion techniques address the issue of multimodality by combining data from different sources and improving the results of monitoring systems. The current paper presents a detailed review of state-of-the-art data fusion solutions, on data storage and indexing from various types of sensors, feature engineering, and multimodal data integration. The review aims to serve as a guide for the early stages of an analytic pipeline of manufacturing prognosis. The reviewed literature showed that in fusion and in preprocessing, the methods chosen to be applied in this sector are beyond the state-of-the-art. Existing weaknesses and gaps that lead to future research goals were also identified.

Keywords: smart manufacturing; data fusion; feature extraction; industrial prognosis

Citation: Tsanousa, A.; Bektsis, E.; Kyriakopoulos, C.; González, G.A.; Leturiondo, U.; Gialampoukidis, I.; Karakostas, A.; Vrochidi, S.; Kompatsiaris, I. Review of Multisensor Data Fusion Solutions in Smart Manufacturing: Systems and Trends. *Sensors* **2022**, *22*, 1734. https://doi.org/10.3390/s22051734

Academic Editor: Natividad Duro Carralero

Received: 22 January 2022
Accepted: 18 February 2022
Published: 23 February 2022

1. Introduction

The advances in Information Communication Technologies (ICTs), along with the era of Industry 4.0 have brought industries closer to the adoption of automation in their processes. Industry 4.0 is formed [1] by the integration of manufacturing operation systems and ICTs. Business models are reframed by adopting the digitisation and Internet of Things (IoT) concepts. The IoT refers to the devices and sensors that are smart, are addressed using communication protocols, are autonomous, and also adapt to the operating conditions [2]. The availability and variability of sensors assist in this transition to automation, by monitoring the function of machinery and tools, as well as measuring the product quality. This intends to improve the failure ratio of machines and products, maximizing the productivity of a company and a good machine function. Along with the increasing usage of sensors in manufacturing procedures comes the need to exploit the large amounts of heterogeneous data they produce and, subsequently, the application of fusion techniques that combine these data with different methods and improve the knowledge extracted from them.

Fusion can be described as the method of combining data from different sources. In the field of machine learning, there are two broad categories of fusion: early and late fusion. Early fusion is implemented in the first stages of an application and refers to the combination of the raw data (data fusion) or the extracted features (feature fusion). Late fusion refers to the combination of the results of different algorithms and is implemented as the last step of the analysis' pipeline. Late fusion can be accomplished with techniques that combine the predicted class probabilities of each algorithm or the predicted classes of each algorithm. More advanced techniques such as stacking and boosting use the output of other classifiers as the input to a new ensemble algorithm.

This paper provides a review of data fusion solutions implemented in the manufacturing sector, with a special emphasis on the sensors employed in the industry, as well as the preprocessing stages of data preparation and feature extraction. Furthermore, the types of data storage and techniques for data extraction used for industrial prognosis are explored. The aim of this literature review is to provide the reader with an introductory guide for the processing of industrial data, with a special focus on the types of fusion employed in industrial prognosis in recent years.

The methodology followed in order to collect the papers included searching in some popular databases such as Scopus, ISO, and Google Scholar. Using keywords, which are mentioned below, combined with the term "manufacturing" or more specific terms relevant to manufacturing applications, e.g., chatter detection, some papers were collected and screened for relatedness to the scope of the current literature review. Papers published in recent years, describing the architecture or the analytical framework of manufacturing problems, were selected. The majority of papers reviewed were published after 2015, with some exceptions that were published in early 2000. Following is the list of keywords used for some of the main sections of the paper:

- "Vibration analysis, current signature, acoustic analysis, thermography, infrared radiation, thermocouple, acoustic emission, NDT, defect inspection, ultrasound, induced current, radiography, penetrating liquids, lubricant analysis, ferrography, spectrometric, chromatography, vibration processing, motor current signature analysis, feature extraction, feature selection, dimensionality reduction, PCA, mRMR, acoustic camera, 3d camera, artificial vision" combined with the term "manufacturing", for the section of data preprocessing;
- Data fusion solutions/applications in manufacturing, ensemble learning in manufacturing applications, fusion for chatter detection, fusion for tool wear detection, and data fusion in industrial prognosis, were some of the keywords used for the section of data fusion solutions in manufacturing.

The analysis of data collected from different sensors of a typical manufacturing process comprises certain steps, similar to the steps of any analytic task that involves prediction (Figure 1). In the beginning, raw data need to be collected from sensors and stored. In order for the data to be analysed, some cleaning and filtering functions are usually applied. Following is the important step of filtering, which removes signal noise, and afterwards, features are usually extracted. Feature engineering is crucial in signal processing, and the transformation methods adopted, as well as the features extracted vary according to the type of sensor and the goal of the analysis. The large amount of features produced is then reduced by applying feature selection methods, which select smaller sets of variables for input in a classification or regression algorithm. Fusion can be applied in different stages of the aforementioned pipeline, whether in the beginning combining the raw data, or after the step of feature extraction to combine the features, or even in the final stage for the combination of the prediction algorithm's results. In the current work, more attention was given to the initial layers of the analytic pipeline, reviewing the types of preprocessing used in industrial prognosis problems and the common useful features produced. Furthermore, the fusion techniques adopted in manufacturing prognosis were reviewed. Thus, the scope of the paper is depicted in Figure 1), responding to the sections included in the dotted lines, while the figure overall describes an analytic pipeline.

Similar literature reviews regarding fusion in the era of Industry 4.0 have been published in recent years. A multisensor data fusion review was presented in [3], with a special focus on multisensor monitoring technology and the architecture of multisensor monitoring systems. An analytical survey on machine learning and fusion applications for industrial prognosis was presented in [4]. The authors provided a categorisation of methods according to the cause of the failure to be predicted. This review referred to the general industry section, while our current literature review paper presents the fusion trends in manufacturing applications. In [5], a review of the techniques used for data fusion for material data was presented. The paper reviewed the methods and underlying theory for material data fusion, explaining the different levels of fusion and how these can be adapted in the combination of heterogeneous material data. The authors also gave insight into the differences of the terms "data integration" and "data fusion", which are often met in manufacturing applications.

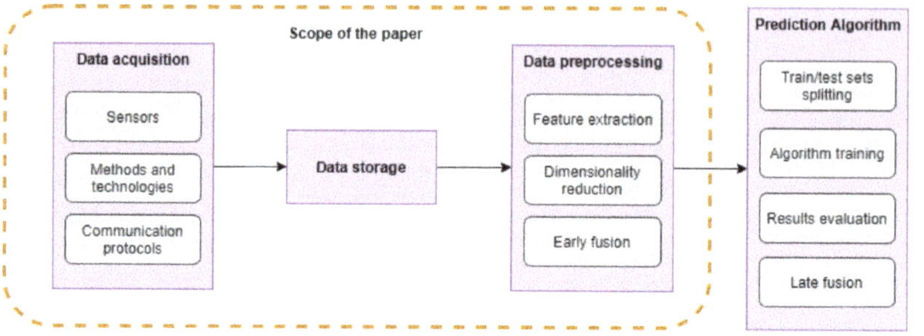

Figure 1. Flowchart that presents the typical analytic pipeline followed in manufacturing applications.

The rest of the paper is organised as follows: Section 2 describes the methods for data storage and extraction, along with the communication protocols enabling the fusion systems. Section 3 refers to the type of sensors employed in manufacturing, data preprocessing techniques, and feature extraction, as well as feature selection methods. Section 4 reviews the fusion methods adopted in different systems and in the manufacturing sector specifically. Finally, the paper is concluded with the Discussion Section.

2. Data Acquisition and Storage

The evolution of technology and especially the Internet of Things (IoT) has led to a new kind of manufacturing known as Smart Manufacturing. Smart Manufacturing is an application of the IoT that focuses on using inexpensive, small-sized, and smart devices that are all interconnected so that they can increase productivity and improve the health of the machines. Big Data in Smart Manufacturing systems are continuously generated data in high volumes produced by said smart devices [6] and are available in various forms, e.g., log files, signal streams, or sensor data. A Big Data analysis system should be able to use these data in real time, as well as save them for historical analysis and long-term pattern detection. This section of the paper deals with the ways data are acquired and stored from such smart devices.

2.1. Data Acquisition Methods and Technologies

In [7], a comprehensive review of Big Data analytics throughout the product lifecycle was made. Most notably, regarding the data acquisition phase of the lifecycle, it was acknowledged that manual-based data acquisition methods are still used in various stages of the lifecycle process, thus making the acquired data from these approaches inaccurate and untimely, and as a consequence, the decisions based on them are usually ineffective. The authors [8] suggested some challenges of data acquisition that need further research to be resolved and used smart mobile devices to provide an example on how IoT technologies can be embedded into the physical world and be able to gather data throughout the whole product lifecycle. A detailed hierarchical architecture of a smart factory was described by [9], emphasizing the need for Wireless Sensor Networks (WSNs) in a smart factory for data monitoring, acquisition, and logging. ZigBee and Bluetooth were also mentioned, on top of Radio-Frequency Identification (RFID), for real-time data acquisition and were described as good choices when it comes down to the cost of the industrial automation of wireless technologies. Furthermore, the devices responsible for data acquisition should be easy to set up and connect with interfaces capable of scaling up.

One methodology employed by [10] was a monitoring tool organised in a WSN. Part of the monitoring tool is a Data Acquisition Device (DAQ) that uses split-core current transformers, closed-loop Hall effect sensors, and a camera to create an easy and not intrusive way to collect data by monitoring the status of the machines. The researchers of the paper used multiple DAQs on the shop floor and used a central gateway to collect and organise the data into packets before they were transmitted. The WSN that was used for data extraction utilised the DIGI XBee ZigBee Radio-Frequency (RF) module. As the paper described, an OPC Unified Architecture (OPC-UA) was used. The OPC-UA provided an extensible data model, which provided the data schema. A NoSQL database was used, as it proved to be more flexible than a Structured Query Language database (SQL) because of the heterogeneity and the different data that were being generated. The authors of [11] create a Cyber–Physical System (CPS) that uses a semantic sensor network. Focus was given on the way data are gathered from the physical sensors. To manage the large volume and velocity of the data, the authors proposed an architecture in which the data are collected through an OPC-UA, an industrial M2M communication protocol. There exists a considerable body of literature on data flow models, but most notably, Reference [12] suggested frameworks that allow users to model their application via visual editors. These programs enable the users to receive data from external sources such as IoT devices and smart sensors.

In [13], the authors suggested an architecture design for a smart manufacturing system. Furthermore, they provided detailed considerations of the way a smart Manufacturing Execution System (MES) should be designed. For real-time data acquisition from the shop floor, the OPC-UA technology was proposed. The authors of [14] expanded on the topic of data transmission with the introduction of WiFi direct, 4G LTE, and Z-Wave. There was also the mention of the protocols being used with these wireless technologies, which include IPv6, MQTT, SOAP, and REST, among others. The authors also mentioned a series of compatible with Supervisory Control And Data Acquisition (SCADA) communication networks such as OPC, Open Database Connectivity (ODBC), RS232, and Dynamic Data Exchange (DDE), as well as some wireless communication standards such as the Highway Addressable Remote Transducer (HART) and ISA100.11a. The most prominent of these protocols are the MQTT and REST API. The MQTT protocol is used to acquire and transmit data from large industrial environments to the cloud where the processes producing the data can be monitored and controlled. The REST Application Programming Interface (API) provides a way to securely collect data from the IoT devices, where the data are collected in formal message arrays and the receivers split those individual messages so the producing device can be identified. SCADA systems provide a fully connected system that a manufacturer can use not only to acquire the data, but also to handle, manage, and archive them long term. Some known SCADA systems are SIMATIC SCADA Systems from Siemens, AVEVA™ Plant from Schneider Electric, Proficy HMI/SCADA from General Electric, and HMI/SCADA from ABB.

A demonstration of real-time data acquisition using the MQTT protocol was described in [15]. The authors implemented a system using temperature and humidity sensors so they could generate data to work with the protocol. For the test, they generated data for 60 s and compared the ability between the Hypertext Transfer Protocol (HTTP) and the MQTT protocol to transfer from the hardware (i.e., sensors) to the server and store them in a MySQL database. To minimise the error and loss of data, each transmission had a sequential ID so the completeness of data could be checked. A conclusion was made that the use of the MQTT protocol proved to be up to six-times faster than HTTP at sending data. On the more technical side, it was reported in [16] that depending on the application and the network coverage required to send data different, protocols may need to be used. Low-Energy (LE) Bluetooth, Near-Field Communications (NFC), and RFID, among others, are technologies used for short-range communication. As a result, industrial applications that require a broader field to be deployed need solutions that can be both energy saving while maintaining a significant coverage area. Such a technology is the Low-Power Wide-Area Network (LPWAN), which includes Sigfox, LoRa, and the Narrowband IoT (NB-IoT) [17].

A Big Data pipeline for data streaming in Industry 4.0 was described in [14,18]. Moreover, data collection and data storing tools were compared and presented. Such tools are Apache Kafka, RabbitMQ, and Amazon Kinesis, which are considered for pushing a high volume of messages that are produced from data producers (i.e., sensors) and even Apache Storm to process and discard "useless data", which were tagged as less important or out of context.

In [19], a system that is capable of data monitoring and acquisition of a Computerised Numerical Control (CNC) machine tool in intelligent manufacturing was proposed and developed. Most notably, the authors compared the different data acquisition methods from a CNC machine, not only on the different data types that can be collected, but also the technical difficulty and implementation costs. For data acquisition, the MTConnect protocol was selected, while for the database, a system that uses the ODBC method was the choice. The machine tool networking was based on an industrial Ethernet and Transmission Control Protocol/Internet Protocol (TCP/IP) technology. Working also with CNC machines, Reference [20] provided a thorough explanation of the way a CNC machine generates data and how these data are acquired, transmitted, and stored. CNC data can be split into two main sources: controller data and external sensor data. As is usual for the sensors, the collected data contain noise from external interference, and for that reason, a necessary step is to clean and preprocess the data. For machines such as a CNC system, a high amount real-time data in the controller is required. Some of the most-used technologies in the field are Ethernet, Profinet, and EtherCAT, but in order for them to work seamlessly in the system, the sensors need to be equipped with acquisition cards.

Previous research showed that data acquisition in real time is based on the configuration of the smart environment [21–23]. Specifically, it was described that the first part of the data acquisition is the data collection, during which raw data are gathered using various technologies depending on the application. Moreover was proposed that Ultra-High-Frequency (UHF) RFID readers can be used to track and collect data in real time from the manufacturing process. Regarding the transmission of the collected data, it was mentioned that for real-time data such as temperature, vibration, and pressure, the Internet, wireless, and 4G methods were used. As far as non-real-time data (e.g., maintenance history) transmission are concerned, tools such Sqoop are preferred. A more in-depth look at the way RFID technology is used to collect data from the shop floor was provided by [24,25], where the authors explained step by step the system architecture they created. Data flow starts from RFID tags, which are attached to the input and output of their manufacturing section. This allows for real-time monitoring of the manufacturing process and can update individual data from each part. The authors concluded that such an architecture (RFID-based IoT solution) with the ability to closely monitor the manufacturing sections leads to improvements in the traceability, quality, and tool wear prediction.

Some authors [26,27] have also suggested Industrial Internet of Things (IIoT) architectures, where the data collection method was thoroughly described. Specifically, all kinds of manufacturing data (e.g., equipment status data, product data, or measurements) can be gathered using wired or wireless methods. The wireless methods include, mostly, as previously mentioned, RFID readers that obtain the raw data. Hive and HBase have been introduced as a distributed data storage system. One way that is considered for data transmission is the Flume interface, which forwards the data to a selected storage system.

2.2. Data Storage Software Solutions

Concerning data storage, a distributed approach is usually the choice where a Distributed Database System (DDBS) is used to store structured data and the Hadoop Distributed File System (HDFS) or NoSQL databases for unstructured data. Other alternatives are the C Open Source Managed Operating System (COSMOS) and Haystack. Especially for distributed file systems, the Google File System (GFS) was one of the first systems developed to handle data-intensive applications.

MongoDB, a NoSQL database, is one of the most popular databases at the moment, and the authors of [10] used it to store the sensor data that were gathered by the DAQs.

As was mentioned by [13], relational databases such as MySQL are not an option due to the complicated nature of the manufacturing data. Subsequently, the authors considered a distributed database as an appropriate option because of its high performance, efficiency, and scalability.

Regarding data storage solutions being proposed, the authors in [18] listed, described, and compared commonly used technologies such as Hadoop Hive and MongoDB. A distinction was made between the data models that were used: firstly, the file system data model, for data stored in a schema-less manner and read in a structured manner with a processing time based on the processing needs of the application; secondly, document-based data model; lastly, a column-based schema. A reference was made regarding the recent data storage technologies and their capabilities to process data for critical real-time applications.

Prior research [28] suggested that shop floor data on a manufacturing site can be gathered using what is called a SCADA system. SCADA provides a singular interface, where all the gathered data collected from different smart devices can be transmitted. Among SCADA, Reference [29] mentioned the Protocol Data Unit (PDU) as an alternative data acquisition tool. It was also suggested that the combination of IoT and cloud services gives the ability for different equipment to be connected and collect huge amounts of data. To organise data in a methodical and effectual way, Database Management Systems (DBMSs) have been created. According to [22], these tools can be split into two categories, relational DBMS and non-relational DBMS. The first category includes SQL databases, meaning databases that usually store data in tables of records. Commercial solutions for SQL databases are Microsoft SQL, PostgreSQL, Oracle, and MySQL. Regarding NoSQL databases, it is possible to use various types of data such as text, binary, and records. One of the benefits of NoSQL over SQL is that it is scalable and can support huge volumes of data, making it perfect for managing data coming from sensors and smart devices. In [30,31], a straightforward software solution explaining the pros and cons of each one was provided. The most commonly used solution is Apache Hadoop, but the authors [30] also proposed other options such as: Redis, SimpleDB, CouchDB, and MongoDB, just to name a few. More research [32] described the criteria of the data model and used them to identify different models where each of the previously mentioned software solutions apply the best.

Authors such as [33] provided a more technical side to the way data being acquired and stored. Even though it was mentioned that the data were collected manually, there was an in-depth review of the storage methods. As has been previously reported by the literature [34], the most adequate NoSQL databases for real-time data storage are HBase and Cassandra. However, it was demonstrated by [35] that Online Transaction Processing (OLTP)-oriented NoSQL databases can lack the support for fast sequential access over a significant amount of data, which sometimes can prove to be a hurdle when it comes to data analytics. Hadoop BDW is proposed as the opposing solution that can handle fast sequential access.

A data storage framework was presented in [36] that can deal with various types of data collected from different devices, for instance RFID readers, monitors, or thermometers. Due to the heterogeneity and volume of data, there is no perfect method for efficiently storing and accessing them. The proposed architecture by the authors included several modules. In more detail, HDFS was used for unstructured file storage, while a database module using NoSQL and relational databases was used to manage the structured data. The authors also investigated a data storage framework capable of being a feasible solution to challenges such as a large volume of data, different data types, rapid generation of data, and the complicated requirements of data management. In detail, for structured data, a database management model was created that combined and extended multiple databases. For unstructured data, a common solution was followed. The authors explained in great detail how the framework extended HDFS for multitenant data isolation. Concluding, it was mentioned that for remote and cross-platform data access, a RESTful-service-generating mechanism was integrated, to provide a platform-independent HTTP interface. Furthermore, the authors of [37] reported that in large-scale manufacturing systems, tens of thousands of data streams flow into the storage at various rates. For that reason, there is a need for improving the bottlenecks that cause the data to arrive irregularly. The authors mentioned Blueflood as a solution that attempts to achieve high scalability. Blueflood is a combination of Cassandra, which handles data storage, and Elasticsearch, which handles indexing. A similar alternative to Blueflood is OpenTSDB, which uses HBase for storage instead of Cassandra. A summary of the software tools used for data management can be found in Table 1.

The authors of [38] separated the requirement and solution components that are assigned to the data storage processes. In [39], it was suggested that due to comprehensive process transparency, structured, semi-structured, and unstructured data should be stored and made available for application-specific processing. Furthermore, in [38], operational data storage and a long data storage system were proposed. The first one requires an edge device unit, which must be able to store and manage real-time data. For operational data storage, the edge devices need to store data efficiently and reliably. For that reason, SQL was recommended. The preferred Relational Database Management Systems (RDBMSs) are MySQL and PostgreSQL. Concerning long-term data storage, a Big Database system is required. Consequently, NoSQL databases provide the best storage solutions as they are able to efficiently store large volumes of unstructured datasets, compared to relational databases [30,40].

A focus on data storage issues and recommendations for a new solution to organise and manage data was given by [41]. In particular, the cloud storage system they presented uses a Document-Oriented Storage System (DO-SS) for the storage of all the information derived from the monitoring systems. The integration between the data collection and storage subsystem occurs with the help of a software module (parser), so that the data can be converted before being stored in MongoDB. The authors also implemented an Object-Oriented Storage System (OO-SS), a widely used object storage system. The main benefit over other solutions, is that the data are protected by being stored as multiple copies, so in case a node fails, there is another one active where the data are stored. This design makes the OO-SS ideal when there is a need for performance and scalability.

In [42], a hybrid framework was conceptualised for an industrial platform that ensures efficient and accurate communication, concerning data transfer among software applications and devices. The framework was characterised as hybrid, because it contains two different technologies for data storage and exploits the best features from each of them. In the proposed framework, structured data are stored in relational database systems, while sensorial data, which most of the time tend to be unstructured, are stored in NoSQL databases. The real-time sensor data are published to an MQTT broker that is suitable to be used to connect with remote locations, and the Raw Data Handler subscribes to MQTT and acquires the generated data. Later, the data are carried to the sensorial repository where they are indexed and can easily be filtered and accessed by timestamps. In general, the authors proposed a hybrid framework that is capable of shop floor data collection and application in industrial environments.

A scalable data storage framework for smart manufacturing was introduced by [43]. In more detail, it is a Software-Defined Hybrid Cloud (SDHC) for saving the equipment-generated data. The main challenge the authors faced was the different data types and formats. With the use of software-defined technology, control data and manufacturing data were separated. The manufacturing data derived from the sensors and controllers can be saved in a key-value database only if the data save request is allowed. Lastly, two improvements were proposed, the first one to deal with the data saving efficiency, which will improve the response time, and the second one with the data-save permission, which will improve the system's robustness. A software architecture was designed by [44]. The framework is highly scalable, so a fleet of IoT boards and sensors can be easily configurable. For data collection, an Arancino board was used that was provided with an AI module that can manage on-board fault prediction. InfluxDB was the database that was selected as it is non-relational and is suitable for industrial scenarios where sensors send data at different rates.

Table 1. Applications of software tools for data management.

Software Tool	Application	Reference
Apache Hadoop	Hadoop is a framework that allows for the distributed processing of large datasets across clusters of computers using simple programming models. It has been used for different kinds of applications such as frameworks that can optimise and organise the way bit data can be searched and accessed. There are also applications regarding storing data derived from sensors that monitor the environmental air pollution. Lastly, tuning systems have been designed to improve the performance of Hadoop and MapReduce.	[45–47]
Apache Storm	One of the most capable software solutions for Big Data is Apache Storm. Several applications exist that employ it. Some of them use it as a data streaming and real-time processing platform, while others create frameworks for dynamically scaling for the analysis of streaming data. Finally, there are multisensor data fusion frameworks that employ Apache Storm due to its high reliability and good processing mode.	[48–50]
Apache Flume	Flume is a distributed, reliable, and available system for efficiently collecting, aggregating, and moving large amounts of event data. It has been used for various kinds of applications such as healthcare and manufacturing. Frameworks have been designed so that the computational scalability of sensor network data can be achieved.	[51–53]
Apache Spark	The aim of Spark is to make data analytics programs run faster by offering a general execution model that optimises arbitrary operator graphs and supports in-memory computing. Most applications use it for sensor analytics. It has been deployed on both industrial and non-industrial applications and can be integrated into pre-existing frameworks.	[54–56]
Apache Kafka	Kafka is well suited for the situations where users need to process real-time data and analyse them. There are papers that focused on learning how to reliably transfer data and studied its application in collaboration with other software solutions.	[57–59]

2.3. Communication Protocols

The smart manufacturing sector benefits from data fusion systems. Next, the required underlying communication technologies enabling the fusion systems are presented. These are related to the IoT and the corresponding protocols. Specifically, networking concepts that are based on software are incorporated into the lower communication layers. Furthermore, adaptivity offers advanced performance since the nature of modern networking systems is dynamic.

Software-Defined Networking (SDN) is the main conceptual networking model [60] under modern IoT fusion environments. It brings the programmable networking logic into the lower architectural layers. This process allows better control and management of networking data flows in a transparent way from the higher-layer networking applications. It is a centralised architecture that defines a stable ground to be used for building networking applications. An open implementation of the SDN networking concept is the

OpenFlow protocol [61], which is widely adopted. A networking foundation is currently maintaining the specification. The whole concept relies on central computing logic, represented by the SDN controller, controlling data flows between core networking components such as switches (i.e., the Southbound API). Fusion techniques over IoT environments are facilitated with the exploitation of the SDN concept.

Achieving high-speed transmissions in IoT environments requires efficient and dynamic channel assignment. Conventional fixed assignment techniques are not adequate for modern environments, in which, due to their dynamic nature, requirements must constantly adapt to the runtime conditions. The IoT over SDN, when combined with deep learning techniques, improves transmission quality. Therefore, a traffic load prediction algorithm that is based on deep learning [62] has been proposed, forecasting network traffic and congestion. Next, a deep-learning-based algorithm that assigns channels has been introduced. Its main role relates to link channel allocation using intelligence in the SDN-IoT network environment.

Since communication between smart devices in the IoT manufacturing sector can be peculiar, event-based data fusion for communication is needed [63]. This is a message exchange system between participating devices that initiates when events occur. Fusion is required since different devices generate heterogeneous notifications, along with data source trust issues that may arise. The contribution of this work consisted of an event-based protocol covering the communication issues of resource-limited sensors and heterogeneous data sources, and it considered the trust degree of the fused data.

There is a vast spectrum of IoT applications that require security and privacy for realistic deployment in the modern era. Trust and data integrity are prerequisites in the IoT ecosystem, otherwise applications will lose high demand and also their potential. In the current case of cellular and sensor networks, special security challenges emerge and correlate with authentication issues, privacy, management, and information storage [64]. Programmable Logic Controllers (PLCs) are an integral part of the industrial control systems [65]. Communication issues between PLCs and the engineering stations or field devices concerning security must be confronted. Modern database systems use communication systems to deploy as cloud-based solutions [66]. Different DBs support various security technologies, though most non-relational solutions overlook modern Big Data applications.

Communication requires a credible authentication model, which guarantees data integrity and secrecy. For that purpose, an IoT node-roaming authentication protocol was introduced [67]. A heterogeneous fusion mechanism comprises the protocol's functionality. Every roaming device communicates with a server, which provides authentication functionality. This process renders attacking attempts from external malicious nodes difficult.

In a smart manufacturing environment, multiple protocols are required for transmitting data efficiently. SDN forms the basis for a heterogeneous network architecture [68] for forwarding multisource manufacturing data and, at the same time, utilising network resources optimally. The core algorithm of the proposed protocol is based on cross-network fusion and scheduling. It was shown that the efficiency was improved for the fusion processes, especially for intelligent manufacturing equipment.

3. Data Preprocessing and Feature Extraction

3.1. Types of Sensors and Variables Measured

In the context of Industry 4.0, the use of sensors has been widely extended, as the capacity to store and use the acquired data has been enhanced [69]. The heterogeneity of the machinery in the manufacturing plants and the specific needs of each sector generate the need to use different kinds of measurements to monitor the machines, the manufacturing processes, and the parts that are produced. For that purpose, a large number of sensors are available on the market. Many of them are already integrated in the machines, even if it is usual to include other sensing equipment in order to complete the measurement chain available for further analysis.

In [70], some general guidelines were given for the condition monitoring and diagnostics of machines. The parameters that can be measured were identified and classified according to the type of machine that was considered. The main parameters that are useful for all kinds of machines are: temperature, noise, vibration, acoustic emissions, ultrasonics, oil pressure, and thermography. Some others will be specific for some particular types of machines. In the next paragraphs, this is explained in more detail.

Vibration analysis is a commonly used technique [44,71,72] as the vibration signature of a component may change as a fault develops. Machines vibrate in their normal operation, some of the vibratory phenomena related being to events that occur periodically (such as the rotation of shafts, the mesh of gear teeth, or the generation of rotating electric fields). Thus, the frequency of that vibration often gives an indication of the source. That is why it is important to establish a baseline for the standard vibration response of a machine and detect any anomaly from that baseline when a fault occurs [73]. For that purpose, different kinds of vibration transducers can be used: proximity probes (measuring the relative distance between the probe and another surface), velocity transducers (providing a signal proportional to the absolute velocity of the element on which the transducer is mounted), accelerometers (giving a signal proportional to the absolute acceleration), dual vibration probes (allowing the measurement of absolute motion), and laser vibrometers (transducers based on the Doppler principle that do not load the measurement object) [73], accelerometers being the most commonly used. In the range of the audible spectrum, acoustic analysis via microphones [74] allows the localisation of an internal material transformation or a noise source, but it is hard to reach to determine the magnitude of a failure, given that manufacturing plants can be noisy environments.

Electrical current is usually measured in electrical machines such as motors or generators [44,75]. This kind of signal has a low implementation cost, as the equipment is simple and economical. Other variables to be measured in such rotational equipment are the torque, the speed, and the temperature. As the temperature changes or reaches a limit, the temperature can be a signal of failure, and the analysis of the temperature is another approach [76]. One of the most-used techniques is thermography, which consists of the determination of the surface temperature by means of measuring the infrared radiation [77]. This technique can be used, for example, to monitor the wear of cutting tools [78]. Moreover, thermocouples are commonly installed in machinery thanks to their low cost, providing the value of the temperature in a single point.

The measurement of Acoustic Emission (AE) [79,80] allows capturing the propagating waves generated by the rapid release of energy from localised sources, as a result of crack propagation, impact, or leakage, among other phenomena [81]. In the manufacturing context, this technique can be useful to monitor the state of the structure of large machinery subjected to cyclic loads or to test the quality of a product. Similarly, ultrasound analysis can be used to detect, identify, and evaluate the size of surface and sub-surface failure or to measure material thickness. Thus, it is mainly used in material or product control stages [82]. Induced currents are useful to test electrically conductive materials [83]. They are a means to evaluate surface and sub-surface failure, as well as to measure material thickness generated in processes such as thermal treatments and coatings.

Radiographies are used to detect internal cracks and the lack of homogeneity of materials [84] (e.g., in pieces manufactured by melting or in welded joints); however, access to the analysed object from all possible perspectives is needed, and this is an expensive technique. Penetrating liquids are also used to detect surface discontinuities.

Lubricant analysis is used to determine the chemical composition of lubricants, as well as to find particles in them. Some of these techniques are ferrography, spectrometric identification, and chromatography. Ferrography looks for iron particles in oils in order to identify the component that is wearing out and to determine the wear level [85]. Spectrometric identification looks for iron, metallic, or non-metallic particles by means of the atomic emission spectrometer, which is useful to identify early-stage failure [86]. Chromatography measures changes in the properties of the lubricants (viscosity, pH, water content, etc.) [73].

The aforementioned sensors and variables measured refer mainly to one-dimensional data. However, higher-dimensional data are also relevant in the context of smart manufacturing. Vision techniques can be used to inspect manufactured products [87]. Three-dimensional vision techniques are also useful in inspection environments, such as the analysis of the quality of an assembly process [88]. Vision techniques can be combined with sound measurements using acoustic cameras, imaging devices with an included microphone array. They process sound signals to form a representation of the originating location. They are used for the identification of noise sources in machinery such as conveyor belts [89].

Even if the aforementioned variables are general-purpose, there are many others that can be used in industrial equipment, which depend on the kind of machinery. For example, those equipment related to pneumatic or hydraulic subsystems (valves, cylinders, and so on) may take advantage of the measurement of pressure, flow, position, etc. Moreover, other equipment suffering from cyclic or transient loads may have installed load cells or strain gauges to measure the force or the displacement, which can be useful to calculate the mechanical stress suffered by a component.

3.2. Data Preprocessing

Data preprocessing is required in order to treat the data and obtain clean data that can be useful for feature extraction. Even if in some scenarios, features are extracted directly from the raw signals, the high complexity of the industrial systems and the nonlinear processes involved ask for the application of signal processing strategies [90]. These strategies may include the removal of the effects of the operating conditions from the acquired signals, the removal of noise in noisy signals, the identification and removal of outliers that may lead to non-significant features, and the transformation of the signals to other domains from which key features can be extracted, among others. There are many common techniques for some of the variables that can be measured and were introduced in Section 3.1. However, some of them are dependent on the nature of the signal. In this subsection, the focus is put on two of the main data sources: vibration signals and current signatures. This selection was made based on the widely extended use in manufacturing machinery in the case of vibration signals, the ease of data acquisition, the fact of being a usual data source in the manufacturing industry, and its usefulness for the detection of faults and inefficiencies in machines in the case of current signatures.

Vibration data preprocessing is one of the most-studied areas in signal preprocessing. ISO 13373-2:2016 [91] summarises some procedures for general purpose applications. Thus, different effects and techniques are described, such as time domain averaging, Fourier transform, spectrograms, order tracking, octave and fractional octave analysis, and cepstrum analysis. According to the standard, there are other techniques for the in-depth analysis of vibration signals. This is the case of the semi-automated bearing diagnostic procedure [73], which consists of five steps for the processing of vibration signals corresponding to rolling element bearings, with different objectives: (i) removal of speed fluctuation, based on order tracking [92]; (ii) removal of discrete frequencies by means of discrete random separation, self-adaptive noise cancellation, or linear prediction [93,94]; (iii) removal of the smearing effect of the signal transfer path using minimum entropy deconvolution [95]; (iv) determination of the optimum band for filtering and demodulation via spectral kurtosis [96]; (v) transformation of the domain by means of envelope analysis [97]. VDI 3832 [98] also provides some guidelines on the analysis of rolling element bearings in driving and driven machines and power transmission elements. Other advanced techniques comprise empirical mode decomposition [99], Hilbert–Huang transform [100], adaptive local iterative filtering [101], or fast iterative filtering decomposition [102], among others.

Regarding the analysis of current signals, Motor Current Signature Analysis (MCSA) is a common practice to diagnose different rotating machinery, as is the case of induction machines [103], permanent magnet synchronous motors [104], or rolling element bearings [105]. This strategy is also complemented with other signal processing techniques such as adaptive filters such as the Wiener filter [106] or time-synchronous averaging [107].

3.3. Feature Extraction

Feature extraction is a key step as this process creates representations of data that increase the effectiveness of the posterior analysis. Thus, the extraction of relevant features for the specific machine or process and the fact of generating features with reduced missing data are important for the accuracy of the models that can be trained afterwards [108]. It should be highlighted that some advanced artificial intelligence models do not need this feature extraction step to be performed previously, as they operate directly with the acquired time series, but these techniques usually require very large data volumes [69]. The feature extraction techniques can be categorised depending on the domain in which the features are extracted [109]. Thus, the features that can be extracted in different domains are presented next.

Time domain features are extracted from the signal itself. Some of them are based on simple statistic indicators, such as the mean, standard deviation, skewness, kurtosis, peak, and root mean square, among others. Time series models can also be used, such as AR, MA, or the combination of both in an ARMA representation. Dynamic features are also employed depending on the characteristics of the signal, for example: overshoot, settling time, or rise time.

Frequency domain features are obtained from the frequency representation of the signal, most commonly after its Fourier transformation. The same statistic indicators as before can be obtained from the signal in this domain, as well as signal energy or some specific frequency peaks' amplitudes. Their main limitation is that the signal considered needs to be stationary. To overcome this limitation, time–frequency features can be obtained, which can be obtained through a short-time Fourier transformation, where the analysis of the frequency domain is performed at several time windows [110] or other types of transformations, such us the Wigner–Ville transformation [111] or wavelets [71].

Two-dimensional data such as images require specific feature extraction. These features can be focused on the colour, the texture, the intensity, or the shape, among others [112]. They can also be related to text recognition, as they are useful for traceability purposes in manufacturing.

A summary of the features can be found in Table 2.

Table 2. Summarised features according to the domain.

Domain	Features	Reference
Time domain	Mean, maximum, minimum, amplitude, variance, standard deviation, skewness, kurtosis, root mean square, peak-to-peak, autoregressive coefficients, overshoot, settling time, rise time	[73,98,109]
Frequency domain	Spectral statistic moments, highest peak amplitude, sum of peak amplitudes	[73,98,105]
Time-frequency domain	Wavelet energy, Wigner–Ville	[71,110, 111]
Two-dimensional domain	RGB, LUV, HSV, HMMD, homogeneity, entropy, contrast, correlation, body shape, length, width, height	[112]

3.4. Dimensionality Reduction and Feature Selection

The feature selection step is crucial as the dimensionality of the data acquired in manufacturing plants and the amount of features extracted from them are usually high. Ideally, the users of the features would prefer a reduced number of significant features for conducting predictions, which are at the same time interpretable. However, the trade-off between the predictive performance and interpretability is infrequent [108]. Thus, the common solution is to apply some feature selection techniques to reach a compromise. This will also help avoid overfitting [113], which is a general problem found in machine learning applications, related to a model that performs very well in the training set, but is not easily generalised to a different set.

In that sense, techniques can be categorised as supervised or unsupervised, depending on the usage of the labels of the target variables [69]. In supervised techniques, the performance of adding or removing a variable is assessed for predicting the target variable, whereas unsupervised techniques use statistical tests to determine if the new features are similar (redundant) or bring new important information. If no labels are available, the second group should be used, whereas if labels are available, its use would be beneficial not to remove features that may not contain useful information in a particular condition, but may be critical for fault detection.

A very common example inside the unsupervised techniques is Principal Component Analysis (PCA) [114], which was used in [115,116]. PCA can be considered both in the feature extraction and in the feature selection groups, as it involves feature transformation (extraction), but when used, it usually intends to also reduce the dimensionality of the features involved, keeping only the transformed features explaining the most variance. Some nonlinear variants of PCA also exist and have also been used, the most common one being kernel PCA [117]; another one can be found in [118].

Another technique is the minimal Redundancy–Maximal Relevancy (mRMR), developed by [119]. The goal of this method is to find those indicators that minimise the redundancy of the data, as the removal of one feature from highly mutually dependent sets will not lead to a change in the information given by them; at the same time, the method must maximise the relevance to the target classes. There is an unsupervised version of this algorithm (UmRMR) that has been used for predictive maintenance in rotating machinery [120] and in structural health monitoring [121].

4. Data Fusion Solutions

4.1. Fusion Solutions for Various IoT Environments

The smart manufacturing sector requires underlying enabling technologies in various environments (Table 3). Furthermore, fusion techniques being deployed in these environments comprise the higher computing layer. Next, these environments are described in detail and can be either distributed, heterogeneous, nonlinear, or object-tracking, distributed in terms of the lack of central control, heterogeneous in terms of variety in the device types that should communicate under the same common protocols, non-linear in terms of the existence of time-varying sensing processes, and finally, object-tracking in terms of correct object localisation and identification.

Data sensing technologies are associated with challenges related to distributed environments [122] as in WSNs, which comprise an integral part of the IoT. Since there is heterogeneity in the nature of IoT environments, nonlinear and tracking issues emerge such as Multi-Target Tracking (MTT), cost effectiveness, errors requiring mitigation, and other asynchronous and track-to-track (T2T) problems. Data fusion methods confront all these issues and challenges.

Fusion in distributed environments is a concept that has been confronted in the last two decades. Concerning IoT environments, practical deployment is still an on-going process. Sensors and microprocessors provide source data for the sub-branches of parallel algorithms in such distributed environments. These are the main components used for tracking environmental changes in remote areas. At the same time, microprocessors provide fusion capabilities to them.

Next, classification of fusion algorithms in WSN environments (subset of the IoT) follows. The Kalman Filter (KF) [123] is an estimation algorithm (prediction corrector) and is utilised due to its scalability. It is useful during state propagation and the update of primitive input data. The KF is a Bayesian fusion algorithm and is practically utilised in weather and environmental monitoring, surveillance areas, as well as in intelligent triage systems for tracking sensitive patient data. Since WSNs are dynamic in nature, during operation, power failure issues may arise, connectivity may be unstable due to environmental factors, and sensors can undergo geographical mobility, altering the way of the network's logical topology. On the other hand, most fusion algorithms exhibit static behaviour.

Table 3. Fusion environments.

Environment	Subfield	Reference
Distributed	Data sensing	[122]
	Kalman filtering	[123]
	Energy/cost efficiency	[124–127]
Heterogeneous	Data correlation	[128]
	Distributed filtering	[129]
	Heterogeneous data	[130]
	Fussy logic/Kalman filter	[131]
	Canonical correlation analysis	[132,133]
	Multimodal fusion	[134]
Non-linear	Multisensor data fusion	[122]
	Sensor-dense IoT networks	[135]
	Fusion based on fuzzy logic	[136]
Object-tracking	Assembly line	[137]
	Transportation network	[138,139]
	Online multi-object tracking	[140]
	Energy efficiency for target tracking	[141]

For WSNs to be cost efficient, they must also be energy efficient. Having a large number of sensors operating together, power instability can cause functional instability at the algorithmic level. A Cuckoo-Based Particle Approach (CBPA) [124] can provide optimisation capabilities in a distributed WSN. This approach is utilised for node deployment in a static cluster. Furthermore, data are aggregated and forwarded to the base station when the cluster heads are selected. This is a generalised swarm algorithm. Next, the Generalised Particle Model Algorithm (GPMA) [125] assists in confronting the energy consumption problem. Specifically, it reduces the complexity by optimising the process of cluster formation with the goal of allocating optimal paths (concerning the reduction of energy consumption) to the base station.

A typical method for reducing energy consumption in many different network architectures (including WSNs) is putting a number of non-passive components into sleep mode when their provision of computing resources is not required. Specifically, sleep mode [126] is used for multimedia sensors in order to save energy. These are only activated by some scalar sensors, which always remain active. The system recognises objects using image analysing techniques, and the results are fused for increasing the recognition performance of sensors.

A cluster formation scheme for energy cost reduction of the data fusion process has been proposed [127]. Due to the limited resources of multisensor schemes, scheduling is a key point for energy efficiency, so a novel hybrid technique has been presented for performing clustering and selection simultaneously. The next step after the selection is the partitioning and processing of data. Blind broadcast messages, along with signal overhead, are reduced during the formation. Next, routing is applied according to the layered architecture with the goal of path elongation minimisation.

Multi-hop routing is an energy-efficient scheme since it utilises traffic grooming techniques for efficient allocation of resources such as bandwidth. Non-passive components are reduced this way along the routing paths. Another technique is to use progressive data fusion, which hops through sensors. This results in low energy consumption. Finally, network planning (predetermination of energy using integer linear programming formulae) also results in reduction. Channel state information and prior knowledge of the routing tree is required during planning.

The basic trait of most fusion IoT systems is the presence of heterogeneity. Various types of devices or sensors interoperate under common communication protocols to assist in fulfilling a specific functionality. An important issue these systems face is the presence of different feature spaces of datasets. Several different spaces comprise the datasets. A challenging task becomes the analysis of the correlations of different data, even when semantic dependencies exist among them.

A framework was proposed [128], aiming at unifying multiple entity views with the purpose of learning embeddings for entity alignments. The views of entity names and their relations and attributes assist in embedding entities, under different combination strategies. The alignment of two different Knowledge Graphs (KGs) is performed by cross-KG inference methods. KG construction and fusion are facilitated by entity matching or resolution. Therefore, entities in different KGs with the same identity are found.

Distributed filtering for the state space models in networking has also been studied [129]. A Bayesian model has been formulated, which handles outliers and heavy-tailed noises, with the purpose of improving the robustness of the filter. A centralised algorithm has been proposed that is based on variational Bayesian methods, providing robust filtering. Next, it has been extended to include the Alternating Direction Method of Multipliers (ADMM). The purpose is to estimate the states and noise covariances.

Schemes have been proposed for managing heterogeneous bio-medical data [130]. A real-time health monitoring system collecting data from the body using sensors has been proposed. A predictive model that is trained on clinical data is applied for detecting malfunctions and generating warnings accordingly. Heterogeneous sensor systems benefit from hybrid fuzzy logic-based algorithms with the Kalman filter. Underground risk has been assessed by a proposed hierarchical fuzzy logic model [131]. For that purpose, two new rule designing and determination methods were presented and evaluated, i.e., the Average Rules Based (ARB) and Max Rules Based (MRB).

Canonical Correlation Analysis (CCA) assists in the analysis of two heterogeneous sets of variables for extracting features that are correlated. In this method [132], there is performance improvement and reduction in computational cost, by optimally predicting the dimension of multimodal information. Next, it was verified that the different types of canonical correlation analyses are special cases of the proposed method, leading to a unified framework. The performance of CCA concerning the fusion output is affected negatively by noisy datasets. When there is dependency between two sets of variables, the Partial Least-Squares (PLS) regression method can be applied. Furthermore, it is applied when the first set contains variables explaining the ones in the other set. In [133], PLS was used to overcome multicollinearity by performing feature extraction. Since PLS is a supervised learning task, it finds orthogonal directions using response values.

In the work of [134], the multimodal fusion of visual and textual similarities was explored from the perspective of an unsupervised framework. These similarities are based on visual features and concepts, as well as textual metadata with the purpose of integrating nonlinear graph-based fusion and PLS regression. A multimodal contextual similarity matrix was constructed, along with the nonlinear combination of relevance scores using query-based similarity vectors.

Multisensor data fusion is affected by nonlinear time-varying sensing processes, producing less-accurate predictions. Mitigating such conditions, optimised algorithms are required for each separate fusion channel. The Extended Kalman Filter (EKF) is extensively used in nonlinear environments. Accuracy improves by using the unscented Kalman filter [122], which is enabled by utilising approximation statistics. The main disadvantages the EKF relate to the linearisation, which is responsible for producing filter instability, and it also requires the existence of the Jacobean matrix. Finally, due to the difficulty of deriving Jacobean matrices, linearisation is a process that is not easily implementable.

The main traits that contribute to the popularity of KF methods relate to simplicity and easy implementation. Outliers inhibit the performance of these filters when sensor-dense IoT networks are operating, resulting in KF breakdowns. Accuracy is not always feasible when AI methods are utilised due to the heuristic nature of the functionality they provide, especially if a large number of sensors is present.

An important research issue is confronting the problem from a Bayesian probabilistic perspective. Using the measurement data, statistically tracking the mean and covariance leads to the estimation of unknown state variables [135]. In such nonlinear networking environments, fusion methods based on fuzzy logic are useful due to the ability to fuse raw data coming from sensors according to predefined rules. These systems can function with high bandwidth and accuracy with sensitive applications. There is also complexity for the update functionality in nonlinear systems [136].

The estimation of a process with an uncertain state with the measurement of noise covariances in nonlinear systems was the research goal of the authors in [142]. For that purpose, an adaptive fusion algorithm with Bayesian cubature was proposed, which jointly estimates system states and covariances with cubature sampling. The goal was to derive the variational Bayesian cubature KF.

The assembly line was assisted by computer vision systems based on object recognition techniques in [137]. These techniques ease the process of identifying complex components and dictate to the operating entity the right task to initiate next.

Data fusion techniques are widely used for the purpose of object tracking. Typical related applications are in robotics and in military equipment and transportation. Specifically, correct target localisation in the military field is facilitated by object tracking techniques. Furthermore, this process involves the detection of threats, the detection of moving objects in restricted areas, as well as timely decision-making.

Object tracking challenges consist of MTT, Track-to-Track issues (T2T), error mitigation, and cost effectiveness. MTT methods rely on data associations that are scene-adaptive. Features of specific targets are accurately determined by exploiting adaptive traits. Data associations take place in hierarchical spaces with different layers.

An example of multi-targeting tracking for an end-to-end transportation network was exhibited in [138]. In this work, an optimal set of trajectories was modelled by generating a graph when exploring deep features. Statistical similarity metrics were utilised for finding the transition cost between nodes. MILP approaches aid in solving the optimisation problem. Another work [139] by the previous authors was related to a hierarchical feature model in the same research field. The proposed algorithm relies on deep features for modelling the appearance of the targets. Next, the tracking problem is confronted by utilizing the unsupervised dimensionality reduction of sparse representation of the feature vectors. Finally, the target is associated by discrete combinatorial optimisation using a Bayesian filter.

The Gaussian mixture probability hypothesis density filter was used for designing an online multi-object tracking method in [140]. In this work, two modules formed the basis for the hierarchical data association, i.e., detection-to-track and track-to-track. These can recover lost tracks and switched IDs. Occlusion problems were also addressed.

Energy efficiency for target tracking in sensor networks is an evolving research field. The authors of [141] presented an algorithm for prediction-based opportunistic sensing in this network type. Sensor nodes self-adapt to target trajectories. They used prediction methods to detect the arrival of targets in close range, so as to put the devices in high consumption mode. When the target was undetected, low-power mode was enabled.

The generic enabling technologies were reviewed for performing fusion in IoT environments. Distributed environments are required when centralised control is not feasible due to the lack of computing and communicating resources. Heterogeneity cannot be discarded since the CapEx and OpEx should remain low. Furthermore, nonlinear and object tracking environments should be confronted as well. In the next section, data fusion for manufacturing environments is described in detail. Some of the techniques used in the general IoT domain can be also found in manufacturing applications.

4.2. Data Fusion Solutions in the Manufacturing Sector

In this section, a review of the fusion methods used to combine the heterogeneous sensors employed in industrial prognosis is presented. As mentioned earlier, in machine learning applications, fusion refers to ensemble learning; it is used to optimise the results of a prediction algorithm, and it is categorised according to the stage of the pipeline that it is implemented. Data fusion combines data at an early level; feature fusion is used to describe the combination of extracted features; late fusion describes the combination of the results of different models.

A thorough overview of data fusion systems in the industrial sector can be found in [4]. The authors categorised the industrial scenarios according to the target of the analysis, which affects the features extracted and the type of fusion adopted. The workflow of "data-driven industrial prognosis", also referred to as the Cross-Industry Standard Process for Data Mining (CRISP), mentioned data fusion and preparation as a middle step in the analysis' pipeline, before proceeding with the modelling and evaluation. Data fusion exploits additional sources of information other than those that already existing in an industry unit, such as environmental factors, which can provide important added value.

In [71], information fusion theory was adopted to combine features in order to deal with the nonlinear problem of machine condition recognition. The sensors used were a dynamometer for cutting force signals and an accelerometer for vibration signals. The abnormal conditions that needed to be recognised were tool wear and workpiece deformation. Before applying feature fusion, signal processing took place. Fast Fourier transformation and then discrete wavelet transformation were used to process the signals and extract features. Approximation and detail components were extracted from the signals and used as the input for the analysis. Cutting parameters and signal characteristics were fused in one characteristic vector with concatenation, and then, Support Vector Machine (SVM) models were applied to recognise the different machine conditions. In [143], the authors performed tool wear prediction and fused multi-domain features with the use of deep neural networks. Cutting force and vibration sensors were used to monitor the tool wear condition. Time, frequency, and time–frequency domain features were extracted from these sensors. The ground truth about the condition of the tools was obtained through a microscope in offline mode. The authors normalised the values of the extracted features and the target values, before applying the Deep Convolutional Neural Network (DCNN) model.

Surface quality control was performed in [72]. Four methods for signal processing were used, and a multisensor data fusion framework was presented. The cutting forces, acoustic emission, and vibration sensors were used to monitor the quality of the surface. Time Direct Analysis (TDA), Power Spectral Density (PSD), Singular Spectrum Analysis (SSA), and Wavelet Packet Transform (WPT) were the four feature extraction methods applied to the sensor data to extract statistical and non-statistical parameters. The multisensor fusion framework refers to the combination of parameters of all available sensors and of cutting forces and vibration sensors only. The framework was examined separately for each feature extraction method and for TDA and PSD together.

A more complex approach for data fusion in two stages was proposed in [144]. The authors dealt with the problem of condition monitoring. The fusion approach was based on Bayesian inference. In the first stage of the fusion framework, the data were combined at a local level, which in the particular application translates into individual health components. The obtained results of this level were used as the inputs in the second stage, which was the global one, and the condition of the whole system was assessed. The authors extracted a variety of time and frequency domain features from the time waveform of the signals. A two-stage Bayesian method combined with PCA was used as a sensor fusion scheme in [115]. The authors combined vibration, electric, and acoustic signals for mechanical and electrical fault diagnosis in induction motors. PCA was used to remove correlated features extracted from the three types of sensors. At the first stage of the Bayesian approach, the principal components of the features were combined with a Gaussian Naive Bayes (GNB) classifier, and the fusion of the local stages' results followed in the second (global) stage of the approach.

In [79], the authors performed CNC machining monitoring with the use of three sensors and also proposed a data fusion framework that improved the monitoring. Microphones were placed at three different locations in order to receive measurements for cutting parameters. The fusion framework comprised the typical procedure of signal extraction, filtering with the use of band-pass filters, and normalisation. Afterwards, using the autocorrelation coefficient, preference weights were calculated, which added a bigger weight to the preferred sensor. Finally, a signal estimate was formed by the sensor with the highest weight and the maximum likelihood estimate. This proposed fusion scheme is suitable for periodic transient signals.

In [44], Deep Neural Networks (DNNs) were used for sensor data fusion and analysis. The data under analysis comprised heterogeneous sensors attached to a replica industrial plant. The goal of the analysis was to assess the working conditions of the industrial plan with a fault prediction algorithm. The sensors involved were a vibration, proximity sensor, temperature, audio (measuring the noise produced by the machine replica), and current sensor. Each of these sensors measures a different condition of the plant. The authors faced the problem of monitoring all of these conditions as a multiclass classification problem, where each of the classes refers to the state of a condition. The fusion of the four sensors, excluding temperature, was performed with a DNN. The sensors had different sampling frequencies. In the preprocessing stage, the outliers were removed. On the audio signals, fast Fourier transformation was applied to extract frequency features. From the current and vibration sensors, the maximum values were extracted using time windows of different lengths for each sensor. No features were extracted from the proximity sensor. All these features were combined using a concatenation layer in the DNN. Deep fusion and feature level fusion were combined in [145] to perform fault severity diagnosis under various operating conditions. Multiple sensors, such as torque sensors and accelerometers, were installed on an experimental conditioning platform, to test the proposed framework. The signals provided by the sensors were segmented, and the features were extracted by each segment separately. The features of different segments were combined and formed a single feature vector. Afterwards, with the use of the DNN, the features were deeply fused in the layers of the network. After the deep fusion stage, the training and testing of classifiers followed.

A DCNN was found in [146]. The authors proposed an adaptive fusion method based on DCNNs that addressed the issues of multisensor feature extraction and the selection of a suitable fusion method in terms of the stage in which it is performed. The capability of DNNs to fuse data into different layers and stages was the reason the authors adopted them in their approach. The adapted network fuses the input data at lower levels and then extracts basic features, which are then, in the middle layer, fused into high-level features and decisions, and finally, at the higher levels, the features and decisions are combined again to obtain the final prediction.

A quite recent paper [74] dealing with the problem of milling chatter detection proposed a multisensor data fusion scheme to combine accelerometer and microphone inputs. After creating chatter features with Wavelet Package Decomposition (WPD), the authors performed parameter optimisation, then identified the WPD coefficients that were correlated to the resonant frequencies, and extracted time–frequency features. The fused features of both sensors were selected with the application of the Recursive Feature Elimination (RFE) method.

In [80], the authors applied a random-forest-based fusion to combine vibration and acoustic signals for fault diagnosis. An accelerometer was used to provide the vibration signals and an acoustic emission sensor for the acoustic signals. The sensors were operating at the same time to monitor the condition of a gearbox. The authors applied wavelet packet transform to extract the statistical parameters of the two sensors, which were later fed into two deep Boltzmann machines, respectively, to extract the deep features. Then, the features were fused using random forests.

In an older paper [147], fault diagnosis was again performed with acoustic and vibration signals, which were combined using Dempster–Shafer theory. This fusion was performed at the decision level. The signals from the two sensors were preprocessed using wavelet analysis, and the extracted data were used as the input in artificial neural networks for classification. The results of the networks were then combined using Dempster–Shafer rules, and the fusion improved the individual performance of the sensors per 10%. Late fusion was also used in [148] to improve the diagnosis of bearing defects in induction motors. The authors first combined the features extracted from the vibration sensors, after removing the redundant ones. Afterwards, they tried four different classifiers, which were later combined using ensemble methods, such as voting and stacking.

Feature fusion with the use of kernel techniques was presented in [149]. The authors referred to the term "fusion" along with feature selection as a technique for dimension reduction. The authors combined kernel theory with factor analysis and proposed a Probabilistic Kernel Factor Analysis framework (PKFA), employed for feature selection and fusion in tool condition monitoring. After the acquisition of data from vibration and force sensors involved in the machine processes, feature extraction was applied to extract statistical and frequency features, which measure the health condition of tools. The proposed PKFA was then applied to select and fuse features. For the prediction of tool wear condition, support vector regression was used.

The above mentioned methods are summarized in Table 4, categorized according to the stage of the analysis they are performed.

Table 4. Summary of the fusion methods.

Fusion Level	Fusion Method	Reference
Feature	Information theory	[71]
Feature/late	Bayesian-based fusion	[115,144]
Feature	D(C)NNs	[44,143,145,146]
Feature	Feature elimination/concatenation	[72,74,79,149]
Late	Dempster–Schafer theory	[147,148]
Feature	Random forest based	[80]

5. Discussion

In this paper, an overview of recent trends in sensor data analysis and fusion applications in the manufacturing sector was provided. The reviewed literature was categorised according to the stages of the analysis. First, data storage solutions were presented, including acquisition methods and communication protocols. Afterwards, preprocessing methods employed in the manufacturing sector were presented. The preprocessing steps of sensor data analysis demand much effort. Considering that most of the acquired data in industrial prognosis are sensor signals, filtering and feature extraction should be applied to eliminate noise.

The current literature review presented first of all the data storage and extraction solutions used in the sector. The vast variety of solutions that were presented are used in order to create a pipeline capable of handling the data from acquiring to storing. Most of them involve a wireless method for gathering data and a database solution capable of storing the data efficiently. Due to the nature and volume of the data, distributed databases are preferred.

Fusion methods are increasingly involved in sensor data analysis in manufacturing applications. From the literature reviewed, it can be stated that data or feature level fusion are more frequently applied in industrial prognosis. It can be also noted that the term data fusion was found to be used broadly to describe the tasks of feature extraction and feature selection, where the features of different sensors were combined using mainly concatenation. Fewer were the applications of late fusion methods, although there is a wider variety of them compared to early fusion ones. DNNs are a popular fusion method in the manufacturing sector, combining different sensors by addressing them in different layers and also offering the advantage of skipping the step of feature extraction.

Regarding signal processing and feature extraction, the computational efforts required by the algorithms imply some handicaps in the implementation of these strategies. In contrast, the use of extremely straightforward techniques can lead to low-quality information for the purposes of the manufacturing industry. Therefore, a compromise is required between the duration of the running time of the processing and feature extraction algorithms and the quality of the data being used for posterior analysis.

There is still room for experimenting with different fusion methods in the field of manufacturing, considering the more narrow variety of fusion methods adopted, compared to other fields. According to this review, as well as other reviews regarding sensor data analysis in the manufacturing sector, the combination of heterogeneous sensor data is mostly achieved through concatenation or DNNs. Real-time applications of prognosis, such as chatter or tool wear detection, definitely affect the type of fusion adopted, while off-line implementations allow for more experimentation. There are of course numerous late fusion methods that are not time demanding and can be employed in a real-time application. The implementation of more sophisticated fusion methods is one possible research direction for further investigation. Researchers should take into account the issues of large data volumes extracted from sensors and the time complexity of feature extraction and of algorithm implementation.

In order to apply efficient fusion solutions in the manufacturing sector, the underlying network technologies should be considered as well. Production environments rely on IoT connectivity, among other technologies, for energy efficiency and performance. Spectrum allocation is a process where machine learning tools emerge for improving adaptivity to the runtime conditions, when embedded in the computational logic and communication protocols. These tools are applied upon the enabling technologies of IoT networks. An overview of the aforementioned enabling technologies was performed, as well as the correlation with distributed, heterogeneous, nonlinear, or object-tracking environments.

It should be highlighted that the conception of all the technologies described in the paper followed its own path and that their strengths were combined in order to generate analysis workflows as the one described in the Introduction. Having said that, the adoption of these technologies by the industry does not follow a homogeneous trend. The differences between sectors (in terms of competitiveness, required quality standards, etc.) generate different needs when it comes to digital transformation and, specifically, multisensor data fusion. Moreover, some companies opt for the application of a subset of the aforementioned workflow, by the use of portable equipment with sensing, signal processing, and feature extraction capabilities for non-skilled workers that can be enough to monitor some of the critical points of the manufacturing plants. Apart from that, the differences with respect to investment capabilities and skills of the personnel of the companies in the specific technologies generate also a heterogeneous adoption of these technologies.

Thus, the purpose of this literature review was to present the recent trends in data fusion solutions for industrial prognosis, specifically for the manufacturing sector. This paper aims to serve as a practical guide for those interested in learning about methodologies and trends in analysing manufacturing sensor data, as well as the more advanced technique of fusion.

Author Contributions: Conceptualisation, A.T. and A.K.; methodology, A.T. and A.G.G.; investigation, A.T., E.B., C.K., A.G.G. and U.L.; writing—original draft preparation, A.T., E.B., C.K., A.G.G. and U.L.; Introduction, A.T.; data acquisition and storage, E.B.; communication protocols, fusion in IoT systems, C.K.; data preprocessing, A.G.G. and U.L.; fusion in manufacturing, A.T.; writing—review and editing, I.G., A.K. and S.V.; supervision, I.G., S.V. and I.K. All authors have read and agreed to the published version of the manuscript.

Funding: This research was supported by the i4Q project, funded by the European Union's Horizon 2020 Research and Innovation Program under Grant Agreement No. 958205.

Institutional Review Board Statement: Not applicable

Informed Consent Statement: Not applicable

Data Availability Statement: Not applicable

Conflicts of Interest: The authors declare no conflict of interest.

Abbreviations

The following abbreviations are used in this manuscript:

ICTs	Information Communication Technologies
IoT	Internet of Things
WSNs	Wireless Sensor Networks
RFID	Radio Frequency Identification
DAQ	Data Acquisition Device
RF	Radio Frequency
OPC-UA	Open Platform Communications Unified Architecture
SQL	Structured Query Language database
CPS	Cyber–Physical System
MES	Smart Manufacturing Execution System
IPv6	Internet Protocol Version 6
MQTT	MQ Telemetry Transport
SOAP	Simple Object Access Protocol
REST	Representational State Transfer
SCADA	Supervisory Control And Data Acquisition
ODBC	Open Database Connectivity
DDE	Dynamic Data Exchange
RS232	Recommended Standard 232
HART	Highway Addressable Remote Transducer
API	Application Programming Interface
HTTP	Hypertext Transfer Protocol
LE	Low Energy
NFC	Near-Field Communications
LPWAN	Low-Power Wide-Area Network
LoRa	Low Range
NB-IoT	Narrowband IoT
CNC	Computerised Numerical Control
TCP/IP	Transmission Control Protocol/Internet Protocol
UHF	Ultra-High Frequency
IIoT	Industrial Internet of Things
DDBS	Distributed Database System
UHF	Ultra-High Frequency
IIoT	Industrial Internet of Things
DDBS	Distributed Database System
HDFS	Hadoop Distributed File System
COSMOS	C Open Source Managed Operating System
GFS	Google File System
PDU	Protocol Data Unit
DBMS	Database Management Systems
OLTP	Online Transaction Processing
RDBMS	Relational Database Management Systems
DO-SS	Document-Oriented Storage System
OO-SS	Object-Oriented Storage System
SDHC	Software-Defined Hybrid Cloud
ISO	International Organization for Standardization
VDI	Verein Deutscher Ingenieure
AE	Acoustic Emission

MCSA	Motor Current Signature Analysis
PCA	Principal Component Analysis
mRMR	minimal Redundancy–Maximum Relevance
CRISP	Cross-Industry Standard Process for Data Mining
SVM	Support Vector Machine
DCNN	Deep Convolutional Neural Network
GNB	Gaussian Naive Bayes
DNN	Deep Neural Networks
WPD	Wavelet Package Decomposition
RFE	Recursive Feature Elimination
PKFA	Probabilistic Kernel Factor Analysis

References

1. Dalenogare, L.S.; Benitez, G.B.; Ayala, N.F.; Frank, A.G. The expected contribution of Industry 4.0 technologies for industrial performance. *Int. J. Prod. Econ.* **2018**, *204*, 383–394. [CrossRef]
2. Shafique, K.; Khawaja, B.A.; Sabir, F.; Qazi, S.; Mustaqim, M. Internet of things (IoT) for next-generation smart systems: A review of current challenges, future trends and prospects for emerging 5G-IoT scenarios. *IEEE Access* **2020**, *8*, 23022–23040. [CrossRef]
3. Kong, L.; Peng, X.; Chen, Y.; Wang, P.; Xu, M. Multisensor measurement and data fusion technology for manufacturing process monitoring: A literature review. *Int. J. Extrem. Manuf.* **2020**, *2*, 022001. [CrossRef]
4. Diez-Olivan, A.; Del Ser, J.; Galar, D.; Sierra, B. Data fusion and machine learning for industrial prognosis: Trends and perspectives towards Industry 4.0. *Inf. Fusion* **2019**, *50*, 92–111. [CrossRef]
5. Zhou, J.; Hong, X.; Jin, P. Information fusion for multi-source material data: Progress and challenges. *Appl. Sci.* **2019**, *9*, 3473. [CrossRef]
6. Lee, J.; Kao, H.A.; Yang, S. Service innovation and smart analytics for industry 4.0 and Big Data environment. *Procedia CIRP* **2014**, *16*, 3–8. [CrossRef]
7. Ren, S.; Zhang, Y.; Liu, Y.; Sakao, T.; Huisingh, D.; Almeida, C.M. A comprehensive review of Big Data analytics throughout product lifecycle to support sustainable smart manufacturing: A framework, challenges and future research directions. *J. Clean. Prod.* **2019**, *210*, 1343–1365. [CrossRef]
8. Zheng, P.; Lin, T.J.; Chen, C.H.; Xu, X. A systematic design approach for service innovation of smart product-service systems. *J. Clean. Prod.* **2018**, *201*, 657–667. [CrossRef]
9. Chen, B.; Wan, J.; Shu, L.; Li, P.; Mukherjee, M.; Yin, B. Smart factory of industry 4.0: Key technologies, application case, and challenges. *IEEE Access* **2017**, *6*, 6505–6519. [CrossRef]
10. Mourtzis, D.; Vlachou, E.; Milas, N. Industrial Big Data as a result of IoT adoption in manufacturing. *Procedia CIRP* **2016**, *55*, 290–295. [CrossRef]
11. Obitko, M.; Jirkovský, V. Big data semantics in industry 4.0. In Proceedings of the International Conference on Industrial Applications of Holonic and Multi-Agent Systems, Valencia, Spain, 2–3 September 2015; Springer: Berlin/Heidelberg, Germany, 2015; pp. 217–229.
12. Gokalp, M.O.; Kayabay, K.; Akyol, M.A.; Eren, P.E.; Koçyiğit, A. Big data for industry 4.0: A conceptual framework. In Proceedings of the 2016 International Conference on Computational Science and Computational Intelligence (CSCI), Las Vegas, NV, USA, 15–17 December 2016; pp. 431–434.
13. Jeon, B.W.; Um, J.; Yoon, S.C.; Suk-Hwan, S. An architecture design for smart manufacturing execution system. *Comput.-Aided Des. Appl.* **2017**, *14*, 472–485. [CrossRef]
14. Saqlain, M.; Piao, M.; Shim, Y.; Lee, J.Y. Framework of an IoT-based industrial data management for smart manufacturing. *J. Sens. Actuator Netw.* **2019**, *8*, 25. [CrossRef]
15. Atmoko, R.; Riantini, R.; Hasin, M. IoT real time data acquisition using MQTT protocol. *J. Phys. Conf. Ser.* **2017**, *853*, 012003. [CrossRef]
16. Xu, J.; Yao, J.; Wang, L.; Ming, Z.; Wu, K.; Chen, L. Narrowband internet of things: Evolutions, technologies, and open issues. *IEEE Internet Things J.* **2017**, *5*, 1449–1462. [CrossRef]
17. Mekki, K.; Bajic, E.; Chaxel, F.; Meyer, F. A comparative study of LPWAN technologies for large-scale IoT deployment. *ICT Express* **2019**, *5*, 1–7. [CrossRef]
18. Sahal, R.; Breslin, J.G.; Ali, M.I. Big data and stream processing platforms for Industry 4.0 requirements mapping for a predictive maintenance use case. *J. Manuf. Syst.* **2020**, *54*, 138–151. [CrossRef]
19. Guo, Y.; Sun, Y.; Wu, K. Research and development of monitoring system and data monitoring system and data acquisition of CNC machine tool in intelligent manufacturing. *Int. J. Adv. Robot. Syst.* **2020**, *17*, 1–12. [CrossRef]
20. Xiao, Y.; Liu, Q. Application of Big Data processing method in intelligent manufacturing. In Proceedings of the 2019 IEEE International Conference on Mechatronics and Automation (ICMA), Tianjin, China, 4–7 August 2019; pp. 1895–1900.
21. Zhang, Y.; Ren, S.; Liu, Y.; Si, S. A Big Data analytics architecture for cleaner manufacturing and maintenance processes of complex products. *J. Clean. Prod.* **2017**, *142*, 626–641. [CrossRef]

22. Dai, H.N.; Wang, H.; Xu, G.; Wan, J.; Imran, M. Big data analytics for manufacturing internet of things: Opportunities, challenges and enabling technologies. *Enterp. Inf. Syst.* **2020**, *14*, 1279–1303. [CrossRef]
23. Majeed, A.; Zhang, Y.; Ren, S.; Lv, J.; Peng, T.; Waqar, S.; Yin, E. A Big Data-driven framework for sustainable and smart additive manufacturing. *Robot.-Comput.-Integr. Manuf.* **2021**, *67*, 102026. [CrossRef]
24. Srinivasan, M.; Prince, E.; Padmanabhan, R. IoT architecture for advanced manufacturing technologies. *Mater. Today Proc.* **2020**, *22*, 2359–2365.
25. Meng, Z.; Wu, Z.; Gray, J. RFID-based object-centric data management framework for smart manufacturing applications. *IEEE Internet Things J.* **2018**, *6*, 2706–2716. [CrossRef]
26. Wan, J.; Tang, S.; Shu, Z.; Li, D.; Wang, S.; Imran, M.; Vasilakos, A.V. Software-defined industrial internet of things in the context of industry 4.0. *IEEE Sens. J.* **2016**, *16*, 7373–7380. [CrossRef]
27. Sanghavi, D.; Parikh, S.; Raj, S.A. Industry 4.0: Tools and implementation. *Manag. Prod. Eng. Rev.* **2019**, *10*, 3–13.
28. Frank, A.G.; Dalenogare, L.S.; Ayala, N.F. Industry 4.0 technologies: Implementation patterns in manufacturing companies. *Int. J. Prod. Econ.* **2019**, *210*, 15–26. [CrossRef]
29. Zhang, X.; Ming, X.; Liu, Z.; Qu, Y.; Yin, D. An overall framework and subsystems for smart manufacturing integrated system (SMIS) from multi-layers based on multi-perspectives. *Int. J. Adv. Manuf. Technol.* **2019**, *103*, 703–722. [CrossRef]
30. Gölzer, P.; Cato, P.; Amberg, M. *Data Processing Requirements of Industry 4.0-Use Cases for Big Data Applications.* In Proceedings of the Twenty-Third European Conference on Information Systems (ECIS), Münster, Germany, 26–29 May 2015; Paper 61.
31. Lade, P.; Ghosh, R.; Srinivasan, S. Manufacturing analytics and industrial internet of things. *IEEE Intell. Syst.* **2017**, *32*, 74–79. [CrossRef]
32. Pokorny, J. NoSQL databases: A step to database scalability in web environment. *Int. J. Web Inf. Syst.* **2013**, *9*, 69–82. [CrossRef]
33. Santos, M.Y.; e Sá, J.O.; Andrade, C.; Lima, F.V.; Costa, E.; Costa, C.; Martinho, B.; Galvão, J. A Big Data system supporting bosch braga industry 4.0 strategy. *Int. J. Inf. Manag.* **2017**, *37*, 750–760. [CrossRef]
34. Costa, C.; Santos, M.Y. Reinventing the energy bill in smart cities with NoSQL technologies. In *Transactions on Engineering Technologies;* Springer: Berlin/Heidelberg, Germany, 2016; pp. 383–396.
35. Costa, C.; Santos, M.Y. The SusCity Big Data warehousing approach for smart cities. In Proceedings of the 21st International Database Engineering & Applications Symposium, Bristol, UK, 12–14 July 2017; pp. 264–273.
36. Jiang, L.; Da Xu, L.; Cai, H.; Jiang, Z.; Bu, F.; Xu, B. An IoT-oriented data storage framework in cloud computing platform. *IEEE Trans. Ind. Inform.* **2014**, *10*, 1443–1451. [CrossRef]
37. Yen, I.L.; Zhang, S.; Bastani, F.; Zhang, Y. A framework for IoT-based monitoring and diagnosis of manufacturing systems. In Proceedings of the 2017 IEEE Symposium on Service-Oriented System Engineering (SOSE), San Francisco, CA, USA, 6–9 April 2017; pp. 1–8.
38. Vater, J.; Harscheidt, L.; Knoll, A. A reference architecture based on edge and cloud computing for smart manufacturing. In Proceedings of the 2019 28th International Conference on Computer Communication and Networks (ICCCN), Valencia, Spain, 29 July–1 August 2019; pp. 1–7.
39. Krumeich, J.; Werth, D.; Loos, P. Prescriptive control of business processes. *Bus. Inf. Syst. Eng.* **2016**, *58*, 261–280. [CrossRef]
40. Raghav, R.; Amudhavel, J.; Dhavachelvan, P. A survey of NoSQL database for analysing large volume of data in big data platform. *Int. J. Eng. Technol. (UAE)* **2018**, *7*, 181–186. [CrossRef]
41. Fazio, M.; Celesti, A.; Puliafito, A.; Villari, M. Big data storage in the cloud for smart environment monitoring. *Procedia Comput. Sci.* **2015**, *52*, 500–506. [CrossRef]
42. Grevenitis, K.; Psarommatis, F.; Reina, A.; Xu, W.; Tourkogiorgis, I.; Milenkovic, J.; Cassina, J.; Kiritsis, D. A hybrid framework for industrial data storage and exploitation. *Procedia CIRP* **2019**, *81*, 892–897. [CrossRef]
43. Wang, H.Y.; Tsung, C.K. Scalable Data-Storage Framework for Smart Manufacturing. In *Proceedings of the International Conference on Frontier Computing, Ischia, Italy, 8–10 May 2018*; Springer: Berlin/Heidelberg, Germany, 2018; pp. 307–313.
44. De Vita, F.; Bruneo, D.; Das, S.K. On the use of a full stack hardware/software infrastructure for sensor data fusion and fault prediction in industry 4.0. *Pattern Recognit. Lett.* **2020**, *138*, 30–37. [CrossRef]
45. Ghaemi, Z.; Farnaghi, M.; Alimohammadi, A. Hadoop-based distributed system for online prediction of air pollution based on support vector machine. *Int. Arch. Photogramm. Remote Sens. Spat. Inf. Sci.* **2015**, *1*, 215–219. [CrossRef]
46. Abdelouarit, K.A.; Sbihi, B.; Aknin, N. Towards an approach based on hadoop to improve and organize online search results in Big Data environment. In *ICCMIT 2016, Communication, Management and Information Technology;* CRC Press: Cosenza, Italy, 2016; pp. 557–564.
47. Ding, X.; Liu, Y.; Qian, D. Jellyfish: Online performance tuning with adaptive configuration and elastic container in hadoop yarn. In Proceedings of the 2015 IEEE 21st International Conference on Parallel and Distributed Systems (ICPADS), Melbourne, Australia, 14–17 December 2015; pp. 831–836.
48. Iqbal, M.H.; Soomro, T.R. Big data analysis: Apache storm perspective. *Int. J. Comput. Trends Technol.* **2015**, *19*, 9–14. [CrossRef]
49. Van Der Veen, J.S.; Van Der Waaij, B.; Lazovik, E.; Wijbrandi, W.; Meijer, R.J. Dynamically scaling apache storm for the analysis of streaming data. In Proceedings of the 2015 IEEE First International Conference on Big Data Computing Service and Applications, Washington, DC, USA, 30 March–2 April 2015; pp. 154–161.
50. Yan, L.; Shuai, Z.; Bo, C. Multisensor data fusion system based on Apache Storm. In Proceedings of the 2017 3rd IEEE International Conference on Computer and Communications (ICCC), Chengdu, China, 13–16 December 2017; pp. 1094–1098.

51. Rashid, M.; Singh, H.; Goyal, V.; Parah, S.A.; Wani, A.R. Big data based hybrid machine learning model for improving performance of medical Internet of Things data in healthcare systems. In *Healthcare Paradigms in the Internet of Things Ecosystem*; Elsevier: Amsterdam, The Netherlands, 2021; pp. 47–62.
52. Manogaran, G.; Lopez, D. Health data analytics using scalable logistic regression with stochastic gradient descent. *Int. J. Adv. Intell. Paradig.* **2018**, *10*, 118–132. [CrossRef]
53. Makeshwar, P.; Kalra, A.; Rajput, N.; Singh, K. Computational scalability with Apache Flume and Mahout for large scale round the clock analysis of sensor network data. In Proceedings of the 2015 National Conference on Recent Advances in Electronics & Computer Engineering (RAECE), Roorkee, India, 13–15 February 2015; pp. 306–311.
54. Shi, W.; Zhu, Y.; Huang, T.; Sheng, G.; Lian, Y.; Wang, G.; Chen, Y. An integrated data preprocessing framework based on apache spark for fault diagnosis of power grid equipment. *J. Signal Process. Syst.* **2017**, *86*, 221–236. [CrossRef]
55. Shyam, R.; HB, B.G.; Kumar, S.; Poornachandran, P.; Soman, K. Apache spark a Big Data analytics platform for smart grid. *Procedia Technol.* **2015**, *21*, 171–178. [CrossRef]
56. Jayaratne, M.; Alahakoon, D.; De Silva, D.; Yu, X. Apache spark based distributed self-organizing map algorithm for sensor data analysis. In Proceedings of the IECON 2017-43rd Annual Conference of the IEEE Industrial Electronics Society, Beijing, China, 29 October–1 November 2017; pp. 8343–8349.
57. Wu, H.; Shang, Z.; Wolter, K. Learning to reliably deliver streaming data with apache kafka. In Proceedings of the 2020 50th Annual IEEE/IFIP International Conference on Dependable Systems and Networks (DSN), Valencia, Spain, 29 June–2 July 2020; pp. 564–571.
58. Kato, K.; Takefusa, A.; Nakada, H.; Oguchi, M. A study of a scalable distributed stream processing infrastructure using Ray and Apache Kafka. In Proceedings of the 2018 IEEE International Conference on Big Data (Big Data), Boston, MA, USA, 11–14 December 2017; pp. 5351–5353.
59. Wu, H.; Shang, Z.; Peng, G.; Wolter, K. A reactive batching strategy of apache kafka for reliable stream processing in real-time. In Proceedings of the 2020 IEEE 31st International Symposium on Software Reliability Engineering (ISSRE), Coimbra, Portugal, 12–15 October 2020; pp. 207–217.
60. Das, T.; Sridharan, V.; Gurusamy, M. A survey on controller placement in SDN. *IEEE Commun. Surv. Tutor.* **2019**, *22*, 472–503. [CrossRef]
61. Alsaeedi, M.; Mohamad, M.M.; Al-Roubaiey, A.A. Toward adaptive and scalable OpenFlow-SDN flow control: A survey. *IEEE Access* **2019**, *7*, 107346–107379. [CrossRef]
62. Tang, F.; Fadlullah, Z.M.; Mao, B.; Kato, N. An intelligent traffic load prediction-based adaptive channel assignment algorithm in SDN-IoT: A deep learning approach. *IEEE Internet Things J.* **2018**, *5*, 5141–5154. [CrossRef]
63. Esposito, C.; Castiglione, A.; Palmieri, F.; Ficco, M.; Dobre, C.; Iordache, G.V.; Pop, F. Event-based sensor data exchange and fusion in the Internet of Things environments. *J. Parallel Distrib. Comput.* **2018**, *118*, 328–343. [CrossRef]
64. Hassija, V.; Chamola, V.; Saxena, V.; Jain, D.; Goyal, P.; Sikdar, B. A survey on IoT security: Application areas, security threats, and solution architectures. *IEEE Access* **2019**, *7*, 82721–82743. [CrossRef]
65. Ghaleb, A.; Zhioua, S.; Almulhem, A. On PLC network security. *Int. J. Crit. Infrastruct. Prot.* **2018**, *22*, 62–69. [CrossRef]
66. Samaraweera, G.D.; Chang, J.M. Security and privacy implications on database systems in Big Data era: A survey. *IEEE Trans. Knowl. Data Eng.* **2019**, *33*, 239–258. [CrossRef]
67. Wan, Z.; Xu, Z.; Liu, S.; Ni, W.; Ye, S. An internet of things roaming authentication protocol based on heterogeneous fusion mechanism. *IEEE Access* **2020**, *8*, 17663–17672. [CrossRef]
68. Wan, J.; Yang, J.; Wang, S.; Li, D.; Li, P.; Xia, M. Cross-network fusion and scheduling for heterogeneous networks in smart factory. *IEEE Trans. Ind. Inform.* **2019**, *16*, 6059–6068. [CrossRef]
69. Watson, N.J.; Bowler, A.L.; Rady, A.; Fisher, O.J.; Simeone, A.; Escrig, J.; Woolley, E.; Adedeji, A.A. Intelligent Sensors for Sustainable Food and Drink Manufacturing. *Front. Sustain. Food Syst.* **2021**, *5*, 408. [CrossRef]
70. International Organization for Standardization. *ISO 17359:2018 Condition Monitoring and Diagnostics of Machines-General Guidelines*; Technical Report; ISO: Geneva, Switzerland, 2018.
71. Liu, C.; Li, Y.; Zhou, G.; Shen, W. A sensor fusion and support vector machine based approach for recognition of complex machining conditions. *J. Intell. Manuf.* **2018**, *29*, 1739–1752. [CrossRef]
72. Garcia Plaza, E.; Nunez Lopez, P.; Beamud Gonzalez, E. Multisensor data fusion for real-time surface quality control in automated machining systems. *Sensors* **2018**, *18*, 4381. [CrossRef] [PubMed]
73. Randall, R. *Vibration-Based Condition Monitoring: Industrial, Automotive and Aerospace Applications*; Wiley: Hoboken, NJ, USA, 2021.
74. Tran, M.Q.; Liu, M.K.; Elsisi, M. Effective Multisensor Data Fusion for Chatter Detection in Milling Process. *ISA Trans.* 2021, *in press*.
75. le Roux, W.; Harley, R.G.; Habetler, T.G. Detecting Rotor Faults in Low Power Permanent Magnet Synchronous Machines. *IEEE Trans. Power Electron.* **2007**, *22*, 322–328. [CrossRef]
76. Alkahtani, A.; Nordin, F.; Sharrif, Z.; Ramasamy, A.; Norzeli, S.; Alkharasani, W.; Md Din, N.; Tiong, S.; Ekanayake, J. Condition monitoring through temperature, vibration and radio frequency emission. *Test Eng. Manag.* **2019**, *81*, 5621–5636.
77. Alfredo Osornio-Rios, R.; Antonino-Daviu, J.A.; de Jesus Romero-Troncoso, R. Recent Industrial Applications of Infrared Thermography: A Review. *IEEE Trans. Ind. Inform.* **2019**, *15*, 615–625. [CrossRef]
78. Brili, N.; Ficko, M.; Klančnik, S. Tool Condition Monitoring of the Cutting Capability of a Turning Tool Based on Thermography. *Sensors* **2021**, *21*, 6687 [CrossRef]

79. Duro, J.A.; Padget, J.A.; Bowen, C.R.; Kim, H.A.; Nassehi, A. Multisensor data fusion framework for CNC machining monitoring. *Mech. Syst. Signal Process.* **2016**, *66*, 505–520. [CrossRef]
80. Li, C.; Sanchez, R.V.; Zurita, G.; Cerrada, M.; Cabrera, D.; Vásquez, R.E. Gearbox fault diagnosis based on deep random forest fusion of acoustic and vibratory signals. *Mech. Syst. Signal Process.* **2016**, *76*, 283–293. [CrossRef]
81. International Organization for Standardization. *ISO 22096:2007 Condition Monitoring and Diagnostics of Machines-Acoustic Emission*; Technical Report; ISO: Geneva, Switzerland, 2007.
82. Sonia, P.; Jain, J.K.; Saxena, K.K. Influence of ultrasonic vibration assistance in manufacturing processes: A Review. *Mater. Manuf. Process.* **2021**, *36*, 1451–1475. [CrossRef]
83. Yuan, F.; Yu, Y.; Li, L.; Tian, G. Investigation of DC Electromagnetic-Based Motion Induced Eddy Current on NDT for Crack Detection. *IEEE Sens. J.* **2021**, *21*, 7449–7457. [CrossRef]
84. Bloomfield, N.; Hughes, B. Digital radiography in NDT-Advances and drivers. In Proceedings of the 56th Annual Conference of the British Institute of Non-Destructive Testing, Telford, UK, 5–7 September 2017.
85. Liu, X.; Wang, J.; Sun, K.; Cheng, L.; Wu, M.; Wang, X. Semantic segmentation of ferrography images for automatic wear particle analysis. *Eng. Fail. Anal.* **2021**, *122*, 105268. [CrossRef]
86. Tan, C.K.; Irving, P.; Mba, D. A comparative experimental study on the diagnostic and prognostic capabilities of acoustics emission, vibration and spectrometric oil analysis for spur gears. *Mech. Syst. Signal Process.* **2007**, *21*, 208–233. [CrossRef]
87. Ullah, U.; Bhatti, F.A.; Maud, A.R.; Asim, M.I.; Khurshid, K.; Maqsood, M. IoT-enabled computer vision-based parts inspection system for SME 4.0. *Microprocess. Microsyst.* **2021**, *87*, 104354. [CrossRef]
88. Zhu, X.; Manamasa, H.; Jiménez Sánchez, J.L.; Maki, A.; Hanson, L. Automatic assembly quality inspection based on an unsupervised point cloud domain adaptation model. *Procedia CIRP* **2021**, *104*, 1801–1806. [CrossRef]
89. Bortnowski, P.; Nowak-Szpak, A.; Król, R.; Ozdoba, M. Analysis and Distribution of Conveyor Belt Noise Sources under Laboratory Conditions. *Sustainability* **2021**, *13*, 2233. [CrossRef]
90. Pisa, I.; Morell, A.; Vilanova, R.; Vicario, J.L. Industrial Control under Non-Ideal Measurements: Data-Based Signal Processing as an Alternative to Controller Retuning. *Sensors* **2021**, *21*, 1237. [CrossRef]
91. International Organization for Standardization. *ISO 13373-2:2016 Condition Monitoring and Diagnostics of Machines-Vibration Condition Monitoring—Part 2: Processing, Analysis and Presentation of Vibration Data*; Technical Report; ISO: Geneva, Switzerland, 2016.
92. Borghesani, P.; Pennacchi, P.; Chatterton, S.; Ricci, R. The velocity synchronous discrete Fourier transform for order tracking in the field of rotating machinery. *Mech. Syst. Signal Process.* **2014**, *44*, 118–133. [CrossRef]
93. Antoni, J.; Randall, R. Unsupervised noise cancellation for vibration signals: Part I—Evaluation of adaptive algorithms. *Mech. Syst. Signal Process.* **2004**, *18*, 89–101. [CrossRef]
94. Antoni, J.; Randall, R. Unsupervised noise cancellation for vibration signals: Part II—A novel frequency-domain algorithm. *Mech. Syst. Signal Process.* **2004**, *18*, 103–117. [CrossRef]
95. Barszcz, T.; Sawalhi, N. Fault detection enhancement in rolling element bearings using the minimum entropy deconvolution. *Arch. Acoust.* **2012**, *37*, 131–141. [CrossRef]
96. Antoni, J. Fast computation of the kurtogram for the detection of transient faults. *Mech. Syst. Signal Process.* **2007**, *21*, 108–124. [CrossRef]
97. Randall, R.B.; Antoni, J. Rolling element bearing diagnostics—A tutorial. *Mech. Syst. Signal Process.* **2011**, *25*, 485–520. [CrossRef]
98. Verein Deutscher Ingenieure. *VDI 3832 Measurement of Structure-Borne Sound of Rolling Element Bearings in Machines and Plants for Evaluation of Condition*; Technical Report; Verein Deutscher Ingenieure: Düsseldorf, Germany, 2013.
99. Rabi, J.; Balusamy, T.; Raj Jawahar, R. Analysis of vibration signal responses on pre induced tunnel defects in friction stir welding using wavelet transform and empirical mode decomposition. *Def. Technol.* **2019**, *15*, 885–896. [CrossRef]
100. Yang, J.; Huang, D.; Zhou, D.; Liu, H. Optimal IMF selection and unknown fault feature extraction for rolling bearings with different defect modes. *Measurement* **2020**, *157*, 107660. [CrossRef]
101. An, X.; Zeng, H.; Yang, W.; An, X. Fault diagnosis of a wind turbine rolling bearing using adaptive local iterative filtering and singular value decomposition. *Trans. Inst. Meas. Control* **2017**, *39*, 1643–1648. [CrossRef]
102. Xu, Y.; Fan, F.; Jiang, X. A fast iterative filtering decomposition and symmetric difference analytic energy operator for bearing fault extraction. *ISA Trans.* **2021**, *108*, 317–332. [CrossRef]
103. Bhole, N.; Ghodke, S. Motor Current Signature Analysis for Fault Detection of Induction Machine—A Review. In Proceedings of the 2021 4th Biennial International Conference on Nascent Technologies in Engineering (ICNTE), Navi Mumbai, India, 15–16 January 2021; pp. 1–6. 10.1109/ICNTE51185.2021.9487715.
104. Karami, M.; Mariun, N.B.; Ab-Kadir, M.Z.A.; Misron, N.; Radzi, M.A.M. Motor Current Signature Analysis-based Non-invasive Recognition of Mixed Eccentricity Fault in Line Start Permanent Magnet Synchronous Motor. *Electr. Power Components Syst.* **2021**, *49*, 133–145. [CrossRef]
105. Bravo, I.; Leturiondo, U.; Arnaiz, A.; Salgado, O. Fault diagnosis of rolling element bearings from current and vibration measurements. In Proceedings of the PHM Society European Conference, Turin, Italy, 6–8 July 2016. [CrossRef]
106. Deekshit, K.K.C.; Rao, M.V.G.; Rao, R.S. Fault Indexing Parameter Based Fault Detection in Induction Motor via MCSA with Wiener Filtering. *Electr. Power Compon. Syst.* **2020**, *48*, 2048–2062. [CrossRef]
107. Sabir, H.; Ouassaid, M.; Ngote, N. Early Fault Estimation of Inter-turn Short-circuit in Rotor Winding of WRIM using ANN-based Combined TSA and MCSA Technique. In Proceedings of the 2020 International Symposium on Advanced Electrical and Communication Technologies (ISAECT), Saidia, Lebanon, 13–15 April 2020; pp. 1–6. [CrossRef]

108. Kuhn, M.; Johnson, K. *Feature Engineering and Selection: A Practical Approach for Predictive Models*; Chapman & Hall/CRC Data Science Series; CRC Press: Boca Raton, FL, USA, 2019.
109. Bleakie, A.; Djurdjanovic, D. Feature extraction, condition monitoring, and fault modelling in semiconductor manufacturing systems. *Comput. Ind.* **2013**, *64*, 203–213. [CrossRef]
110. Chandra, N.H.; Sekhar, A. Fault detection in rotor bearing systems using time frequency techniques. *Mech. Syst. Signal Process.* **2016**, *72*, 105–133. [CrossRef]
111. Baydar, N.; Ball, A. A comparative study of acoustic and vibration signals in detection of gear failures using Wigner–Ville distribution. *Mech. Syst. Signal Process.* **2001**, *15*, 1091–1107. [CrossRef]
112. Salau, A.O.; Jain, S. Feature extraction: A survey of the types, techniques, applications. In Proceedings of the 2019 International Conference on Signal Processing and Communication (ICSC), Noida, India, 7–9 March 2019; pp. 158–164.
113. Srivastava, N.; Hinton, G.; Krizhevsky, A.; Sutskever, I.; Salakhutdinov, R. Dropout: A simple way to prevent neural networks from overfitting. *J. Mach. Learn. Res.* **2014**, *15*, 1929–1958.
114. Jolliffe, I. Principal component analysis. In *Encyclopedia of Statistics in Behavioral Science*; John Wiley & Sons: Chichester, UK, 2005.
115. Stief, A.; Ottewill, J.R.; Baranowski, J.; Orkisz, M. A PCA and two-stage Bayesian sensor fusion approach for diagnosing electrical and mechanical faults in induction motors. *IEEE Trans. Ind. Electron.* **2019**, *66*, 9510–9520. [CrossRef]
116. Yang, B.S.; Han, T.; Yin, Z.J. Fault diagnosis system of induction motors using feature extraction, feature selection and classification algorithm. *JSME Int. J. Ser. Mech. Syst. Mach. Elem. Manuf.* **2006**, *49*, 734–741. [CrossRef]
117. Navi, M.; Meskin, N.; Davoodi, M. Sensor fault detection and isolation of an industrial gas turbine using partial adaptive KPCA. *J. Process Control.* **2018**, *64*, 37–48. [CrossRef]
118. Lee, J.M.; Yoo, C.; Choi, S.W.; Vanrolleghem, P.A.; Lee, I.B. Nonlinear process monitoring using kernel principal component analysis. *Chem. Eng. Sci.* **2004**, *59*, 223–234. [CrossRef]
119. Peng, H.; Long, F.; Ding, C. Feature selection based on mutual information criteria of max-dependency, max-relevance, and min-redundancy. *IEEE Trans. Pattern Anal. Mach. Intell.* **2005**, *27*, 1226–1238. [CrossRef]
120. Hamaide, V.; Glineur, F. Unsupervised Minimum Redundancy Maximum Relevance Feature Selection for Predictive Maintenance: Application to a Rotating Machine. *Int. J. Progn. Health Manag.* **2021**, *12*. [CrossRef]
121. Zugasti, E.; Mujica, L.E.; Anduaga, J.; Martinez, F. Feature Selection—Extraction Methods Based on PCA and Mutual Information to Improve Damage Detection Problem in Offshore Wind Turbines. *Key Eng. Mater.* **2013**, *569–570*, 620–627. [CrossRef]
122. Alam, F.; Mehmood, R.; Katib, I.; Albogami, N.N.; Albeshri, A. Data fusion and IoT for smart ubiquitous environments: A survey. *IEEE Access* **2017**, *5*, 9533–9554. [CrossRef]
123. Jondhale, S.R.; Deshpande, R.S. Kalman filtering framework-based real time target tracking in wireless sensor networks using generalized regression neural networks. *IEEE Sens. J.* **2018**, *19*, 224–233. [CrossRef]
124. Khabiri, M.; Ghaffari, A. Energy-aware clustering-based routing in wireless sensor networks using cuckoo optimization algorithm. *Wirel. Pers. Commun.* **2018**, *98*, 2473–2495. [CrossRef]
125. Verma, N.; Singh, D. Data redundancy implications in wireless sensor networks. *Procedia Comput. Sci.* **2018**, *132*, 1210–1217. [CrossRef]
126. Koyuncu, M.; Yazici, A.; Civelek, M.; Cosar, A.; Sert, M. Visual and auditory data fusion for energy-efficient and improved object recognition in wireless multimedia sensor networks. *IEEE Sens. J.* **2018**, *19*, 1839–1849. [CrossRef]
127. Din, S.; Ahmad, A.; Paul, A.; Rathore, M.M.U.; Jeon, G. A cluster-based data fusion technique to analyse Big Data in wireless multisensor system. *IEEE Access* **2017**, *5*, 5069–5083. [CrossRef]
128. Zhang, Q.; Sun, Z.; Hu, W.; Chen, M.; Guo, L.; Qu, Y. Multi-view knowledge graph embedding for entity alignment. *arXiv* **2019**, arXiv:1906.02390.
129. Hua, J.; Li, C. Distributed robust Bayesian filtering for state estimation. *IEEE Trans. Signal Inf. Process. Over Netw.* **2018**, *5*, 428–441. [CrossRef]
130. Mishra, A.; Mohapatro, M. An IoT framework for bio-medical sensor data acquisition and machine learning for early detection. *Int. J. Adv. Technol. Eng. Explor.* **2019**, *6*, 112–125. [CrossRef]
131. Fayaz, M.; Ullah, I.; Kim, D.H. Underground risk index assessment and prediction using a simplified hierarchical fuzzy logic model and kalman filter. *Processes* **2018**, *6*, 103. [CrossRef]
132. Gao, L.; Qi, L.; Chen, E.; Guan, L. Discriminative multiple canonical correlation analysis for information fusion. *IEEE Trans. Image Process.* **2017**, *27*, 1951–1965. [CrossRef] [PubMed]
133. Ferrer-Cid, P.; Barcelo-Ordinas, J.M.; Garcia-Vidal, J.; Ripoll, A.; Viana, M. Multisensor data fusion calibration in IoT air pollution platforms. *IEEE Internet Things J.* **2020**, *7*, 3124–3132. [CrossRef]
134. Gialampoukidis, I.; Moumtzidou, A.; Liparas, D.; Tsikrika, T.; Vrochidis, S.; Kompatsiaris, I. Multimedia retrieval based on nonlinear graph-based fusion and partial least squares regression. *Multimed. Tools Appl.* **2017**, *76*, 22383–22403. [CrossRef]
135. Fang, H.; Tian, N.; Wang, Y.; Zhou, M.; Haile, M.A. Nonlinear Bayesian estimation: From Kalman filtering to a broader horizon. *IEEE/CAA J. Autom. Sin.* **2018**, *5*, 401–417. [CrossRef]
136. Jiao, Z.; Wu, R. A new method to improve fault location accuracy in transmission line based on fuzzy multisensor data fusion. *IEEE Trans. Smart Grid* **2018**, *10*, 4211–4220. [CrossRef]
137. Cohen, Y.; Naseraldin, H.; Chaudhuri, A.; Pilati, F. Assembly systems in Industry 4.0 era: A road map to understand Assembly 4.0. *Int. J. Adv. Manuf. Technol.* **2019**, *105*, 4037–4054. [CrossRef]

138. Ullah, M.; Cheikh, F.A. Deep feature based end-to-end transportation network for multi-target tracking. In Proceedings of the 2018 25th IEEE International Conference on Image Processing (ICIP), Athens, Greece, 7–10 October 2018; pp. 3738–3742.
139. Ullah, M.; Mohammed, A.K.; Cheikh, F.A.; Wang, Z. A hierarchical feature model for multi-target tracking. In Proceedings of the 2017 IEEE International Conference on Image Processing (ICIP), Beijing, China, 17–20 September 2017; pp. 2612–2616.
140. Song, Y.m.; Yoon, K.; Yoon, Y.C.; Yow, K.C.; Jeon, M. Online multi-object tracking with gmphd filter and occlusion group management. *IEEE Access* **2019**, *7*, 165103–165121. [CrossRef]
141. Hare, J.Z.; Gupta, S.; Wettergren, T.A. POSE: Prediction-based opportunistic sensing for energy efficiency in sensor networks using distributed supervisors. *IEEE Trans. Cybern.* **2017**, *48*, 2114–2127. [CrossRef]
142. Dong, X.; Chisci, L.; Cai, Y. An adaptive variational Bayesian filter for nonlinear multisensor systems with unknown noise statistics. *Signal Process.* **2021**, *179*, 107837. [CrossRef]
143. Huang, Z.; Zhu, J.; Lei, J.; Li, X.; Tian, F. Tool wear predicting based on multi-domain feature fusion by deep convolutional neural network in milling operations. *J. Intell. Manuf.* **2020**, *31*, 953–966. [CrossRef]
144. Jaramillo, V.H.; Ottewill, J.R.; Dudek, R.; Lepiarczyk, D.; Pawlik, P. Condition monitoring of distributed systems using two-stage Bayesian inference data fusion. *Mech. Syst. Signal Process.* **2017**, *87*, 91–110. [CrossRef]
145. Liu, J.; Hu, Y.; Wang, Y.; Wu, B.; Fan, J.; Hu, Z. An integrated multisensor fusion-based deep feature learning approach for rotating machinery diagnosis. *Meas. Sci. Technol.* **2018**, *29*, 055103. [CrossRef]
146. Jing, L.; Wang, T.; Zhao, M.; Wang, P. An adaptive multisensor data fusion method based on deep convolutional neural networks for fault diagnosis of planetary gearbox. *Sensors* **2017**, *17*, 414. [CrossRef]
147. Khazaee, M.; Ahmadi, H.; Omid, M.; Moosavian, A.; Khazaee, M. Classifier fusion of vibration and acoustic signals for fault diagnosis and classification of planetary gears based on Dempster–Shafer evidence theory. *Proc. Inst. Mech. Eng. Part J. Process Mech. Eng.* **2014**, *228*, 21–32. [CrossRef]
148. Agahi, H.; Mahmoodzadeh, A. Decision fusion scheme for bearing defects diagnosis in induction motors. *Electr. Eng.* **2020**, *102*, 2269–2279. [CrossRef]
149. Wang, J.; Xie, J.; Zhao, R.; Mao, K.; Zhang, L. A new probabilistic kernel factor analysis for multisensory data fusion: Application to tool condition monitoring. *IEEE Trans. Instrum. Meas.* **2016**, *65*, 2527–2537. [CrossRef]

MDPI AG
Grosspeteranlage 5
4052 Basel
Switzerland
Tel.: +41 61 683 77 34

Sensors Editorial Office
E-mail: sensors@mdpi.com
www.mdpi.com/journal/sensors

www.ingramcontent.com/pod-product-compliance
Lightning Source LLC
Chambersburg PA
CBHW061614220926
51850CB00005B/507
* 9 7 8 3 7 2 5 8 1 5 8 9 0 *